Smart Nanomaterials Technology

Series Editors

Azamal Husen, Wolaita Sodo University, Wolaita, Ethiopia

Mohammad Jawaid, Department of Chemical and Petroleum Engineering, United Arab Emirates University, Al Ain, United Arab Emirates

AF790316

Nanotechnology is a rapidly growing scientific field and has attracted a great interest over the last few years because of its abundant applications in different fields like biology, physics and chemistry. This science deals with the production of minute particles called nanomaterials having dimensions between 1 and 100 nm which may serve as building blocks for various physical and biological systems. On the other hand, there is the class of smart materials where the material that can stimuli by external factors and results a new kind of functional properties. The combination of these two classes forms a new class of smart nanomaterials, which produces unique functional material properties and a great opportunity to larger span of application. Smart nanomaterials have been employed by researchers to use it effectively in agricultural production, soil improvement, disease management, energy and environment, medical science, pharmaceuticals, engineering, food, animal husbandry and forestry sectors.

This book series in Smart Nanomaterials Technology aims to comprehensively cover topics in the fabrication, synthesis and application of these materials for applications in the following fields:

- Energy Systems—Renewable energy, energy storage (supercapacitors and electrochemical cells), hydrogen storage, photocatalytic water splitting for hydrogen production
- Biomedical—controlled release of drugs, treatment of various diseases, biosensors,
- Agricultural—agricultural production, soil improvement, disease management, animal feed, egg, milk and meat production/processing,
- Forestry—wood preservation, protection, disease management
- Environment—wastewater treatment, separation of hazardous contaminants from wastewater, indoor air filters.

Antonio Juárez-Maldonado ·
Adalberto Benavides-Mendoza ·
Damaris Leopoldina Ojeda-Barrios ·
Gonzalo Tortella Fuentes · Amedea B. Seabra
Editors

Plant Biostimulation with Nanomaterials

Springer

Editors
Antonio Juárez-Maldonado
Department of Botany
Antonio Narro Agrarian Autonomous University
Saltillo, Coahuila, Mexico

Adalberto Benavides-Mendoza
Department of Horticulture
Antonio Narro Agrarian Autonomous University
Saltillo, Mexico

Damaris Leopoldina Ojeda-Barrios
Faculty of Agrotechnological Sciences
Autonomous University of Chihuahua
Chihuahua, Mexico

Gonzalo Tortella Fuentes
Center of Excellence in Biotechnological Research Applied to the Environment
CIBAMA-BIOREN
Universidad de La Frontera
Temuco, Chile

Amedea B. Seabra
Center for Natural and Human Sciences
Federal University of ABC (UFABC)
Bairro Santa Terezinha, Brazil

ISSN 3004-8273 ISSN 3004-8281 (electronic)
Smart Nanomaterials Technology
ISBN 978-981-96-4650-0 ISBN 978-981-96-4648-7 (eBook)
https://doi.org/10.1007/978-981-96-4648-7

© The Editor(s) (if applicable) and The Author(s), under exclusive license to Springer Nature Singapore Pte Ltd. 2025, corrected publication 2025

This work is subject to copyright. All rights are solely and exclusively licensed by the Publisher, whether the whole or part of the material is concerned, specifically the rights of translation, reprinting, reuse of illustrations, recitation, broadcasting, reproduction on microfilms or in any other physical way, and transmission or information storage and retrieval, electronic adaptation, computer software, or by similar or dissimilar methodology now known or hereafter developed.
The use of general descriptive names, registered names, trademarks, service marks, etc. in this publication does not imply, even in the absence of a specific statement, that such names are exempt from the relevant protective laws and regulations and therefore free for general use.
The publisher, the authors and the editors are safe to assume that the advice and information in this book are believed to be true and accurate at the date of publication. Neither the publisher nor the authors or the editors give a warranty, expressed or implied, with respect to the material contained herein or for any errors or omissions that may have been made. The publisher remains neutral with regard to jurisdictional claims in published maps and institutional affiliations.

This Springer imprint is published by the registered company Springer Nature Singapore Pte Ltd.
The registered company address is: 152 Beach Road, #21-01/04 Gateway East, Singapore 189721, Singapore

If disposing of this product, please recycle the paper.

Preface

Nanoscience and nanotechnology have advanced exponentially in recent years, as the number of scientific publications and patents related to using and applying nanomaterials has increased significantly worldwide. For this reason, all the implications of using nanomaterials in different areas of knowledge, ranging from molecular biology to the ecosystem level, must be considered.

In particular, the application of nanomaterials to plants has been a very relevant topic of study for humans since a large part of food is obtained from agricultural production systems, thus impacting food security. If we add to the growing need for food by an increasing population, as well as the environmental impacts of climate change that put pressure on agricultural production systems. In that case, it becomes vital to use new technologies to improve the productivity of plants as well as the quality of the products obtained from them.

In this sense, using biostimulants in agriculture as novel tools to address current global challenges has become increasingly important. In both extensive crops and intensive production, chemical, biological, or physical biostimulation allows for increased production and improved crop quality, accompanied by the mitigation of environmental impacts in the form of reduced use of pesticides and greater efficiency in using fertilizers and water. Moreover, biostimulants are agents that increase stress tolerance in crops and, therefore, have potential value in the current context of climate change. Nanomaterials have recently been considered biostimulants, with a dual category of physical and chemical biostimulation-inducing agents. The physical process of biostimulation occurs through interactions at the interfaces of the nanomaterial corona with cell walls and membranes. The chemical process depends on the composition of the corona and core of the nanomaterial.

It was not long known that most of the research on the biological impact of nanomaterials was directed toward their toxicological aspects. However, fortunately, the dual nature of biostimulation by nanomaterials described above has been recognized, and it has been concluded that, in adequate amounts, nanomaterials have a favorable impact on crop plants. Research on plant biostimulation with nanomaterials has been encouraged; however, much research is still needed to reach the large-scale application of nanotechnology in agriculture. First, the interactions between nanomaterials and plant cells differ depending on the corona's size, shape, specific surface area, and composition, in addition to the dynamic biological contexts of development and interactions with other environmental factors. Second, there needs to be more certainty about the environmental fate of nanomaterials and their long-term ecological impact. Therefore, crop, community, or ecosystem responses are often difficult to predict.

The editors consider that, in this book, crop biostimulation via nanomaterials has been addressed in detail, considering the different spheres of organization, from a molecular vision to ecosystems. The 16 chapters of this book cover topics such as hormetic responses caused by nanomaterials; the impact of nanomaterials at the molecular, biochemical, and physiological levels in plants; the induction of tolerance to biotic and abiotic stresses; the induction of metabolites in plants; the improvement of the nutritional quality of crops; the synthesis of nanomaterials through different approaches; and the possible impacts of nanomaterials on ecosystems.

The editors hope that through this book, readers will obtain a comprehensive overview of this dynamic and attractive topic.

Saltillo, Mexico — Antonio Juárez-Maldonado
Saltillo, Mexico — Adalberto Benavides-Mendoza
Chihuahua, Mexico — Damaris Leopoldina Ojeda-Barrios
Temuco, Chile — Gonzalo Tortella Fuentes
Bairro Santa Terezinha, Brazil — Amedea B. Seabra

Contents

Hormesis and Nanomaterials. From Biostimulation to Toxicity

Elena A. Erofeeva

Abstract Currently, more and more evidence is accumulating in favor of plant hormesis (i.e., low-dose stimulation and high-dose inhibition) when exposed to various nanomaterials (NMs) (metal NMs, carbon-based NMs, polymeric NMs, and hybrid NMs). However, the mechanisms underlying the stimulating and inhibitory effects of NMs have not been sufficiently studied. Here, the regularities of plant hormesis, including its mechanisms, are discussed in terms of the concept of environmental hormesis. The qualitative features of nanomaterial effects on plants and their relationship with direct and indirect hormetic stimulation of plants by NMs in low doses are considered. It has been shown that along with direct stimulation, any nanomaterials activate non-specific and specific defense mechanisms of plants, which largely determine their hormetic effects. The concept of environmental hormesis can be useful for understanding the patterns and mechanisms underlying the effects of NMs on plants, as well as for improving their efficiency and environmental safety.

Keywords Stress · Plant defense · Priming · Trade-off · Hormetic mechanisms

1 Introduction

It is known that various low-dose stressors (pollutants, abiotic factors, allelochemicals and elicitors) cause plant hormesis, that is, low-dose stimulation and high-dose inhibition [9, 16, 27, 28]. This indicates the ability of moderate stressors of various qualities to cause horemtic stimulation and demonstrates the universality of plant hormesis. According to the concept of environmental hormesis, any low-dose stressors can induce hormetic stimulation, including new environmental challenges [1, 27].

Currently, significant data have been accumulated on hormesis induced by low doses of various nanomaterials that can cause direct and indirect stimulation in plants

E. A. Erofeeva (✉)
Department of Ecology, Institute of Biology and Biomedicine, Lobachevsky State University of Nizhni Novgorod, Nizhni Novgorod, Russian Federation
e-mail: erofeeva@unn.ru

© The Author(s), under exclusive license to Springer Nature Singapore Pte Ltd. 2025

A. Juárez-Maldonado et al. (eds.), *Plant Biostimulation with Nanomaterials*, Smart Nanomaterials Technology, https://doi.org/10.1007/978-981-96-4648-7_1

[4, 34, 62]. However, the contribution of these mechanisms to hormetic effects and their interaction remain insufficiently studied. Here, these issues are discussed in terms of the concept of environmental hormesis. The qualitative features of the effects of nanoparticles in plants and their relationship with direct and indirect hormetic stimulation of plants with nanomaterials in low doses are considered.

1.1 Plant Hormetic Responses

Hormesis is a biphasic adaptive response of plants and other organisms characterized by low-dose stimulation and high-dose inhibition (Fig. 1) [17]. Hormetic stimulation is considered as an overcompensation response to an initial moderate disruption of homeostasis, which is caused by excessive activity of protective systems [9, 16].

The range of doses that cause hormetic stimulation is the hormetic zone of the dose–response curve (Fig. 1). Based on studies of the hormetic effects of various low-dose stressors, it has been shown that maximal stimulating effect (MAX) is typically 130–160% of the control. However, in some cases, the maximal stimulation can (rarely) reach 200% of the control [17]. The zone of inhibitory doses begins after the maximal dose of a stressor that does not have damaging effects (NOAEL—no-observed adverse effect level) [15]. It is shown that MAX and the distance between the dose causing MAX and NOAEL are time-dependent and depend on the duration of exposure to a low-dose stressor.

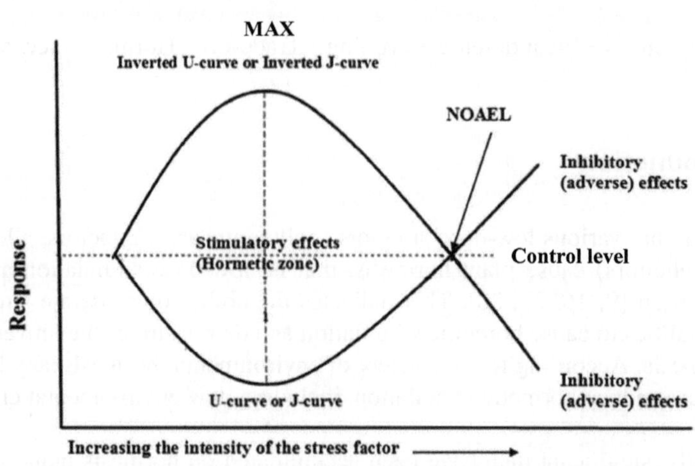

Fig. 1 Two types of hormetic dose–response curves (modified from Erofeeva [30]. The inverted U-curve (or inverted J-curve) demonstrates low-dose stimulation and high-dose inhibition and is the most frequently observed. The U-curve (or J-curve) demonstrates less damage at low doses and more damage at high doses. NOAEL, the no-observed adverse-effect level, is the maximal dose of a stressor that does not cause any damage. MAX is the maximum stimulating effect of a stressor [9, 17]

Hormesis may be accompanied by preconditioning (priming) when a low-dose stressor increases the resistance of living systems to the subsequent more severe effects of this stressor or other strong stressors [9, 17]. Thus, in some cases, preconditioning can induce cross-tolerance or cross-adaptation, since resistance to one low-dose stressor increase tolerance to other strong stressors [26]. Apparently, this allows living beings to cope with the instability of the environment and new challenges [28].

However, hormetic stimulation is not always accompanied by a synchronous improvement of various indicators, since hormetic trade-offs may take place [13, 22, 31]. Hormetic trade-offs are phenomena when the same dose of a low-dose stressor stimulates some indicators and worsens other indicators in the living system (trade-off 1) or hormetic stimulation of some indicators occurs, while others remain unchanged (trade-off 2) [28, 31].

In plants, trade-offs 1 and 2 are found both between different functional traits (e.g., growth and defense in plants), and between the endpoints of the same functional trait (e.g., seed weight and seed number in plants). It is assumed that hormetic trade-offs can be observed in two cases: (1) plant indicators have different responses (hormesis, inhibition or zero reaction) in the same dose range; (2) plant indicators have hormetic responses with different hormetic zones [31]. It is shown that trade-offs can have a positive, negative or zero effect on preconditioning, offspring, and fitness of the population [31].

It is suggested that hormetic trade-offs can be associated with the energy redistribution between different functional traits in plants due to the limited resources of the organism, which is based on the antagonism of molecular pathways regulating various functional traits. This allows the organism to make regulatory "decisions" that fine-tune the phenotype in response to various environmental challenges и and thus improve phenotypic plasticity [31]. For instance, some phytohormones regulating growth and protection have an antagonistic effect on plants under stress caused by herbivores. In particular, gibberellin and jasmonic acid pathways act as reciprocal suppressors via an interaction of their respective transcription factors and suppressor proteins JASMONATE ZIM DOMAIN (JAZ) and DELLA. This leads to growth inhibition and defense activation (synthesis of secondary metabolites, increased antioxidant defense, etc.) in response to damage caused by herbivores [31, 32, 75].

Currently, hormetic responses have been found in various plant groups, such as higher plants (woody and herbaceous, monocotyledonous and dicotyledonous) and various algae [2, 9, 26, 30], as well as at all development stages of higher plants from seeds and seedlings to mature plants [28].

Plant hormesis occurs in response to a wide range of qualitatively different low-dose stressors, including those that plants have not previously encountered during evolution [28]. Among the natural stressors, hormetic effects are manifested for various abiotic factors such as mineral nutrition, temperature, soil moisture, light, etc. In addition, various pollutants (pesticides and nanopesticides, engine fuels, micro/nanoplastics, pharmaceuticals, etc.), and their mixtures (agrochemicals, antibiotics or polybrominated diphenyl ethers) can induce plant hormesis [28]. Plant hormesis caused by biotic stressors remains poorly understood. However, some compounds involved in biotic interactions (allelochemicals, elicitors, etc.) [34, 55, 69], as well as moderate damage to plants caused by phytopathogens [31, 47], can cause hormetic responses.

These facts demonstrate that various abiotic, biotic or anthropogenic stressors regardless of the control level (zero as for many pollutants or optimal as in the case of abiotic factors when the response is described by the Shelford curve) and factor type (natural or anthropogenic) can induce plant hormesis [28]. Thus, the concept of environmetal hormesis, which postulates that any low-dose factor can cause hormesis [1, 28], is increasingly being confirmed.

Hormetic responses in plants have been found at various biological levels such as cells, populations, and even communities [1, 30, 31]. At the same time, hormesis at lower biological levels can be the basis for hormetic responses at higher levels [29]. Some indirect evidence shows that the hormetic responses of plant communities to low-dose impacts can stimulate ecosystem indicators. For example, the stimulating effects of moderate climate change on the state of forest ecosystems have been found [30].

1.2 Mechanisms of Plant Hormesis

Currently, the mechanisms underlying hormetic stimulation of plants have not been sufficiently studied. Nevertheless, experimental studies of plant hormesis show that three types of hormetic mechanisms are possible:

(1) *Direct stimulation of plants*. Some low-dose stressors can cause direct stimulation by interacting with biomolecules/structures of plant cells and stimulating their functioning. For instance, low concentrations of the herbicide phosphinotricin activate chloroplastic glutamine synthetase (a key enzyme for nitrogen assimilation) isoforms and nitrogen assimilation, thereby stimulating plant growth [23, 27].

(2) *Indirect stimulation of non-specific defense in plants*. Hormetic stimulation is also based on the moderate activation of the same non-specific defense systems that protect plants from any severe stressors and include, in particular, active chemical compounds, stress hormones and antioxidant defense, as well as the synthesis of stress proteins (i.e., heat shock proteins) [27, 28]. For instance, various studies demonstrate moderate activation of antioxidant enzymes (catalase, peroxidases, superoxide dismutase—SOD, etc.) and increased levels of non-enzyme antioxidants (alpha-tocopherol, ascorbic acid, flavonoids, phenolic compounds, reduced glutathione) with hormetic stimulation of growth and photosynthetic indicators by heavy and rare earth metals (Zn, Cd, Pb, W, La, La + Cd), ground-level ozone, carbon nanotubes, and aquaculture wastewater [28], as well as changes of phytohormones levels (abscisic acid—ABA, cytokinins, auxins, etc.) and their ratios with growth hormesis [28].

(3) *Indirect stimulation of specific defense mechanisms in plants*. Hormetic stimulation can also be observed for specific adaptive responses determined by the nature of the stressor and the 'points' of its impact on plant metabolism and, respectively, activation of the specific defenses. For example, low doses of heavy metals that stimulate growth increase the level of chelator compounds (phytochelatins, some organic acids, etc.) that are involved in specific heavy metal detoxification in plants [27].

Thus, it is suggested that hormesis depends on both indirect stimulation of general (universal for any stressor) and specific (due to the characteristics of the stressor) protective mechanisms of plants, as well as on the ability of the stressor to cause direct stimulation the functions of proteins and other biomolecules [27, 28].

1.3 Types of Nanomaterials and Their Applications

Nanomaterials are defined as materials where at least one of their dimensions is in the nanoscale, i.e. smaller than 100 nm [39]. There are four main types of nanomaterials such as metal nanomaterials, carbon-based nanomaterials, polymeric nanomaterials, and hybrid nanomaterials. Fe, Mg, Co, Mg, Zn, Cu, Au, Ce, Ag, Ni, Mn, and their oxides are the most frequently synthesized metal nanomaterials. Fe, Co, Mg, and Mn are commonly used to synthesize magnetic nanomaterials [53].

Various carbon-based materials include graphene, carbon nanotubes, carbon nanofibers, graphite, fullerenes, and diamonds [39, 53]. The term 'polymeric nanoparticle' is applied to nanospheres and nanocapsules, and a variety of polymer-based nanomaterials. Hybrid nanomaterials are conjugates of organic/inorganic structures. In particular, metal nanomaterials, carbon-based nanomaterials, and polymeric nanomaterials are mixed together to produce hybrid nanomaterials, thus enhancing the applicability and controlling the biocompatibility of the materials [39, 53].

Due to larger specific areas, lower modification of temperature, shorter distance of interparticle diffusion, large number of associated adsorption sites and differing surface chemistry compared to other materials, nanomaterials are considered to be a marvelous remarkable absorbents and catalysts [60]. Therefore, innovative engineered nanomaterials find application in several production fields, including electrical engineering, chemistry, material sciences, biomedicine, textile, cosmetic and food industry [37]. In agriculture, nanomaterials are used as nanopesticides and fertilizers. For hydrophobic pesticides, nanoencapsulation may be a tool to provide greater stability, dispersion in aqueous media, and allowing a controlled release of the active compound. In nanofertilizers, micro- or macronutrients can be encapsulated by nanomaterials which allow to release of nutrients into the soil gradually and in a controlled way maintaining soil fertility, thus preventing eutrophication and pollution of water resources [14]. At the same time, the environmental safety of the use of nanomaterials remains sufficiently assessed. After discharge into the environment, nanomaterials are tedious for detection and if the remediation is not secured, it leads to several environmental issues [60].

1.4 Features of the Nanomaterial Effect on Plants

The peculiarities of the biological effects of nanomaterials are due to their high adsorption activity (i.e., binding of various molecules, ions on their surface), catalytic activity in some types of nanomaterials, the ability to integrate into cellular structures due to nanoscale dimensions, as well as the products of dissociation and biotransformation of these substances. For example, metal nanomaterials can dissociate with the release of metal ions, which largely determine their biological effects and toxicity.

The high adsorption activity of nanomaterials is due to the following reasons: (a) dispersed nanomaterials have a very large surface area and high particle number per mass unit, (b) the fraction of atoms at the surface in nanomaterials is increased, and (c) the atoms situated at the surface in nanomaterials have fewer direct neighbors [39]. As a consequence of each of these differences, the chemical and physical properties of nanomaterials change compared to their larger-dimension counterparts [39].

Nanoparticles are able to penetrate plant tissues through the leaves and root system. Some studies show that nanoparticles are able to penetrate through the pores of the cell wall and even change the size of its pores. Nanoparticles penetrate through the cell membrane by both clathrin-dependent endocytosis and clathrin-independent endocytosis [45, 51]. It is believed that different types of nanoparticles have such a mechanism of entry into plant cells [42, 71]. At the same time, some nanomaterials can disrupt the plasma membrane, inducing the formation of pores for crossing into the cell [56].

Small nanoparticles penetrate into the aboveground parts of plants more effectively than large ones. The penetration efficiency of nanoparticles depends on plant species, as well as on the size and surface charge of the nanoparticles. It is known that positively charged nanoparticles are actively absorbed only by the roots of plants, while negatively charged nanoparticles are also able to actively move from the roots to the stems and leaves [45].

Nanoparticles have two ways to move through plant tissues: the apoplastic and the symplastic transports. Apoplastic transport takes place outside the plasma membrane through the extracellular spaces, cell walls of adjacent cells and xylem vessels, whereas symplastic transport involves movement of water and substances between the cytoplasm of adjacent cells through specialized structures called plasmodesmata [43, 56, 73].

The toxicity of nanoparticles to plant cells depends on the size, while the toxicity increases with decreasing size. This, in particular for metal nanoparticles, is associated with an increase in the ability to dissociate and, accordingly, the entry of metal ions into the cell, which have a toxic effect [73]. Dendric and spindle-shaped nanoparticles have higher cytotoxicity than spherical particles. High toxicity is also characteristic of cationic nanoparticles, compared with anionic and neutral ones.

Studies based on the use of animal blood have shown that nanoparticles may interact with bio-molecules such as proteins, nucleic acids, lipids and even biological metabolites due to their nano-size and large surface-to-mass ratio. The formation of nanoparticle-protein complexes is commonly referred to as the nanoparticle-protein corona which can influence the biological reactivity of the nanoparticles [59].

It is suggested that some plant organs such as fruits, grains, flowers, and young leaves can accumulate nanoparticles. However, some nanoparticles could be transformed or degraded by plants [56].

Some metal nanoparticles have metal atoms with a variable degree of oxidation on their surface, which determines their ability to be electron donors and participate in cell redox processes in a similar way to some antioxidant enzymes (catalase, superoxidismutase). For instance, cerium (Ce) is a rare-earth element which belongs to the lanthanide series. In solution Ce may exist as Ce3+ and Ce4+ and depending on environmental conditions can cycle between both oxidation states [57, 61]. It has been shown that nanoceria displays superoxide-dismutase mimetic activity, catalase mimetic activity or the capacity to scavenge nitric oxide radicals. However, nanoceria has also been found to display oxidase activities and to generate damaging oxygen radicals in a range of organisms and cell systems [57].

In addition, it has been shown that enzyme-like activities are possessed by various nanoparticles where $nCeO_2$, nFe_3O_4, nCo_3O_4 nanoparticles imitate catalase; $nCeO_2$, nFe_3O_4, nCo_3O_4, $nMnO_2$, nCuO, and nAu mimic peroxidase; $nCeO_2$, nPt, and fullerene mimic superoxidismutase activity [61]. This indicates that not only metal nanoparticles have catalytic properties and the mechanisms of this phenomenon may be different.

It should be noted that high concentrations of nanoparticles have genotoxic effects in plants. Mechanisms of genotoxicity can be divided into direct genotoxicity or indirect genotoxicity. Direct genotoxicity comes from the physical interaction between nanoparticles and DNA, while the indirect genotoxicity may be due to the decrease of DNA repair or the increase of reactive oxygen species content [33].

In addition, nanoparticle stress is manifested by the differential expression of genes, which participate in a series of biological reactions, such as redox, detoxification, hormone signal pathways and signal transduction [33]. The mechanisms underlying the effects of nanomaterials on gene expression have not yet been sufficiently studied. Perhaps, in this case, nanomaterials can also directly affect expression regulatory proteins and DNA and indirectly cause these effects through the stress reaction of plants.

Nanomaterials undergo biotransformation in plant cells, the mechanisms of which are still poorly understood [54, 64]. For instance, transformation of metal-based nanoparticles in many plants such as wheat, soybean, and maize, as well as in germinated roots are reported. Some studies have shown that the nanoparticles toxicity to plant cells is likely to be dependent on the transformation's mechanisms and the behavior of these particles in transformed status [24].

Thus, the biological effects of nanomaterials may depend on their following features:

1. All hanomaterials have high adsorption activity based mainly on weak interactions between molecules. This allows them to interact with various biomolecules and modulate their functions moderately, which is especially important in the case of proteins. The nanoparticle surface can modify the structure and therefore the function of the adsorbed protein [59]. Weak interactions are widely represented in cellular structures, and perhaps this determines the low toxicity of nanomaterials for cells compared, for example, with metal ions [53], which significantly disrupt the conformation of proteins, forming chelates with them.
2. Some nanomaterials can participate directly in the redox processes in plant cells.
3. Nanomaterials can directly affect the permeability of the cell wall through the modulation of the pore size and plasmalemma of plant cells.
4. Biological effects depend on the biotransformation of nanomaterials and its products, including the dissociation of metal nanoparticles to form metal ions.

Thus, these facts demonstrate the high ability of nanoparticles to have various direct effects on plant cells. At the same time, it should take into account that any direct effects of nanoparticles that disrupt cell structures and functions will induce stress responses in plants.

1.5 Plant Hormesis and Nanomaterials

Currently, significant evidence has been accumulated for the stimulating effects of low doses/concentrations of various nanomaterials in plants, especially for metal nanoparticles [20, 36, 40, 64, 67] (Table 1). At the same time, high concentrations of nanomaterials are usually toxic to plants and cause inhibitory effects [33, 53, 64]. These facts indicate that different types of nanomaterials can cause biphasic hormetic reactions in plants.

Plant stimulation was found for various metal nanomaterials containing metals and their oxides, carbon-based nanomaterials, hybrid nanomaterials (e.g., PEG-coated Ag-nanoparticles and Ag-nanoparticles with carbon coated, Zn-chitosan nanoparticles), as well as polymer nanomaterials (example, nanoparticles of the bioactive polymer chitosan) (Table 1).

Moreover, nanomaterials have a stimulating effect on various groups of plants, from algae to higher plants, both herbaceous and woody forms, monocotyledonous and dicotyledonous species at different stages of development from seeds and seedlings to vegetative and generative stages [46, 58, 67, 71] (Table 1).

Numerous studies have shown that various nanomaterials stimulate seed germination and plant growth-related indicators, as well as yield. In addition, nanoparticles in low concentrations increase the content of photosynthetic pigments and the rate of photosynthesis (Iavicoli et al. 2017), [11, 38, 58] (Table 1). It has been shown that some nanomaterials have direct stimulating effects on the photosynthetic apparatus of plants and can, in particular, improve the photochemical reactions of photosystem II [4]. For instance, the addition of 2–4 mg L^{-1} of Cu nanocarboxylates or 0.4–4 mg L^{-1}

Table 1 Examples of plant hormesis caused by various nanomaterials

Nanomaterials (NMs) and their types	Plant species/Stimulatory effects	References
Ag-nanoparticles (metal NMs)	*Triticum aestivum*/growth, chlorophyll content, nutrient use efficiency, yield	[38, 58]
TiO_2—nano particles (metal NMs)	*Zea mays*/Growth, yield	[40, 67]
ZnO—nanoparticles (metal NMs)	*Oryza sativa L.*/Growth	[11, 40]
γFe_2O_3-nanoparticles (metal NMs)	*Glycine max* (L.) Merr.)/photosynthetic rate, growth	[11]
Zn—chitosan nanoparticles (polimeric NMs)	Zea mays (seeds, seedlings)/antifungal resistence, growth, yield, activity of antioxidant enzimes (superoxide dismutase, peroxidase)	[20]
Cu—nanoparticles (metal NMs)	*Phaeodactylum tricornutum*/ algal growth, photosynthetic pigment content	[74], Iavicoli et al. (2017)
Factory-synthesized multi-walled-carbon nanotubes (carbon-based NMs)	*Zea mays*/water content, growth	[64], Iavicoli et al. (2017)
PEG-coated and carbon-coated Ag-nanoparticles (hybrid NMs)	Poplars (*Populus deltoides* × nigra) and *Arabidopsis thaliana*/growth	[71], Iavicoli et al. (2017)
Chitosan nanoparticles (polimeric NMs)	*Triticum aestivum*/germination, growth, auxin-related gene expression, indole-3-acetic acid content	[46]
Chitosan nanoparticles (polimeric NMs)	*Coffea arabica*/pigment content, rate of photosynthesis, nutrient uptake	[68]
$MnFe_2O_4$ magnetic nanoparticles	*Hordeum vulgare*/seeds, growth, chlorophyll content	[66]
Magnetic nickel ferrite ($NiFe_2O_4$) nanoparticles (metal NMs)	*Hordeum vulgare* (three weeks seedlings)/ calcium, magnesium, potassium, sodium and manganese content, chlorophyll and carotenoid content, soluble protein content, catalase activity	[66]
Silicon nanoparticles (polimeric NMs)	*Triticum aestivum*/growth, chlorophyll content, yield	[36, 61]

of Se nanocarboxylates caused the evident initial increase in such chlorophyll a fluorescence parameters as maximal quantum yield of photosystem II photochemistry (F v/F m) and the quantum yield of photosystem II photochemistry in the light-adapted state (F v′/F m′) in green algae *Chlorella vulgaris*, which is also accompanied by the accumulation of biomass [50].

Various nanomaterials have stimulating effects on plant defense systems, in particular, by increasing the activity of antioxidant enzymes [20, 53, 66] and the content of non-enzymatic antioxidants [44] (Table 1). For instance, in *Vicia faba*, TiO_2 nanoparticles (0.01%) significantly increased shoot length, leaf area and root dry weight of plants under normal conditions. These growth promoting effects were simultaneous with increased levels of chlorophyll b, soluble sugars and proline, and enhanced activities of antioxidant enzymes [44].

It should be noted that stimulation of plant antioxidant defense can be caused by any nanoparticles (Table 1), and not only nanoparticles "mimicking" the activity of antioxidant enzymes. This indicates that indirect activation of plant defense mechanisms by low doses of nanoparticles plays an important role in their effects on plant condition. It is considered that nanomaterials regulate reactive oxygen species production (ROS) in stressed plant cells by mimicking ROS quenching antioxidants or eliciting antioxidants production [62].

In addition, low doses of nanomaterials increase the uptake of nutrients, in particular the content of calcium, magnesium, potassium, sodium and manganese in the leaves [66, 68] (Table 1).

Some studies have shown that nanomaterials in stimulating doses affect gene expression and phytohormone content (Table 1) [46]. For example, the application of 5 μg/mL chitosan nanoparticles induced the auxin-related gene expression in *Triticum aestivum*, accelerated indole-3-acetic acid (IAA) biosynthesis and transport, and reduced IAA oxidase activity resulting in the increase of IAA concentration in wheat shoots and roots [46].

The intensity of the hormetic stimulation may depend on the way plants are treated with nanomaterials. For instance, in *Glycine max*, Fe_2O_3 nanoparticles more significantly stimulated plant indicators (photosynthetic rate, growth) with foliar treatment compared to root treatment [11].

Low doses of nanomaterials of various types can cause plants to be preconditioned to various strong stressors—abiotic and pollutants. For example, in *Fragaria ananassa* Duch, Fe_3O_4 nanoparticles (0.8 ppm) improved morphological and growth parameters, as well as drought resistance [49]. Similar data were obtained for Se nanoparticles, which stimulated growth and high temperature resistance in *Lycopersicum esculentum* [35]. In *Glycine max* (L.) Merr., CeO nanoparticles (0, 10, 100 and 500 mg kg^{-1}) increased the photosynthetic rate, RuBisCo carboxylase, and water use efficiency, which is accompanied by enhanced salinity tolerance [18]. Silicon dioxide nanoparticles (100 μM) improved Cu and Mn resistance in *Arundinaria pygmaea* and promoted growth, photosynthesis and the action of protective enzymes [25].

In addition, it has been shown that an increase in plant resistance to heavy metals caused by nanoparticles may also be associated with the adsorption of ions of these metals on the surface of nanoparticles and thus a decrease in the bioavailability of these pollutants, that is, direct effects of nanomaterials are observed here, and not mediated through stimulation of plant defense mechanisms [4].

Some nanoparticles also can increase plant resistance to various phytopathogens (bacteria, fungi, viruses). It is believed that this is due to their direct toxic effects on phytopathogens [41]. However, low concentrations of some nanomaterials can stimulate plant immunity. For instance, in *Camellia sinensis*, chitosan nanoparticles treatment of leaves produced significant improvement in the plant's innate immune response through induction of defense enzyme activity, upregulation of defense related genes including that of several antioxidant enzymes as well as elevation of the levels of total phenolics [19]. It is shown that bio-Cu nanoparticles can suppress watermelon bacterial fruit blotch by triggering the expressions of genes related to stomatal closure (*ClMPK9, ClLOX1, ClPR1, ClPR2,* and *RbohF*) to activate innate immune response, and thus restricting pathogen entry in plants [52]. Similarly, other studies have demonstrated that bio-chitosan-Fe nanocomposites also activated immune response to suppress bacterial leaf blight in rice plants by modulating the expressions of defensive genes (*OsNPR1, OSPR1, OsPR5, OsPR9,* and *OsEDS1*) [10, 52]. In addition, low concentrations of copper nanoparticles stimulate plant growth, various biochemical parameters, phytohormone levels, as well as plant immunity and resistance to phytopathogens [48]. This demonstrates the hormetic preconditioning of plants by nanomaterials to phytopathogens by activating plant immunity.

The hormesis of nanoparticles can be accompanied by hormetic trade-offs similar to the hormetic effects of other types of low-dose stressors. For instance, in seedling of barley (*Hordeum vulgare* L.), magnetic nanoparticles in low concentration (250 mg L^{-1}) significantly stimulated growth, but did not affect chlorophyll level and antioxidant defense, which demonstrate hormetic trade-off 2. At the same time, high concentration of nanoparticles (1000 mg L^{-1}) inhibited plant growth, but stimulated antioxidant defense (catalase activity) and chlorophyll a content (hormetic trade-off 1). $NiFe_2O_4$ nanoparticles (125, 250, 500, and 1000 mg/L) caused hormetic trade-off 2 inbarley when they significant increased some physiological parameters, such as chlorophyll content (50%), carotenoid levels (51%), and soluble protein (35%). However, plant biomass and growth indicators did not change up to 500 mg/L level [65].

Recent assessment of the quantitative aspects of hormetic stimulation caused by various nanomaterials in higher plants and algae has shown that they are very similar to indicators of hormesis induced by other stressors in low doses. The median maximum stimulating response (MAX) was 123% and is usually below 200% of the control response to the hormesis of nanomaterials, similar to the stimulating effect of other stress factors [4]. In addition, MAX was also lower in algae than in plants, and occurred commonly at concentrations of nanomaterials <100 mg L^{-1}. The no-observed-adverse-effect level (NOAEL) to MAX ratio was 2.4 for algae and 1.7 for plants, and the two distributions differed significantly. Ag nanoparticles induced higher MAX than TiO_2 and ZnO nanoparticles. The MAX varied upon nanomaterial application methods, growth stage of application (seed versus vegetative), type of endpoint and time window. While nanomaterial size did not affect significantly the MAX, sizes $\leq$50 nm appeared to have lower NOAEL:MAX ratio than sizes $\geq$100 nm, suggesting higher risks from incorrect application [4].

1.6 Mechanisms Underlying Nanomaterial Hormesis in Plants

Numerous data show that the hormetic stimulating effects of various nanomaterials are very similar to the stimulation caused by other stressors in low doses. For example, stimulation of growth, rate and other indicators of photosynthesis, chlorophyll levels, antioxidant defense is observed when exposed to moderate abiotic stressors (temperature, mineral nutrition, humidity, light) [6, 26], various pollutants (heavy metals, herbicides, pharmaceuticals, hydrocarbons, etc.) [3, 5, 7, 12, 27, 28], and even chemical agents involved in biotic interactions of plants with other species [34, 55, 69], as well as moderate damage to plants by phytopathogens [47]. It should be noted that, like the hormesis of nanomaterials, the hormetic effects of these low-dose stressors could be accompanied by changes in levels of phytohormones and gene expression (Agathokleous et al. 2020c) [28].

These facts indicate the existence of universal (i.e., non-specific) mechanisms of hormetic stimulation in plants, which do not depend on the quality of a low-dose stressor and, accordingly, the mechanisms of its effect on plant cells [27].

It is believed that, in particular, the antioxidant system, heat shock proteins, various phytohormones involved in responses to any stressors, primarily abscisic acid, belong to the non-specific defense in plants [7, 27–29].

It is assumed that any stressors in low doses cause moderate damage to plant cells and thus induce activation of non-specific defense systems of plants, which leads to hormetic overcompensation, and, accordingly, stimulating effects (Fig. 2). In this case, the type of damage caused by stressors of different quality may vary. Moreover, it may be a stressor that plants have not previously encountered in the process of evolution. Nevertheless, plants, like other groups of organisms, are able to successfully cope with new environmental challenges. It is suggested that plants can withstand most environmental challenges, including new ones, due to the systems maintaining all the parameters of the plant body and cells. These systems are activated by appropriate specific stress damage [27].

Moderate damaging effects of nanomaterials may be based on the following features:

1. Nanomaterials are characterized by high adsorption activity and can bind to proteins, thus affecting their conformation and disrupting functions.
2. Due to the nanoscale size, these materials can be embedded in various cellular structures, disrupting their functions.
3. Metal nanomaterials are capable of dissociating with the release of metal ions, including heavy metals, which act as protein poisons, forming chelates with proteins and thus disrupting their functions.
4. Some nanoparticles contain elements of variable oxidation state, which can be involved in redox processes in such a way as to cause a disruption of the redox processes in cells.

In turn, the specificity of the damaging effects of nanomaterials on cells determines the specifics of activation of the corresponding protective mechanisms of

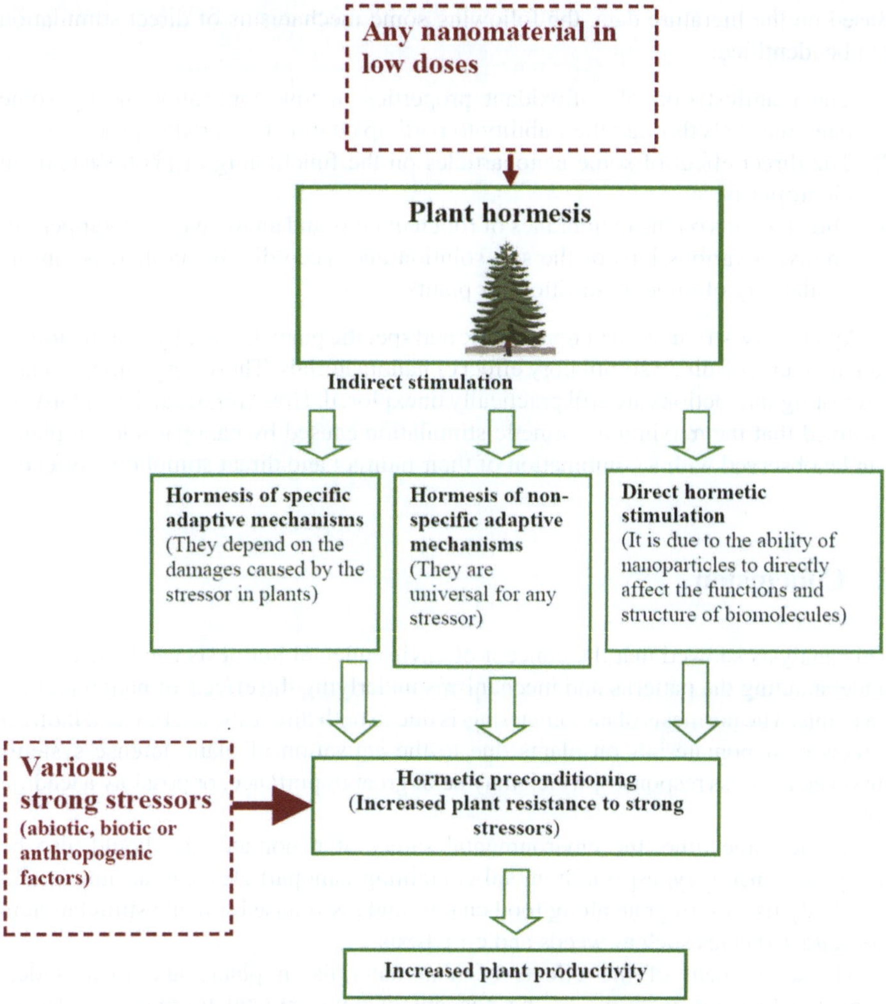

Fig. 2 Mechanisms of hormetic stimulation of plants caused by low doses of various nanomaterials (explanations are provided in the text of this section) [27–29], with significant changes)

cells (Fig. 2). For example, one should expect an increase in the level of protective chelating compounds (peptides, organic acids, etc.) that bind metal ions released from metal nanoparticles, causing their detoxification. This is a specific protection against metal nanomaterials, but not other types of nanoparticles, the effects of which are not related to metal ions.

At the same time, some nanomaterials have features that indicate their ability, along with nonspecific and specific indirect activation of plant defense systems, to cause direct stimulating effects due to the peculiarities of their effect on plant cells (Fig. 2).

Based on the literature data, the following some mechanisms of direct stimulation can be identified:

1. The manifestation of antioxidant properties in low concentrations by some nanomaterials through their ability to participate directly in redox processes.
2. The direct effect of some nanoparticles on the functioning of photosystems in chloroplasts.
3. Direct effects on the membranes of root hair cells and an increase in their permeability to various ions of the soil solution and, accordingly, an increase in the availability of mineral nutrition for plants.

Apparently, stimulation of nonspecific and specific plant defense by nanomaterials can interact with direct stimulatory effect of nanomaterials. These very important and interesting interactions are still practically unexplored. However, it can be tentatively assumed that the maximum hormetic stimulation caused by nanoparticles in plants can be observed with a combination of their indirect and direct stimulating effects.

2 Conclusion

This analysis showed that the concept of environmental hormesis can be useful for understanding the patterns and mechanisms underlying the effects of nanomaterials on plants. The hormesis of nanomaterials is due to both direct stimulation and indirect effects of nanomaterials on plants due to the activation of plant defense systems involved in stress responses, which may be of great importance, or possibly a leading role.

At the same time, the environmental safety of nanomaterials should also be analyzed, since they, especially metal containing nanoparticles, can accumulate in soil, body tissues, migrate along food chains, and also cause hormetic stimulation in non-target species such as weeds and crop pests.

The assessment of the effects of nanomaterials in plants and their underlying mechanisms in terms of the concept of environmetal hormesis is also of applied importance, in particular for improving the effectiveness of preparations based on nanomaterials (nanoherbicides, nano fertilizers, etc.) and improving their environmental safety.

The assessment of the impact of nanomaterials on plants and their underlying mechanisms from the point of view of the concept of environmental hormesis is also of practical importance, in particular, to increase the effectiveness of nanomaterials (e.g., nanoherbicides, nanofertilizers, etc.) and increase their environmental safety.

References

1. Agathokleous E (2018) Environmental hormesis, a fundamental non-monotonic biological phenomenon with implications in ecotoxicology and environmental safety. Ecotoxicol Environ Saf 148:1042–1053. https://doi.org/10.1016/j.ecoenv.2017.12.003
2. Agathokleous E (2021) The rise and fall of photosynthesis: hormetic dose response in plants. J For Res 32:889–898. https://doi.org/10.1007/s11676-020-01252-1
3. Agathokleous E, Kitao M, Calabrese EJ (2018) Human and veterinary antibiotics induce hormesis in plants: scientific and regulatory issues and an environmental perspective. Environ Int 120:489–495. https://doi.org/10.1016/j.envint.2018.08.035
4. Agathokleous E, Feng ZZ, Iavicoli I, Calabrese EJ (2019) The two faces of nanomaterials: a quantification of hormesis in algae and plants. Environ Int 131:105044. https://doi.org/10.1016/j.envint.2019.105044
5. Agathokleous E, Belz RG, Calatayud V, De Marco A, Hoshika Y, Kitao M, Saitanis CJ, Sicard P, Paoletti E, Calabrese EJ (2019) Predicting the effect of ozone on vegetation via linear non-threshold (LNT), threshold and hormetic dose-response models. Sci Total Environ 649:61–74. https://doi.org/10.1016/j.scitotenv.2018.08.264
6. Agathokleous E, Kitao M, Harayama H, Calabrese EJ (2019) Temperature-induced hormesis in plants. J For Res 30:13–20. https://doi.org/10.1007/s11676-018-0790-7
7. Agathokleous E, Barceló D, Tsatsakis A, Calabrese EJ (2020) Hydrocarbon-induced hormesis: 101 years of evidence at the margin? Environ Pollut 265(Pt B):114846. https://doi.org/10.1016/j.envpol.2020.114846
8. Agathokleous E, Kitao M, Calabrese EJ (2020) Hormesis: highly generalizable and beyond laboratory. Trends Plant Sci 25(11):1076–1086. https://doi.org/10.1016/j.tplants.2020.05.006
9. Agathokleous E, Calabrese EJ (2020) A global environmental health perspective and optimisation of stress. Sci Total Environ 704:135263. https://doi.org/10.1016/j.scitotenv.2019.135263
10. Ahmed T, Noman M, Jiang H, Shahid M, Ma C, Wu Z, Nazir MM, Ali MA, White JC, Chen J, Li B (2022) Bioengineered chitosan-iron nanocomposite controls bacterial leaf blight disease by modulating plant defense response and nutritional status of rice (*Oryza sativa* L.). Nano Today 45, Article 101547. https://doi.org/10.1016/j.nantod.2022.101547
11. Alidoust D, Isoda A (2013) Effect of γFe_2O_3 nanoparticles on photosynthetic characteristic of soybean (Glycine max (L.) Merr.): foliar spray versus soil amendment. Acta Physiol Plant 35:3365–3375. https://doi.org/10.1007/s11738-013-1369-8
12. Belz RG (2020) Low herbicide doses can change the responses of weeds to subsequent treatments in the next generation: metamitron exposed PSII-target-site resistant *Chenopodium album* as a case study. Pest Manag Sci 76(9):3056–3065. https://doi.org/10.1002/ps.5856
13. Berry R, López-Martínez G (2020) A dose of experimental hormesis: when mild stress protects and improves animal performance. Comp Biochem Physiol Part A: Mole Integr Physiol 242:110658. https://doi.org/10.1016/j.cbpa.2020.110658
14. Bratovcic A, Hikal WM, Said-Al Ahl HAH, Tkachenko KG, Baeshen RS, Sabra AS, Sany H (2021) Nanopesticides and nanofertilizers and agricultural development: scopes, advances and applications. Open J Ecol 11:301–316. https://doi.org/10.4236/oje.2021.114022
15. Calabrese EJ, Blain RB (2009) Hormesis and plant biology. Environ Pollut 157(1):42–8. https://doi.org/10.1016/j.envpol.2008.07.028
16. Calabrese EJ (2018) Hormesis: path and progression to significance. Int J Mol Sci 19:2871. https://doi.org/10.3390/ijms19102871
17. Calabrese EJ (2008) Hormesis: why it is important to toxicology and toxicologists. Environ Toxicol Chem 27(7):1451–74. https://doi.org/10.1897/07-541
18. Cao Z, Stowers C, Rossi L, Zhang W, Lombardini L, Ma X (2017) Physiological effects of cerium oxide nanoparticles on the photosynthesis and water use efficiency of soybean (*Glycine max* (L.) Merr.). Environ Sci Nano 4:1086–1094. https://doi.org/10.1039/C7EN00015D

19. Chandra S, Chakraborty N, Dasgupta A, Sarkar J, Panda K, Acharya K (2015) Chitosan nanoparticles: a positive modulator of innate immune responses in plants. Sci Rep 5:15195. https://doi.org/10.1038/srep15195
20. Choudhary RC, Kumaraswamy RV, Kumari S, Sharma SS, Pal A, Raliya R, Biswas P, Saharan V (2019) Zinc encapsulated chitosan nanoparticle to promote maize crop yield. Int J Biol Macromol 127:126–135. https://doi.org/10.1016/j.ijbiomac.2018.12.274
21. Cocozza C, Perone A, Giordano C, Salvatici MC, Pignattelli S, Raio A, Schaub M, Sever K, Innes JL, Tognetti R, Cherubini P (2019) Silver nanoparticles enter the tree stem faster through leaves than through roots. Tree Physiol 39(7):1251–1261. https://doi.org/10.1093/treephys/tpz046
22. Costantini D (2019) Hormesis promotes evolutionary change. Dose Response 17(2):1559325819843376. https://doi.org/10.1177/1559325819843376
23. Dragićević M, Platiša J, Nikolić R, Todorović S, Bogdanović M, Mitić N, Simonović A (2012) Herbicide phosphinothricin causes direct stimulation hormesis. Dose Response 11(3):344–360. https://doi.org/10.2203/dose-response.12-039.Simonovic
24. Elbehiry F, Elbasiouny H, El-Ramady H (2022) Bioaccumulation and biotransformation of metal-based nanoparticles in plants. In: Rajput VD, Minkina T, Sushkova S, Mandzhieva SS, Rensing C (eds) Nanomaterial-plant interactions, toxicity of nanoparticles in plants, vol 5. Academic, pp 299–315. https://doi.org/10.1016/B978-0-323-90774-3.00001-5
25. Emamverdian A, Ding Y, Mokhberdoran F, Ahmad Z, Xie Y (2020) Determination of heavy metal tolerance threshold in a bamboo species (*Arundinaria pygmaea*) as treated with silicon dioxide nanoparticles. Glob Ecol Cons 24:e01306. https://doi.org/10.1016/j.gecco.2020.e01306
26. Erofeeva EA (2021) Plant hormesis and Shelford's tolerance law curve. J For Res 32:1789–1802. https://doi.org/10.1007/s11676-021-01312-0
27. Erofeeva EA (2022) Environmental hormesis of non-specific and specific adaptive mechanisms in plants. Sci Total Environ 804:150059. https://doi.org/10.1016/j.scitotenv.2021.150059
28. Erofeeva EA (2022) Hormesis in plants: its common occurrence across stresses. Curr Opin Toxicol 30:100333. https://doi.org/10.1016/j.cotox.2022.02.006
29. Erofeeva EA (2022) Environmental hormesis: from cell to ecosystem. Curr Opin Environ Sci Health 29:100378. https://doi.org/10.1016/j.coesh.2022.100378
30. Erofeeva EA (2023) Hormetic effects of abiotic environmental stressors in woody plants in the context of climate change. J For Res 34:7–19. https://doi.org/10.1007/s11676-022-01591-1
31. Erofeeva EA (2023) Environmental hormesis in living systems: the role of hormetic trade-offs. Sci Total Environ 901:166022. https://doi.org/10.1016/j.scitotenv.2023.166022
32. Figueroa-Macías JP, García YC, Núñez M, Díaz K, Olea AF, Espinoza L (2021) Plant growth-defense trade-offs: molecular processes leading to physiological changes. Int J Mol Sci 22(2):693. https://doi.org/10.3390/ijms22020693
33. Gao M, Chang J, Wang Z, Zhang H, Wang T (2023) Advances in transport and toxicity of nanoparticles in plants. J Nanobiotechnol 21:75. https://doi.org/10.1186/s12951-023-01830-5
34. Godínez-Mendoza PL, Rico-Chávez AK, Ferrusquía-Jimenez NI, Carbajal-Valenzuela IA, Villagómez-Aranda AL, Torres-Pacheco I, Guevara-González RG (2023) Plant hormesis: revising of the concepts of biostimulation, elicitation and their application in a sustainable agricultural production. Sci Total Environ 894:164883. https://doi.org/10.1016/j.scitotenv.2023.164883
35. Haghighi M, Abolghasemi R, da Silva JAT (2014) Low and high temperature stress affect the growth characteristics of tomato in hydroponic culture with Se and nano-Se amendment. Sci Hortic 178:231–240. https://doi.org/10.1016/j.scienta.2014.09.006
36. Hussain A, Rizwan M, Ali Q, Ali S (2019) Seed priming with silicon nanoparticles improved the biomass and yield while reduced the oxidative stress and cadmium concentration in wheat grains. Environ Sci Pollut Res 26:7579–7588. https://doi.org/10.1007/s11356-019-04210-5
37. Iavicoli I, Leso V, Fontana L, Calabrese EJ (2018) Nanoparticle exposure and hormetic dose–responses: an Update. Int J Mol Sci 19:805. https://doi.org/10.3390/ijms19030805

38. Jhanzab HM, Razzaq A, Jilani G, Rehman A, Hafeez A, Yasmeen F (2015) Silver nanoparticles enhance the growth, yield and nutrient use efficiency of wheat. Int J Agron Agric Res 7:15–22
39. Joudeh N, Linke D (2022) Nanoparticle classification, physicochemical properties, characterization, and applications: a comprehensive review for biologists. J Nanobiotechnol 20(1):262. https://doi.org/10.1186/s12951-022-01477-8
40. Jyothi TV, Hebsur NS (2017) Effect of nanofertilizers on growth and yield of selected cereals—a review. Agric Rev 38(2):112–120. https://doi.org/10.18805/ag.v38i02.7942
41. Kanakari E, Dendrinou-Samara C (2023) Fighting phytopathogens with engineered inorganic-based nanoparticles. Materials 16:2388. https://doi.org/10.3390/ma16062388
42. Khodakovskaya M, Dervishi E, Mahmood M, Xu Y, Li Z, Watanabe F, Biris AS (2009) Carbon nanotubes are able to penetrate plant seed coat and dramatically affect seed germination and plant growth. ACS Nano 3:3221–3227. https://doi.org/10.1021/nn900887m
43. Koelmel J, Leland T, Wang H, Amarasiriwardena D, Xing B (2013) Investigation of gold nanoparticles uptake and their tissue level distribution in rice plants by laser ablation-inductively coupled-mass spectrometry. Environ Pollut 174:222–228. https://doi.org/10.1016/j.envpol.2012.11.026
44. Latef AAHA, Srivastava AK, El-sadek MSA, Kordrostami M, Tran L-SP (2018) Titanium dioxide nanoparticles improve growth and enhance tolerance of broad bean plants under saline soil conditions. Land Degrad Dev 29(4):1065–1073. https://doi.org/10.1002/ldr.2780
45. Li H, Ye X, Guo X, Geng Z, Wang G (2016) Effects of surface ligands on the uptake and transport of gold nanoparticles in rice and tomato. J Hazard Mater 314:188–196. https://doi.org/10.1016/j.jhazmat.2016.04.043
46. Li R, He J, Xie H, Wang W, Bose SK, Sun Y, Hu J, Yin H (2019) Effects of chitosan nanoparticles on seed germination and seedling growth of wheat (*Triticum aestivum* L.). Int J Biol Macromol 126:91–100. https://doi.org/10.1016/j.ijbiomac.2018.12.118
47. López Sánchez A, Pascual-Pardo D, Furci L, Roberts MR, Ton J (2021) Costs and benefits of transgenerational induced resistance in *Arabidopsis*. Front Plant Sci 12:644999. https://doi.org/10.3389/fpls.2021.644999
48. Ma C, Borgatta J, Hudson BG, Tamijani AA, La Torre-Roche D, ZuverzaMena N, Shen Y, Elmer W, Xing B, Mason SE (2020) Advanced material modulation of nutritional and phytohormone status alleviates damage from soybean sudden death syndrome. Nat Nanotechnol 15(12):1033–42
49. Mozafari A, Havas F, Ghaderi N (2018) Application of iron nanoparticles and salicylic acid in in vitro culture of strawberries (*Fragaria ananassa* Duch.) to cope with drought stress. Plant Cell, Tissue Organ Culture 132:511–523
50. Mykhaylenko NF, Zolotareva EK (2017) The effect of copper and selenium nanocarboxylates on biomass accumulation and photosynthetic energy transduction efficiency of the green algae *Chlorella vulgaris*. Nanoscale Res Lett 12:147. https://doi.org/10.1186/s11671-017-1914-2
51. Navarro E, Baun A, Behra R, Hartmann NB, Filser J, Miao A-J, Quigg A, Santschi PH, Sigg L (2008) Environmental behavior and ecotoxicity of engineered nanoparticles to algae, plants, and fungi. Ecotoxicology 17:372–386. https://doi.org/10.1007/s10646-008-0214-0
52. Noman M, Ahmed T, White JC, Nazir MM, Li D, Song F (2023) *Bacillus altitudinis*-stabilized multifarious copper nanoparticles prevent bacterial fruit blotch in watermelon (*Citrullus lanatus* L.): direct pathogen inhibition, in planta particles accumulation, and host stomatal immunity modulation. Small 19(15):2207136. https://doi.org/10.1002/smll.202207136
53. Omar RA, Afreen S, Talreja N, Chauhan D, Ashfaq M (2019) Impact of nanomaterials in plant systems. In: Prasad R (eds) Plant nanobionics: nanotechnology in the life sciences. Springer, Cham. https://doi.org/10.1007/978-3-030-12496-06
54. Pagano L, Rossi R, White JC, Marmiroli N, Marmiroli M (2023) Nanomaterials biotransformation: in planta mechanisms of action. Environ Pollut 318:120834. https://doi.org/10.1016/j.envpol.2022.120834
55. Park M-H, Kim K, Hwang S-J (2020) Differential effects of the allelochemical juglone on growth of harmful and non-target freshwater algae. Appl Sci 10:2873. https://doi.org/10.3390/app10082873

56. Pérez-de-Luque A (2017) Interaction of nanomaterials with plants: what do we need for real applications in agriculture? Front Environ Sci 5:12. https://doi.org/10.3389/fenvs.2017.00012
57. Pulido-Reyes G, Rodea-Palomares I, Das S, Sakthivel TS, Leganes F, Rosal R, Seal S, Fernández-Piñas F (2015) Untangling the biological effects of cerium oxide nanoparticles: the role of surface valence states. Sci Rep 5:15613. https://doi.org/10.1038/srep15613
58. Razzaq A, Ammara R, Jhanzab HM, Mahmood T, Hafeez A, Hussain S (2016) A novel nanomaterial to enhance growth and yield of wheat. J Nanosci Technol 2:55–58
59. Saptarshi SR, Duschi A, Lopata AL (2013) Interaction of nanoparticles with proteins: relation to bio-reactivity of the nanoparticle. J Nanobiotechnology 11:26
60. Saravanan A, Kumar PS, Karishma S, Vo D-VN, Jeevanantham S, Yaashikaa PR, George CS (2021) A review on biosynthesis of metal nanoparticles and its environmental applications. Chemosphere 264(2):128580. https://doi.org/10.1016/j.chemosphere.2020.128580
61. Sarraf M, Vishwakarma K, Kumar V, Arif N, Das S, Johnson R, Janeeshma E, Puthur JT, Aliniaeifard S, Chauhan DK, Fujita M, Hasanuzzaman M (2022) Metal/metalloid-based nanomaterials for plant abiotic stress tolerance: an overview of the mechanisms. Plants 11:316. https://doi.org/10.3390/plants11030316
62. Saswati B, Sumanti G, Jayita S (2023) Nanoparticles regulate redox metabolism in plants during abiotic stress within hormetic boundaries. Funct Plant Biol 50:850–869. https://doi.org/10.1071/FP23068
63. Shang L, Nienhaus K, Nienhaus GU (2014) Engeneering nanoparticles interacting with cells: size matters. J Nanobiotechnol 12:5
64. Tiwari DK, Dasgupta-Schubert N, Villaseñor Cendejas LM, Villegas J, Carreto Montoya L, Borjas Garcıa SE (2014) Interfacing carbon nanotubes (CNT) with plants: enhancement of growth, water and ionic nutrient uptake in maize (*Zea mays*) and implications for nanoagriculture. Appl Nanosci 4:577–591
65. Tombuloglu H, Slimani Y, Tombuloglu G, Almessiere M, Baykal A, Ercan I, Sozeri H (2019) Tracking of NiFe2O4 nanoparticles in barley (*Hordeum vulgare* L.) and their impact on plant growth, biomass, pigmentation, catalase activity, and mineral uptake. Environ Nanotechnol Monit Manag 11:100223. https://doi.org/10.1016/j.enmm.2019.100223
66. Tombuloglu H, Tombuloglu G, Slimani Y, Ercan I, Sozeri H, Baykal A (2018) Impact of manganese ferrite (MnFe2O4) nanoparticles on growth and magnetic character of barley (*Hordeum vulgare* L.). Environ Pollut 243:872–881. https://doi.org/10.1016/j.envpol.2018.08.096
67. Trela-Makowej A, Orzechowska A, Szymańska R (2024) Less is more: the hormetic effect of titanium dioxide nanoparticles on plants. Sci Total Environ 910:168669. https://doi.org/10.1016/j.scitotenv.2023.168669
68. Van SN, Minh HD, Anh DN (2013) Study on chitosan nanoparticles on biophysical characteristics and growth of Robusta coffee in green house. Biocatal Agric Biotechnol 2(4):289–294. https://doi.org/10.1016/j.bcab.2013.06.001
69. Vargas-Hernandez M, Macias-Bobadilla I, Guevara-Gonzalez RG, Romero-Gomez SJ, Rico-Garcia E, Ocampo-Velazquez RV, Alvarez-Arquieta LL, Torres-Pacheco I (2017) Plant hormesis management with biostimulants of biotic origin in agriculture. Front Plant Sci 8:1762. https://doi.org/10.3389/fpls.2017.01762
70. Wang W-N, Tarafdar JC, Biswas P (2013) Nanoparticle synthesis and delivery by an aerosol route for watermelon plant foliar uptake. J Nanopart Res 15:1417. https://doi.org/10.1007/s11051-013-1417-8
71. Wang J, Koo Y, Alexander A, Yang Y, Westerhof S, Zhang Q, Schnoor JL, Colvin VL, Braam J, Alvarez PJ (2013) Phytostimulation of poplars and *Arabidopsis* exposed to silver nanoparticles and Ag+ at sublethal concentrations. Environ Sci Technol 47:5442–5449
72. Wang X, Xie H, Wang P, Yin H (2023) Nanoparticles in plants: uptake, transport and physiological activity in leaf and root. Materials 16:3097. https://doi.org/10.3390/ma16083097
73. Zhai G, Walters KS, Peate DW, Alvarez PJ, Schnoor JL (2014) Transport of gold nanoparticles through plasmodesmata and precipitation of gold ions in woody poplar. Environ Sci Technol Lett 1:146–151. https://doi.org/10.1021/ez400202b

74. Zhu Y, Xu J, Lu T, Zhang M, Ke M, Fu Z, Pan X, Qian H (2017) A comparison of the effects of copper nanoparticles and copper sulfate on *Phaeodactylum tricornutum* physiology and transcription. Environ Toxicol Pharmacol 56:43–49
75. Züst T, Agrawal AA (2017) Trade-offs between plant growth and defense against insect herbivory: an emerging mechanistic synthesis. Annu Rev Plant Biol 68:513–534. https://doi.org/10.1146/annurev-arplant-042916-040856

Biostimulation of Crops with Nanomaterials from a Transcriptomic Perspective

Oscar Sariñana-Aldaco, Rebeca Casique-Valdés, Luz Leticia Rivera-Solís, Verónica Soriano-Puente, Guadalupe Magdaleno-García, Adalberto Benavides-Mendoza, and Susana González-Morales

Abstract Nanotechnology is playing an increasingly important role in modern agriculture, as it offers new opportunities to improve the efficiency and sustainability of agricultural systems, promoting food security. The role of nanomaterials (NMs) as biostimulants in plants has been demonstrated through numerous investigations. The application of NMs to plants can induce positive changes in the metabolism, composition, and nutraceutical quality of the plants, as well as their tolerance to stress or toxicity. The interaction between the surface charges of NMs and plant cells leads to a sequence of events of perception, transduction, signaling, effector action and modification of genetic expression, which modifies metabolism or cellular characteristics. However, at the molecular level, there is little information on the changes in gene expression induced by NMs in plants. The objective of this review is to describe the impact of NMs as biostimulants in plants from the perspective of the analysis of the transcriptional response (changes in gene expression), which will provide a greater understanding of the molecular mechanisms involved in plant cells exposed to NMs.

Keywords Biostimulants · Gene expression · Protein kinase · Transcription factor

O. Sariñana-Aldaco · R. Casique-Valdés
SECIHTI-Department of Horticulture, Antonio Narro Autonomous Agrarian University, Saltillo, Coahuila, Mexico

L. L. Rivera-Solís · V. Soriano-Puente · G. Magdaleno-García
Doctorate of Science in Protected Agriculture, Antonio Narro Autonomous Agrarian University, Saltillo, Coahuila, Mexico

A. Benavides-Mendoza
Department of Horticulture, Antonio Narro Autonomous Agrarian University, Saltillo, Coahuila, Mexico

S. González-Morales (✉)
SECIHTI-Antonio Narro Autonomous Agrarian University, Saltillo, Coahuila, Mexico
e-mail: sgonzalezmo@secihti.mx; qfb_sgm@hotmail.com

© The Author(s), under exclusive license to Springer Nature Singapore Pte Ltd. 2025

A. Juárez-Maldonado et al. (eds.), *Plant Biostimulation with Nanomaterials*, Smart Nanomaterials Technology, https://doi.org/10.1007/978-981-96-4648-7_2

1 Introduction

Faced with continuous significant changes at the global level, the agricultural sector faces a series of emerging challenges in its quest to guarantee a sustainable diet for the growing world population (9.7 billion by 2050). In addition, climate change constitutes a significant threat to food security and sustainable environmental management [1]. Given this scenario, it is essential to move forward urgently in the implementation of techniques that improve the efficiency of agricultural, livestock and forestry activities and that fundamentally reduce the ecological impact associated with these practices [2]. An alternative to the aforementioned methods would be techniques that, without the need to increase the surface area of cultivated soil and efficiently use water, fertilizers and pesticides, could increase agricultural productivity [3]. An example of a technique for improving the sustainability of modern agriculture is the introduction of biostimulants to crops.

The definition of biostimulant according to Regulation 2019/1009 of the European Parliament [4] states that it is a product that stimulates nutritional processes in plants, regardless of the nutrient content of the product, with the aim of improving one or more of the following characteristics of the plant or the plant rhizosphere: (1) efficiency in the use of nutrients, (2) tolerance to abiotic stress, (3) quality traits and (4) availability of nutrients confined in the soil or rhizosphere. There are a wide variety of environmental factors capable of inducing biostimulation; however, in this chapter, we will refer exclusively to the NMs of metals, semimetals and their oxides, organic, inorganic and carbon [5].

Nanotechnology can be used as an efficient tool in the design, development and distribution of a new generation of biostimulants [1]. This approach encompasses a set of techniques designed to use the physicochemical properties of distinguished materials such as NMs, which manifest particular characteristics when their dimensions are in the range of 0.4–100 nm. It is important to note that this interval represents arbitrary formal agreement to establish limits between NMs and materials with micrometric dimensions, since the characteristic properties of NMs are continuously manifested in a spectrum of dimensions ranging from less than 1–1000 nm [6]. The physicochemical properties of NMs can be modified by changing their size, shape (spherical, rod, hyperbranched, laminar or layered structures), chemical composition (organic, inorganic, hybrid or composite) and surface properties (surface charge and coating, functional groups or target end groups) [7].

The applications of NMs in agriculture include as nanofertilizers, fertilizer nanocarriers, nanopesticides, pesticide nanocarriers, nanosensors, as well as nanobiostimulants [8, 9]. These applications are due to multiple characteristics such as size, shape, surface area and reactivity, compared to those of bulk materials [10]. The action of NMs as biostimulants is based on the impact of surface charges and the electrical double layer on their interaction with plant cells. This suggests the possibility that surface interactions are part of the explanation for the ability of NMs to provoke beneficial responses in plants [5].

The interaction of plants with NMs leads to changes in behavior at the subcellular, cellular, tissue, organ or whole-organism levels. Adaptive responses are triggered through a sequence of events that encompass perception, transduction, and signaling, which, in turn, modify gene expression or metabolism [11]. An additional factor that contributes to the complexity of the interactions between plant cells and NMs is the formation of the superficial crown. The surface of NMs adsorbs biomolecules, giving rise to the formation of a crown whose composition is characteristic of each type of NM and the environment in which said NMs are found [5]. The initial interaction between NMs and cells occurs between the crown of NMs and receptors, as well as between the surface of cell walls and membranes [12]. After the first contact of NMs with plant cells, biochemical signals, such as abscisic acid (ABA), salicylates or other hormones and metabolites, are generated in the root or leaf cells. These signals travel through vascular structures to the rest of the plant, where they induce biostimulation, which manifests through adjustments in metabolism and gene expression [13].

The progress of omics sciences, driven by the development of genome sequencing technologies and the decrease in costs associated with these techniques, has revolutionized the understanding of the identification of metabolic pathways in plants [14]. Through transcriptomic studies, it is feasible to identify molecular markers that are related to common responses in diverse crop species [15]. These molecular markers can be used to understand the molecular mechanism of action of NMs, their interaction with the environment and the genotype of plants.

The objective of this chapter is to present the state of the art of the genes that are expressed in different plant species when they are subjected to a biostimulation process with NMs. The following sections are included in this manuscript: (1) NMs of metals, semimetals and their oxides, (2) inorganic NMs, (3) carbon NMs, and (4) organic NMs. In each section, we present papers that describe the genes whose level of transcription changes in plants in response to the application of NMs. In total, 44 recent papers (published in the last decade) from the four categories of NMs were included.

2 NMs of Metals, Semimetals and Their Oxides

The synthesis and use of NMs metallic, semimetals and their oxides has become of great interest because of their differences from conventional size materials [5]. These NMs have unique physical and chemical properties, such as optical, electrical and thermal properties, which is why they have been used in agriculture as biostimulants to improve nutritional efficiency, agronomic characteristics, functional quality and combat stress in crop plants [16, 17]. Like any other material (nonmetallic, carbon or organic), these NMs can affect the cell wall or even DNA, the function of which is very diverse since they can modify all omic processes in plants through its physical and chemical properties [5, 18].

Among these NMs, copper (Cu), zinc (Zn), iron (Fe), silicon (Si), titanium (Ti), silver (Ag) and their oxides stand out [19]. When metallic NMs and their oxides interact with walls and membranes, a second phase occurs in which they are transformed into ionic forms of the element and can be stored in the apoplast. These extracellular ions, depending on their concentration, oxidation–reduction potential and complexation with biomolecules, cause membrane modification by binding to receptors, which results in the expression of genes affected by this biostimulation [20, 21].

Metallic NMs, because of their different structures and charge in their crown, can modify the expression of genes in plants, as mentioned above; however, when they enter the cytoplasm and pass through ionic forms, their expression can be modified in a different way. For example, when Cu passes from the nanoform to the ionic form, during photosynthesis, it can accept electrons from plastoquinone and transfer them to oxygen, which will generate superoxide anions (O_2^-), with the ability to initiate signaling through activated protein kinases by mitogens (MAPKs) culminating in gene expression [22]. Mosa et al. [23] reported that the application of Cu–NMs (50, 100 and 200 mg L^{-1}) in nutrient solution induced the expression of the Cu–Zn *SOD* gene in cucumber shoots and roots. This gene encodes the enzyme superoxide dismutase (SOD), which is responsible for the dismutation of O_2^- to oxygen and hydrogen peroxide (H_2O_2), which means the passage from a highly toxic species to a less toxic one, protecting cells from oxidative stress [24].

Zn-NMs and their oxides are another of the most studied in recent years, due to their benefits as plant biostimulants. Pejam et al. [25] showed that ZnO–NMs (3 mg L^{-1}) differentially expressed the *R2R3MYB* and *MKK2* genes in tomato. The *R2R3MYB* gene encodes a transcription factor (TF) that is responsible for regulating the biosynthesis of secondary metabolites, mainly phenolic compounds in plants [26]. The *MKK2* gene encodes a MAPKK protein that is responsible for positively regulating the resistance of plants to low temperatures [27]. Most of the metallic NMs applied are based on Fe and its oxides. In this sense, Hu et al. [28] reported that the application of γ-Fe_2O_3–NMs (20, 50, and 100 mg L^{-1}) in *Citrus maxima* (Burm.) Merr. repressed the *FRO2* and *Nramp3* genes. These genes are normally activated when plants lack available Fe, which is why the low expression of these genes can be explained since the plants were not exposed to nutrient deficiency. The *FRO2* gene encodes an iron chelate reductase, an enzyme responsible for the reduction of Fe^{3+} to Fe^{2+}, for its subsequent absorption and assimilation in plants [29]. *Nramp3* is a protein that is located in the vacuolar membrane and is responsible for transporting Fe^{2+} to the cytoplasm under deficient conditions [30].

Among the metallic and semimetal NMs and their oxides, those based on beneficial elements such as Si and Ti stand out. Chandra et al. [31] indicate that in priming with Si–NMs (3 and 5%) in seeds of *Cicer arietinum* L. mitigates stress caused by Al (3 mM) by inducing the expression of the *SOD*, *CAT* and *APX* genes, which encode antioxidant enzymes responsible for reducing the H_2O_2 in water and oxygen [32]. The difference between these enzymes is that APX uses the antioxidant ascorbate to carry out the reduction process [33]. In the same way, Omar et al. [34] applied TiO_2–NMs (10 and 20 mg L^{-1}) in broad bean plants by the nutrient solution and under saline stress (100 and 200 mM NaCl), and show in their results a very marked expression of the Cu–Zn *SOD*, *Fe-SOD*, *CAT*, *GR*, *HSP17.9* and *HSP70* genes. The

GR gene encodes the enzyme glutathione reductase, which catalyzes the reduction of oxidized glutathione to reduced glutathione; this enzyme can directly reduce RONSS or can bind to glutathione peroxidase and reduce oxidizing species [35]. *HSP* genes encode heat shock proteins and act as molecular chaperones; that is, they protect other proteins from misfolding during abiotic stress [36].

Among the NMs belonging to this category, those of Ag also stand out, which in recent years have shown encouraging effects as plant biostimulants, mainly due to their antimicrobial action. However, some studies have indicated that they can also mitigate abiotic stress. Yadu et al. [37] primed seeds of *Cajanus cajan* L. with Ag–NMs (1.2 nM) under sodium fluoride (75 mg L^{-1}) stress, and the results indicated that the *P5CS* gene was differentially expressed when Ag–NMs were applied with no stress. The *P5CS* gene encodes the enzyme pyrroline-5-carboxylate synthetase, which is responsible for the synthesis of the amino acid proline, which acts as an osmotic agent, protecting plants from dehydration [38].

It is important to mention that metals and semimetals have physical and chemical properties that promote changes between them in different reactions. For example, by losing electrons, cations can be formed; when reacting with oxygen, some cations can form oxides, and the combination of an alkali metal with water can produce a metallic hydroxide [39]. In addition, there are elements belonging to this category that can act directly as antioxidants due to their electronic configuration [40]. All these properties mean that they can have varied effects on crop plants; however, when they are in nano form, their effects are more pronounced, and better results can be achieved. Table 1 shows the studies with the levels of transcription of genes with different NMs belonging to this category.

3 Inorganic NMs

In this section, inorganic NMs that are nonmetals or other compounds with characteristics different from those of metals and semimetals will be specifically discussed. Within this category, the use of NMs of selenium (Se), sulfur (S), phosphorus (P) and clay stands out because these NMs have shown some important effects at the transcriptomic level in crops.

Currently, Se is one of the most studied elements in agricultural areas due to its potential as a biostimulant and biofortifier [53]. It has been shown that Se can replace S in some physiological, biochemical and molecular functions due to its similar chemical characteristics [54]. However, when Se is applied in its nano form, its effects are better at lower concentrations than when Se is applied in its ionic form since it is less toxic [55]. Among the beneficial effects of Se-NMs, the improvements in agronomic variables, yield, quality and tolerance to biotic and abiotic stress stand out through the positive regulation of different genes [56]. Kang et al. [57] mention that one of the main biomolecules that gives melon the characteristic bitter taste is cucurbitacin, which plays an important role in repelling insects and herbivores and in conferring resistance to bacterial and fungal diseases. Accordingly, the authors

Table 1 Description of genes whose transcription level change in response to the application of metals, semimetals and their oxides nanomaterials in plants

NMs	Size (nm)/ Shape	Zeta potential (mV)	Plant species	Application method and rate	Differentially expressed genes	References
Ag	20/ Spherical	NP (not presented)	*C. box*	Seed priming/ 1.2 nM	*P5CS*	[37]
Au	38.5/ Polygonal and spherical	17.2	*Oryza sativa* ssp. *japonica*	Nutrient solution/50, 100, and 200 μM	*OsHMA2, OsHMA3, OsIRT1, OsIRT2, OsNramp1, OsNramp5* and *OsLCT1*	[41]
CaO	10–24/ Spherical	NP	*Hordeum vulgare* L	Nutrient solution/ 25 mg L^{-1}	*HvPHT1.1, HvPHT1.3, HvPHT1.4* and *HvPHT1.6*	[42]
CeO_2	24.4/ Spherical	9.6	Brassica chinensis L	Soil/0.7, 7, 70, and 350 mg kg^{-1}	*PetH* and *Psb27*	[43]
Cu–Fe	60.1	24.2	*Lactuca sativa* L	Foliar/1, 10, and 100 μg mL^{-1}	*MAPK, ZOG, ZOX* and *PAL*	[44]
Cu	25/ Spherical	NP	*Solanum lycopersicum* L	Root area/ 10 mg	*SOD, CAT* and *JA*	[45]
Cu	10–30	NP	*Cucumis sativus* L	Nutrient solution/50, 100, and 200 mg L^{-1}	Cu–Zn *SOD*	[23]

(continued)

Table 1 (continued)

NMs	Size (nm)/ Shape	Zeta potential (mV)	Plant species	Application method and rate	Differentially expressed genes	References
CuO	40	NP	*Glycine max* (L.) Merr	Root suspension/2, 5, and 10 mg L^{-1}	*GmCYCB1;2*, *GmCYCU4;1* and *GmCYCD3;1*	[46]
K	3–8	NP	*Cucurbita pepo* L	Foliar/0.5 g L^{-1}	*SOD*, *CAT*, *APX*, *GR* and *PrxQ*	[47]
MgO	100	NP	*S. lycopersicum*	Drench/0.1, 0.5, 0.7, and 1%	*PR1*, *LoxA*, *Osm* and *GluA*	[48]
MnO	50/ Spherical	29.2	*Capsicum annum* L	Seed priming/ 0.1, 0.5, and 1 mg L^{-1}	*Mn-SOD*	[49]
Pd	50/ Spherical	NP	*Brassica napus* L	Leaf infiltration/17, 35, and 70 mg L^{-1}	*PR1*	[50]
Yes	138–160	–23.3	*C. arietinum*	Seed priming/ 3 and 5%	*SOD*, *CAT* and *APX*	[31]
SiO_2	126/ Spherical	NP	*Zea mays* L	Foliar/0.25 g L^{-1}	*Osmotin-34*, *psbD* and *AQPs*	[51]
TiO_2	5–10	NP	*Vicia faba* L	Nutrient solution/10 and 20 mg L^{-1}	Cu–Zn *SOD*, *Fe-SOD*, *CAT*, *GR*, *HSP17.9* and *HSP70*	[34]

(continued)

Table 1 (continued)

NMs	Size (nm)/ Shape	Zeta potential (mV)	Plant species	Application method and rate	Differentially expressed genes	References
ZnO	10–30	–20	*S. lycopersicum*	Foliar/3 mg L^{-1}	*R2R3MYB* and *MKK2*	[25]
ZnO	20–45	11.29	*S. lycopersicum*	Foliar/20 and 100 mg L^{-1}	*TIP3-2*, *TPI2-1* and *TIP1-1*	[52]
γ-Fe_2O_3	20.2/ Spherical	–11.7	*C. maximum*	Foliar/20, 50, and 100 mg L^{-1}	*FRO2* and *Nramp3*	[28]

evaluated the effect of Se-NMs (2.5, 5 and 10 mg L^{-1}) in the melon crop, and their results indicate a differential expression of genes that code for cucurbitacin (*CmBt*, *CmBi*, *Cm170*, *Cm180*, *Cm490*, *Cm710* and *Cm890*), which was positively correlated with the concentration of the metabolite.

Se-NMs applied at high concentrations can cause negative effects on the morphology, physiology and biochemistry of plants, and they can cause genotoxicity; however, at adequate concentrations, Se-NMs can cause positive effects [58]. El-Badri et al. [56] treated canola seeds with Se-NMs (150 μmol L^{-1}) under saline stress (150 and 200 mM NaCl), and the results indicated that the expression of the *GST-U4*, *SULT1A* and *GoLS3* genes was positively regulated. The *GST-U4* gene encodes the enzyme glutathione S transferase, which is responsible for the detoxification of xenobiotics through conjugation with reduced glutathione [59]. The *SULT1A* gene is a member of the sulfotransferase family, which is involved in the selenation pathway [60]. *GoLS3* gene encode to the galactinol synthase, this is a key regulator of oligosaccharide synthesis in the raffinose family that promotes cold stress tolerance in plants [61].

S is considered an essential element for plants since it is involved in the synthesis of proteins and is directly related to the expression of genes [62]. It also plays a crucial role in the protection of cells against dehydration caused by abiotic stress, coupled with its antifungal function, because various secondary metabolites have S in their structure, which is what gives it its bioactive function [63]. Accordingly, some studies have shown that the use of S in nano form has positive impacts on gene expression in crops. Yuan et al. [64] mention that the use of S-NMs (300 mg L^{-1}) in canola increased tolerance to stress by mercury (Hg) (10 mg L^{-1}) by increasing the transcriptional level of the *BnGST*, *BnPOD*, *BnAPX*, *Bn SOD*, *BnST* and *BnPHO1* genes. The *BnST* gene encodes a sulfate transporter, which is a membrane protein [65]. *BnPHO1* encodes an inorganic phosphate exporter that contributes to its load in vascular tissue and can be induced by deficiency of this element [66]. Similarly, Cao et al. [67] applied S-NMs (1 mg L^{-1}) to tomato plants under *Fusarium oxysporum* stress and showed that the use of S-NMs suppressed infection by the pathogen by increasing the expression of the *PR1* and *NPR1* genes. *PR1* encodes a protein that inhibits the growth and development of pathogens [68]. These proteins are synthesized and accumulated de novo in plants as a consequence of attack by pathogens and some abiotic stresses [69]. *NPR1* promotes the synthesis of a protein that regulates acquired systemic resistance by interacting with TFs of the TGA type, which allows the expression of the *PR* genes [70].

Within this category, the use of hydroxyapatite NMs (HAp-NMs) as sources of phosphorus for plants also stands out. In addition, unlike other NMs, they are recognized for their biocompatibility and intrinsic biodegradability, which makes them suitable for use in agriculture since they do not cause problems in human or environmental health [71]. HAp-NMs are characterized by mitigating heavy metal stress in plants by improving the soil pH, which reduces the solubility and mobility of metals [72]. Wu et al. [73] reported that the use of HAp-NMs (20, 40, 60 and 80 mg L^{-1}) in rice plants mitigated cadmium (Cd) stress (1 mg L^{-1}) by regulating the synthesis pathways of jasmonic acid and gibberellins and the differential expression of the *OsZIP1* and *OsNramp1* genes. The *OsZIP1* gene plays an important role in Zn transport and metal homeostasis in plants [74]. *OsNramp1* is a macrophage protein associated with natural resistance that plays an important role in the transport and homeostasis of metal ions such as Cd [75].

Within this category, clay NMs (clay-NMs) also stand out and have been used as nanocarriers for the continuous release of agrochemicals and other biostimulant compounds, including beneficial microorganisms from plants [76]. Aseel et al. [77] applied clay-NMs (1%) to potato plants under stress by the virus Y, and the results showed that the addition of clay-NMs improved potato growth and decreased disease symptoms, as did the increase in antioxidant enzyme activity and expression of the *PAL-1*, *PR-5* and *CHI-2* genes. The *PAL-1* gene encodes the enzyme phenylalanine ammonium lyase, which is responsible for the synthesis of phenolic compounds, which are antioxidants with high reducing power in plant cells [78]. The *CHI-2* gene encodes a chitinase enzyme that is highly important for defending plants against fungal pathogens [79]. Phytopathogenic fungi contain chitin in their cell wall, so the action of chitinases is directed toward this area to cause lysis of the fungus [80]. The use of clay-NMs also helps soils increase their fertility and retain water, and the microbiome and plants develop better [81]. Within this category, the use of nanozeolite also stands out; however, no studies have reported on the transcription of genes in crop plants. Table 2 shows the studies with the levels of transcription of genes with different NMs belonging to this category.

4 Carbon NMs

The use of nanotechnology in agriculture relies not only on the application of metallic and nonmetallic NMs but also on the use of carbon NMs (CNMs). These NMs are easily absorbed by plant cells, causing positive effects on plant growth and development [86]. Among its most marked effects are the antifungal and bactericidal effects, stimulation of seed germination and antioxidant activity of plants [86, 87]. The most studied CNMs in agriculture are carbon nanotubes (CNT), graphene (GPH) and fullerene (FUll) [87]. In addition, research shows that they improve the aforementioned aspects by regulating the expression of stress-sensitive genes in plants, as mentioned by Subotić et al. [88], they priming cherry tomato seeds with FUll (200 mg L^{-1}), and detected increases in foliar pigments, lycopene in fruits and the transcription levels of the *PIP1;3*, *PIP1;5* and *PIP2;4* genes. These genes are important under conditions of osmotic stress since they are responsible for the encoding of structural membrane proteins, whose function is essential for the transport of water and solutes between the cell and its environment in response to osmotic changes [89]. Martínez-Ballesta et al. [90] evaluated the application of CNT (10 mg L^{-1}) in protoplasts of *Brassica oleracea* var. italica under salinity stress (80 mM NaCl), the results revealed that under standard growth conditions, the expression of the *PIP1* and *PIP2* genes increased, while under stress conditions, their expression was not modified. Treatment with CNT (20 mg L^{-1}) was evaluated in *B. napus* in vitro under

Table 2 Description of genes whose transcription level changed in response to the application of inorganic nanomaterials in plants

NMs	Size (nm)/Shape	Zeta potential (mV)	Plant species	Application method and rate	Differentially expressed genes	References
Clay	120–320/ Hexagonal and tabular	NP	*Solanum tuberosum* L	Foliar/1%	*PAL-1*, *PR-5* and *CHI-2*	[77]
HAp	60–80	NP	*Or sativa*	Nutrient solution/ 20, 40, 60, and 80 mg L^{-1}	*OsZIP1* and *OsNramp1*	[73]
S	37–61/Spherical	10.09	*B. napus*	In vitro/300 mg L^{-1}	*BnGST*, *BnPOD*, *BnAPX*, *Bn SOD*, *BnST* and *BnPHO1*	[64]
S	47	–16.26	*Or sativa*	Soil/100, 200, 300, and 400 mg kg^{-1}	*OsSultr1;2*, *OsSultr2;2*, *OsSiR* and *OsLsi1*	[82]
S	30/ Orrthorhombic	25.7	*S. lycopersicum*	Foliar/1 mg L^{-1}	*PR1* and *NPR1*	[67]
Se	NP	NP	*Cucumis melo* L	Foliar/2.5, 5, and 10 mg L^{-1}	*CmBt*, *CmBi*, *Cm170*, *Cm180*, *Cm490*, *Cm710* and *Cm890*	[57]
Se	120–260/ Spherical	–32.4	*B. napus*	Seed priming/ 150 μmol L^{-1}	GST-U4, *SULT1A* and *GoLS3*	[56]
Se	10–45/Spherical	NP	*C annuum*	In vitro 0.5, 1, 10, and 30 mg L^{-1}	*bZIP1* and *WRKY1*	[83]
Se	50–200/Spherical	NP	*C. melo*	Foliar/5 mg L^{-1}	*GLU*, *CHT1*, *CHT2*, *SOD*, *CAT*, *APX*, *POD*, *ODC* and *CPA*	[84]
Se	61.9/Spherical	–36.3	*Raphanus sativus* L. var. Radculus pers	Foliar/1, 5, and 10 mg L^{-1}	*ARF6*, *ARF16*, *ARF17*, *DEWAX1* and *RsDEWAX2*	[85]

salinity (125 mM NaCl), and the results showed that the defense system of plants was activated by the expression of the genes *BnSOS1*, *BnKT1*, *BnNHX1*, *BnAPX* and *BnMnSOD* [91]. *BnSOS1* is a Na^+/H^+ ion membrane anticarrier, and *BnNHX1* is a Na^+/H^+ vacuolar exchanger [92]. *BnKT1* is a K^+ transporter [93]. All these transporters under conditions of Na^+ stress are responsible for the regulation of Na^+ and K^+ within the cell, always trying to keep Na^+ within the vacuole or in the apoplast, and K^+ at the cytoplasmic level [92].

Currently, the use of GPH NMs is one of the most studied since it can promote plant growth without toxic effects, as demonstrated by Zhao et al. [94] who applied GPH (100 μg mL^{-1}) to soybean plants under drought stress. The results showed that the transcript levels of the genes *GmP5CS*, *GmGOLS*, *GmDREB1* and *GmNCED1* increased substantially, which provided the crop with some tolerance to drought stress. The *GmGOLS* gene is responsible for the expression of the enzyme galactinol synthase, which is responsible for the synthesis of galactinol, a sugar that functions as an osmolyte in plants under drought and salinity stress [95]. The *DREB1* gene encodes a protein that binds to dehydration response elements and belongs to a family of TFs that regulate the expression of genes sensitive to osmotic stress in plants [96]. The *NCED1* gene encodes the enzyme 9-cis-epoxycarotenoid dioxygenase 1, which is responsible for the synthesis of ABA [97]. ABA is a phytohormone called a stress hormone that regulates the closing and opening of stomata under conditions of osmotic stress [98].

In industry, there are also solutions that are based on CNMs, as indicated by Li et al. [99] who applied a nanocarbon solution (NC) (3.75, 11.25 and 18.75 L ha^{-1}) to *B. oleracea* to evaluate the metabolism of glucosinolates and found that the content of glucoraphanin and glucoerucine increased, which was directly related to the expression of the genes *MAM1*, *IPMI2* and *CYP79F1*, which encode different aliphatic glucosinolates. Glucosinolates are derived from the secondary metabolism of the *Brassicaceae* family and have defensive functions against pathogens through their transformation to isothiocyanates and thiocyanates, which are responsible for antifungal activity. Additionally, they also act directly as antioxidants by reducing oxidative species [100]. One of the main functions of CNMs is antifungal activity, which is highly important since they can be used to reduce the use of pesticides and improve the quality of products without harming the environment [86]. Table 3 shows the studies with the levels of transcription of genes with different NMs belonging to this category.

5 Organic NMs

Due to the greater biocompatibility of biomaterials, there is a growing inclination toward biomaterial research using natural products [104]. Organic NMs consist of carbon-based components, which is why their morphology is highly versatile. Two main types of organic building blocks are recognized: traditional amphiphiles with

Table 3 Description of genes whose transcription level changed in response to the application of carbon nanomaterials in plants

NMs	Size (nm)/Shape	Zeta potential (mV)	Plant species	Application method and rate	Differentially expressed genes	References
CNT	6–13/Cylindrical	NP	*B. oleracea*	Protoplast incubation/10 mg L^{-1}	PIP1 and *PIP2*	[90]
CNT	6–12/Cylindrical	NP	*B. napus*	In vitro/20 mg L^{-1}	*BnSOS1*, *BnKT1*, *BnNHX1*, *BnAPX* and *BnMnSOD*	[91]
FUll	6–18/Spherical	NP	*S. lycopersicum*	Soil/200 mg L^{-1}	*PIP1;3*, *PIP1;5* and *PIP2;4*	[88]
GPH	<100/Hexagonal	NP	*G. max*	Drench/ 100 µg mL^{-1}	*GmP5CS*, *GmGOLS*, *GmDREB1* and *GmNCED1*	[94]
GPH	<100/Hexagonal	NP	*Melissa officinalis* L	Foliar/40 and 60 mM	*RAS*, *PAL1* and *TAT*	[101]
GPH	40/Laminate	NP	*S. lycopersicum*	Nutrient solution/50 and 100 mg L^{-1}	SlExt1, *LeCTR1* and *SlIAA3*	[102]
NC	<6	NP	*B. oleracea*	Nutrient solution/ 3.75, 11.25, and 18.75 L ha^{-1}	*MAM1*, *IPMI2* and *CYP79F1*	[99]
NC	40/Spherical	NP	*Triticum aestivum* L	Soil/ 0.3%	*TaNRT2.2*, *TaNRT2.3*, *TaGS1* and *TaGS2*	[103]

a "head and tail" structure, such as liposomes, and polymer-based building blocks, such as chitosan [7].

Ali et al. [105] studied *Catharanthus roseus* (L.) plants subjected to drought stress (100 and 50% of the field capacity) and with the foliar application of chitosan NMs (1%), the results showed that the stress was mitigated and the alkaloid content increased. This activity is associated with the induction of the expression of the genes *STR* (strictosidine synthase), *DAT* (deacetylvindoline-4-O-acetyltransferase), *PRX1* (peroxidase 1) and *GS* (geissoschizine synthase), which are associated with the biosynthesis of indole terpenoid alkaloids. In the same crop and with the foliar application of chitosan NMs (1%), tolerance to stress due to salinity (150 mM NaCl) was evaluated, which the expression of the *MAPK3*, *GS* (geissoschizine synthase) and *ORCA3 genes* increased (the AP2 domain is sensitive to the octadecanoid derivative) [106]. The MAPK kinase pathway represents a crucial signaling pathway linked to all aspects of growth, development, performance and adaptation to abiotic and biotic stresses in plants [107]. The *GS* and *ORCA3* (TF) genes participate in the biosynthesis of indole terpene alkaloids [106]. Li et al. [108] applied NMs of chitosan (1–100 μg mL^{-1}) in *T. aestivum* seeds, obtaining a positive effect on both germination and growth promotion in seedlings and the induction of genetic expression related to auxin biosynthesis. Abdelkhalek et al. [109] inoculated *Nicotiana glutinosa* L. with alfalfa mosaic virus and NMs of chitosan/dextran (100 μg mL^{-1}) were applied via a foliar approach, resulting in a reduction in the severity of the disease and an increase in the transcription of the *POD, PR-1* and *PAL* genes (mentioned above).

Hegedűs et al. [110] evaluated the application of supercritical botanical extracts encapsulated in nanoscale liposomes (10–240 g ha^{-1}) in *H. vulgare* plants where crop yield was increased and the transcriptional profile showed positive regulation of genes related to pathogenesis (PR) and other genes, associated with defense pathways. While [111] studied the foliar application of garlic extract with supercritical carbon dioxide encapsulated in nanoscale liposomes composed of lipids of plant origin (240 g ha^{-1}) in *T. aestivum* plants, the expression of the *NCED3, HVA22, PUB19, LEA* (related to the ABA pathway), *PR1, PR2, PR3, PR4, PR5* and *PR6* (genes related to pathogenesis) genes and the TFs that respond to the biotic and abiotic stresses *HOX, DREB, ZAT* and *MYB* were increased. In *Nicotiana tabacum* L. infected with tobacco mosaic virus (TMV) and subjected to foliar application of quercetin nanoliposomes, the expression levels of the genes *Nbhsp70er-1* and *Nbhsp70c-A* were reduced, which resulted in the inhibition of TMV proliferation. The *hsp70* gene encodes a key protein involved in TMV replication [112].

For many years, the role of salicylic acid as a stimulant in agriculture has been investigated due to its multiple functions as a signaling molecule and second messenger [113]. However, its effects have now been investigated in nanometric form. Consistent with this, Salem et al. [114] applied NMs of salicylic acid (0.05 and 0.1 mM) in *C. roseus* under drought stress. The results revealed increases in plant height, chlorophylls, carotenoids and alkaloids. The expression of *LEA*, *MPK1* and *MPK6* genes also were increased. The *LEA* gene (late embryogenesis abundant proteins) encodes a protein that accumulate in large concentrations due to the effects of dehydration in plants; these effects are caused by stress due to drought, salinity, heat and cold [115]. The *MPK1* and *MPK6* genes code for protein kinases, which are the signal transduction pathways that control processes that include growth, differentiation, and stress response in plants [107]. Table 4 shows the studies with the levels of transcription of genes with different NMs belonging to this category.

Based on the information presented in this chapter, a set of possible molecular markers is proposed in Fig. 1 that can be used to evaluate the potential of NMs as biostimulants in plants. Specific genes that act as transcription factors and regulatory and functional proteins related to the biostimulation process were selected.

6 Conclusion and Perspectives

The use of NMs in plants has great potential for improving agriculture and addressing challenges related to food security and sustainability. The biostimulant effect of NMs on plants depends on various factors, such as the type of NM, shape, size, dose, application and type of crop.

The genes described in this chapter encode transcription factors, functional proteins, and regulatory proteins. The most studied genes belonged to the metabolic pathways of antioxidant metabolism, the salicylic acid pathway, aquaporins, the abscisic acid pathway, the phenylpropanoid pathway, protein kinases, osmoregulation and heat shock proteins. The genes of the metabolic pathways mentioned above can act as molecular markers of biostimulation in plants following the application of NMs. The inclusion of these genes in future studies will allow a more precise and specific evaluation of the effects of NMs on the growth, development and stress tolerance of plants.

Most of the studies presented in this chapter are related to the use of metallic or semimetal NMs and their oxides; thus, there is an opportunity to focus future research efforts on the analysis of inorganic, organic and carbon NMs in different plant species.

Table 4 Description of genes whose transcription level changed in response to the application of organic nanomaterials in plants

NMs	Size (nm)/Shape	Zeta potential (mV)	Plant species	Application method and rate	Differentially expressed genes	References
Botanical extracts encapsulated in liposomes	100–200/Multi-laminar	NP	*H. vulgare*	Foliar/ 10–240 g ha^{-1}	*PR*	[110]
Chitosan	NP	NP	*C. roseus*	Foliar/1%	*STR, DAT, PRX1* and *GS*	[105]
Chitosan	NP	NP	*C. roseus*	Foliar/1%	*MAPK3*, GS and ORCA3	[106]
Chitosan/Dextran	91.68/Spherical	NP	*N. glutinosa*	Foliar/ 100 μg mL^{-1}	*POD, PR-1* and *PAL*	[109]
Chitosan	80–200	40.9	*T. aestivum*	Seed priming/ 1–100 μg mL^{-1}	*YUCCA9, AUX1* and *ARF*	[108]
Garlic extract encapsulated in liposomes	149.2	−40.66 and −40.53	*T. aestivum*	Foliar/240 g ha^{-1}	*NCED3, HVA22, PUB19, LEA, PR1, PR2, PR3, PR4, PR5, PR6, HOX, DREB, ZAT* and *MYB*	[111]
Salicylic acid	5.71–17.3/Spherical	3.99	*C. roseus*	Foliar/0.05 and 0.1 mM	*LEA, MPK1* and *MPK6*	[114]
Quercetin nanoliposomes	117	−48.5, −42.4, and + 43.5	*N. tabacum*	Foliar/0.3 g L^{-1}	*Nbhsp70er-1* and *Nbhsp70c-A*	[112]

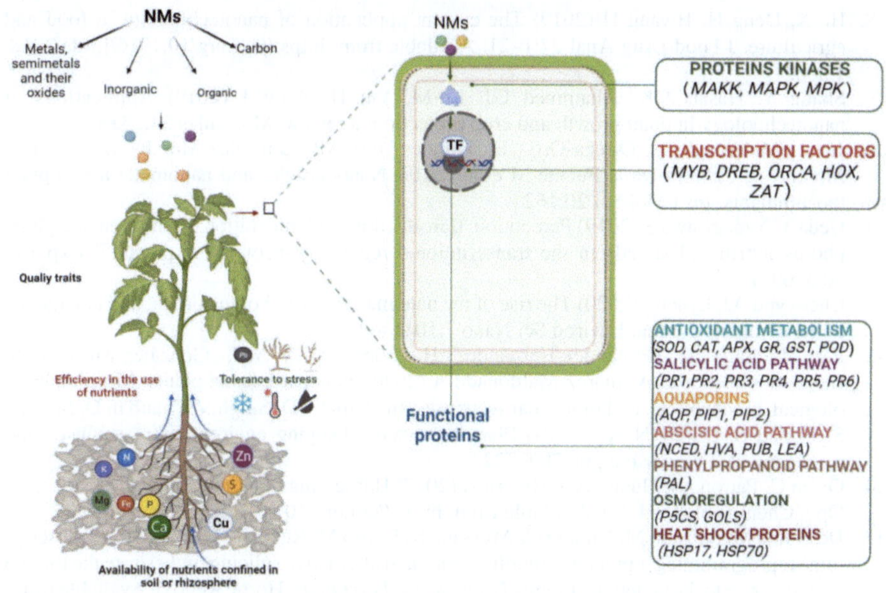

Fig. 1 Biostimulant effects of the application of NMs in plants and their impact on the transcriptional response

References

1. Magnabosco P, Masi A, Shukla R, Bansal V, Carletti P (2023) Advancing the impact of plant biostimulants to sustainable agriculture through nanotechnologies. Chem Biol Technol Agric 10:117. Available from: https://doi.org/10.1186/s40538-023-00491-8
2. Evans JR, Lawson T (2020) From green to gold: agricultural revolution for food security. J Exp Bot 71:2211–5
3. González-Morales S, Cárdenas-Atayde PA, Garza-Alonso CA, Robledo-Olivo A, Benavides-Mendoza A (2022) Plant biostimulation with nanomaterials: a physiological and molecular standpoint. In: Fernandes Fraceto L, Pereira de Carvalho HW, de Lima R, Ghoshal S, Santaella C (eds) Inorganic nanopesticides and nanofertilizers: a view from the mechanisms of action to field applications. Springer, Cham, pp 153–185
4. European Union (EU) (2019) Regulation of the European parliament and of the council laying down rules on the making available on the market of EU fertilizing products ond amending Regulations (EC) No 1069/2009 and (EC) No 1107/2009 and repealing Regulation (EC) No 2003/2003. Available from: https://eur-lex.europa.eu/legal-content/EN/TXT/?uri=OJ:L:2019:170:TOC
5. Juárez-Maldonado A, Tortella G, Rubilar O, Fincheira P, Benavides-Mendoza A (2021) Biostimulation and toxicity: the magnitude of the impact of nanomaterials in microorganisms and plants. J Adv Res 31:113–26
6. Miernicki M, Hofmann T, Eisenberger I, von der Kammer F, Praetorius A (2019) Legal and practical challenges in classifying nanomaterials according to regulatory definitions. Nat Nanotechnol 14:208–216. Available from: https://doi.org/10.1038/s41565-019-0396-z
7. Lombardo D, Calandra P, Pasqua L, Magazù S (2020) Self-assembly of organic nanomaterials and biomaterials: the bottom-up approach for functional nanostructures formation and advanced applications. Materials 13:1048

8. He X, Deng H, Hwang H (2019) The current application of nanotechnology in food and agriculture. J Food Drug Anal 27:1–21. Available from: https://doi.org/10.1016/j.jfda.2018.12.002
9. Shang Y, Hasan MK, Ahammed GJ, Li M, Yin H, Zhou J (2019) Applications of nanotechnology in plant growth and crop protection: a review. Molecules 24:2558
10. Juárez-Maldonado A, Ortega-Ortíz H, Morales-Díaz AB, González-Morales S, Morelos-Moreno Á, Cabrera-De la Fuente M et al (2019) Nanoparticles and nanomaterials as plant biostimulants. Int J Mol Sci 20:162
11. Ueda Y, Yanagisawa S (2019) Perception, transduction, and integration of nitrogen and phosphorus nutritional signals in the transcriptional regulatory network in plants. J Exp Bot 70:3709–17
12. Chetwynd AJ, Lynch I (2020) The rise of the nanomaterial metabolite corona, and emergence of the complete corona. Environ Sci Nano 7:1041–60
13. Pérez-Labrada F, Hernández-Hernández H, López-Pérez MC, González-Morales S, Benavides-Mendoza A, Juárez-Maldonado A (2020) Nanoparticles in plants: Morphophysiological, biochemical, and molecular responses. In: Tripathi D, Singh V, Chauhan D, Sharma S, Prasad S, Dubey N et al (eds) Plant life under changing environment: responses and management. Elsevier Inc., pp 289–322
14. Owen C, Patron NJ, Huang A, Osbourn A (2017) Harnessing plant metabolic diversity. Curr Opin Chem Biol 40:24–30. Available from: https://doi.org/10.1016/j.cbpa.2017.04.015
15. De Palma M, Salzano M, Villano C, Aversano R, Lorito M, Ruocco M et al (2019) Transcriptome reprogramming, epigenetic modifications and alternative splicing orchestrate the tomato root response to the beneficial fungus *Trichoderma harzianum*. Hortic Res 6:5. Available from: https://doi.org/10.1038/s41438-018-0079-1
16. Rastogi A, Tripathi DK, Yadav S, Chauhan DK, Živčák M, Ghorbanpour M et al (2019) Application of silicon nanoparticles in agriculture. 3 Biotech 9:90. Available from: https://doi.org/10.1007/s13205-019-1626-7
17. Seleiman MF, Al-Selwey WA, Ibrahim AA, Shady M, Alsadon AA (2023) Foliar applications of ZnO and SiO_2 nanoparticles mitigate water deficit and enhance potato yield and quality traits. Agronomy 13:466
18. Kumari M, Pandey S, Mishra SK, Giri VP, Agarwal L, Dwivedi S et al (2020) Omics-based mechanistic insight into the role of bioengineered nanoparticles for biotic stress amelioration by modulating plant metabolic pathways. Front Bioeng Biotechnol 8:242
19. Tighe-Neira R, Carmora E, Recio G, Nunes-Nesi A, Reyes-Diaz M, Alberdi M et al (2018) Metallic nanoparticles influence the structure and function of the photosynthetic apparatus in plants. Plant Physiol Biochem 130:408–417. Available from: https://doi.org/10.1016/j.plaphy.2018.07.024
20. Lv J, Christie P, Zhang S (2019) Uptake, translocation, and transformation of metal-based nanoparticles in plants: recent advances and methodological challenges. Environ Sci Nano 6:41–59
21. López-Moreno ML, Cedeño-Mattei Y, Bailón-Ruiz SJ, Vazquez-Nuñez E, Hernandez-Viezcas JA, Perales-Pérez OJ et al (2018) Environmental behavior of coated NMs: physicochemical aspects and plant interactions. J Hazard Mater 347:196–217
22. León Morales JM, Sepúlveda-Jiménez G (2012) El daño por oxidación causado por cobre y la respuesta antioxidante de las plantas. Interciencia 37:805–11
23. Mosa KA, El-Naggar M, Ramamoorthy K, Alawadhi H, Elnaggar A, Wartanian S et al (2018) Copper nanoparticles induced genotoxicty, oxidative stress, and changes in superoxide dismutase (*SOD*) gene expression in cucumber (*Cucumis sativus*) plants. Front Plant Sci 9:872
24. Che Y, Zhang N, Zhu X, Li S, Wang S, Si H (2020) Enhanced tolerance of the transgenic potato plants overexpressing Cu/Zn superoxide dismutase to low temperature. Sci Hortic 261:108949. Available from: https://doi.org/10.1016/j.scienta.2019.108949
25. Pejam F, Ardebili ZO, Ladan-Moghadam A, Danaee E (2021) Zinc oxide nanoparticles mediated substantial physiological and molecular changes in tomato. PLoS One 16:e0248778. Available from: https://doi.org/10.1371/journal.pone.0248778

26. Cao Y, Li K, Li Y, Zhao X, Wang L (2020) MYB transcription factors as regulators of secondary metabolism in plants. Biology 9:61
27. Gao J, Zhang S, He W-D, Shao X-H, Li C-Y, Wei Y-R et al (2017) Comparative phosphoproteomics reveals an important role of MKK2 in banana (*Musa* spp.) cold signal network. Sci Rep 7:40852. Available from: https://doi.org/10.1038/srep40852
28. Hu J, Guo H, Li J, Wang Y, Xiao L, Xing B (2017) Interaction of γ-Fe_2O_3 nanoparticles with *Citrus maxima* leaves and the corresponding physiological effects via foliar application. J Nanobiotechnology 15:51
29. Montejano-Ramírez V, Valencia-Cantero E (2023) Cross-talk between iron deficiency response and defense establishment in plants. Int J Mol Sci 24:6236
30. Agafonov O, Selstø CH, Thorsen K, Xu XM, Drengstig T, Ruoff P (2016) The organization of controller motifs leading to robust plant iron homeostasis. PLoS ONE 11:e0147120
31. Chandra J, Chauhan R, Korram J, Satnami ML, Keshavkant S (2020) Silica nanoparticle minimizes aluminium imposed injuries by impeding cytotoxic agents and over expressing protective genes in *Cicer arietinum*. Sci Hortic 260:108885. Available from: https://doi.org/10.1016/j.scienta.2019.108885
32. Gomes MP, Kitamura RSA, Marques RZ, Barbato ML, Zámocký M (2022) The role of H_2O_2-scavenging enzymes (Ascorbate Peroxidase and Catalase) in the tolerance of *Lemna minor* to antibiotics: implications for phytoremediation. Antioxidants 11:151
33. Raza A, Sharif Y, Chen K, Wang L, Fu H, Zhuang Y et al (2022) Genome-wide characterization of ascorbate peroxidase gene family in peanut (*Arachis hypogea* L.) revealed their crucial role in growth and multiple stress tolerance. Front Plant Sci 13:962182
34. Omar SA, Elsheery NI, Pashkovskiy P, Kuznetsov V, Allakhverdiev SI, Zedan AM (2023) Impact of titanium oxide nanoparticles on growth, pigment content, membrane stability, DNA damage, and stress-related gene expression in *Vicia faba* under saline conditions. Horticulturae 9:1030
35. Cohen A, Hacham Y, Welfe Y, Khatib S, Avice J-C, Amir R (2020) Evidence of a significant role of glutathione reductase in the sulfur assimilation pathway. Plant J 102:246–61
36. González-Gordo S, Palma JM, Corpas FJ (2023) Small Heat Shock Protein (*sHSP*) Gene Family from Sweet Pepper (*Capsicum annuum* L.) fruits: involvement in ripening and modulation by Nitric Oxide (NO). Plants 12:389
37. Yadu B, Chandrakar V, Korram J, Satnami ML, Kumar M, Keshavkant S (2018) Silver nanoparticle modulates gene expressions, glyoxalase system and oxidative stress markers in fluoride stressed *Cajanus cajan* L. J Hazard Mater 353:44–52. Available from: https://doi.org/10.1016/j.jhazmat.2018.03.061
38. Zuo S, Li J, Gu W, Wei S (2022) Exogenous proline alleviated low temperature stress in maize embryos by optimizing seed germination, inner proline metabolism, respiratory metabolism and a hormone regulation mechanism. Agriculture 12:548
39. Rudakiya DM, Patel Y (2021) Bioremediation of metals, metalloids, and nonmetals. In: Panpatte DG, Jhala YK (eds) Microbial rejuvenation of polluted environment microorganisms for sustainability. Springer Nature, Singapore, pp 33–49
40. Pisoschi AM, Pop A (2015) The role of antioxidants in the chemistry of oxidative stress: a review. Eur J Med Chem 97:55–74. Available from: https://doi.org/10.1016/j.ejmech.2015.04.040
41. Jiang M, Dai S, Wang B, Xie Z, Li J, Wang L et al (2021) Gold nanoparticles synthesized using melatonin suppress cadmium uptake and alleviate its toxicity in rice. Environ Sci Nano 8:1042–56
42. Nazir MM, Li Q, Noman M, Ulhassan Z, Ali S, Ahmed T et al (2022) Calcium oxide nanoparticles have the role of alleviating arsenic toxicity of barley. Front Plant Sci 13:843795
43. Hong J, Jia S, Wang C, Li Y, He F, Gardea-Torresdey JL (2023) Transcriptome reveals the exposure effects of CeO_2 nanoparticles on pakchoi (*Brassica chinensis* L.) photosynthesis. J Hazard Mater 444:130427
44. Wu H, Wan X, Niu J, Xu H, Zhang Y, Xue X et al (2023) Enhancing lettuce yield via Cu/Fe-layered double hydroxide nanoparticles spraying. J Nanobiotechnology 21:417. Available from: https://doi.org/10.1186/s12951-023-02178-6

45. Hernández-Hernández H, Juárez-Maldonado A, Benavides-Mendoza A, Ortega-Ortiz H, Cadenas-Pliego G, Sánchez-Aspeytia D et al (2018) Chitosan-PVA and copper nanoparticles improve growth and overexpress the *SOD* and *JA* genes in tomato plants under salt stress. Agronomy 8:175
46. Liu C, Yu Y, Liu H, Xin H (2021) Effect of different copper oxide particles on cell division and related genes of soybean roots. Plant Physiol Biochem 163:205–214. Available from: https://doi.org/10.1016/j.plaphy.2021.03.051
47. Rady MM, Mossa A-TH, Youssof AMA, Osman AS, Ahmed SMA, Mohamed IAA (2023) Exploring the reinforcing effect of nano-potassium on the antioxidant defense system reflecting the increased yield and quality of salt-stressed squash plants. Sci Hortic 308:111609
48. Imada K, Sakai S, Kajihara H, Tanaka S, Ito S (2016) Magnesium oxide nanoparticles induce systemic resistance in tomato against bacterial wilt disease. Plant Pathol 65:551–60
49. Ye Y, Cota-Ruiz K, Hernández-Viezcas JA, Valdés C, Medina-Velo IA, Turley RS et al (2020) Manganese nanoparticles control salinity-modulated molecular responses in *Capsicum annuum* L. Through priming: a sustainable approach for agriculture. ACS Sustain Chem Eng 8:1427–1436
50. Maryška L, Jindřichová B, Siegel J, Záruba K, Burketová L (2023) Impact of palladium nanoparticles on plant and its fungal pathogen. A case study: *Brassica napus-Plenodomus lingam.* AoB Plants 15:1–11
51. Sharf-Eldin AA, Alwutayd KM, El-Yazied AA, El-Beltagi HS, Alharbi BM, Eisa MAM et al (2023) Response of maize seedlings to silicon dioxide nanoparticles (SiO_2NPs) under drought stress. Plants 12:2592
52. Sun L, Wang Y, Wang R, Wang R, Zhang P, Ju Q et al (2020) Physiological, transcriptomic and metabolomic analyses reveal zinc oxide nanoparticles modulate plant growth in tomato. Environ Sci Nano 7:3587–604
53. Antoshkina M, Golubkina N, Poluboyarinov P, Skrypnik L, Sekara A, Tallarita A et al (2023) Effect of sodium selenate and selenocystine on savoy cabbage yield, morphological and biochemical characteristics under *Chlorella* Supply. Plants 12:1020
54. González-Morales S, Pérez-Labrada F, García-Enciso EL, Leija-Martínez P, Medrano-Macías J, Dávila-Rangel IE et al (2017) Selenium and sulfur to produce *Allium* functional crops. Molecules 22:558
55. García Márquez V, Morelos Moreno Á, Benavides Mendoza A, Medrano MJ (2020) Ionic selenium and nanoselenium as biofortifiers and stimulators of plant metabolism. Agronomy 10:1399
56. El-Badri AM, Hashem AM, Batool M, Sherif A, Nishawy E, Ayaad M et al (2022) Comparative efficacy of bio-selenium nanoparticles and sodium selenite on morpho-physiochemical attributes under normal and salt stress conditions, besides selenium detoxification pathways in *Brassica napus* L. J Nanobiotechnology 20:163. Available from: https://doi.org/10.1186/s12951-022-01370-4
57. Kang L, Wu Y, Zhang J, An Q, Zhou C, Li D et al (2022) Nano-selenium enhances the antioxidant capacity, organic acids and cucurbitacin B in melon (*Cucumis melo* L.) plants. Ecotoxicol Environ Saf 241:113777. Available from: https://doi.org/10.1016/j.ecoenv.2022.113777
58. Hernández-Hernández H, Quiterio-Gutiérrez T, Cadenas-Pliego G, Ortega-Ortiz H, Hernández-Fuentes AD, Cabrera De La Fuente M et al (2019) Impact of selenium and copper nanoparticles on yield, antioxidant system, and fruit quality of tomato plants. Plants 8:355
59. Zhou Y, Tang Q, Wu M, Mou D, Liu H, Wang S et al (2018) Comparative transcriptomics provides novel insights into the mechanisms of selenium tolerance in the hyperaccumulator plant *Cardamine hupingshanensis*. Sci Rep 8:2789. Available from: https://doi.org/10.1038/s41598-018-21268-2
60. Hasan MK, Liu C-X, Pan Y-T, Ahammed GJ, Qi Z-Y, Zhou J (2018) Melatonin alleviates low-sulfur stress by promoting sulfur homeostasis in tomato plants. Sci Rep 8:10182
61. Jing Q, Chen A, Lv Z, Dong Z, Wang L, Meng X et al (2023) Systematic analysis of galactinol synthase and raffinose synthase gene families in potato and their expression patterns in development and abiotic stress responses. Genes 14:1344

62. Narayan OP, Kumar P, Yadav B, Dua M, Johri AK (2022) Sulfur nutrition and its role in plant growth and development. Plant Signal Behav 2030082. Available from: https://doi.org/10.1080/15592324.2022.2030082
63. Künstler A, Gullner G, Ádám AL, Nagy JK, Király L (2020) The versatile roles of sulfur-containing biomolecules in plant defense—a road to disease resistance. Plants 9:1705. Available from: https://doi.org/10.1016/j.copbio.2019.01.009
64. Yuan H, Liu Q, Guo Z, Fu J, Sun Y, Gu C et al (2021) Sulfur nanoparticles improved plant growth and reduced mercury toxicity via mitigating the oxidative stress in *Brassica napus* L. J Clean Prod 318:128589
65. Xu Z-R, Cai M-L, Chen S-H, Huang X-Y, Zhao F-J, Wang P (2021) High-affinity sulfate transporter Sultr1;2 Is a major transporter for Cr(VI) uptake in plants. Environ Sci Technol 55:1576–84
66. Wang Y, Wang F, Lu H, Liu Y, Mao C (2021) Phosphate uptake and transport in plants: an elaborate regulatory system. Plant Cell Physiol 62:564–72
67. Cao X, Wang C, Luo X, Yue L, White JC, Elmer W et al (2021) Elemental sulfur nanoparticles enhance disease resistance in tomatoes. ACS Nano 15:11817–27
68. Ali S, Ganai BA, Kamili AN, Bhat AA, Mir ZA, Bhat JA et al (2018) Pathogenesis-related proteins and peptides as promising tools for engineering plants with multiple stress tolerance. Microbiol Res 212–213:29–37. Available from: https://doi.org/10.1016/j.micres.2018.04.008
69. Russo R, Sicilia A, Caruso M, Arlotta C, Di Silvestro S, Gmitter FG et al (2021) De Novo transcriptome sequencing of rough lemon leaves (*Citrus jambhiri* Lush.) in Response to *Plenodomus tracheiphilus* Infection. Int J Mol Sci 22:882
70. Shimizu K, Suzuki H, Uemura T, Nozawa A, Desaki Y, Hoshino R et al (2022) Immune gene activation by *NPR* and *TGA* transcriptional regulators in the model monocot *Brachypodium distachyon.* Plant J 110:470–81
71. Marchiol L, Filippi A, Adamiano A, Esposti LD, Iafisco M, Mattiello A et al (2019) Influence of hydroxyapatite nanoparticles on germination and plant metabolism of tomato (*Solanum lycopersicum* L.): Preliminary evidence. Agronomy 9:161
72. Zhu F, He S, Shang Z (2019) Effect of vegetables and nano-particle hydroxyapatite on the remediation of cadmium and phosphatase activity in rhizosphere soil through immobilization. Int J Phytoremediation 21:610–6
73. Wu H, Tong J, Jiang X, Wang J, Zhang H, Luo Y et al (2023) More effective than direct contact: Nano hydroxyapatite pre-treatment regulates the growth and Cd uptake of rice (*Oryza sativa* L.) seedlings. J Hazard Mater 463:132889
74. Ma Y, Wen Y, Wang C, Wu Z, Yuan X, Xiong Y et al (2023) *ZIP* genes are involved in the retransfer of zinc ions during the senescence of zinc-deficient rice leaves. Int J Mol Sci 24:13989
75. Yang Y, Zheng J, Liang Y, Wang X, Li K, Chen L et al (2023) Natural resistance-associated macrophage protein (*Nramp*) Family in Foxtail Millet (*Setaria italica*): characterization, expression analysis and relationship with metal content under Cd stress. Agronomy 13:2000
76. Pavlicevic M, Abdelraheem W, Zuverza-Mena N, O'Keefe T, Mukhtar S, Ridge G et al (2022) Engineered nanoparticles, natural nanoclay and biochar, as carriers of plant-growth promoting bacteria. Nanomaterials 12:4474
77. Aseel DG, Abdelkhalek A, Alotibi FO, Samy MA, Al-Askar AA, Arishi AA et al (2022) Foliar application of nanoclay promotes potato (*Solanum tuberosum* L.) growth and induces systemic resistance against potato virus Y. Viruses 14:2151
78. Chen X, Wang P, Gu M, Hou B, Zhang C, Zheng Y et al (2022) Identification of *PAL* genes related to anthocyanin synthesis in tea plants and its correlation with anthocyanin content. Hortic Plant J 8:381–394. Available from: https://doi.org/10.1016/j.hpj.2021.12.005
79. Xu J, Zhang N, Wang K, Xian Q, Dong J, Qi X et al (2022) Chitinase Chi 2 positively regulates cucumber resistance against *Fusarium oxysporum* f. sp. cucumerinum. Genes 13:62
80. Alonso de Robador JM, Ortega Pérez N, Sanchez-Ballesta MT, Tello Mariscal ML, Pintos López B, Gómez-Garay A (2023) Plant defence induction by *Meyerozyma guilliermondii* in *Vitis vinifera* L. Agronomy 13:2780

81. Chugh G, Siddique KHM, Solaiman ZM (2021) Nanobiotechnology for agriculture: smart technology for combating nutrient deficiencies with nanotoxicity challenges. Sustainability 13:1781
82. Meselhy AG, Sharma S, Gou Z, Singh G, Yuan H, Tripathi RD et al (2021) Nanoscale sulfur improves plant growth and reduces arsenic toxicity and accumulation in rice (*Oryza sativa* L.). Environ Sci Technol 55:13490–13503
83. Sotoodehnia-Korani S, Iranbakhsh A, Ebadi M, Majd A, Oraghi Ardebili Z (2020) Selenium nanoparticles induced variations in growth, morphology, anatomy, biochemistry, gene expression, and epigenetic DNA methylation in *Capsicum annuum*; an *in vitro* study. Environ Pollut 265:114727. Available from: https://doi.org/10.1016/j.envpol.2020.114727
84. Kang L, Wu Y, Jia Y, Chen Z, Kang D, Zhang L et al (2023) Nano-selenium enhances melon resistance to *Podosphaera xanthii* by enhancing the antioxidant capacity and promoting alterations in the polyamine, phenylpropanoid and hormone signaling pathways. J Nanobiotechnology 21:377. Available from: https://doi.org/10.1186/s12951-023-02148-y
85. Cheng B, Wang C, Chen F, Yue L, Cao X, Liu X et al (2022) Multiomics understanding of improved quality in cherry radish (*Raphanus sativus* L. var. radculus pers) after foliar application of selenium nanomaterials. Sci Total Environ 824:153712
86. González-García Y, Cadenas-Pliego G, Alpuche-Solís ÁG, Cabrera RI, Juárez-Maldonado A (2021) Carbon nanotubes decrease the negative impact of *Alternaria solani* in tomato crop. Nanomaterials 11:1080
87. González-García Y, López-Vargas ER, Cadenas-Pliego G, Benavides-Mendoza A, González-Morales S, Robledo-Olivo A et al (2019) Impact of carbon nanomaterials on the antioxidant system of tomato seedlings. Int J Mol Sci 20:5858
88. Subotić A, Jevremović S, Milošević S, Trifunović-Momčilov M, Đurić M, Koruga Đ (2022) Physiological response, oxidative stress assessment and aquaporin genes expression of cherry tomato (*Solanum lycopersicum* L.) exposed to hyper-harmonized fullerene water complex. Plants 11:2810
89. Tayade R, Rana V, Shafiqul M, Nabi RBS, Raturi G, Dhar H et al (2022) Genome-wide identification of aquaporin genes in Adzuki Bean (*Vigna angularis*) and expression analysis under drought stress. Int J Mol Sci 23:16189
90. Martínez-Ballesta MC, Chelbi N, Lopez-Zaplana A, Carvajal M (2020) Discerning the mechanism of the multiwalled carbon nanotubes effect on root cell water and nutrient transport. Plant Physiol Biochem 146:23–30. Available from: https://doi.org/10.1016/j.plaphy.2019.11.008
91. Zhao G, Zhao Y, Lou W, Su J, Wei S, Yang X et al (2019) Nitrate reductase-dependent nitric oxide is crucial for multi-walled carbon nanotubes-induced plant tolerance against salinity. Nanoscale 11:10511–23
92. Keisham M, Mukherjee S, Bhatla SC (2018) Mechanisms of sodium transport in plants—progresses and challenges. Int J Mol Sci 19:647
93. Gupta BK, Sahoo KK, Anwar K, Nongpiur RC, Deshmukh R, Pareek A et al (2021) Silicon nutrition stimulates Salt-Overly Sensitive (SOS) pathway to enhance salinity stress tolerance and yield in rice. Plant Physiol Biochem 166:593–604. Available from: https://doi.org/10.1016/j.plaphy.2021.06.010
94. Zhao L, Wang W, Fu X, Liu A, Cao J, Liu J (2022) Graphene oxide, a novel nanomaterial as soil water retention agent, dramatically enhances drought stress tolerance in soybean plants. Front Plant Sci 13:810905
95. Martins CPS, Fernandes D, Guimarães VM, Du D, Silva DC, Almeida AAF et al (2022) Comprehensive analysis of the *GALACTINOL SYNTHASE* (*GolS*) gene family in citrus and the function of *CsGolS6* in stress tolerance. PLoS ONE 17:e0274791
96. Akhtar M, Jaiswal A, Taj G, Jaiswal JP, Qureshi MI, Singh NK (2012) DREB1/CBF transcription factors: their structure, function and role in abiotic stress tolerance in plants. J Genet 91:385–95
97. Nykiel M, Gietler M, Fidler J, Graska J, Rybarczyk-Płońska A, Prabucka B et al (2022) Differential water deficit in leaves is a principal factor modifying barley response to drought stress. Int J Mol Sci 23:15240

98. Liu H, Song S, Zhang H, Li Y, Niu L, Zhang J et al (2022) Signaling transduction of ABA, ROS, and Ca^{2+} in plant stomatal closure in response to drought. Int J Mol Sci 23:14824
99. Li Z, Liu G, He H, Liu Y, Han F, Liu W (2022) Effects of nanocarbon solution treatment on the nutrients and glucosinolate metabolism in broccoli. Food Chem X 15:100429. Available from: https://doi.org/10.1016/j.fochx.2022.100429
100. Poveda J, Eugui D, Velasco P (2020) Natural control of plant pathogens through glucosinolates: an effective strategy against fungi and oomycetes. Phytochem Rev 19:1045–1059. Available from: https://doi.org/10.1007/s11101-020-09699-0
101. Soraki RK, Gerami M, Ramezani M (2021) Effect of graphene/metal nanocomposites on the key genes involved in rosmarinic acid biosynthesis pathway and its accumulation in *Melissa officinalis*. BMC Plant Biol 21:260. Available from: https://doi.org/10.1186/s12870-021-03052-z
102. Guo X, Zhao J, Wang R, Zhang H, Xing B, Naeem M et al (2021) Effects of graphene oxide on tomato growth in different stages. Plant Physiol Biochem 162:447–455. Available from: https://doi.org/10.1016/j.plaphy.2021.03.013
103. Yang M, Dong C, Shi Y (2023) Nano fertilizer synergist effects on nitrogen utilization and related gene expression in wheat. BMC Plant Biol 23:26. Available from: https://doi.org/10.1186/s12870-023-04046-9
104. Virlan MJR, Miricescu D, Radulescu R, Sabliov CM, Totan A, Calenic B et al (2016) Organic nanomaterials and their applications in the treatment of oral diseases. Molecules 21:207
105. Ali EF, El-Shehawi AM, Ibrahim OHM, Abdul-Hafeez EY, Moussa MM, Hassan FAS (2021) A vital role of chitosan nanoparticles in improvisation the drought stress tolerance in *Catharanthus roseus* (L.) through biochemical and gene expression modulation. Plant Physiol Biochem 161:166–175
106. Hassan FAS, Ali E, Gaber A, Fetouh MI, Mazrou R (2021) Chitosan nanoparticles effectively combat salinity stress by enhancing antioxidant activity and alkaloid biosynthesis in *Catharanthus roseus* (L.) G. Don. Plant Physiol Biochem 162:291–300
107. Raja V, Majeed U, Kang H, Andrabi KI, John R (2017) Abiotic stress: Interplay between ROS, hormones and MAPKs. Environ Exp Bot 137:142–57. Available from: https://doi.org/10.1016/j.envexpbot.2017.02.010
108. Li R, He J, Xie H, Wang W, Bose SK, Sun Y et al (2019) Effects of chitosan nanoparticles on seed germination and seedling growth of wheat (*Triticum aestivum* L.). Int J Biol Macromol 126:91–100
109. Abdelkhalek A, Qari SH, Abu-Saied MAA-R, Khalil AM, Younes HA, Nehela Y et al (2021) Chitosan nanoparticles inactivate alfalfa mosaic virus replication and boost innate immunity in *Nicotiana glutinosa* plants. Plants 10:2701
110. Hegedűs G, Kutasy B, Kiniczky M, Decsi K, Juhász Á, Nagy Á et al (2022) Liposomal formulation of botanical extracts may enhance yield triggering *PR* Genes and phenylpropanoid pathway in barley (*Hordeum vulgare*). Plants 11:2969
111. Kutasy B, Kiniczky M, Decsi K, Kálmán N, Hegedűs G, Alföldi ZP et al (2023) 'Garlic-lipo'4Plants: liposome-encapsulated garlic extract stimulates ABA pathway and *PR* Genes in Wheat (*Triticum aestivum*). Plants 12:743
112. Wang J, Hao K, Yu F, Shen L, Wang F, Yang J et al (2022) Field application of nanoliposomes delivered quercetin by inhibiting specific *hsp70* gene expression against plant virus disease. J Nanobiotechnology. 20:16
113. Sariñana-Aldaco O, Sánchez-Chávez E, Troyo-Diéguez E, Tapia-Vargas LM, Díaz-Pérez JC, Preciado-Rangel P (2020) Foliar aspersion of salicylic acid improves nutraceutical quality and fruit yield in tomato. Agriculture 10:482
114. Salem D, El-Garhy HAS, Ismail IA, Dessoky ES, Samra BN, Shoala T (2022) Nanobiotechnological approaches to enhance drought tolerance in *Catharanthus roseus* plants using salicylic acid in bulk and nanoform. Molecules 27:5112
115. Ozturk M, Turkyilmaz Unal B, García-Caparrós P, Khursheed A, Gul A, Hasanuzzaman M (2021) Osmoregulation and its actions during the drought stress in plants. Physiol Plant 172:1321–35

Changes in the Activity of Plant Enzymes and Proteins Associated with Biostimulation using Nanomaterials

Carlos Alberto Garza-Alonso and Yolanda González-García

Abstract Nanotechnology is a widely used tool in various fields of everyday life, such as industry, medicine, and agriculture. In addition, nanomaterials (NMs) have been shown to be useful for fertilizer, pesticide, and biostimulant applications. Plant biostimulation is an area of agricultural science in which there has been widespread interest in the multiple advantages of cultivable species, such as increased productivity, resistance to biotic and abiotic stress, and more efficient use of water and minerals. Several studies have shown that different NMs can potentially be applied as plant biostimulants, resulting in positive responses in crops. Among these responses are modifications in gene expression in plant cells, which positively or negatively alter certain proteins involved in the different metabolic processes of plants. On the other hand, the application of NMs also alters the enzymatic activity of a wide variety of plant enzymes. Changes in both the structure and production of proteins, as well as in enzymatic activity, will result in greater or lesser efficiency in physiological processes such as photosynthesis, respiration, and assimilation of nutrients, among others. Therefore, the aim of this chapter was to review in detail how different NMs induce biostimulation in plant species, in addition to their effects on proteins and enzymatic activity, emphasizing the physiological processes mentioned above.

Keywords Nanoparticles · Nanotechnology · Plant physiology · Proteomics

C. A. Garza-Alonso (✉)
Universidad Autónoma de Nuevo León, Facultad de Agronomía, Gral. Escobedo, Nuevo León, México
e-mail: carlos.garza.alonso@gmail.com

Y. González-García
Instituto Nacional de Investigaciones Forestales Agrícolas y Pecuarias, Campo Experimental Todos Santos, La Paz, Baja California Sur, México

© The Author(s), under exclusive license to Springer Nature Singapore Pte Ltd. 2025
A. Juárez-Maldonado et al. (eds.), *Plant Biostimulation with Nanomaterials*, Smart Nanomaterials Technology, https://doi.org/10.1007/978-981-96-4648-7_3

1 Introduction

The need for food and other products obtained through agricultural activities is constantly increasing, mainly due to the excessive increase in the world population and the reduction in arable land. Some projections suggest that the world's population will suffer uncontrolled increases by 2050, adding to people's longer life expectancy [105]. However, the production of food and other products can be affected by various environmental changes caused by climate change, affecting the growth, development and productivity of crops [79].

Nanotechnology has gained importance in recent years due to its applications in industry, medicine and agriculture, the latter involving various applications, such as pesticides or fertilizers. In general, nanotechnology considers materials with dimensions smaller than 100 nm; however, other authors include larger materials [77]. NMs encompass a wide variety of products, ranging from metal nanoparticles (NPs) (e.g., ZnO and Fe_2O_3) to carbon materials (carbon nanotubes, graphene, and fullerene), nanochitosan, and nanozeolites [122].

In addition to fulfilling the aforementioned roles as pathogen control methods and means of plant nutrition, NMs also have the ability to induce biostimulation in plants, improve overall growth and development, and provide greater tolerance to biotic and abiotic stressors [69]. Among the mechanisms that explain the biostimulant effect of NMs are modifications in the protein structures of plants. In the literature, several studies have demonstrated the alterations produced by NMs in the expression of various proteins for specific metabolic processes, such as photosynthesis, cellular respiration; access, transport and metabolism of nutrients; the antioxidant defense system; and germination [57].

Changes in protein composition and activity have been documented by proteomic studies, which consist of the complete analysis of proteins in plants and comparison with existing databases, as well as transcriptomic studies, which consist of the study of the expression of genes involved in protein coding. In addition, each metabolic pathway in which they intervene was classified [27].

On the other hand, some studies have reported direct effects on protein composition and the activity of certain enzymes. An example of this is the research of Wu et al. [132], who studied the application of carbon nanotubes (CNTs) in *Malus domestica* plants and reported greater activity of some enzymes involved in cellular respiration, more specifically in the Krebs cycle. Similarly, Sun et al. [119] reported that ZnO NPs modified the activity of the enzymes sucrose synthase, invertase, hexokinase, fructokinase, ADP-glucose pyrophosphorylase, phospho-glucosidase, and phosphoglucosidase in *Zea mays* plants. Other reports have demonstrated changes in a wide variety of enzymes involved in the antioxidant system of plants [44], as well as in the modification of specific proteins for mineral access and transport in plant cells [120].

However, the changes mentioned above are dependent on different factors, including aspects of the plants (plant species, phenological stage), characteristics of the NMs (size, shape, electric charge) and interactions between the two (form of application, dose, exposure time) [122]. Therefore, understanding how certain NMs can be used to alter the protein composition of plants is difficult. The study of modifications in proteins and enzymes is highly relevant since it could have applications in various branches of crop production, such as plant breeding, fertilization programs, and biotic and abiotic stress management.

Based on the above, the objective of this chapter was to review in detail the mechanisms by which NMs induce changes in the groups of enzymes and plant proteins involved in the main processes of primary and secondary metabolism in plant species, as well as the current reports that specifically indicate how proteins are altered when plants are subjected to the application of NMs.

2 Classifications of Plant Enzymes and Proteins

Proteins are macromolecules that perform a wide variety of functions, either specific or multifunctional, in plants [129]. According to the functions they perform in plants, proteins can be classified as follows: structural, transport, involved in defense, enzymatic, signal recognition and transduction, and reserve proteins (Fig. 1).

Structural proteins correspond to the most abundant group, as these are the proteins responsible for forming the different tissues and vegetative organs. On the other hand, transport proteins include all those that form the different transmembrane channels that allow the access of water, minerals and other substances, as well as proteins involved in the transport of all types of compounds in an up/down direction in plants [41]. In the group of defense proteins, we can mention as an example the heat shock proteins or proline, which are produced when plants are under some condition of

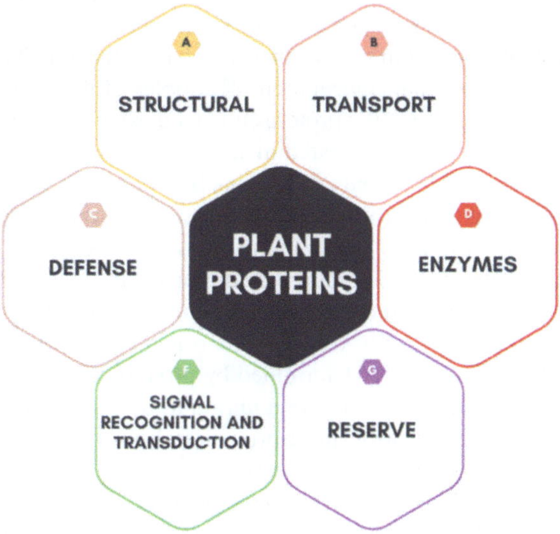

Fig. 1 Classification of principal proteins according with they function in plants

biotic/abiotic stress [128]. On the other hand, the group of enzymatic proteins is also one of the most important and numerous groups in plants. Among the main enzymes and proteins those involved in the process of photosynthesis include Rubisco, an enzyme responsible for carbon fixation in the photosynthetic process [126, 139], the coenzyme NADPH, which is used in the CO_2 fixation phase of the Calvin-Benson-Bassham (CBB) cycle [22, 86], and the carbonic anhydrase (CA), which is responsible for catalyzing the conversion of CO_2 and bicarbonate (HCO_3) and is extremely important for the efficient delivery of CO_2 to Rubisco [93], and nitrate reductase (NR), which reduces nitrate to nitrite, which enters the chloroplast and is used for photosynthesis reactions [36].

The enzymes involved in cellular respiration include hexokinase (HK), which catalyzes the phosphorylation of glucose to glucose 6-phosphate via the consumption of one molecule of ATP [85], fructokinase (FRK), which is the gateway to fructose metabolism, as this enzyme directly regulates the flow of fructose into cells [138], and the enzymes responsible for sucrose synthesis and degradation, sucrose synthase (SUSY), sucrose-phosphate synthase (SPS) and neutral invertase (NINV) [145], and enzymes involved in glycogenesis, such as ADP-glucose pyrophosphorylase, phospho-glucosidase, and phospho-glucosidase [13]. In addition, some enzymes, such as aconitase and succinate dehydrogenase (SDH), are vital for the Krebs cycle [35].

The proteins and enzymes are involved in the antioxidant defense system, especially superoxide dismutase (SOD), peroxidase (POD/POX), catalase (CAT), the ascorbate–glutathione (AsA-GSH) cycle enzyme ascorbate peroxidase (APX), monodehydroascorbate reductase (MDHAR), dehydroascorbate reductase (DHAR), glutathione reductase (GR), peroxiredoxins (PRX), glutathione peroxidases (GPX) and glutathione S-transferases (GST) [32, 112], which are produced in various compartments and organelles of the cell to prevent and/or neutralize the formation of free radicals and maintain homeostasis within the cell [112].

The proteins responsible for signal recognition encompass different families of proteins located in membranes and cell walls, which are the first to perceive stimuli from the external environment. Examples of these proteins include G proteins [99]. On the other hand, the proteins involved in signal transduction correspond to families of receptor proteins located inside cells, as well as to different families of transcription factors (TFs) that are responsible for receiving previously detected signals and producing various alterations in genetic transcription [58]. The next section (Sect. 3) explains in greater detail the process by which cells perceive stimuli by NMs and how proteins in this group play an important role in the response of cells.

The plant proteome represents the total set of proteins synthesized by an individual cell, a type of tissue, or an organ system at a specific time or state in plants [100]. These can be determined by proteomics, which involves the study of protein expression and function, in addition to protein interactions, cell locations, differential abundance, posttranscriptional and posttranslational modifications and protein stimulation/repression [17, 62].

When plants perceive a nanomaterial, they modify their development program through different changes, including the production of different enzymes and proteins that are involved in different functions in the plant [130]. The following lines explain the mechanisms by which NMs modify the proteins and enzymes involved in the main physiological processes of plants.

3 Mechanism of Action of NMs on Protein Modification and Enzyme Activity

The mechanism through which NMs induce biostimulation in plant cells is complex and is currently poorly understood. In general, this process can be divided into two major stages. In the first stage, NMs interact with cell membranes, which are recognized by cells and subsequently trigger a signaling system. On the other hand, the second stage consists of the specific effects of the elements that make up the nucleus of the NMs, which are released once the NMs enter the plant cells [69].

3.1 First Phase of the Biostimulation Mechanism

As previously mentioned, the first stage consists of contact between the NMs and the plant cells. The main factors involved in this stage are the physicochemical characteristics of the NMs, mainly their dimensions, shape, roughness, electric charge, and hydrophobicity, among others [124]. In addition, the dose, exposure time, and mode of application play a role [127]. First, the size of the NMs is one of the most important factors, as it tells us that the specific surface of contact, they will have with cell membranes. In general, small NMs have a larger specific contact surface area than large NMs [70]. This surface area represents an increase in the number of contacts with cell membranes and different organelles. In addition to the contact surface, another advantage of small NMs is that they more easily enter cells because they can cross membranes directly through pores or induce the formation of new pores, while larger NMs must enter through other cellular mechanisms, such as endocytosis [12]. Another physical factor of NMs is shape, which can vary depending on the method of synthesis, where spherical, hemispherical, star, or irregular shapes predominate [12]. In other materials such as carbon NMs, cylindrical shapes (carbon nanotubes) and completely flat shapes can be found with a thickness of just one atom (graphene) [26]. The way in which shape intervenes with cells is due to initial contact, where irregularly shaped cells may induce a greater response due to damage to membranes [42]. A similar phenomenon occurs with the roughness of NMs, where rough materials produce some friction upon contact with membranes, which also modifies their final net charge [9]. On the other hand, the electric charge interferes with the way in which NMs are attracted or rejected by the membranes of the cells, which also have a specific net charge, where an electric double layer is formed that allows or stops the attraction of materials [69]. Such a double layer usually has a net negative charge. When NMs are positively charged, they strongly attract to cell membranes, which is not the case for NMs with opposite (negative) charges. However, NMs can attract other compounds into the medium, such as proteins or protein residues, to form an outer layer known as the corona, which can result in a net positive charge and be attracted by the electrical double layers of cell membranes [141] (Fig. 2).

A related factor is hydrophobicity, which causes the repulsion of NMs when they are in an aqueous medium [29]. On the other hand, the dose, exposure time and method of application are closely related to each other. In general, small and irregularly shaped NMs have greater biostimulant effects at low doses since increasing the dose runs the risk of producing toxicity in cells. However, larger NMs of regular shapes (e.g., spherical) should have an effect at a higher dose, as well as an increase in exposure time to produce a response in the cells. Similarly, the form of application plays an important role since the plant tissues of the leaves, roots, stems and seeds are completely different. In the case of leaves, NMs can enter plants more easily

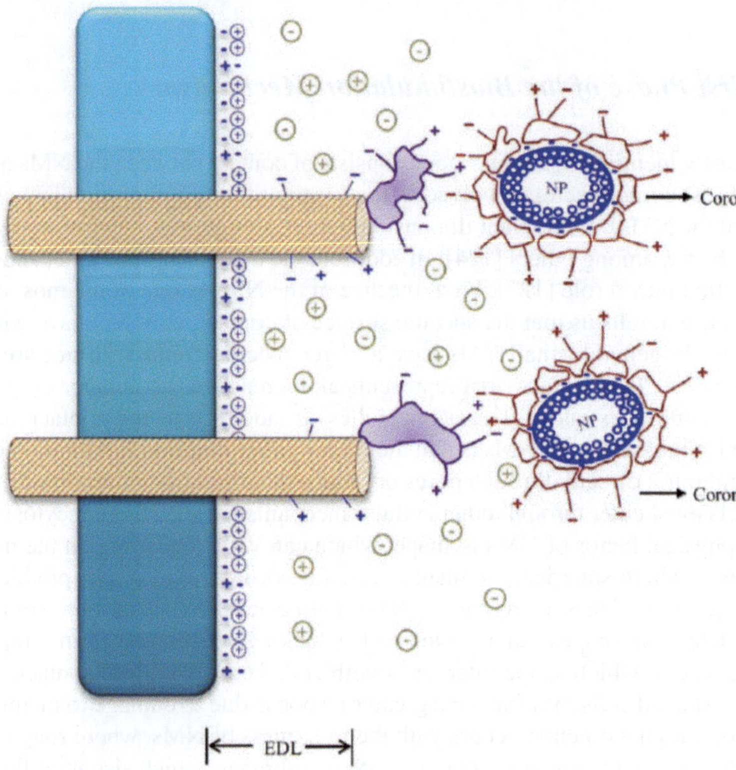

Fig. 2 Representation of the electrical double layer (EDL) and attraction of NPs coated with a protein crown. The net charge of the LDS is mostly negative, while the crowns of the NPs are positively charged, which allows them to be attracted. Figure from Juárez-Maldonado et al. [69]

through stomatal openings, while when applied to roots, they must follow a wider path [42].

Once we know all of the above, the next step is to explain how plant cells perceive the stimuli generated by NMs. As previously mentioned, NMs have a larger contact surface area than do other nonnanometric materials, so these characteristics result in various alterations in cell membranes. One of these alterations can be direct damage, which can be determined through the production of malondialdehyde (MDA), a compound that indicates lipid peroxidation, which means damage to the lipid structures that make up membranes [7]. Another alteration may be the change in membrane stability and permeability [108]. All of the above results in the production of different free radicals, which represent a cell signaling mechanism that is subsequently perceived by other internal structures in the cell [123]. These radicals consist of reactive oxygen species (ROS), such as superoxide (O_2^-), hydrogen peroxide (H_2O_2), and hydroxyl (OH^-) species, reactive nitrogen species (RNS), such as nitric oxide (NO), and reactive sulfur species (RSS), such as sulfuric acid (H_2S) [92]. For practical purposes, these species are grouped as reactive oxygen, nitrogen and sulphur species (RONSS). In addition to RONSS, the perception of cell membranes can also produce hormonal signals, such as salicylic acid (SA) and jasmonic acid (JA), among others [45]. In addition, stimulation by NMs in cell membranes induces the accumulation of Ca^{2+} ions, which also play an important role in the mechanism of cell signaling [76].

These cell signalers can act together or separately in the next phase, which involves the transduction of signals into the cell. In this process, one of the most important mechanisms is the MAP-kinse pathway (MAPK), which is activated through the previously mentioned signaling pathways and consists of proteins that are found freely in the cytoplasm [45]. This phosphorylation represents changes in the protein structures that will later be assembled to form new structures that will be very different from those that would have been formed if the stimulus had not been perceived by the NMs. Protein modifications are known as posttranscriptional modifications (PTMs) because they occur after the transcription and assembly of these proteins [52]. However, an alternative route to CVA is direct modification of the genetic material of cells to give rise to completely different proteins [50]. In this case, the previously produced cellular signalers are detected by protein structures known as transcription factors (TFs), where the most important families are *MYB*, *NAC*, *WRKY* and *HLH* [45]. These TFs are activated by cellular signalers, producing changes in the genetic material and resulting in an alteration in the transcription of proteins, which in turn will produce new protein structures in different plant cells, as well as the assembly of different families of plant enzymes that will intervene in the different metabolic processes of plants [50] (Fig. 3).

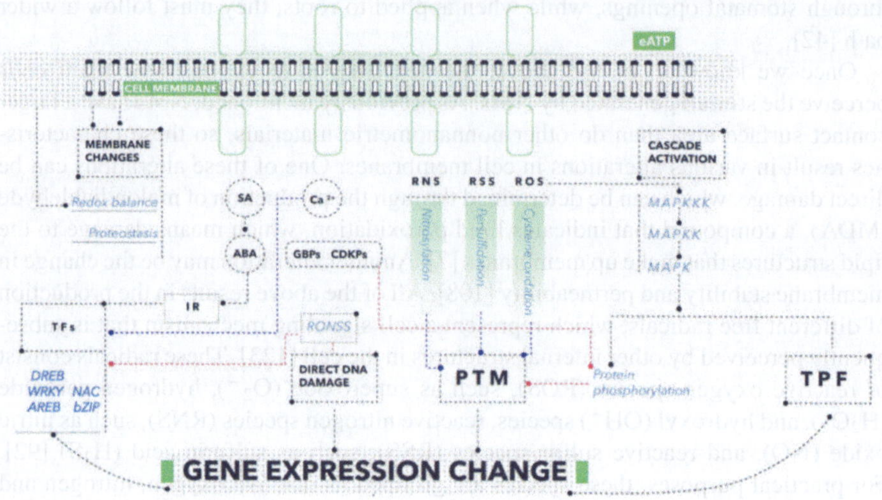

Fig. 3 Mechanisms involved in cell signaling after the first contact of NMs with plant cells. Figure from Garza-Alonso et al. [45]

3.2 Second Phase of the Biostimulation Mechanism

As previously mentioned, the second stage of the biostimulation process with NMs is dependent on the elements that make up the nucleus (e.g., Zn, Fe, Cu, and Mo) [69]. Unlike non-nanometric materials, NMs can more easily enter plant cells and be translocated to different plant organs, where they will later be biotransformed, releasing the elements found in the core of the materials [68]. In this case, these elements will structurally fulfill specific functions for proteins and enzymes, as well as activators or inactivators [123]. For example, Zn^{2+} ions are indispensable for the transcription of mRNAs for protein assembly, so more Zn^{2+} in the medium represents greater efficiency in this process [96]. Another example is the presence of Mo, a fundamental element for molybdoenzymes such as nitrate reductase, xanthine dehydrogenase, sulfite oxidase, aldehyde oxidase, and the reducing component amidoxime [59].

In general, both stages of the biostimulation process result in changes in the levels and structures of plant proteins, as well as in the assembly of more complex structures and enzyme activity that will subsequently fulfill specific roles in plant physiology. In the following sections, the effects of NMs on proteins and enzymes involved in the main metabolic processes of plants, such as photosynthesis, respiration, the antioxidant system and the assimilation of mineral elements, are explained in greater detail.

4 Effect of NMs on Proteins and Enzyme Activity

4.1 Effects on Proteins and Enzymes Involved in Photosynthesis

Photosynthesis is a biochemical process carried out by plants, cyanobacteria, and algae where sunlight, water and CO_2 are used to produce the chemical energy necessary to produce organic matter [80]. In plants, this process occurs in chloroplasts, which are located mainly in the spongy parenchyma and palisade tissues of leaves [75]. These organelles are united in a double membrane and have structures called grana, which are arranged in the shape of a disk, forming the thylakoids where they carry out light-dependent reactions [89].

The dark phase of this process, the Calvin–Benson–Bassham (CBB) cycle, occurs in the stroma [86]. The photosynthetic process includes the capture of light in the electron transport chain, initial CO_2 fixation in the mesophyll, decarboxylation in the bundle sheath, refixation of CO_2 by ribulose-1,5-bisphosphate carboxylase/oxygenase (Rubisco) in the bundle sheath, and CBB cycle activity,additionally, the partitioning of triose phosphate is carried out by multiple protein complexes and enzymes that catalyze different reactions of this complex process [91].

Since photosynthesis is one of the main determinants of yield in crop plants, increasing the efficiency of these processes is an interesting challenge for increasing agricultural productivity and ensuring global food security [80]. The use of NPs can boost plant production by accelerating growth, as they can improve various photosynthetic indices, such as the modulation of stomatal aperture,regulate photophosphorylation, the carbon assimilation pathway, the electron transport chain, and the intercellular CO_2 concentration; increase chlorophyll and carotenoid biosynthesis; increase chloroplast stability; improve the electron transport chain and increase mineral elements; increase ROS scavenging [31], increase the level of carboxylation and oxygenation of Rubisco; and promote the synthesis of photosynthesis-promoting enzymes and coenzymes [43] (Table 1).

Several publications have reported that the application of different nanomaterials can increase the efficiency of Rubisco activity in various plant species, for example, Wu et al. [131] by infiltrating 100 mg L^{-1} CeO_2 NPs into the mesophyll of *Arabidopsis thaliana* leaves, Xiao et al. [134] when 100 mg L^{-1} amine-functionalized graphene (G-NH_2) NPs are applied to *Brassica napus*, Gao et al. [42] when 200 mg L^{-1} $CaCO_3$ NPs are applied to hydroponic solution on *Triticum aestivum* plants, de Sousa et al. [30] when *Zea mays* plants are grown in soil containing 12 or 25 mgkg^{-1} SiO_2 NPs, and Wang et al. [125] when *Vigna radiata* plants are grown by seed priming 0.02, 0.04 or 0.12 mg mL^{-1} carbon dots (CDs).

ATP synthase is another enzyme that can be stimulated by the application of NMs. For example, foliar spray of Fe_2O_3 at a concentration of 3 mML^{-1} increased ATP synthase activity in *Trachyspermum ammi* plants, streamlining the photosynthesis process and improving plant yields [47]. Similarly, TiO_2 NPs applied to nutrient solution at concentrations of.01 and 1 g L^{-1} improved the activity of ATP synthase

Table 1 Effects of NMs on plant proteins and enzymes involved in photosynthesis

Nanomaterials	Plant species	Dose/Application route	Increased enzyme	References
CDs	*Oryza sativa*	0.56 mg mL^{-1} In vitro	Rubisco	Li et al. [81]
CeO_2	*Arabidopsis thaliana*	50 mg L^{-1} Leaf infiltration	Rubisco	Wu et al. [131]
CeO_2	*Glycine max*	100 mg kg^{-1} soil	Rubisco	Cao et al. [18]
Fe_2O_3	*Trachyspermum ammi*	3 mM L^{-1} Foliar sprayed	ATP synthase	Ghassemi-Golezani and Abdoli [47]
TiO_2, SiO_2, ZnO and Fe_2O_3	*Linum usitatissimum*	50 mg L^{-1} Foliar sprayed	Carbonic anhydrase	Singh et al. [113]
ZnO	*Solanum lycopersicum*	10, 50, 100 and 200 mg L^{-1} Foliar sprayed	Carbonic anhydrase	Faizan and Hayat [38]
CeO_2	*Glycine max*	100 mg kg^{-1} Soil	Rubisco	Cao et al. [19]
ZnO	*Solanum lycopersicum*	50 mg L^{-1} Soil	Carbonic anhydrase	Faizan et al. [37]
$CaCO_3$	*Triticum aestivum*	200 mg L^{-1} Hydroponic solution	Rubisco	Gao et al. [42]
G-NH_2	*Brassica napus*	100 mg L^{-1} Seed priming	Rubisco	Xiao et al. [134]
SiO_2	*Zea mays*	12 and 25 mg kg^{-1} Soil	Rubisco	de Sousa et al. [30]
Grafeno	*Medicago sativa*	0.5% 1% and 2% Seed priming	Ferredoxin	Chen et al. [25]
CuO	*Brassica juncea*	2, 4, 8 and 16 mg L^{-1} Seed priming	Carbonic anhydrase	Faraz et al. [39]
TiO_2	*Mentha piperita*	50, 100, 150 and 250 mg L^{-1} Foliar sprayed	Carbonic anhydrase	Ahmad et al. [4]
TiO_2	*Solanum lycopersicum*	0.01 and 1 g L^{-1} Nutrient solution	Rubisco, NADPH and ATP synthase	Li et al. [83]
TiO_2	*Pisum sativum*	10, 50 and 100 mg L^{-1} Nutrient solution	Rubisco	Skiba et al. [115]
CDs	*Vigna radiata*	0.02, 0.04 and 0.12 mg mL^{-1} Seed priming	Rubisco	Wang et al. [125]

and other enzymes related to the process of photosynthesis, such as Rubisco and the coenzyme nicotinamide adenine nucleotide phosphate (NADPH), in *Solanum lycopersicum* plants.

It has been reported that Carbonic anhydrase can also be stimulated by the application of NMs and make photosynthesis more efficient since the *Solanum lycopersicum* plants it was increased with the application of 50 mg L^{-1} of ZnO NPs applied via soil [37], in *Linum usitatissimum* leaves with the application of 50 mg L^{-1} of TiO_2, SiO_2, ZnO and Fe_2O_3 NPs applied via foliar sprayed [113] and in *Brassica juncea* plants with the application of 2, 4, 8, and 16 mg L^{-1} of CuO NPs in seed priming [39].

4.2 Effects on Proteins and Enzymes Involved in Cellular Respiration

Respiration is the process by which cells transform carbohydrates into energy available for all physiological processes in plants [63]. NMs invert proteins and enzymes involved in respiration in a variety of ways, where increases in their production and activity are usually found. An example of this is the work carried out by Wu et al. [133], who studied the application of CNTs in *Malus domestica* plants. The results showed an increase in the activity of several enzymes involved in respiration, such as hexokinase (HK), fructokinase (FRK), aldose-6-phosphate reductase (A6PR), sucrose synthase (SUSY), sorbitol dehydrogenase (SDH), sucrosephosphate synthase (SPS) and neutral invertase (NINV). A similar trend was reported for *Withania somnifera* plants that were subjected to different nanoparticles (Zn, Ni, Ag and CdSe), in which genes involved in the activity of the enzymes neutral invertase (NInv), fructokinase (FK), UDP-glucose pyrophosphorylase (UGPase), hexokinase (HxK), phosphoglucomutase (PGM), soluble acid invertase (SAInv), sucrose-6F-P phosphohydrolase (SPP), sucrose phosphate synthase (SPS), sucrose: fructan 6-fructosyltransferase (6-SFT), sucrose synthase (SuS), and glucose-6-phosphate isomerase (GPI) were overexpressed [114]. A contrary effect was reported by Zhang et al. [146], where the application of TiO_2 NPs to *Oryza sativa* plants resulted in reductions in various compounds and enzyme activity involved in the tricarboxylic acid (TCA) cycle.

In another study carried out by [116], a proteomic study was carried out in *Brassica napus* plants subjected to application of ZnO NPs,the authors reported the overexpression of various groups of proteins involved in hydrolase activity, hydrolyzing o-glycosyl compounds, triose-phosphate isomerase activity, monooxygenase activity, ribulose-bisphosphate carboxylase activity, glucan endo-1,3-β-D-glucosidase activity, hydrolase activity, and hydrolyzing O-glycosyl compounds. In other reports, it has been shown that the application of ZnO NPs to *Zea mays* plants increases the activity of some enzymes involved in cellular respiration, such as sucrose synthase, invertase, hexokinase, fructokinase, ADP-glucose pyrophosphorylase, phospho-glutamyl complementase and phospho-glutamylisomerase [119]. On the other hand, the application of Si-NPs to *Solanum lycopersicum* plants also increased the activity of the enzymes hexokinase and fumarase [8]. In another study, the application of ZnO NPs to *Solanum lycopersicum* plants resulted in the overexpression of several genes related to the enzymatic activity of the respiratory process, specifically those that encode glyceraldehyde-3-phosphate dehydrogenase (*Solyc12g094635.1*), phosphoglycerate mutase (*Solyc05g050800.3*), phosphoglycerate mutase-like protein AT74 (*Solyc01g098590.3*), and pyruvate decarboxylase (*Solyc02g077240.3*) [120].

The activity of the enzymes aconitase and succinate dehydrogenase (SDH), which are involved in the Krebs cycle, increased when Fe NPs were applied to *Brassica napus* plants [109]. On the other hand, research by Stałanowska et al. [118] showed that the application of ZnO NPs to *Pisum sativum* and *Triticum aestivum* plants increased the concentration of organic acids such as citric and malic acid, as well as increasing the concentrations of sucrose and homoserine, which could be explained by the greater activity of the enzymes involved in the TCA cycle.

4.3 Effects on Proteins and Enzymes Involved in the Antioxidant Defense System

During each stage of development, plants face constantly changing environmental conditions, which produce numerous biotic and abiotic stress factors that cause the overproduction of free radicals, mainly ROS [112]. Additionally, various metabolic processes, such as respiration, photorespiration, and photosynthesis, produce ROS, mainly in chloroplast peroxisomes and mitochondria [21], and their accumulation generates oxidative stress, reducing performance and causing programmed cell death (PCD) [71]. However, ROS not only are toxic molecules that negatively affect plant growth but can also act as signaling molecules to activate acclimatization and defense mechanisms in plant systems [102]. The ability of plants to utilize these free radicals as signaling molecules depends on the ability of the cells to scavenge them via enzymatic and nonenzymatic antioxidant defense systems, which counteract reactive species and maintain homeostasis within the cell [112].

In particular, the enzyme-type antioxidant defense system comprises the following enzymes: superoxide dismutase (SOD), peroxidase (POD/POX), catalase (CAT), the ascorbate–glutathione (AsA-GSH) cycle enzyme ascorbate peroxidase (APX), monodehydroascorbate reductase (MDHAR), dehydroascorbate reductase (DHAR), glutathione reductase (GR), peroxiredoxins (PRX), glutathione peroxidases (GPX) and glutathione S-transferases (GST) [32, 112]. Upon the perception of free radicals by plant cells, the enzyme SOD is considered the first line of defense against oxidative stress, leading to the dismutation of O_2^- to H_2O_2, subsequently, H_2O_2 derived from this or other reactions can be reduced to H_2O via the intervention of CAT, GPX, APX, MDHAR, DHAR, GR, PRX, GPX and GST [103, 104, 133].

Improving the enzymatic defense system in plants is of immense importance for coping with stress, improving crop yields, and ensuring environmental and agricultural sustainability [90]. At appropriate doses and through various routes of application, nanomaterials have the potential to enhance the enzymatic antioxidant defense system (Table 2). Several studies have confirmed that foliar spraying of NMs can promote the production of several enzymes, e.g., Abdelkhalek and Al-Askar [2] reported that ZnO NPs at a dose of 100 $\mu g\ mL^{-1}$ increased the SOD content in *Solanum lycopersicum.* Li et al. [82] showed that doses of 0.01, 0.05, and 0.1% Zn and $ZnSO_4$ NPs promoted the production of SOD, POD and CAT in *Brassica rapa, respectively.* Rasouli et al. [107] applied 1000 and 2000 mg L^{-1} ZnO NPs to *Capsicum annum* and observed increases in GPX, SOD and GR. Emamverdian et al. [33] reported increases in SOD, POD, CAT, and GR in *Bambusa vulgaris* upon the application of 60 mgL^{-1} Co NPs. Alhaithloul et al. [11] reported increases in SOD, POD, and CAT in *Triticum aestivum* following the application of 25, 50, 75 and 100 mgL^{-1} Cu NPs.

Additionally, when applied directly to the soil or substrate, NMs promote the activation of the plant enzyme defense system, as in species such as *Coriandrum sativum;* additionally, the application of 500, 1000, or 1500 mg L^{-1} Si NPs increases the production of POD, CAT, and SOD [40]. In *Pisum sativum,* a concentration of 3 mM SiO_2 NPs promoted CAT production [61]. in *Solanum lycopersicum* 10 and 50 mg L^{-1} of multiwallet carbon nanotubes (MWCNTs) increased the APX, GPX and CAT contents [49].

Similarly, when applied via seed priming, the activation of antioxidant enzymes in different plant species has been demonstrated; for example, Se NPs can promote the production of APX, DHAR, MDHAR, GR, and GPX enzymes in *Glycine max* plants at a concentration of 25 mg L^{-1} [10]. Gohari et al. [48] reported increases in SOD, CAT, APX and GPX in *Dracocephalum moldavica* upon the application of 50 mg L^{-1} TiO_2 NPs. Banerjee et al. [14] evaluated the application of 60 mg L^{-1} SiO_2 NPs on *Oryza sativa* plants and observed increases in GPX and GR. Gautam et al. [46] reported an increase in CAT activity upon application of 150 $\mu g\ mg^{-1}$ MgO NPs in *Spinacia oleracea.*

Additionally, when NMs are applied in vitro by dosing the culture media Murashige and Skoog (MS), NMs stimulate the plant enzyme defense system, e.g., in *Pleioblastus pygmaeu* plants. Emamverdian et al. [34] showed that dosing the culture media with 50, 100, 150, or 200 mM L^{-1} Zn NPs increased the SOD, CAT, APX and GS contents. Şener and Sayğı [111] reported an increase in SOD and CAT in *Rubus loganobaccus* following the dosing of culture media with 0.1 and 0.2 mg L^{-1} Ag NPs, respectively. Behbahani et al. [106] observed an increase in CAT activity after the application of 1, 4, 10, 30, or 50 mg L^{-1} Se NPs in *Momordica charantia* to culture media.

Table 2 Effects of NMs on plant proteins and enzymes involved in the antioxidant defense system

Nanomaterials	Plant species	Dose/Application route	Increased enzyme	References
Zn	*Pleioblastus pygmaeu*	50, 100, 150, and 200 Mm L^{-1} In vitro	SOD, CAT, APX, GS	Emamverdian et al. [34]
Zn	*Oryza sativa*	20, 50 and 100 mg L^{-1} Seed priming	SOD	Yan et al. [136]
ZnO	*Capsicum annum*	1000 and 2000 mg L^{-1} Foliar sprayed	GPX, SOD, GR	Rasouli et al. [107]
ZnO	*Solanum lycopersicum*	100 µg mL^{-1} Foliar sprayed	POD	Abdelkhalek and Al-Askar [2]
Zn and $ZnSO_4$	*Brassica rapa*	0.01%, 0.05% and 0.1% Foliar sprayed	SOD, POD, CAT	Li et al. [82]
SiO_2	*Oryza sativa*	50 mg L^{-1} Seed priming	CAT, SOD, POD	Jiang et al. [66]
TiO_2	*Vitex agnus –Castus*	200, 400, 600 and 800 mg L^{-1} Foliar sprayed	CAT, POD	Moshirian Farahi et al. [94]
CeO_2	*Cucumis sativus*	1 mM mL^{-1} Foliar sprayed	SOD, POD, CAT	Chen et al. [24]
Ag	*Rubus loganobaccus*	0.1 and 0.2 mg L^{-1} In vitro	SOD, CAT	Şener and Sayğı [111]
Nano Biochar	*Solanum lycopersicum*	3% Foliar sprayed	CAT, SOD, POD	Mubashir et al. [95]
Co	*Bambusa vulgaris*	60 mg L^{-1} Foliar sprayed	SOD, POD, CAT, GR	Emamverdian et al. [33]
Si	*Coriandrum sativum*	500, 1000 and 1500 mg L^{-1} Soil	POD, CAT, SOD	Fatemi et al. [40]
Ag	*Lactuca sativa*	25, 50 and 75 mg L^{-1} Foliar sprayed	GST, GR	Akhoundnejad et al. [6]
Fe_3O_4	*Moringa oleífera*	40 mg L^{-1} Foliar sprayed	POX, SOD	Tawfik et al. [121]
Se	*Momordica charantia*	1, 4, 10, 30 and 50 mg L^{-1} In vitro MS	CAT	Rajaee Behbahani et al. [106]
Cu	*Triticum aestivum*	25, 50, 75 and 100 mg L^{-1} Foliar Sprayed	SOD, POD, CAT	Alhaithloul et al. [11]

(continued)

Table 2 (continued)

Nanomaterials	Plant species	Dose/Application route	Increased enzyme	References
TiO_2	*Dracocephalum moldavica*	50 mg L^{-1} Seed priming	SOD, CAT, APX, GPX	Gohari et al. [48]
SiO_2	*Oryza sativa*	60 mg L^{-1} Seed priming	GTS, GPX	Banerjee et al. [14]
SiO_2	*Pisum sativum*	3 Mm Soil	CAT	Ismail et al. [61]
Se	*Fragaria × ananassa*	10 and 20 mg L^{-1} Foliar Sprayed	SOD, POD	Zahedi et al. [143]
Se	*Glycine max*	25 mg L^{-1} Seed priming	APX, DHAR, MDHAR, GR, GPX	Albqmi et al. [10]
MgO	*Spinacia oleracea*	150 μg mg^{-1} Seed priming	CAT	Gautam et al. [46]
MWCNT	*Solanum lycopersicum*	10 and 50 mg L^{-1} Soil	APX, GPX, CAT	González-García et al. [49]
Cu, Fe, Zn	*Pyrus communis*	250 mg L^{-1} Foliar sprayed	SOD, CAT, APX, POD	Hernández-Fuentes et al. [56]
MWCNT	*Vitis vinifera*	30, 60, 90, 150, and 180 μg mL^{-1} Soil	POD, SOD, CAT, DHAR, APX, GR, MDHAR	Li et al. [84]

4.4 *Effects on Proteins and Enzymes Involved in Mineral Assimilation and Nutrient Metabolism*

Several reports have shown changes in proteins related to the assimilation and metabolism of mineral elements, as well as in the activity of some enzymes involved in these processes. In a study by [47], foliar applications of Fe_2O_3 NPs promoted the activity of the transport proteins H^+-ATPase and H^+-pyrophosphatase (H^+-PPase), which are located in vacuole membranes (tonoplasts) and are responsible for the entry and exit of cations such as H^+, Na^+ and K^+. This experiment was carried out on Trachyspermum ammi L. plants, where increases in the concentrations of minerals such as Ca, Mg, Fe and Zn were also found.

On the other hand, the application of ZnO and CuO NPs in barley plants results in modifications of the expression of genes responsible for the coding of several families of transport proteins, such as Zn-regulated transporter (*ZRT*), Fe-regulated transporter, (IRT)-like proteins (*ZIPs*), P-type ATPases heavy metals transporters (HMA), metal tolerance proteins (MTPs) and copper transporters (*COPTs*) [67]. Specifically, the application of ZnO NPs resulted in the overexpression of the *ZIP1, ZIP3, ZIP6* and *ZIP14* genes, which encode specific proteins involved in the access and transport of ions such as Zn, Fe, Mn and Cu. In addition, the combined application of ZnO + CuO NPs increased the expression of the *PAA1*, *PAA2* and *RAN1* genes, which are involved in the transport of Cu^{2+} in different organelles. The *COPT5* and *MTP1* genes, which are responsible for the access of cations such as Zn, Cu, Cd^{2+}, Co^{2+} and Ni^{2+} to vacuoles, were also overexpressed. In another study carried out by [116], a proteomic study was carried out in *Brassica napus.* plants subjected to applications of ZnO NPs, in which the various groups of proteins involved in the transmembrane transport of protons, protein channels and integral components of membranes were overexpressed. Similarly, a proteomic study of *Triticum aestivum* plants subjected to TiO_2 NPs revealed the overexpression of proteins involved in mineral transport [110].

In a study conducted by Sun et al. [120], ZnO NPs were applied to *Solanum lycopersicum* plants, and positive changes were detected in terms of the expression of genes encoding the following ion transporters: NH_4^+: ammonium transporter 2 (*Solyc10g076480.2*) and ammonium transporter 1 member 1 (*Solyc09g090730.2*), K^+: transporter genes KAT3-like (*Solyc01g010080.3*) and SKOR (*Solyc11g011500.2*); Ca^{2+}: transporter genes cation/H^+ antiporter 18-like (*Solyc02g071070.3*); calcium-transport ATPase 13 (*Solyc09g082870.2*); autoinhibited Ca^{2+}-transport ATPase 10 (*Solyc07g008320.3*); SO_4^-: high-affinity sulfate transporter type 1 (*Solyc10g047170.2*); Zn^{2+}: transporter genes Solyc06g005620.3 and *Solyc07g043230.3*; and IRT1 (*Solyc02g069200.3*), Nramp5-like (*Solyc11g018530.2*) and IRT2 (*Solyc02g069190.3*).

On the other hand, Jhanzab et al. [64] applied AgNPs to *Triticum aestivum.* plants and found the overexpression of proteins involved in metal transport, such as the ER membrane protein complex, Synaptotagmin-like mitochondrial lipid-binding proteins and copper transport proteins. On the other hand, foliar applications of SiO_2 NPs to *Citrus* x *sinensis* plants resulted in an increase in the expression of several families of Na^+ cotransporters, including *sSOS1, CsSOS2, CsSOS3* and *CsNHX1* [88]. A similar trend was reported by Nicolas-Espinosa et al. [98], where the application of B-encapsulation in *Ipomea batatas* favored the overexpression of the genes *BOR2*, *BOR4* and *BOR7-1*, which are involved in the access and transport of B in plants. On the other hand, the application of Se NPs to *Oryza sativa* plants promoted the overexpression of the *OsSULTR1* gene 2 and *OsPT2*, which encode S and P transporter proteins, respectively [137]. In another study, Soliman et al. [117] reported the overexpression of genes encoding the proteins sodium/proton exchanger (*NHX1*), salt overly sensitive 1 (*SOS1*), Ca^{2+} protein exchanger (*CAX1*) and plasma membrane H^+-ATPase (*ATPase*) in *Triticum aestivum* plants subjected to Se NP application. Similarly, the application of CuO NPs to *Lactuca sativa* plants resulted in an increase in certain heavy metal-associated isoprenylated plant proteins (*HIPPs*), specifically *HIPP01*, *HIPP09*, *HIPP05*, *HIPP39*, *HIPP37*, *HIPP03*, *HIPP36*, *HIPP21*, *HIPP32*, *HIPP07*, *HIPP31* and *HIPP26*, as well as an increase in the proteins *RAN1*, *SKU5*, *MTP4*, *MTP11* and *NRAMP6*, all of which are involved in metal transport [135].

In a study carried out by Abbasifar et al. [1], Mo-NPs were applied to *Spinacia oleracea* L plants, where an increase in the activity of the enzyme nitrate reductase (NR), which is responsible for the reduction of NO_3^- to NO_2^-, was found.

The increase in the activity of this enzyme was also proven by the decrease in the concentration of NO_3^- in plant tissues. All of the above can be explained by the dual mechanism of action of the nanocomposites (see Sect. 3), where the increase in NR activity can be attributed to the interaction of the NPs with cell membranes, in addition to the use of Mo atoms available in cells, as these atoms are important elements for the activation of this enzyme [60]. On the other hand, the application of ZnO and Fe_2O_3 NPs to Eleusine coracana plants increased the expression of certain Zn and Fe transport proteins, such as *EcFER1*, *EcYSL2*, *EcIRT2*, *EcZTP29* and *EcZIP1* [20]. Another study carried out by Zeeshan et al. [144] revealed that, in *Glycine max* plants subjected to the application of SeNPs, there was an increase in the expression of proteins in the sulfate transporter (SLC26 family), MATE efflux family, ATP-binding cassette transporter (ABC transporter), Fe^{2+}/Zn^{2+}-regulated transporter and Mn^{2+} and Fe^{2+} transporters of the NRAMP family.

NMs also have the ability to inhibit several metal transporters, such as the repression of the *HvPHT1* gene; *HvPHT1*; *HvPHT1*; and *HvPHT1*, which encode proteins involved in As transport in plants. This was reported by Nazir et al. [97] for *Hordeum vulgare* plants subjected to the application of CaO NPs. A similar trend was reported by Huang et al. [60], who applied Fe_2O_3 NPs to *Oryza sativa* plants, they found that the inhibition of the proteins *OsCd1*, *OsNRAMP5*, *OsIRT2* and *OsIRT1* was responsible for the access and transport of Cd^{2+} and Fe^{2+} in cells. In the same study, an increase in the activity of *OsYSL15* proteins, which are responsible for Fe^{3+} access, was also found. On the other hand, the application of Au NPs to *Oryza sativa* plants suppressed genes related to metal access in the roots, such as *OsIRT2*, *OsNramp1*, *OsIRT1*, *OsHMA2* and *OsHMA3* [65]. A similar effect was reported by Lai et al. [78], who applied Si–TiO_2 NPs to *Oryza sativa* plants and reported a reduction in the expression of the *OsHMA3* and *OsNramp1* transporters involved in Cd access.

An important group of proteins is the aquaporins, which are transmembrane channels that allow water and nutrients to enter plant cells. These aquaporins can be modified by the application of nanomaterials; therefore, various reports in the literature on these materials and plant species exist [15]. Recent research has demonstrated how nanomaterials positively or negatively modify the expression of genes involved in aquaporin synthesis (Table 3).

4.5 *Effects on Proteins and Enzymes Involved in Germination and Other Physiological Processes*

There are some reports on the effects of NMs on enzymes and proteins involved in the process of seed germination. The main mechanism that explains the positive effects of NMs on the germination and initial growth of plants is through an increase in the activity of hydrolytic enzymes (e.g., α-amylase, β-amylase, and protease), which are responsible for the hydrolysis of seed reserves [72]. An example of this is the study carried out by Mahakham et al. [87], where the application of Ag NPs to *Oryza*

Table 3 Recent studies on the modification of aquaporins in plant species subjected to various nanomaterials

Nanomaterials	Plant species	Alterations in aquaporins	References
Mo NPs	*Nicotinia tabacum*	NtTIP1, NtPIP1, NtPIP1; 1, NtPIP2; 1	Chen et al. [23]
SeNPs	*Glycine max*	Major intrinsic proteins family (MIP)	Zeeshan et al. [144]
Se NPs	*Oryza sativa*	NIP2; 1	Yang et al. [137]
Se NPs	*Triticum aestivum*	Nodulin intrinsic protein (NIP), plasma membrane intrinsic protein (PIP1) and tonoplast intrinsic protein (TIP1.1)	Soliman et al. [117]
Fullerene	*Zea mays*	Nodulin 26-like intrinsic protein1-1 (NIP1-1) and tonoplast intrinsic protein2-1 (TIP2-1)	Ozfidan-Konakci et al. [101]
SiO_2 NPs	*Citrus x sinensis*	CsPIP1-1, CsPIP2-3, and CsTIP4-1	Mahmoud et al. [88]
B-nanoencapsulated	*Ipomea potatoes*	PIN1; 2, PIN 1; 3; PIN 4; 1, PIN 4; 2, PIN 5; 1, PIN 6; 1, and NIP7	Nicolas-Espinosa et al. [98]
Al_2O_3 and ZnO NPs	*Hordeum vulgare*	HvTip1; 1 and HvPip1; 1	Akdemir [5]
SiO_2 NPs	*Triticum aestivum*	*Triticum aestivum* plasma membrane intrinsic protein (TaPIP1) and *Triticum aestivum* nodulin 26-like intrinsic protein (TaNIP2)	Younis et al. [140]
La_2O_3 NPs	*Zea mays*	PIP; 1, PIP; 2, PIP2; 4, PIP2; 5, TYPE1; 1, NIP1; 1	Yue et al. [142]

sativa seeds promoted greater activity of the enzymes α-amylase and dehydrogenase in addition to overexpressing the PIP1 gene 1 and PIP2,1, all of which favor the germination characteristics of the seeds. On the other hand, the application of Ag NPs to the seeds of *Cicer arietinum*, *Pisum sativum* and *Vigna radiata* resulted in an overall increase in hydrolytic enzymes [74]. Likewise, the application of Fe-zero-valent NPs promoted the activity of these enzymes in *Oryza sativa* seeds [51]. In another study, the application of FeO NPs to *Oryza sativa* seeds was shown to increase the activity of the enzyme α-amylase, which was also proven by increasing the concentration of soluble sugars in the seed [3]. In another study conducted by Bovand et al. [16], it was reported that TiO_2 NPs increased α-amylase activity in seeds of *Calendula officinalis* and *Melissa officinalis.* The same effect was found in *Stevia rebaudiana* seeds subjected to the application of SiO_2 NPs [55]. On the other hand, Gupta et al. [53] reported increases in the activity of the enzymes α-amylase, β-amylase and dehydrogenase in Cucumis sativus seeds following the application of Fe_3O_4 and ZnO. A similar effect was observed in *Zea mays* seeds, where the application of Cu-chitosan NPs increased the activity of the enzymes α-amylase and protease [28].

On the other hand, it has been shown that the application of zero-valence Fe nanoparticles (nZVIs) in in vitro culture of *Arabidopsis thaliana* increased the activity of H^+-ATPase proteins, specifically the isoform responsible for stomatal opening and closing; this phenomenon was confirmed through the overexpression of the *AHA2* gene and verified by counting stomata on leaves. This increase in H^+-ATPase activity also promoted water access and increased leaf area [73]. In another study, it was reported that the application of CoNPs and AgNPs to *Rosa hybrida* plants decreased the activity of the enzymes pectinase and cellulase, both of which are involved in leaf abscessation [54].

5 Conclusions and Perspectives

The study of plant protein structures is complex; however, in recent years, it has been possible to discover how proteins and enzymes interact with changes in the environment. It has also been possible to clarify the mechanisms by which NMs alter the composition and function of these proteins, positively impacting mainly those proteins and enzymes that intervene in the processes involved in photosynthesis, cellular respiration, antioxidant defense system, mineral assimilation and nutrient metabolism physiological processes. Changes in plant proteins have been mostly investigated for certain physiological processes, such as photosynthesis or the antioxidant defense system; however, additional studies are needed on processes such as cellular respiration. Such studies can result in a certain level of complexity due to the natural conditions of the proteins and enzymes of interest, as well as the methodologies and equipment currently available. Knowing in detail how NMs positively or negatively modify the composition and functions of plant proteins will allow us to make use of different NMs with a biostimulant effect with greater precision and certainty in terms of their dose, form and time of application. With this, it will be possible to establish a biostimulation plan for crops, adjusting it based on the objective of agricultural production, resulting in greater growth, development and productivity of crops and facing the food security problems that are approaching in the coming years.

Acknowledgements The authors thank Facultad de Agronomía of the Universidad Autónoma de Nuevo León and Campo Experimental Todos Santos of the Instituto Nacional de Investigaciones Forestales Agrícolas y Pecuarias for all the facilities provided for the development of this chapter.

References

1. Abbasifar A, ValizadehKaji B, Iravani MA (2020) Effect of green synthesized molybdenum nanoparticles on nitrate accumulation and nitrate reductase activity in spinach. J Plant Nutr 43:13–27. https://doi.org/10.1080/01904167.2019.1659340
2. Abdelkhalek A, Al-Askar AA (2020) Green synthesized ZnO nanoparticles mediated by *Mentha Spicata* extract induce plant systemic resistance against tobacco mosaic virus. Appl Sci 10:5054. https://doi.org/10.3390/app10155054
3. Afzal S, Sharma D, Singh NK (2021) Eco-friendly synthesis of phytochemical-capped iron oxide nanoparticles as nano-priming agent for boosting seed germination in rice (*Oryza sativa* L.). Environ Sci Pollut Res 28:40275–40287. https://doi.org/10.1007/s11356-020-12056-5
4. Ahmad B, Shabbir A, Jaleel H et al (2018) Efficacy of titanium dioxide nanoparticles in modulating photosynthesis, peltate glandular trichomes and essential oil production and quality in *Mentha piperita* L. Curr Plant Biol 13:6–15. https://doi.org/10.1016/j.cpb.2018.04.002
5. Akdemir H (2021) Evaluation of transcription factor and aquaporin gene expressions in response to Al_2O_3 and ZnO nanoparticles during barley germination. Plant Physiol Biochem 166:466–476. https://doi.org/10.1016/j.plaphy.2021.06.018
6. Akhoundnejad Y, Karakas O, Demirci O (2022) Response of lettuce to silver nanoparticles under drought conditions. Iran J Sci Technol Trans A: Sci 46:111–120. https://doi.org/10.1007/s40995-021-01241-x
7. Al-Selwey WA, Alsadon AA, Alenazi MM et al (2023) Morphological and biochemical response of potatoes to exogenous application of ZnO and SiO_2 nanoparticles in a water deficit environment. Horticulturae 9. https://doi.org/10.3390/horticulturae9080883
8. Alam P, Arshad M, Al-Kheraif AA et al (2022) Silicon nanoparticle-induced regulation of carbohydrate metabolism, photosynthesis, and ROS homeostasis in *Solanum lycopersicum* subjected to salinity stress. ACS Omega 7:31834–31844. https://doi.org/10.1021/acsomega.2c02586
9. Alan BO, Barisik M, Ozcelik HG (2020) Roughness effects on the surface charge properties of silica nanoparticles. J Phys Chem C 124:7274–7286. https://doi.org/10.1021/acs.jpcc.0c00120
10. Albqmi M, Yaghoubi Khanghahi M, Selim S et al (2023) Positive interaction of selenium nanoparticles and olive solid waste on vanadium-stressed soybean plant. Agriculture 13:426. https://doi.org/10.3390/agriculture13020426
11. Alhaithloul HAS, Ali B, Alghanem SMS et al (2023) Effect of green-synthesized copper oxide nanoparticles on growth, physiology, nutrient uptake, and cadmium accumulation in *Triticum aestivum* (L.). Ecotoxicol Environ Saf 268:115701. https://doi.org/10.1016/j.ecoenv.2023.115701
12. Azim Z, Singh NB, Singh A et al (2023) A review summarizing uptake, translocation and accumulation of nanoparticles within the plants: current status and future prospectus. J Plant Biochem Biotechnol 32:211–224. https://doi.org/10.1007/s13562-022-00800-6
13. Balan D, Tokas J, Singal HR (2018) UDP-glucose pyrophosphorylase: isolation, purification and characterization from developing thermotolerant wheat (*Triticum aestivum*) grains. Protein Expr Purif 148:68–77. https://doi.org/10.1016/j.pep.2018.04.007
14. Banerjee A, Singh A, Sudarshan M, Roychoudhury A (2021) Silicon nanoparticle-pulsing mitigates fluoride stress in rice by fine-tuning the ionomic and metabolomic balance and refining agronomic traits. Chemosphere 262:127826. https://doi.org/10.1016/j.chemosphere.2020.127826
15. Bárzana G, Garcia-Gomez P, Carvajal M (2022) Nanomaterials in plant systems: smart advances related to water uptake and transport involving aquaporins. Plant Nano Biol 1:100005. https://doi.org/10.1016/J.PLANA.2022.100005
16. Bovand F, Chavoshi S, Ghorbanpour M (2023) Titanium dioxide and silicon dioxide nanoparticles differentially affect germination and biochemical traits in marigold (*Calendula officinalis* L.) and lemon balm (*Melissa officinalis* L.). Nanotechnol Environ Eng 8:281–295. https://doi.org/10.1007/S41204-022-00279-5/METRICS

17. Cao P, Zhao Y, Wu F et al (2022) Multi-omics techniques for soybean molecular breeding. Int J Mol Sci 23:4994. https://doi.org/10.3390/ijms23094994
18. Cao Z, Rossi L, Stowers C et al (2018) The impact of cerium oxide nanoparticles on the physiology of soybean (*Glycine max* (L.) Merr.) under different soil moisture conditions. Environ Sci Pollut Res 25:930–939. https://doi.org/10.1007/s11356-017-0501-5
19. Cao Z, Stowers C, Rossi L et al (2017) Physiological effects of cerium oxide nanoparticles on the photosynthesis and water use efficiency of soybean (*Glycine max* (L.) Merr.). Environ Sci Nano 4:1086–1094. https://doi.org/10.1039/C7EN00015D
20. Chandra AK, Pandey D, Tiwari A et al (2021) Metal based nanoparticles trigger the differential expression of key regulatory genes which regulate iron and zinc homeostasis mechanism in finger millet. J Cereal Sci 100:103235. https://doi.org/10.1016/j.jcs.2021.103235
21. Chauhan J, Prathibha M, Singh P et al (2023) Plant photosynthesis under abiotic stresses: damages, adaptive, and signaling mechanisms. Plant Stress 10:100296. https://doi.org/10.1016/j.stress.2023.100296
22. Chemingui H, Smiri M, Missaoui T, Hafiane A (2019) Zinc oxide nanoparticles induced oxidative stress and changes in the photosynthetic apparatus in fenugreek (*Trigonella foenum graecum* L.). Bull Environ Contam Toxicol 102:477–485. https://doi.org/10.1007/s00128-019-02590-5
23. Chen J, Yin Y, Zhu Y et al (2023) Favorable physiological and morphological effects of molybdenum nanoparticles on tobacco (*Nicotiana tabacum* L.): root irrigation is superior to foliar spraying. Front Plant Sci 14:1–16. https://doi.org/10.3389/fpls.2023.1220109
24. Chen L, Peng Y, Zhu L et al (2022) CeO_2 nanoparticles improved cucumber salt tolerance is associated with its induced early stimulation on antioxidant system. Chemosphere 299:134474. https://doi.org/10.1016/j.chemosphere.2022.134474
25. Chen Z, Niu J, Guo Z et al (2021) Graphene enhances photosynthesis and the antioxidative defense system and alleviates salinity and alkalinity stresses in alfalfa (*Medicago sativa* L.) by regulating gene expression. Environ Sci Nano 8:2731–2748. https://doi.org/10.1039/D1EN00257K
26. Chen Z, Zhao J, Cao J et al (2022) Opportunities for graphene, single-walled and multi-walled carbon nanotube applications in agriculture: a review. Crop Design 1:100006. https://doi.org/10.1016/j.cropd.2022.100006
27. Choe K, Pak U, Pang Y et al (2023) Advances and challenges in spatial transcriptomics for developmental biology. Biomolecules 13:1–20. https://doi.org/10.3390/biom13010156
28. Choudhary RC, Joshi A, Kumari S et al (2017) Preparation of Cu-chitosan nanoparticle and its effect on growth and enzyme activity during seed germination in maize. J Pharmacognosy Phytochem JPP 6:669–673
29. Crandon LE, Boenisch KM, Harper BJ, Harper SL (2020) Adaptive methodology to determine hydrophobicity of nanomaterials in situ. PLoS ONE 15:1–17. https://doi.org/10.1371/journal.pone.0233844
30. de Sousa A, Saleh AM, Habeeb TH et al (2019) Silicon dioxide nanoparticles ameliorate the phytotoxic hazards of aluminum in maize grown on acidic soil. Sci Total Environ 693:133636. https://doi.org/10.1016/j.scitotenv.2019.133636
31. Dilnawaz F, Kalaji MH, Misra AN (2023) Nanotechnology in improving photosynthesis under adverse climatic conditions: cell to Canopy action. Plant Nano Biol 4:100035. https://doi.org/10.1016/j.plana.2023.100035
32. Dumanović J, Nepovimova E, Natić M et al (2021) The significance of reactive oxygen species and antioxidant defense system in plants: a concise overview. Front Plant Sci 11. https://doi.org/10.3389/fpls.2020.552969
33. Emamverdian A, Ghorbani A, Li Y et al (2023) Responsible mechanisms for the restriction of heavy metal toxicity in plants via the Co-foliar spraying of nanoparticles. Agronomy 13:1748. https://doi.org/10.3390/agronomy13071748
34. Emamverdian A, Hasanuzzaman M, Ding Y et al (2022) Zinc oxide nanoparticles improve pleioblastus pygmaeus plant tolerance to arsenic and mercury by stimulating antioxidant defense and reducing the metal accumulation and translocation. Front Plant Sci 13. https://doi.org/10.3389/fpls.2022.841501

35. Eprintsev AT, Fedorin DN, Cherkasskikh MV, Igamberdiev AU (2020) Regulation of expression of the mitochondrial and cytosolic forms of aconitase in maize leaves via phytochrome. Plant Physiol Biochem 146:157–162. https://doi.org/10.1016/j.plaphy.2019.11.018
36. Faizan M, Akhter J, Noureldeen A et al (2021) Ecotoxicology and environmental safety zinc oxide nanoparticles and 24-epibrassinolide alleviates Cu toxicity in tomato by regulating ROS scavenging, stomatal movement and photosynthesis. Ecotoxicol Environ Saf 218:112293. https://doi.org/10.1016/j.ecoenv.2021.112293
37. Faizan M, Faraz A, Mir AR, Hayat S (2021) Role of zinc oxide nanoparticles in countering negative effects generated by cadmium in *Lycopersicon esculentum.* J Plant Growth Regul 40:101–115. https://doi.org/10.1007/s00344-019-10059-2
38. Faizan M, Hayat S (2019) Effect of foliar spray of ZnO-NPs on the physiological parameters and antioxidant systems of *Lycopersicon esculentum.* Polish J Nat Sci 34:87–105
39. Faraz A, Faizan M, Rajput DV et al (2023) CuO nanoparticle-mediated seed priming improves physio-biochemical and enzymatic activities of *Brassica juncea.* Plants 12:803. https://doi.org/10.3390/plants12040803
40. Fatemi H, Esmaiel Pour B, Rizwan M (2021) Foliar application of silicon nanoparticles affected the growth, vitamin C, flavonoid, and antioxidant enzyme activities of coriander (*Coriandrum sativum* L.) plants grown in lead (Pb)-spiked soil. Environ Sci Pollut Res 28:1417–1425. https://doi.org/10.1007/s11356-020-10549-x
41. Gani U, Vishwakarma RA, Misra P (2021) Membrane transporters: the key drivers of transport of secondary metabolites in plants. Plant Cell Rep 40:1–18. https://doi.org/10.1007/S00299-020-02599-9/FIGURES/2
42. Gao M, Chang J, Wang Z et al (2023) Advances in transport and toxicity of nanoparticles in plants. J Nanobiotechnol 21:1–17. https://doi.org/10.1186/s12951-023-01830-5
43. Gao Y, Chen S, Li Y, Shi Y (2023b) Effect of nano-calcium carbonate on morphology, antioxidant enzyme activity and photosynthetic parameters of wheat (*Triticum aestivum* L.) seedlings. Chem Biol Technol Agric 10:31. https://doi.org/10.1186/s40538-023-00404-9
44. Garza-Alonso CA, Cadenas-Pliego G, Juárez-Maldonado A et al (2023) Fe_2O_3 nanoparticles can replace Fe-EDTA fertilizer and boost the productivity and quality of *Raphanus sativus* in a soilless system. Scientia Horticulturae 321. https://doi.org/10.1016/j.scienta.2023.112261
45. Garza-Alonso CA, Olivares-Sáenz E, González-Morales S et al (2022) Strawberry biostimulation: from mechanisms of action to plant growth and fruit quality. Plants 11:1–34. https://doi.org/10.3390/plants11243463
46. Gautam A, Sharma P, Ashokhan S et al (2023) Inhibitory impact of MgO nanoparticles on oxidative stress and other physiological attributes of spinach plant grown under field condition. Physiol Mol Biol Plants. https://doi.org/10.1007/s12298-023-01391-9
47. Ghassemi-Golezani K, Abdoli S (2021) Improving ATPase and PPase activities, nutrient uptake and growth of salt stressed ajowan plants by salicylic acid and iron-oxide nanoparticles. Plant Cell Rep 40:559–573. https://doi.org/10.1007/s00299-020-02652-7
48. Gohari G, Mohammadi A, Akbari A et al (2020) Titanium dioxide nanoparticles (TiO_2 NPs) promote growth and ameliorate salinity stress effects on essential oil profile and biochemical attributes of *Dracocephalum moldavica.* Sci Rep 10:912. https://doi.org/10.1038/s41598-020-57794-1
49. González-García Y, López-Vargas ER, Cadenas-Pliego G et al (2019) Impact of carbon nanomaterials on the antioxidant system of tomato seedlings. Int J Mole Sci 20. https://doi.org/10.3390/ijms20235858
50. González-Morales S, Solís-Gaona S, Valdés-Caballero MV et al (2021) Transcriptomics of biostimulation of plants under abiotic stress. Front Gene 12. https://doi.org/10.3389/fgene.2021.583888
51. Guha T, Gopal G, Das H et al (2021) Nanopriming with zero-valent iron synthesized using pomegranate peel waste: a "green" approach for yield enhancement in *Oryza sativa* L. cv. Gonindobhog. Plant Physiol Biochem 163:261–275. https://doi.org/10.1016/j.plaphy.2021.04.006

52. Guo J, Li S, Brestic M et al (2023) Modulations in protein phosphorylation explain the physiological responses of barley (*Hordeum vulgare*) to nanoplastics and ZnO nanoparticles. J Hazard Mater 443:130196. https://doi.org/10.1016/J.JHAZMAT.2022.130196
53. Gupta N, Jain SK, Tomar BS et al (2022) Impact of foliar application of ZnO and Fe_3O_4 nanoparticles on seed yield and physio-biochemical parameters of cucumber (*Cucumis sativus* L.) seed under open field and protected environment vis a vis during seed germination. Plants 11:1–20. https://doi.org/10.3390/plants11233211
54. Ha NTM, Manh Do C, Hoang TT et al (2020) The effect of cobalt and silver nanoparticles on overcoming leaf abscission and enhanced growth of rose (*Rosa hybrida* L. 'Baby Love') plantlets cultured in vitro. Plant Cell, Tissue Organ Cult 141:393–405. https://doi.org/10.1007/s11240-020-01796-4
55. Hasanaklou NT, Mohagheghi V, Hasanaklou HT et al (2023) Seed nano-priming using silica nanoparticles: effects in seed germination and physiological properties of *Stevia Rebaudiana* Bertoni. Chem Biol Technol Agric 10:1–13. https://doi.org/10.1186/s40538-023-00445-0
56. Hernández-Fuentes AD, Montaño-Herrera A, Pinedo-Espinoza JM et al (2023) Changes of the antioxidant system in pear (*Pyrus communis* L.) fruits by foliar application of copper, selenium, iron, and zinc nanoparticles. J Agric Food Res 14:100885. https://doi.org/10.1016/j.jafr.2023.100885
57. Hossain Z, Yasmeen F, Komatsu S (2020) Nanoparticles: synthesis, morphophysiological effects, and proteomic responses of crop plants. Int J Mole Sci 21. https://doi.org/10.3390/ijms21093056
58. Hrmova M, Hussain SS (2021) Plant transcription factors involved in drought and associated stresses. Int J Mole Sci 22:5662. https://doi.org/10.3390/IJMS22115662
59. Huang G, Pan D, Wang M et al (2022) Regulation of iron and cadmium uptake in rice roots by iron(III) oxide nanoparticles: insights from iron plaque formation, gene expression, and nanoparticle accumulation. Environ Sci Nano 9:4093–4103. https://doi.org/10.1039/D2EN00487A
60. Huang XY, Hu DW, Zhao FJ (2022) Molybdenum: more than an essential element. J Exp Bot 73:1766–1774. https://doi.org/10.1093/jxb/erab534
61. Ismail LM, Soliman MI, Abd El-Aziz MH, Abdel-Aziz HMM (2022) Impact of silica ions and nano silica on growth and productivity of pea plants under salinity stress. Plants 11:494. https://doi.org/10.3390/plants11040494
62. Jain A, Singh HB, Das S (2021) Deciphering plant-microbe crosstalk through proteomics studies. Microbiol Res 242:126590. https://doi.org/10.1016/j.micres.2020.126590
63. Jethva J, Schmidt RR, Sauter M, Selinski J (2022) Try or die: dynamics of plant respiration and how to survive low oxygen conditions. Plants 11:205. https://doi.org/10.3390/PLANTS11020205
64. Jhanzab HM, Razzaq A, Bibi Y et al (2019) Proteomic analysis of the effect of inorganic and organic chemicals on silver nanoparticles in wheat. Int J Mole Sci 20. https://doi.org/10.3390/ijms20040825
65. Jiang M, Dai S, Wang B et al (2021) Gold nanoparticles synthesized using melatonin suppress cadmium uptake and alleviate its toxicity in rice. Environ Sci Nano 8:1042–1056. https://doi.org/10.1039/d0en01172j
66. Jiang Y, Yang J, Li M et al (2022) Effect of silica-based nanomaterials on seed germination and seedling growth of rice (*Oryza sativa* L.). Nanomaterials 12:4160. https://doi.org/10.3390/nano12234160
67. Jośko I, Kusiak M, Xing B, Oleszczuk P (2021) Combined effect of nano-CuO and nano-ZnO in plant-related system: From bioavailability in soil to transcriptional regulation of metal homeostasis in barley. J Hazard Mater 416. https://doi.org/10.1016/j.jhazmat.2021.126230
68. Juárez-Maldonado A (2022) Impact of nanomaterials on plants: what other implications do they have? Biocell 46:651–654. https://doi.org/10.32604/biocell.2022.017350
69. Juárez-Maldonado A, Ortega-Ortíz H, Morales-Díaz AB et al (2019) Nanoparticles and nanomaterials as plant biostimulants. Int J Mol Sci 20:1–19. https://doi.org/10.3390/ijms20010162

70. Juárez-Maldonado A, Tortella G, Rubilar O et al (2021) Biostimulation and toxicity: the magnitude of the impact of nanomaterials in microorganisms and plants. J Adv Res 31:113–126. https://doi.org/10.1016/j.jare.2020.12.011
71. Kerchev P, van der Meer T, Sujeeth N et al (2020) Molecular priming as an approach to induce tolerance against abiotic and oxidative stresses in crop plants. Biotechnol Adv 40:107503. https://doi.org/10.1016/j.biotechadv.2019.107503
72. Khan MN, Fu C, Li J et al (2023) Seed nanopriming: How do nanomaterials improve seed tolerance to salinity and drought? Chemosphere 310. https://doi.org/10.1016/j.chemosphere.2022.136911
73. Kim JH, Oh Y, Yoon H et al (2015) Iron nanoparticle-induced activation of plasma membrane H+-ATPase promotes stomatal opening in *Arabidopsis thaliana*. Environ Sci Technol 49:1113–1119. https://doi.org/10.1021/es504375t
74. Koley R, Mondal A, Mondal NK (2023) Green synthesized silver nanoparticles mediated regulation on hydrolytic enzymes and ROS homeostasis promote growth of three pulses (*Cicer arietinum* L., *Pisum sativum* L., and *Vigna radiata* L.). Energy, Ecol Environ 8:537–555. https://doi.org/10.1007/S40974-023-00293-6/METRICS
75. Kordyum E, Polishchuk O, Akimov Y, Brykov V (2022) Photosynthetic apparatus of hydrocharis morsus-ranae in different solar lighting. Plants 11:2658. https://doi.org/10.3390/plants11192658
76. Köster P, DeFalco TA, Zipfel C (2022) Ca^{2+} signals in plant immunity . EMBO J 41:1–18. https://doi.org/10.15252/embj.2022110741
77. Kumari R, Suman K, Karmakar S et al (2023) Regulation and safety measures for nanotechnology-based agri-products. Front Genome Edit 5:1–12. https://doi.org/10.3389/fgeed.2023.1200987
78. Lai M, Ghouri F, Sarwar S et al (2023) Modulation of metal transporters, oxidative stress and cell abnormalities by synergistic application of silicon and titanium oxide nanoparticles: a strategy for cadmium tolerance in rice. Chemosphere 345:140439. https://doi.org/10.1016/J.CHEMOSPHERE.2023.140439
79. Leal Filho W, Nagy GJ, Setti AFF et al (2023) Handling the impacts of climate change on soil biodiversity. Sci Total Environ 869:161671. https://doi.org/10.1016/J.SCITOTENV.2023.161671
80. Leister D (2023) Enhancing the light reactions of photosynthesis: strategies, controversies, and perspectives. Mol Plant 16:4–22. https://doi.org/10.1016/j.molp.2022.08.005
81. Li H, Huang J, Lu F et al (2018) Impacts of carbon dots on rice plants: boosting the growth and improving the disease resistance. ACS Appl Bio Mater 1:663–672. https://doi.org/10.1021/acsabm.8b00345
82. Li J, Zafar S, Javaid A et al (2023a) Zinc nanoparticles (ZnNPs): high-fidelity amelioration in turnip (*Brassica rapa* L.) production under drought stress. Sustainability 15:6512. https://doi.org/10.3390/su15086512
83. Li X, Ghanizadeh H, Han Z et al (2023) Metabolic profile and physiological mechanisms underlying the promoting effects of TiO_2NPs on the photosynthesis and growth of tomato. Sci Hortic 322:112394. https://doi.org/10.1016/j.scienta.2023.112394
84. Li Y, Liu M, Yang X et al (2022) Multi-walled carbon nanotubes enhanced the antioxidative system and alleviated salt stress in grape seedlings. Sci Hortic 293:110698. https://doi.org/10.1016/j.scienta.2021.110698
85. Liu Y, Jiang Y, Liu X et al (2023) Identification and expression analysis of hexokinases family in *Saccharum spontaneum* L. under drought and cold stresses. Plants 12:1215. https://doi.org/10.3390/plants12061215
86. Ludwig M, Hartwell J, Raines CA, Simkin AJ (2024) The Calvin-Benson-Bassham cycle in C4 and crassulacean acid metabolism species. Semin Cell Dev Biol 155:10–22. https://doi.org/10.1016/j.semcdb.2023.07.013
87. Mahakham W, Sarmah AK, Maensiri S, Theerakulpisut P (2017) Nanopriming technology for enhancing germination and starch metabolism of aged rice seeds using phytosynthesized silver nanoparticles. Sci Rep 7:1–21. https://doi.org/10.1038/s41598-017-08669-5

88. Mahmoud LM, Shalan AM, El-Boray MS et al (2022) Application of silicon nanoparticles enhances oxidative stress tolerance in salt stressed 'Valencia' sweet orange plants. Sci Hortic 295:110856. https://doi.org/10.1016/J.SCIENTA.2021.110856
89. Mandal R, Dutta G (2020) From photosynthesis to biosensing: chlorophyll proves to be a versatile molecule. Sens Int 1:100058. https://doi.org/10.1016/j.sintl.2020.100058
90. Maqsood MF, Shahbaz M, Khalid F et al (2023) Biogenic nanoparticles application in agriculture for ROS mitigation and abiotic stress tolerance: a review. Plant Stress 10:100281. https://doi.org/10.1016/j.stress.2023.100281
91. Marquardt A, Henry RJ, Botha FC (2021) Effect of sugar feedback regulation on major genes and proteins of photosynthesis in sugarcane leaves. Plant Physiol Biochem 158:321–333. https://doi.org/10.1016/j.plaphy.2020.11.022
92. Medrano-Macías J, Flores-Gallegos AC, Nava-Reyna E et al (2022) Reactive oxygen, nitrogen, and sulfur species (RONSS) as a metabolic cluster for signaling and biostimulation of plants: an overview. Plants 11:3203. https://doi.org/10.3390/PLANTS11233203
93. Mitra J, Sharma PD, Paul PK (2019) Do phylloplane microfungi in fluence activity of Rubisco and Carbonic anhydrase. S Afr J Bot 124:118–126. https://doi.org/10.1016/j.sajb.2019.04.033
94. Moshirian Farahi SM, Taghavizadeh Yazdi ME, Einafshar E et al (2023) The effects of titanium dioxide (TiO_2) nanoparticles on physiological, biochemical, and antioxidant properties of Vitex plant (*Vitex agnus - Castus* L). Heliyon 9:e22144. https://doi.org/10.1016/j.heliyon.2023.e22144
95. Mubashir A, Nisa Z, Shah AA et al (2023) Effect of foliar application of nano-nutrients solution on growth and biochemical attributes of tomato (*Solanum lycopersicum*) under drought stress. Front Plant Sci 13:1–11. https://doi.org/10.3389/fpls.2022.1066790
96. Natasha N, Shahid M, Bibi I et al (2022) Zinc in soil-plant-human system: a data-analysis review. Sci Total Environ 808:152024. https://doi.org/10.1016/J.SCITOTENV.2021.152024
97. Nazir MM, Li Q, Noman M et al (2022) Calcium oxide nanoparticles have the role of alleviating arsenic toxicity of barley. Front Plant Sci 13. https://doi.org/10.3389/fpls.2022.843795
98. Nicolas-Espinosa J, Garcia-Gomez P, Rios JJ et al (2022) Nanoencapsulated boron foliar supply increased expression of NIPs Aquaporins and BOR transporters of In vitro *Ipomoea batatas* Plants. Appl Sci (Switzerland) 12. https://doi.org/10.3390/app12041788
99. Nishad R, Ahmed T, Rahman VJ, Kareem A (2020) Modulation of plant defense system in response to microbial interactions. Front Microbiol 11:514909. https://doi.org/10.3389/FMICB.2020.01298/BIBTEX
100. Niu L, Li C, Wang W et al (2023) Cadmium tolerance and hyperaccumulation in plants–a proteomic perspective of phytoremediation. Ecotoxicol Environ Saf 256:114882. https://doi.org/10.1016/j.ecoenv.2023.114882
101. Ozfidan-Konakci C, Alp FN, Arikan B et al (2022) The biphasic responses of nanomaterial fullerene on stomatal movement, water status, chlorophyll a fluorescence transient, radical scavenging system and aquaporin-related gene expression in *Zea mays* under cobalt stress. Sci Total Environ 826:154213. https://doi.org/10.1016/J.SCITOTENV.2022.154213
102. Peláez-Vico MÁ, Fichman Y, Zandalinas SI et al (2022) ROS and redox regulation of cell-to-cell and systemic signaling in plants during stress. Free Radical Biol Med 193:354–362. https://doi.org/10.1016/j.freeradbiomed.2022.10.305
103. Pinto A, Azenha M, Fidalgo F, Teixeira J (2023) 2,4-dichlorophenoxyacetic acid detoxification occurs primarily in tomato leaves by the glutathione S-transferase phi members 4 and 5. Sci Hortic 321:112214. https://doi.org/10.1016/j.scienta.2023.112214
104. Rafique MU, Nahid N, Azeem F et al (2023) Genome wide analysis for the identification and characterization of Superoxide-dismutase gene family in *Rosa chinensis* ascertains the role of salinity-responsive RcMSD1 protein and its interaction with peroxyl radical. Plant Stress 10:100218. https://doi.org/10.1016/j.stress.2023.100218
105. Raftery AE, Ševčíková H (2023) Probabilistic population forecasting: short to very long-term. Int J Forecast 39:73–97. https://doi.org/10.1016/J.IJFORECAST.2021.09.001

106. Rajaee Behbahani S, Iranbakhsh A, Ebadi M et al (2020) Red elemental selenium nanoparticles mediated substantial variations in growth, tissue differentiation, metabolism, gene transcription, epigenetic cytosine DNA methylation, and callogenesis in bittermelon (*Momordica charantia*); an in vitro experiment. PLoS ONE 15:e0235556. https://doi.org/10.1371/journal.pone.0235556
107. Rasouli F, Asadi M, Hassanpouraghdam MB et al (2022) Foliar application of ZnO-NPs influences chlorophyll fluorescence and antioxidants pool in *Capsicum annum* L. under Salinity. Horticulturae 8:908. https://doi.org/10.3390/horticulturae8100908
108. Raza MU, Abasi F, Shahbaz M et al (2023) Phytomediated silver nanoparticles (AgNPs) embellish antioxidant defense system, ameliorating HLB-diseased 'Kinnow' mandarin plants. Molecules 28:2044. https://doi.org/10.3390/molecules28052044
109. Rezayian M, Niknam V, Arabloo M (2023) Iron nanoparticle regulate succinate dehydrogenase activity in canola plants under drought stress. Sci Rep 13:1–14. https://doi.org/10.1038/s41598-023-36105-4
110. Satti SH, Raja NI, Ikram M et al (2022) Plant-Based titanium dioxide nanoparticles trigger biochemical and proteome modifications in *Triticum aestivum* L. under biotic stress of *Puccinia striiformis*. Molecules 27:1–21. https://doi.org/10.3390/molecules27134274
111. Şener S, Sayğı H (2023) The role of silver nanoparticles in response of *in vitro* boysenberry plants to drought stress. Horticulturae 9:1177. https://doi.org/10.3390/horticulturae9111177
112. Singh D (2022) Juggling with reactive oxygen species and antioxidant defense system–a coping mechanism under salt stress. Plant Stress 5:100093. https://doi.org/10.1016/j.stress.2022.100093
113. Singh P, Arif Y, Siddiqui H et al (2021) Nanoparticles enhances the salinity toxicity tolerance in *Linum usitatissimum* L. by modulating the antioxidative enzymes, photosynthetic efficiency, redox status and cellular damage. Ecotoxicol Environ Saf 213:112020. https://doi.org/10.1016/j.ecoenv.2021.112020
114. Singh R, Singh DP, Gupta P et al (2019) Nanoparticles alter the withanolide biosynthesis and carbohydrate metabolism in *Withania somnifera* (Dunal). Ind Crops Prod 127:94–109. https://doi.org/10.1016/j.indcrop.2018.10.049
115. Skiba E, Pietrzak M, Michlewska S et al (2024) Photosynthesis governed by nanoparticulate titanium dioxide. The *Pisum sativum* L. case study. Environ Pollut 340:122735. https://doi.org/10.1016/j.envpol.2023.122735
116. Sohail, Sawati L, Ferrari E et al (2022) Molecular effects of biogenic zinc nanoparticles on the growth and development of *Brassica napus* L. revealed by proteomics and transcriptomics. Front Plant Sci 13. https://doi.org/10.3389/fpls.2022.798751
117. Soliman MH, Alnusairi GSH, Khan AA et al (2023) Biochar and selenium nanoparticles induce water transporter genes for sustaining carbon assimilation and grain production in salt-stressed wheat. J Plant Growth Regul 42:1522–1543. https://doi.org/10.1007/s00344-022-10636-y
118. Stałanowska K, Szablińska-Piernik J, Okorski A, Lahuta LB (2023) Zinc oxide nanoparticles affect early seedlings' growth and polar metabolite profiles of pea (*Pisum sativum* L.) and wheat (*Triticum aestivum* L.). Int J Mole Sci 24. https://doi.org/10.3390/ijms241914992
119. Sun L, Song F, Zhu X et al (2021) Nano-ZnO alleviates drought stress via modulating the plant water use and carbohydrate metabolism in maize. Arch Agron Soil Sci 67:245–259. https://doi.org/10.1080/03650340.2020.1723003
120. Sun L, Wang Y, Wang R et al (2020) Physiological, transcriptomic, and metabolomic analyses reveal zinc oxide nanoparticles modulate plant growth in tomato. Environ Sci Nano 7:3587–3604. https://doi.org/10.1039/d0en00723d
121. Tawfik MM, Mohamed MH, Sadak MS, Thalooth AT (2021) Iron oxide nanoparticles effect on growth, physiological traits and nutritional contents of *Moringa oleifera* grown in saline environment. Bull Natl Res Centre 45:177. https://doi.org/10.1186/s42269-021-00624-9
122. Tortella G, Rubilar O, Pieretti JC et al (2023) Nanoparticles as a promising strategy to mitigate biotic stress in agriculture. Antibiotics 12. https://doi.org/10.3390/antibiotics12020338

123. Venzhik YV, Deryabin AN (2023) Regulation of Pro-/antioxidant balance in higher plants by nanoparticles of metals and metal oxides. Russian J Plant Physiol 70. https://doi.org/10.1134/S1021443722602312
124. Wahab A, Munir A, Saleem MH et al (2023) Interactions of metal-based engineered nanoparticles with plants: an overview of the state of current knowledge, research progress, and prospects. J Plant Growth Regul 42:5396–5416. https://doi.org/10.1007/s00344-023-10972-7
125. Wang H, Zhang M, Song Y et al (2018) Carbon dots promote the growth and photosynthesis of mung bean sprouts. Carbon 136:94–102. https://doi.org/10.1016/j.carbon.2018.04.051
126. Wang R, Song H, Zhang W et al (2023) Structural insights into the functions of Raf1 and Bsd2 in hexadecameric Rubisco assembly. Mol Plant 16:1927–1936. https://doi.org/10.1016/j.molp.2023.10.011
127. Wang X, Xie H, Wang P, Yin H (2023) Nanoparticles in Plants: uptake, transport and physiological activity in leaf and root. Materials 16:1–21. https://doi.org/10.3390/ma16083097
128. Waters ER, Vierling E (2020) Plant small heat shock proteins – evolutionary and functional diversity. New Phytol 227:24–37. https://doi.org/10.1111/NPH.16536
129. Werelusz P, Galiniak S, Mołoń M (2024) Molecular functions of moonlighting proteins in cell metabolic processes. Biochimica et Biophysica Acta (BBA)—Mole Cell Res 1871:119598. https://doi.org/10.1016/j.bbamcr.2023.119598
130. Wu H, Li Z (2022) Nano-enabled agriculture: how do nanoparticles cross barriers in plants? Plant Commun 3:100346. https://doi.org/10.1016/j.xplc.2022.100346
131. Wu H, Tito N, Giraldo JP (2017) Anionic cerium oxide nanoparticles protect plant photosynthesis from abiotic stress by scavenging reactive oxygen species. ACS Nano 11:11283–11297. https://doi.org/10.1021/acsnano.7b05723
132. Wu M, Su H, Li C et al (2023) Effects of foliar application of single-walled carbon nanotubes on carbohydrate metabolism in crabapple plants. Plant Physiol Biochem 194:214–222. https://doi.org/10.1016/j.plaphy.2022.11.023
133. Wu Q, Chen Y, Zou W et al (2023) Genome-wide characterization of sugarcane catalase gene family identifies a *ScCAT1* gene associated disease resistance. Int J Biol Macromol 232:123398. https://doi.org/10.1016/j.ijbiomac.2023.123398
134. Xiao X, Wang X, Liu L et al (2022) Effects of three graphene-based materials on the growth and photosynthesis of *Brassica napus* L. Ecotoxicol Environ Saf 234:113383. https://doi.org/10.1016/j.ecoenv.2022.113383
135. Xiong T, Zhang S, Kang Z et al (2021) Dose-dependent physiological and transcriptomic responses of lettuce (*Lactuca sativa* L.) to copper oxide nanoparticles—insights into the phytotoxicity mechanisms. Int J Mole Sci 22. https://doi.org/10.3390/ijms22073688
136. Yan S, Wu F, Zhou S et al (2021) Zinc oxide nanoparticles alleviate the arsenic toxicity and decrease the accumulation of arsenic in rice (*Oryza sativa* L.). BMC Plant Biol 21:150. https://doi.org/10.1186/s12870-021-02929-3
137. Yang C, Wang C, Khan Z et al (2023) Algal polysaccharides–Selenium nanoparticles regulate the uptake and distribution of selenium in rice plants. Front Plant Sci 14:1–13. https://doi.org/10.3389/fpls.2023.1135080
138. Yang J, Zhan R, Jin Y et al (2020) Functional analysis of the promoter of the *MdFRK2* gene encoding a high-affinity fructokinase in apple (*malus* × *domestica*). Sci Hortic 265:109088. https://doi.org/10.1016/j.scienta.2019.109088
139. Younes NA, Dawood MFA, Wardany AA (2019) Biosafety assessment of graphene nanosheets on leaf ultrastructure, physiological and yield traits of *Capsicum annuum* L. and *Solanum melongena* L. Chemosphere 228:318–327. https://doi.org/10.1016/j.chemosphere.2019.04.097
140. Younis AA, Khattab H, Emam MM (2020) Impacts of silicon and silicon nanoparticles on leaf ultrastructure and tapip1 and tanip2 gene expressions in heat stressed wheat seedlings. Biologia Plantarum 64:343–352. https://doi.org/10.32615/bp.2020.030
141. Yu Y, Dai W, Luan Y (2023) Bio- and eco-corona related to plants: Understanding the formation and biological effects of plant protein coatings on nanoparticles. Environ Pollut 317:120784. https://doi.org/10.1016/J.ENVPOL.2022.120784

142. Yue L, Ma C, Zhan X et al (2017) Molecular mechanisms of maize seedling response to *La2O3* NP exposure: water uptake, aquaporin gene expression and signal transduction. Environ Sci Nano 4:843–855. https://doi.org/10.1039/c6en00487c
143. Zahedi SM, Abdelrahman M, Hosseini MS et al (2019) Alleviation of the effect of salinity on growth and yield of strawberry by foliar spray of selenium-nanoparticles. Environ Pollut 253:246–258. https://doi.org/10.1016/j.envpol.2019.04.078
144. Zeeshan M, Hu YX, Guo XH et al (2023) Physiological and transcriptomic study reveal SeNPs-mediated AsIII stress detoxification mechanisms involved modulation of antioxidants, metal transporters, and transcription factors in *Glycine max* L. (Merr.) roots. Environ Pollut 317:120637. https://doi.org/10.1016/j.envpol.2022.120637
145. Zhang L, Zhu L, Xu Y et al (2023) Genome-wide identification and function analysis of the sucrose phosphate synthase MdSPS gene family in apple. J Integr Agric 22:2080–2093. https://doi.org/10.1016/j.jia.2023.05.024
146. Zhang Y, Liu N, Wang W et al (2020) Photosynthesis and related metabolic mechanism of promoted rice (*Oryza sativa* L.) growth by TiO_2 nanoparticles. Front Environ Sci Eng 14. https://doi.org/10.1007/s11783-020-1282-5

Changes in the Mineral and Secondary Metabolite Compositions of Plants Exposed to Nanomaterials

Oscar Cruz Alvarez, María Teresa Martínez Damián, Adalberto Benavides-Mendoza, Ofelia Adriana Hernández-Rodríguez, and Damaris Leopoldina Ojeda-Barrios

Abstract Over the last 20,000 years, agriculture has become humankinds' primary economic activity. Very recently (in comparison) nanotechnology has become a most promising area of scientific endeavour. This technology is creating exciting new opportunities in widespread areas of industry. These opportunities include in agriculture, where nanomaterials (NMs) offer improvements in nutrient supply efficiency to crop plants, that lead both to increased plant productivity and to increased sustainability in today's sometimes quite unsustainable agroeconomic production systems. The role of NMs as nanofertilisers and as elicitors in plants has been demonstrated through numerous studies. The use of NMs can improve plant mineral composition and increase the synthesis and accumulation of numerous beneficial secondary metabolites. These secondary metabolites not only increase a plant's tolerance of biotic and abiotic stresses of all sorts but also raises its nutraceutical quality and yield. There is a dearth of studies on the impacts of NMs used as nanofertilisers on traceability in the edible organs and on their long-term effects on soil health and water. This chapter offer a review of the effects of foliar, soil and substrate applications of NMs both as nanofertilisers and as elicitors on the mineral compositions and secondary metabolite compositions of food plants.

Keywords Bioactive compounds · Nanofertilisers · Nutrient uptake · Oxidative stress · Phytotoxicity

O. C. Alvarez · O. A. Hernández-Rodríguez · D. L. Ojeda-Barrios (✉)
Facultad de Ciencias Agrotecnológicas, Universidad Autónoma de Chihuahua, Chihuahua, México
e-mail: dojeda@uach.mx

M. T. M. Damián
Departamento de Fitotecnia, Universidad Autónoma Chapingo, Estado de México, Texcoco de Mora, México

A. Benavides-Mendoza
Departamento de Horticultura, Universidad Agraria Antonio Narro. Buenavista, Saltillo, México

© The Author(s), under exclusive license to Springer Nature Singapore Pte Ltd. 2025
A. Juárez-Maldonado et al. (eds.), *Plant Biostimulation with Nanomaterials*, Smart Nanomaterials Technology, https://doi.org/10.1007/978-981-96-4648-7_4

1 Introduction

Agriculture is one of the principle nexuses between human beings and environmental deterioration, being a major cause of soil degradation and ecosystem modification and loss [6]. However, agriculture is also an essential activity for the production of both fresh and processed foods for human consumption. Therefore, there is urgent need for the development of innovative technologies that will improve plant gene expression while at the same time resulting in reduced negative impacts on the environment. But, if they are to be implemented, such technologies must also be both practicable to apply and economic. It is arguable that incremental improvements in existing technologies are unlikely to yield the technological gains required, whereas novel multidisciplinary approaches may yet be able to. In this regard, nanotechnology is a relatively new, multidisciplinary field of research that has resulted in a number of novel and practicable applications in biotechnology, medical science, environmental science, engineering, food technology and agriculture [87]. In plant science these applications are being achieved through the use of nanomaterials (NMs) with particle sizes between 1 and 100 nm. These NMs are synthesised by chemical, physical and biological means, to achieve size control, morphology, scalability, simplicity, and environmental friendliness [90]. The unique properties of NMs can help improve plant performance by minimising the loss of mineral nutrients, by better control of pathogens and by their ability to function as elicitors that trigger the plant's own defence mechanisms against stresses caused by biotic factors such as pathogens, and by abiotic factors both aerial (such as extreme air temperatures) and edaphic (such as severe soil water deficits, excesses of heavy metals and sodicity/salinity [63, 67].

Studies have already demonstrated the efficacy of NMs when applied to adult plants, to seedlings or to seeds, through foliar sprays, applications to the substrate or impregnation [65, 95]. Here, multiple effects of broad technical interest have been demonstrated, including gains in germination percentage, vegetative growth (plant height, leaf area, number of leaves etc.), flowering and fruit set, with consequent gains in both crop yield and quality [24, 71]. There are now numerous reports on the roles that NMs can play in established plants in protecting against stress under conditions of water deficit, excesses of salinity, heavy metals (e.g., cadmium, arsenic, cerium) and extremes of temperature [46, 57]. These take place by mimicking some specific functions of antioxidant enzymes, including superoxide dismutase (SOD), catalase (CAT), peroxidase (POD), ascorbate peroxidase (APX), guiacol peroxidase (GPX) and phenylalanine ammonia lyase (PAL) [118]. It has been reported that applications of high concentrations of NMs can induce symptoms of phytotoxicity by increasing the generation and accumulation of reactive oxygen species (ROS) [75].

The synthesis and accumulation of secondary metabolites (including carotenoids, anthocyanins, phenolic compounds, flavonoids, flavones, chlorophyll, carotenoids, malonaldelhyde, proline and glutathione) in plant tissues and organs (seeds, roots, leaves, shoots and fruits) [44, 112], is linked to modifications of antioxidant enzyme activity and the uptake and accumulations of essential mineral nutrients, including Zn, Mg, Fe, Cu and Mn [2, 94]. Use of NMs can have a range of effects in plants

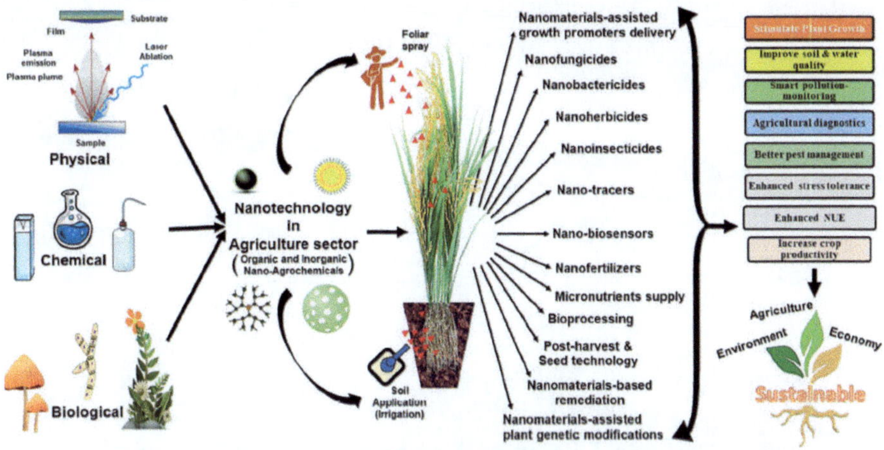

Fig. 1 Different aspects of nanotechnology in agriculture. Adapted from Naaz et al. [86]

including significantly increasing seed germination percentage, improving initial seedling development, increasing leaf area, flowering and fruit set All these especially in plant growing under suboptimal and marginal conditions of climate or whose health is compromised by anthropogenic activities [10, 46, 94]. Such benefits of NMs have been reported in beans [15], grapes [48], wheat [124], safflower [33], tomato [95], asparagus [24] and pecan [89]. All of these properties are affected by the sizes and shapes of the NMs which are both determining factors in their efficacy (Fig. 1).

It is well known that overuse of fertilisers, pesticides, and agricultural implements (mowing, tillage etc.) in horticulture has multiple negative effects on the environment, reducing water availability, loss of local biodiversity, soil compaction and degradation, and accumulation of chemical residues in the soil in soil water and in local flora and fauna [7, 99]. The use NMs offers improvements in agroecosystem care and reductions in cost by improving the efficiency of product delivery, so reducing the required product dose strength and the number of product applications. This benefit applies in particular to fertilisers, fungicides and herbicides formulated as NMs. In agriculture, NMs can be classified according to their area of use or by the problem to be solved. This includes (a) nanostructures for the controlled release of substances such as fertilisers and biocides, (b) NMs for the control of pathogens (e.g., microorganisms) and pests (e.g., insects), (c) NMs for treating and monitoring contaminated water and soil and (d) NM-based food packaging for the transport and distribution of harvested products [20, 119]. Such applications are briefly described in Fig. 2.

The rapid development of nanotechnology and its wide application now in agriculture have spawned a good number of research and review articles dealing with biocompatible NMs (carbon NMs, metal NMs, ceramic NMs, semiconductor NMs, polymeric NMs and lipid-based NMs) and their use with various plant species, both cultivated and wild [55]. However, their ecological aspects including degradation time and the potential issues with the upscaling of their use in agriculture have not

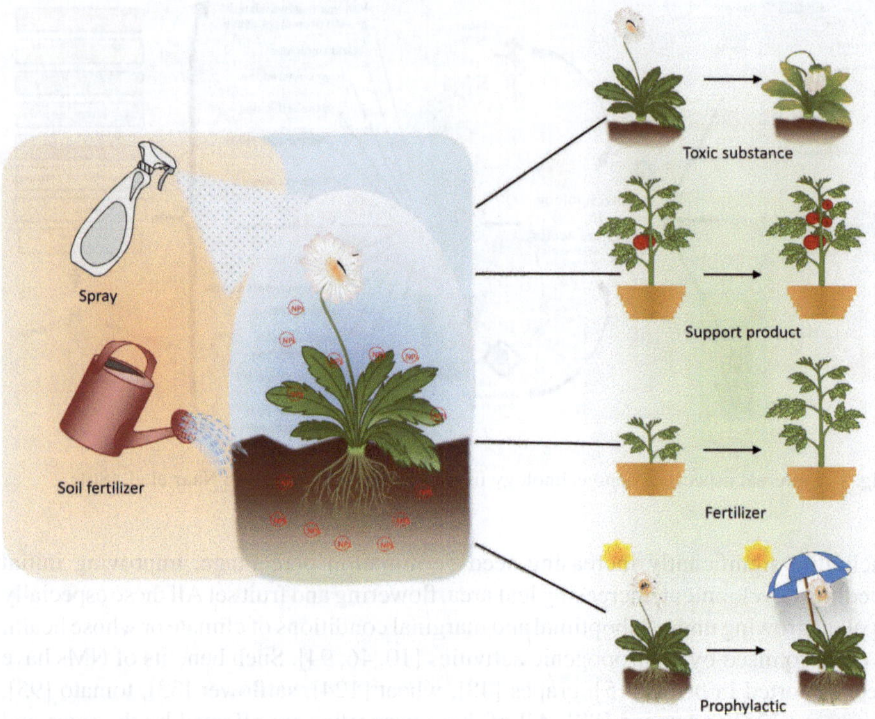

Fig. 2 Possible application of nanomaterials for agriculture. Adapted from Sturikova et al. [111]

been thoroughly examined [67]. Therefore, the integration of NM use with crop plants demands a multidisciplinary approach to properly understand and characterise the mechanisms of action of NMs in crop growth and associated physiological cycles. This chapter explores changes in the mineral and metabolic composition of plants exposed to NMs.

2 Nanomaterials as Fertilisers: Changes in the Mineral Composition and Metabolism of Plants

The optimal expression of the genotype-environment interaction in horticultural production systems by characterising both yield and quality of the harvested products requires continual replenishment of the essential soil mineral nutrients. These nutrients (N, P, K, Ca, Mg, S, Fe, Mn, Cu, Zn, B, Mo, Cl and Ni) are subject to loss due to extraction by the roots and to leaching following rainfall and irrigation. They can also suffer reduced bioavailability through certain unfavourable physicochemical properties of the soil [15]. Hence, attention to soil mineral availability is fundamental if plants and other autotrophic organisms are to take up, process and accumulate the various mineral nutrients essential to the functioning of their physiological and biochemical processes [40, 100]. This recognises that optimal levels of these nutrients vary among plant species, with phenological stage and with the culture medium (Kirkby 2023). On the other hand, there is another group of elements that plants can take up without their being classified as essential, but which nevertheless have beneficial effects on the expression of phenotype, as well helping confer resistance to pests and diseases and to climatic factors that generate abiotic stress [14]. These beneficial nutrients include Na, Si, Co, I, Se, and Al [16, 78].

The use of excessive amounts of these conventional fertilisers in horticulture and agriculture, coupled with the low uptake efficiencies of many crops, is causing significant direct problems in human health as well as indirect ones via environmental pollution as these nutrients interact with the soil, with water bodies, with microbial communities and with the atmosphere [100]. The growing uptake of nanotechnology is helping to solve a range of problems in various science fields including agriculture, by transforming minerals into nanometric-sized particles ($\leq$100 nm) [97]. A major advance relates to the production and commercial availability of N, P, K, Ca, Cu, Mn, Fe, Ni, Mo, Zn, Se and Ti as NMs [35]. These products represent sustainable alternatives for the supply of essential and beneficial nutrients to increase plant productivity and efficient water use under both abiotic and biotic stress conditions [102]. This is because these NMs possess unique characteristics due to their small size (between 1 and 100 nm), their very large surface/volume ratios, their spherical or quasi-spherical morphologies, and their stability in aqueous media at different values of pH [17].

Compared to conventional chemical fertilisers formulated as solids (granules, powders, macro-granules, tablets and sticks) or liquids (suspensions or mixtures and solutions), NM fertilisers have outstanding characteristics including high solubility (dispersibility), high bioavailability due to their efficient uptake properties (slow and controlled release), as well as lower rates of loss through their interactions with the soil, the plant, with microorganisms and in the air [81]. They also have good potential for reducing groundwater contamination and eutrophication, as they are released to the soil only slowly and are rapidly and more completely taken up by plants [84] (see Fig. 3). However, they can induce oxidative stress, challenge membrane integrity, and interact with cell organelles [90]. Here, their levels of toxicity in their interactions with the plant and possible risks to human health when fresh food grown under these mineral nutrition regimes is eventually consumed, remain unclear [123]. Currently, NM fertilisers can be classified into three main groups according to (1) their mode of action (controlled release, precision delivery, plant growth stimulant, control of water and nutrient losses), (2) their composition (organic, inorganic, hybrid and nutrient-loaded) and (3) their consistency (surface coated, synthetic polymer or bio-based and nanocarbon based) [17, 122]. However, the classification of some NM fertilisers is more complex as they can fall into more than one group, due to variations in their morphology, dimensions, uniformity, and agglomeration [79, 128].

Nanomaterials exhibit stability characteristics that allow application in multiple products in modern agriculture including in pesticides, herbicides, and fertilisers

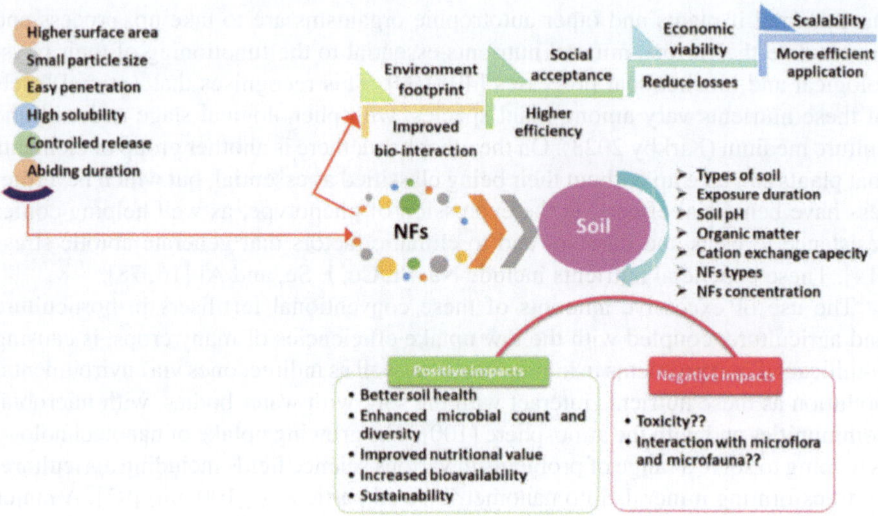

Fig. 3 An overview of the role of nanofertilisers in soil. Adapted from Verma et al. [117]

[27]. The use of NMs can reduce usage of limited of natural resources because they can be applied in a more controlled and targeted manner e.g., through smart delivery systems, where particle size is a determining factor in the interaction with the plant [77]. Furthermore, NMs can be essential components of integrated agronomic management in precision farming systems, where the type and concentration of the input is matched to the different actual needs of plants [88].

Recent research indicates favourable tolerance in established plants to multiple abiotic stress conditions following foliar or substrate spray applications of NM fertilisers of both essential and beneficial nutrients by increasing the concentrations of these minerals and the production of secondary metabolites [22]. However, such responses are strongly affected by genome and crop phenological stage, dosage, the weather, the season and if plants are of the same genotype (i.e., yellow, or white maize) [101]. The different plant responses when conventional bulk or NM fertilisers are used do not depend on the different responses to essential or beneficial nutrients [90]. The different responses may have to do with differences in the product's physicochemical properties including surface:volume ratio, hydrophobicity, water solubility and reactivity [87, 122]. For example, Hussein et al. [56] reported increased percentages of P, K, Na, Ca, Mg and Cu in wheat grains following sprays with 100 ppm NM N fertiliser, while N, Zn, Mn and Fe were increased following sprays with 200 ppm NM N fertiliser. Previous studies have demonstrated the role of nanomaterial fertilisers such as potentially useful tool to lessen the negative effects of global warming on avocado development [94]. They can also improve nutritional value and tolerance to drought, salinity, and heavy metals, raising levels of α-tocopherol, phenols, indoles, ascorbic acid, proline, and malonaldehyde in carom (*Trachyspermum ammi*) [80], basil [30], beans [15], coriander [34]. These secondary metabolites are widely perceived by consumers as good for their health, with elevated levels also being associated with better prices at market.

There are not many studies that allow prediction of the ecological and food safety aspects NMs [117]. It has been reported that plant exposure to NMs of C, Si, Se, Au, Ag, Ce and Ti does not pose a risk to crops [12, 44], due in part to their low concentrations in the soil. However, due to the small size of these materials it is possible to detect them in leaves and roots but not in food products, such as fruits [37]. When applied at high rates to low-growing crops such as lettuce, beans, asparagus, pumpkin, chilli, and tomato, there are reports of toxicities in the plant, but with no significant effect human health when harvested products are consumed [120]. However, when NMs are inhaled or consumed directly they can affect pregnancy, foetal development and infant health through an effect on the synthesis of steroid hormones (oestrogen and progesterone) [26]. Toxicity damage can also occur to various body systems such as the respiratory, nervous, and endocrine systems. Here, the size, shape, surface area, and chemical composition of NMs are key determinants of toxicity [121]. Some metal NMs have been shown to be carcinogenic causing damage to DNA due to ROS, mutation, apoptosis and cell cycle inhibition [70]. However, it is important to distinguish between the low level use of NMs in horticulture and the unintentional release of concentrated amounts into the biosphere (e.g., spills). Due to the high biological efficiency of these products, horticulture distributes low amounts over relatively wide areas. Also, any potential or small health and environmental risks associated with NM use in horticulture must be compared with the known and significant damage currently being caused by overuse of conventional chemical fertilisers which cause soil degradation, atmospheric pollution via volatilisation and water pollution via leaching and runoff [93]. An overview of the effects on the concentrations of minerals and secondary metabolites in plants exposed to NM fertilisers are presented in Table 1.

3 Nanomaterials as Elicitors: Changes in the Composition of Metabolites in Plants

The growing demand for healthy food and the scarcity of key food production resources including quality water and quality soil lead to excessive use of mineral fertilisers. Also, the sometimes stressed plants are especially susceptible to pests and diseases so the need for insecticides, miticides, herbicides, nematicides, fungicides is increased. Synthetic plant growth regulators are also in demand to manage climatological and other stressors. All these chemical interventions used to maximise crop production come also with problems including increasing resistance of pests to pesticides, ecological damage to the surround natural fauna and flora, agro-ecosystem contamination (soil, groundwater and air) and potential damage to human health through consumption of chemically contaminated food [19]. The interactions of all these factors with climate change has created a need to find and implement new more

Table 1 Overview of reported effects on mineral and metabolite concentrations in plants using nanomaterials such as nanofertilisers

Mineral nutrient	Crop	Effect	Particle size and concentration	References
N	Wheat (*Triticum aestivum*) Pomegranate (*Punica granatum*)	– Optimum values of macro- and micro-nutrient concentrations in grains except N, Zn, and Mn – Increased concentration of N, P, K, Mn and Zn	– <100 nm; 100, 200 ppm – < 100 nm; 0.25, 0.50 g N L^{-1}	[22, 56]
P	Orange (*Citrus sinensis*) Maize (*Zea mays*)	– Increased N, P, K, Fe, Zn and Mn in leaves – Increased leaf pigments, and chemical (phenols, indoles) content of yellow sweet corn	– <100 nm; 11.45 mg L^{-1} – < 100 nm; 50, 100 mg L^{-1}	[29, 101]
K	Kaki (*Diospyros kaki*) Wheat (*Triticum aestivum*)	– Reduced leaf concentrations of Chl, carotenes, Na, Cl and K but increased concentrations of Mg, Ca, N and P – Increased content of Chl, carotenoids and seed protein	– 20 nm; 3% – 21–30 nm; 20, 40, 60 ppm	[74, 107]
Ca	Avocado (*Persea americana*) Cherry (Prunus dulcis) Carom (*Trachyspermum ammi* L.) Tomato (*Solanum licopersicon*)	– Increased Ca and P concentrations in fruit and reduced high-temperature stress – Increased fruit Ca content – Increased enzymatic activity of superoxide dismutase (SOD) and peroxidase (POD) under drought stress. Also, increased α-tocopherol and ascorbic acid (AA) content – Increased content of AA and phenols in fruit under NaCl stress	– ≈100 nm; 15, 30, 45 mg L^{-1} – 40 nm; 1.4% – 160 nm; 75, 100 ppm – ≈15 nm; 50, 100, 200 mM	[3, 18, 80, 94]

(continued)

Table 1 (continued)

Mineral nutrient	Crop	Effect	Particle size and concentration	References
Zn	Bean (*Phaseolus vulgaris*) Tomato (*Solanum licopersicun*) Apple (*Malus sylvestris*) Pecan (*Carya illinoinensis*)	– Increased Zn concentration in stems, roots, leaves and fruit. Increased concentration of sucrose and Chl in the plant – Increased lycopene, AA and total phenol content. decreased the absorption and concentration of Zn and B – Increased the content of AA, phenols and total flavonoids in fruit – Reduced leaf concentrations of Fe, Cu, Mn and Zn and increased total carotenoids, K and Mg. Increased crude protein and fatty acids (unsaturated and saturated) in the kernel	– <100 nm; 150 mg L^{-1} – <100 nm; 75, 100, 125 mg L^{-1} – 2–20 nm; 250 ppm – < 100 nm; 200; 2000, 4000 mg L^{-1}	[5, 15, 32, 83, 89]
Mg	Black-eyed pea (*Vigna unguiculata*) Cotton (*Gossypium hirsutum*)	– Increased Fe and Mg uptake – Increased foliar concentrations of N, P, K and Mg	– 20 nm; 0.5 g L^{-1} – 50 nm; 60 mg L^{-1}	[23, 60]
Fe	Basil (*Ocimum basilicum*) Cucumber (*Cucumis sativus*) Maize (*Zea mays*)	– Increased total carbohydrate content, Chl (a and b), phenols and total flavonoids in leaves – Increased concentrations of Fe, AA, total phenols, and flavonoids – Increased foliar Ca and Fe content	– 12.6 and < 100 nm; 0.3 mol/L; 1000, 2000, 4000 ppm – 50 nm; 100 mg L^{-1} – ≈20 nm, 100 ppm	[1, 30, 48, 58]

(continued)

Table 1 (continued)

Mineral nutrient	Crop	Effect	Particle size and concentration	References
B	Pomegranate (*Punica granatum)* Pea (*Pisum sativum)*	– It Increased fruit concentration of Fe, Mn, Cu and Zn – Reduced total sugars and total phenols – Increased plant height, leaf area, enzyme activity of catalase (CAT), SOD, ascorbate peroxidase (APX) and guaicol peroxidase (GPX)	– < 100 nm; 3.25 mg L^{-1} ≈100 nm; 12.5, 25, 50 ppm	[21, 112]
Cu	Tomato (*Solanum licopersicum*) Basil (*Ocimum basilicum*)	– Increased Cu content and total phenols in leaves. Increased AA, glutathione, and total phenols in fruits – Increased concentration of Chl (a) and (b), phenols and total flavonoids in leaves	– ≈50 nm; 250 mg L^{-1} – <100 nm;1000, 2000, 4000 ppm	[1, 92]
Mo	Pea (*Pisum sativum*)	– Increased yield, height, leaf area, nodule number and phenols in pea plants under drought stress. Increased activation of CAT, APX, SOD and GPX enzymes	– < 100 nm;12.5, 25, 50 ppm	[113]

(continued)

Table 1 (continued)

Mineral nutrient	Crop	Effect	Particle size and concentration	References
Ti	Tomato (*Solanum licopersicum*) Indian mustard (*Brassica juncea*) Broad beans (*Vicia faba*) Wheat (*Triticum aestivum*)	– Increased Chl concentration, CAT, POD, glutathione synthetase and transferase activity. Increased distribution of P, S, Mg and Fe in root and leaf – Increased proline content of shoots and leaves but decreased malondialdehyde (MDA) by 50% (shoots) and 9% (leaves) – Under control conditions increased shoot length, leaf area and root dry weight. Under NaCl stress increased the activity of SOD, APX, CAT and the contents of MDA and proline – Increased plant height, dry and fresh weights of roots and shoots, lengths of shoots and roots, number of leaves, relative water consumption (RWC), membrane stability index and Chl content	– 30–50 nm; 0.5, 1, 2, 4 g L^{-1} – ≈25 nm; 200, 500, 1000, 1500 mg L^{-1} – 25 nm; 0·01%, 0·02%, 0·03% – 30–111 nm; 20, 40, 60, 80 mg L^{-1}	[2, 85, 96, 115]

(continued)

Table 1 (continued)

Mineral nutrient	Crop	Effect	Particle size and concentration	References
Se	Pepper (*Capsicum annuum*) Coriander (*Coriandrum sativum*) Cucumber (*Cucumis sativum*); tomato (*Solanum licopersicum*); Eggplant (*Solanum melongena*)	The content of phenols, flavonoids and glutathione in bell pepper fruits was increased. Likewise, showed increased in the activities APX, GPX, CAT and phenylalanine ammonia lyase (PAL) – Reduced uptake of Cd and effective in alleviating Cd stress. Increased shoot and root weights, Chl, and RWC – Prior to heat stress, plants showed a near doubling of leaf area	– 2–20 nm; 10, 50 mg L^{-1} – 10–40 nm; 20, 40, 60 mg L^{-1} – ≈100 nm; 1, 5, 10, 25 μg kg^{-1}	[12, 44, 45]
Si	Coriander (*Coriandrum sativum*) Bean (*Phaseolus vulgaris*)	– Minimised accumulation of MDA and enhance the activities of POD, CAT, and SOD in plants under Pb stress. Likewise, increased the flavonoid and anthocyanin contents – Increased shoot length, leaf number, leaf area, dry weight. Increased yield pod number and green pod yield, photosynthetic pigments (chl a, b, and carotenoids), and gas exchange - transpiration, leaf net photosynthetic and stomatal conductance	– 20–35 nm; 1.5 mM – <100 nm; 2.5, 5, 10 mmol/L	[28, 34]

environmentally friendly interventions that will achieve the same (or similar) results in terms of agroeconomic performance [25] but in the face too of increased frequencies and severities of random climatological events that bring with them increased incidence of abiotic stresses (heat, cold, drought, flooding, high wind etc.) and associated increased occurrences of severe pest and disease problems. A current research strategy is to better understand the roles of organic and inorganic elicitors that can be applied to a crop to trigger its defence mechanisms by stimulating various physiological and biochemical processes and so increase resistance to a range of biotic and abiotic stressors [108].

Among the elicitors are a range of compounds produced by pathogens and others produced by plants in response to pathogens. These include folate, provitamin A, minerals (K, Ca, Mg), flavonoids, phenolic acids, alkaloids, carotenoids, fibres [82, 98]. So far, more than 5000 such compounds have been identified and isolated from fruits, vegetables and grains [73]. Their main functions are to stimulate the plant's defences against insect, animal and pathogen attack or to increase survival in the face of a wide range of other biotic and abiotic stresses [43]. This diversity of bioactive compounds may be classified as the terpenoids (monoterpenes, diterpenes, triterpenes and polyterpenes), the nitrogen-containing alkaloids (tropane, quinolones and amines), the polyphenols (phenolic acids, stilbenes, lignans and flavonoids) and the sulphur compounds (glutathione, phytoalexins, thioniums and glucosinolates) [103]. As well as being of potential use as agrochemicals, these compounds also have a wide range of applications in medicine, pharmacology and cosmetics, also in the food industry. Lastly, many of these compounds create opportunities within the scope of nanotechnology [12].

The activation of multiple signal transduction pathways constitutes one of the most practical and inexpensive strategies for enhancing the synthesis and bioaccumulation of these secondary metabolites in plants. The do this by enhancing the transmission of signals within each cell, or plant, that alter the expressions of various gene networks, thus stimulation many of the well-known secondary metabolic pathways (Shikimic-acid, Malonic-acid, Mevalonic-acid, and Methylerythritol-phosphate) [91, 125]. An efficient way to achieve such effects is through foliar or soil spraying of substances with elicitor functions [98]. The NMs offer one potential option as abiotic elicitors that can induce the production of bioactive compounds in plant cells and tissues through stimulation of the expressions of related genes (Fig. 4) [39]. However, the physiological and biochemical mechanism of many of these processes remains unclear. Recent research proposes that NMs induce the generation of reactive oxygen species (ROS) and secondary signalling messengers leading to timely transcriptional regulation in the primary and secondary metabolisms of plants so as to ameliorate the severity of impact of various stresses [31].

It is well known that plants under stress from drought, salinity, infection, radiation, herbivory, physical and biological damage etc., increase their productions of secondary metabolites, and this is often related to the increased synthesis and accumulation of ROS (redox adjustment). This stress response mechanism can be stimulated by various elicitors which activate the plant's chemical defences (Fig. 5) [116]. Such elicitors can often be formulated as a commercial product (a bio-stimulant) that can help improve plant nutrition, productivity, and performance under stresses [110]. Other avenues for stimulating the synthesis of such bioactive compounds are precursor feeding, metabolic engineering, permeabilisation, immobilisation and bioreactor cultures [25]. Encouragingly, many studies have shown that NMs can be used to induce the production of bioactive compounds in plants (Table 2).

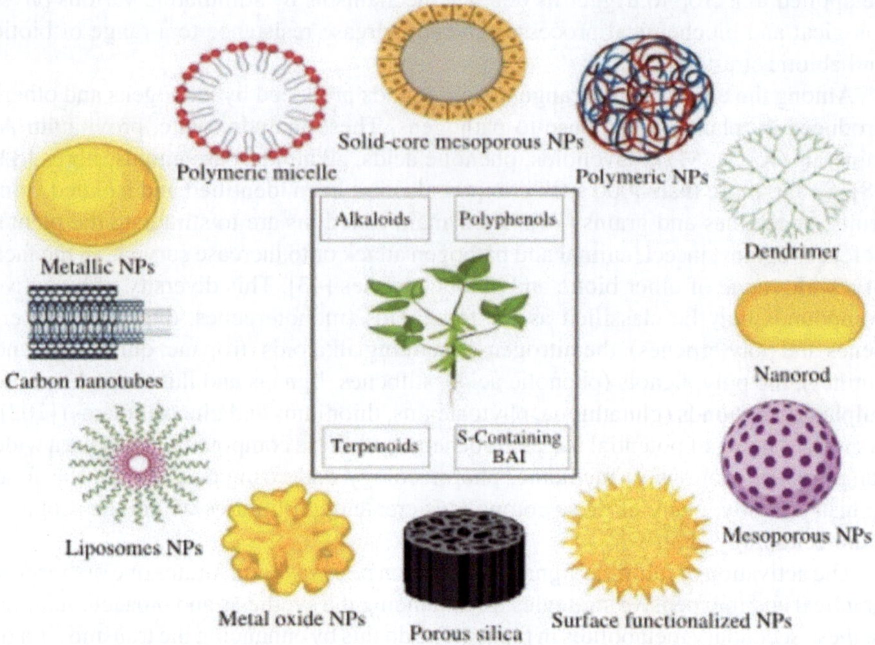

Fig. 4 Graphical representation of different types of nanomaterials which can act as potential elicitors of secondary metabolites in plants. Adapted from Khan et al. [62]

4 Conclusions and Perspectives

Among the benefits offered by the use of nano-fertilisers is their ability to control and slow the release of nutrients to the plant, this both increases the efficiency of plant uptake and reduces pollution of the agro-ecosystem. The materials also increase concentrations of minerals and secondary metabolites in the plant, with these benefits also being reflected in the yield and nutraceutical quality of the harvested products. However, most studies discussed in this chapter relate to metallic or semi-metallic nanofertilisers and their oxides used in relatively low doses and in annual (arable) crop species. Hence, there is a need to increase the number of species studied to include also crop plants having different production cycles (perennials) and growth habits (trees)—so in fruit trees. More comprehensive studies are also required that encompass their roles in crop physiology and biochemistry, and address aspects of traceability in the edible organs, the optimal application methods, doses and timings (taking into account the crop, the soil and the climate), trophic transfer, impacts on the soil microbiome and their long-term effects on soil health.

There is now good scientific evidence that NM use in agriculture as mineral nutrients, metallic or semi-metallic elements (including gold, silver, titanium, lead, silicon, sodium, cerium) and organic compounds (including chitosan, melatonin,

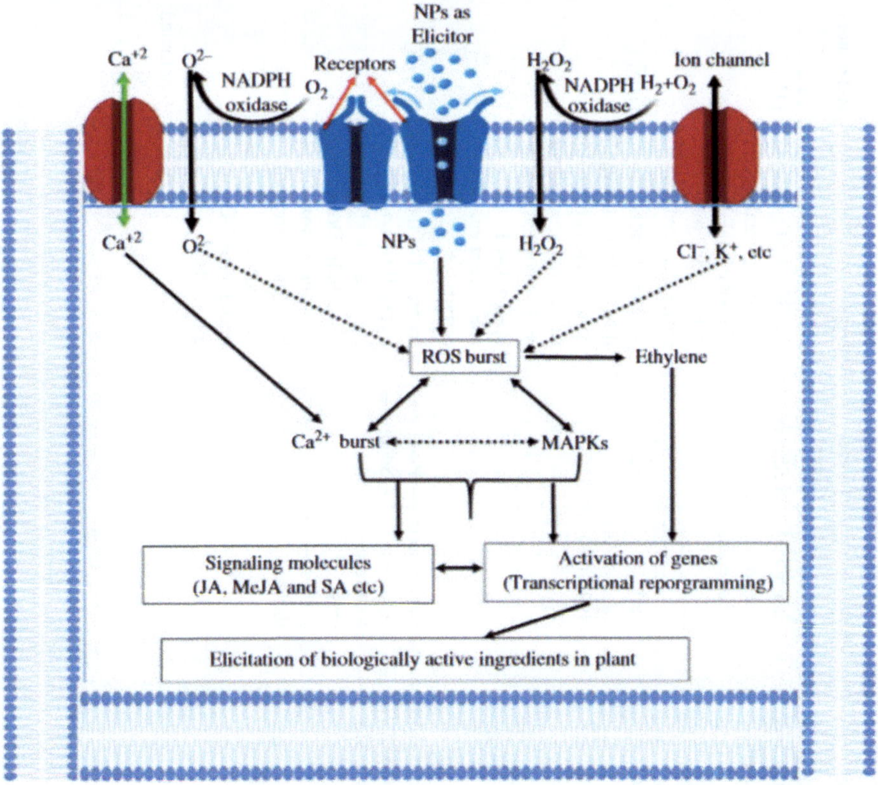

Fig. 5 Representation of a proposed mechanism involved in nano-elicited production of secondary metabolites in various in vitro systems of plants. Adapted from Anjum et al. [9]

humic acids, fulvic acids) can help mitigate negative impacts of various biotic and abiotic stresses in plants (including pests, diseases, drought, salinity, heavy metals) by increasing plant resistance by the activating the production and accumulation of a range secondary metabolites. However, the fat of all these NMs in the biosphere and their health effects when consumed by humans or animals remain largely unclear.

Table 2 Summary associated with nanoparticles (NMs) as elicitors of bioactive compound production in plants

NPs	Size	Concentration	NPs treatment	Plant	Tissue/ Organ	Effects	References
ZnO and CuO	– 20–30 y 25–30 nm, respectively	– 2, 20, 200, 2000 mg L^{-1}	– One time only added to the culture media	– Candy leaf (*Stevia rebaudiana*)	– Shoots	– Increased rebaudioside, stevioside, total phenolic,total flavonoid content, and total antioxidant activity	[4]
ZnO NPs, jasmonate and chitosan	– Nr	– ZnO—20, 60, 100 ppm; Jasmonate—100, 250, 500 μM; Chittosan: 10, 50, 100 μM	– Foliar sprays	– Carla (*Momordica charantia*)	– Shoots	– Rise of phenols, flavonoids, anthocyanin content and carotenoids as well as carbohydrate, proline content and up-regulated antioxidant enzyme activity	[105]
Ag	– Nr-10, 40 and 100 nm	– 0.25, 0.5, 1, 1.5, 2 mg L^{-1} – 0.5, 1.0, 5.0 mg L^{-1}	– One time only added to the culture media – 12, 24, 72 h in MS medium	– Isatis (*Isatis constricta*) – Mouseear cress (*Arabidopsis thaliana*)	– Plantlets – Seedlings	– Increased of indigo and tryptanthrin content – Increased of glutathione disulphide. Decreased of 4-glucosyloxy-benzoate, pinresinol hexoside and kaempferol-3-O-glucoside	[61, 68]

(continued)

Table 2 (continued)

NPs	Size	Concentration	NPs treatment	Plant	Tissue/ Organ	Effects	References
SiO_2 and TiO_2	SiO_2: 5–15 nm and TiO_2: ≈ 25 nm	– 5, 10, 20 ppm after 15 days of incubation	– One time only added to the MS medium supplemented with 4.5 µM 2,4-dichlorophenoxyacetic acid and 5 µM NAA	– Argan (*Argania spinosa*)	– Callus culture (leaves)	Enhance α-tocopherol accumulation	[53]
Iron-Oxide coated with Citral and Chitosan	Nr	– FeO-15, 30 µm L^{-1}; chitosan and citral (15, 30 µm L^{-1}, respectively)	– Foliar sprayed under salinity stress (0, 30, 60 mM NaCl)	– Spearmint (*Mentha spicata* L.)	– Leaves	– Increase in proline, sugar, chl, anthocyanin, RWC, and essential oil content	[51]
SiO_2; TiO_2; ZnO; Fe_3O_4; CuO; CeO_2	≤100 nm	– SiO_2 (50, 100 mg L^{-1}), TiO_2 (20, 60 mg L^{-1}), ZnO (50, 100 mg L^{-1}), Fe3O4 (100, 200 mg L^{-1}), CuO (50, 100 mg L^{-1}) y CeO_2 (50, 100 mg L^{-1})	– Foliar sprayed on plants irrigated with a cd solution	– Lechuga (*Lactuca sativa*)	– Leaves	– It increased ROS and MDA production, improved SOD, POD, CAT, and APX activity, photosynthetic efficiency and K, Mg, Ca, Fe, Zn and Cd uptake	[13]
Chitosan (CNPs)	19–38 nm	– CNPs-30, 60, 90 ppm	– Foliar sprayed on plants under different humidity levels	– Sage (*Salvia abrotanoides*)	– Leaves	– Decreased electrolyte conductivity but improved RWC, Chl, carotenoids, phenol, flavonoid, soluble sugar, proline, and protein	[11]

(continued)

Table 2 (continued)

NPs	Size	Concentration	NPs treatment	Plant	Tissue/ Organ	Effects	References
Ceria NPs and Ce^{3+} ions	<25 nm	– 50, 100, 200 mg Ce kg^{-1}	– Applied to soil for 12 weeks	– Bean (*Phaseolus vulgaris*)	– Roots	– Increase of stomatal conductance, and P, S, Ca, Mn, Fe, Cu, and Zn contents	[76]
Carbon (63% of C, 34% of O, 1.6%of H, and 1.4% of N)	20–130 nm	– 80 mM	– Impregnation of seeds treated with a solution containing 25 mM NaCl for 12 h	– Radish (*Raphanus sativus*)	– Sprouts	– Enhance antioxidant capacity, polyphenols, flavonoids, polyamines, anthocyanin and proline content	[50]
Ag_2O	≤100 nm	– Ag—0.005, 0.01, 0.025, 0.05, 0.1 g L^{-1}; $AlCl_3$—0, 100, 300, 500 μM	– One time only added to the culture media	– Pineapple (*Ananas comosus*)	– Plantlets	– Stimulated a higher concentration of Chl, carotenoids, shoot formation, and assimilation K, Ca, Mg, Fe, Cu, Mn, and Zn	[114]
Graphene oxide (Go NPs) (NbO_2)	≤100 nm	– NaCl (0, 50, 100 mM), Go NPs (0, 50, 100 mg L^{-1}), and proline-functionalized graphene oxide nanoparticles (Go-Pro NPs) (0, 50, 100 mg L^{-1}),	– Foliar sprayed for three times with an interval of five days	– Grape (*Vitis vinifera*)	– Seedlings	– Increases RWC, activity of SOD as well as contents of proline, MDA and hydrogen peroxide (H_2O_2)	[123]

(continued)

Table 2 (continued)

NPs	Size	Concentration	NPs treatment	Plant	Tissue/Organ	Effects	References
Au	≤100 nm	– 1,2, 4 μg mL^{-1}	– Various exposure time (8, 24, 48, and 72 h) on Cell suspension cultures (Gamborg's B5 medium)	– Yew tree (*Taxus baccata*)	– Callus culture	– Activity of antioxidant enzymes including POX, SOD, CAT, and PPO was increased. Likewise, fenols and taxans contents was increased	[42]
Melatonine;Au	≈39 nm	– 50, 100, 200 μM	– The 14 day-old plants were grown in 1/2 MS and 100 μM CdSO with applications Mel-Au NPs	– Rice (*Oriza sativa*)	– Seedling	– Restored chlorophyll biosynthesis and mitigated Cd-induced oxidative stresses, while the antioxidant enzyme activities (OD, POD, and CAT) and proline and MDA contents were significantly increased	[59]

Nr-Not reported

References

1. Abbasifar A, Shahrabadi F, ValizadehKaji B (2020) Effects of green synthesized zinc and copper nano-fertilizers on the morphological and biochemical attributes of basil plant. J Plant Nutr 43(8):1104–1118. https://doi.org/10.1080/01904167.2020.1724305
2. Abdel-Latef AAH, Srivastava AK, El-sadek MSA, Kordrostami M, Tran LSP (2018) Titanium dioxide nanoparticles improve growth and enhance tolerance of broad bean plants under saline soil conditions. Land Degrad Dev 29(4):1065–1073. https://doi.org/10.1002/ldr.2780
3. Abeed AH, AL-Huqail AA, Albalawi S, Alghamdi SA, Ali B, Alghanem SM et al (2023) Calcium nanoparticles mitigate severe salt stress in *Solanum lycopersicon* by instigating the antioxidant defense system and renovating the protein profile. S Afr J Bot 161:36–52. https://doi.org/10.1016/j.sajb.2023.08.005
4. Ahmad MA, Javed R, Adeel M, Rizwan M, Ao Q, Yang Y (2020) Engineered ZnO and CuO nanoparticles ameliorate morphological and biochemical response in tissue culture regenerants of candyleaf (*Stevia rebaudiana*). Molecules 25(6):1356. https://doi.org/10.3390/molecules25061356
5. Ahmed R, Uddin MK, Quddus MA, Samad MYA, Hossain MM, Haque ANA (2023) Impact of foliar application of zinc and zinc oxide nanoparticles on growth, yield, nutrient uptake and quality of tomato. Horticulturae 9(2):162. https://doi.org/10.3390/horticulturae9020162
6. Akanbi-Gada MA, Ogunkunle CO, Vishwakarma V, Viswanathan K, Fatoba PO (2019) Phytotoxicity of nano-zinc oxide to tomato plant (*Solanum lycopersicum* L.): Zn uptake, stress enzymes response and influence on non-enzymatic antioxidants in fruits. Environ Technol Innov 14:100325. https://doi.org/10.1016/j.eti.2019.100325
7. Al-Juthery HW, Lahmod NR, Al-Taee RA (2021) Intelligent, nano-fertilizers: a new technology for improvement nutrient use efficiency (article review). IOP Conf Ser Earth Environ Sci 735(1):012086. https://doi.org/10.1088/1755-1315/735/1/012086
8. Alidoust D, Isoda A (2013) Effect of γFe_2O_3 nanoparticles on photosynthetic characteristic of soybean (*Glycine max* (L.) Merr.): foliar spray versus soil amendment. Acta Physiol Plant 35:3365–3375. https://doi.org/10.1007/s11738-013-1369-8
9. Anjum S, Komal A, Abbasi BH, Hano C (2021) Nanoparticles as elicitors of biologically active ingredients in plants. In: Nanotechnology in plant growth promotion and protection: recent advances and impacts. Wiley, West Sussex, UK. https://doi.org/10.1002/9781119745884.ch9
10. Arumugam M, Manikandan DB, Dhandapani E, Sridhar A, Balakrishnan K, Markandan M, Ramasamy T (2021) Green synthesis of zinc oxide nanoparticles (ZnO NPs) using *Syzygium cumini*: Potential multifaceted applications on antioxidants, cytotoxic and as nanonutrient for the growth of *Sesamum indicum*. Environ Technol Innov 23:101653. https://doi.org/10.1016/j.eti.2021.101653
11. Attaran-Dowom S, Karimian Z, Mostafaei-Dehnavi M, Samiei L (2022) Chitosan nanoparticles improve physiological and biochemical responses of *Salvia abrotanoides* (Kar.) under drought stress. BMC Plant Biol 22(1):364. https://doi.org/10.1186/s12870-022-03689-4
12. Babashpour-Asl M, Farajzadeh-Memari-Tabrizi E, Yousefpour-Dokhanieh A (2022) Foliar-applied selenium nanoparticles alleviate cadmium stress through changes in physio-biochemical status and essential oil profile of coriander (*Coriandrum sativum* L.) leaves. Environ Sci Pollut Res 29(53):80021–80031. https://doi.org/10.1007/s11356-022-19941-1
13. Bano N, Khan S, Hamid Y, Ullah MA, Khan AG, Bano F et al (2023) Effect of foliar application of nanoparticles on growth, physiology, and antioxidant enzyme activities of lettuce (*Lactuca sativa* L.) plants under cadmium toxicity. Environ Sci Pollut Res 30(44):99310–99325. https://doi.org/10.1007/s11356-023-29241-x
14. Barker AV, Pilbeam DJ (2015) Handbook of plant nutrition. CRC Press
15. Bautista-Diaz J, Cruz-Alvarez O, Hernández-Rodríguez OA, Sánchez-Chávez E, Jacobo-Cuellar JL, Preciado-Rangel P et al (2021) Zinc sulphate or zinc nanoparticle applications to leaves of green beans. Folia Hortic 33(2):365–375. https://doi.org/10.2478/fhort-2021-0028

16. Broadley M, Brown P, Cakmak I, Ma JF, Rengel Z, Zhao F (2012) Beneficial elements. In: Marschner's mineral nutrition of higher plants. Academic, Oxford, UK. https://doi.org/10.1016/B978-0-12-384905-2.00008-X
17. Butt BZ, Naseer I (2020) Nanofertilizers. In: Nanoagronomy. Springer International Publishing AG, Cham, Switz. https://doi.org/10.1007/978-3-030-41275-3_8
18. Carrasco-Correa EJ, Mompó-Roselló Ò, Simó-Alfonso EF (2023) Calcium oxide nanofertilizer as alternative to common calcium products for the improvement of the amount of peel fruit calcium. Environ Technol Innov 31:103180. https://doi.org/10.1016/j.eti.2023.103180
19. Chhipa H (2017) Nanofertilizers and nanopesticides for agriculture. Environ Chem Lett 15–22. https://doi.org/10.1007/s10311-016-0600-4
20. Coronado-Partida L, Patron-Soberano A, Rodríguez-González V, Gutiérrez-Martínez P (2023) Antifungal potential of eco-friendly chitosan-sodium benzoate to inhibit the development of *Rhizopus stolonifer* isolated from jackfruit. J Plant Dis Protect 130:905–913. https://doi.org/10.1007/s41348-023-00746-4
21. Davarpanah S, Tehranifar A, Davarynejad G, Abadía J, Khorasani R (2016) Effects of foliar applications of zinc and boron nano-fertilizers on pomegranate (*Punica granatum* cv. Ardestani) fruit yield and quality. Sci Hortic 210:57–64. https://doi.org/10.1016/j.scienta.2016.07.003
22. Davarpanah S, Tehranifar A, Davarynejad G, Aran M, Abadía J, Khorassani R (2017) Effects of foliar nano-nitrogen and urea fertilizers on the physical and chemical properties of pomegranate (*Punica granatum* cv. Ardestani) fruits. HortScience 52(2):288–294. https://doi.org/10.21273/HORTSCI11248-16
23. Delfani M, Baradarn FM, Farrokhi N, Makarian H (2014) Some physiological responses of black-eyed pea to iron and magnesium nanofertilizers. Commun Soil Sci Plant Anal 45(4):530–540. https://doi.org/10.1080/00103624.2013.863911
24. Desoukey SF, Taha Z, El-Shabrawi HM, Sabh AZ (2020) Impact of zinc oxide nanoparticles on *Asparagus officinalis* plant. Plant Arch 20:9323–9338
25. Ding Y, Wang Q, Zhu G, Zhang P, Rui Y (2023) Application and perspectives of nanopesticides in agriculture. J Nanopart Res 25(8):159. https://doi.org/10.1007/s11051-023-05811-5
26. Dugershaw BB, Aengenheister L, Hansen SSK, Hougaard KS, Buerki-Thurnherr T (2020) Recent insights on indirect mechanisms in developmental toxicity of nanomaterials. Part Fibre Toxicol 17:1–22. https://doi.org/10.1186/s12989-020-00359-x
27. Duhan JS, Kumar R, Kumar N, Kaur P, Nehra K, Duhan S (2017) Nanotechnology: the new perspective in precision agriculture. Biotechnol Rep 15:11–23. https://doi.org/10.1016/j.btre.2017.03.002
28. El-Saadony MT, Desoky ESM, Saad AM, Eid RS, Selem E, Elrys AS (2021) Biological silicon nanoparticles improve *Phaseolus vulgaris* L. yield and minimize its contaminant contents on a heavy metals-contaminated saline soil. J Environ Sci 106:1–14. https://doi.org/10.1016/j.jes.2021.01.012
29. El-shereif AR, Zerban SM, Elmaadawy MI (2023) Impact of nano fertilizers and chemical fertilizers on valencia orange (*Citrus sinensis* [L.] Osbeck) growth, yield and fruit quality. Appl Ecol Environ Res 21(2). https://doi.org/10.15666/aeer/2102_13751387
30. Elfeky SA, Mohammed MA, Khater MS, Osman YA, Elsherbini E (2013) Effect of magnetite nano-fertilizer on growth and yield of *Ocimum basilicum* L. Int J Indigenous Med Plants 46(3):1286–11293
31. Elnahal AS, El-Saadony MT, Saad AM, Desoky ESM, El-Tahan AM, Rady MM et al (2022) The use of microbial inoculants for biological control, plant growth promotion, and sustainable agriculture: a review. Eur J Plant Pathol 162(4):759–792. https://doi.org/10.1007/s10658-021-02393-7
32. Estrada-Arellano KL, Vázquez-Vázquez C, Betancourt-Galindo R, Muy-Rangel MD, Valenzuela-Núñez LM, García-Hernández JL et al (2023) Foliar fertilization with ZnO nanoparticles and its effect on production, biophysical and nutraceutical quality of pecan nuts (Carya illinoinensis). Terra Latinoam 41:e1585. https://doi.org/10.28940/terra.v41i0.1585

33. Farhangi-Abriz S, Ghassemi-Golezani K (2021) Changes in soil properties and salt tolerance of safflower in response to biochar-based metal oxide nanocomposites of magnesium and manganese. Ecotoxicol Environ Saf 211:111904. https://doi.org/10.1016/j.ecoenv.2021.111904
34. Fatemi H, Esmaiel Pour B, Rizwan M (2021) Foliar application of silicon nanoparticles affected the growth, vitamin C, flavonoid, and antioxidant enzyme activities of coriander (*Coriandrum sativum* L.) plants grown in lead (Pb)-spiked soil. Environ Sci Pollut Res 28:1417–1425. https://doi.org/10.1007/s11356-020-10549-x
35. Fatollahpour-Grangah M, Rashidi V, Mirshekari B, Khalilvand-Behrouzyar E, Farahvash F (2020) Effects of nanofertilizers on physiological and yield characteristics of pinto bean cultivars under water deficit stress. J Plant Nutr 43(19):2898–2910. https://doi.org/10.1080/01904167.2020.1799000
36. Fazal H, Abbasi BH, Ahmad N, Ali M, Shujait Ali S, Khan A, Wei DQ (2019) Sustainable production of biomass and industrially important secondary metabolites in cell cultures of selfheal (*Prunella vulgaris* L.) elicited by silver and gold nanoparticles. Artif Cells Nanomed Biotechnol 47(1):2553–2561. https://doi.org/10.1080/21691401.2019.1625913
37. Feregrino-Pérez AA, Magaña-López E, Guzmán C, Esquivel K (2018) A general overview of the benefits and possible negative effects of the nanotechnology in horticulture. Sci Hortic 238:126–137. https://doi.org/10.1016/j.scienta.2018.03.060
38. Ganguly S, Dana K, Mukhopadhyay TK, Parya TK, Ghatak S (2011) Organophilic nano clay: a comprehensive review. Trans Indian Ceram Soc 70(4):189–206. https://doi.org/10.1080/0371750X.2011.10600169
39. García-Enciso EL, Robledo-Olivo A, Benavides-Mendoza A, Solís-Gaona S, González-Morales S (2018) Effect of elicitors of natural origin on tomato plants subjected to biotic stress. Revista Mexicana de Ciencias Agrícolas 9(SPE20):4212–4221. https://doi.org/10.29312/remexca.v0i20.991
40. García-Santiago JC, Valdez-Aguilar LA, Cartmill AD, Cartmill DL, Juárez-López P, Díaz-Pérez JC (2019) Subirrigation of container-grown tomato I: decreased concentration of the nutrient solution sustains growth and yield. Water 11(10):2064. https://doi.org/10.3390/w11102064
41. Gohari G, Farhadi H, Panahirad S, Zareei E, Labib P, Jafari H et al (2023) Mitigation of salinity impact in spearmint plants through the application of engineered chitosan-melatonin nanoparticles. Int J Biol Macromole 224:893–907. https://doi.org/10.1016/j.ijbiomac.2022.10.175
42. Golinejad S, Mirjalili MH, Rezadoost H, Ghorbanpour M (2023) Molecular, biochemical, and metabolic changes induced by gold nanoparticles in *Taxus baccata* L. cell culture. Ind Crops Prod 192:115988. https://doi.org/10.1016/j.indcrop.2022.115988
43. González-García Y, Cadenas-Pliego G, Benavides-Mendoza A, Juárez-Maldonado A (2023) Nanomaterials as novel elicitors of plant secondary metabolites. In: Nanotechnology in agriculture and agroecosystems, Elsevier, pp 113–139. https://doi.org/10.1016/B978-0-323-99446-0.00016-7
44. González-García Y, Cárdenas-Álvarez C, Cadenas-Pliego G, Benavides-Mendoza A, Cabrera-de-la-Fuente M, Sandoval-Rangel A et al (2021) Effect of three nanoparticles (Se, Si and Cu) on the bioactive compounds of bell pepper fruits under saline stress. Plants 10(2):217. https://doi.org/10.3390/plants10020217
45. Gudkov SV, Shafeev GA, Glinushkin AP, Shkirin AV, Barmina EV, Rakov II et al (2020) Production and use of selenium nanoparticles as fertilizers. ACS Omega 5(28):17767–17774. https://doi.org/10.1021/acsomega.0c02448
46. Guha T, Gopal G, Kundu R, Mukherjee A (2020) Nanocomposites for delivering agrochemicals: a comprehensive review. J Agric Food Chem 68(12):3691–3702. https://doi.org/10.1021/acs.jafc.9b06982
47. Guillén-Castillo OI, Rojas-Velázquez ÁN, Alcalá-Jáuregui JA, Loredo-Osti C, Ramírez-Tobías HM, Bertolini V, Ortiz JCR (2023) Nitrogen nanocomposite as a slow-release fertilizer in soil columns with lettuce plants. Terra Latinoamericana 41:e1712. https://doi.org/10.28940/terra.v41i0.1712

48. Guillén-Enríquez RR, Sánchez-Chávez E, Fortis-Hernández M, Márquez-Guerrero SY, Espinosa-Palomeque B, Preciado-Rangel P (2023) ZnO nanoparticles improve bioactive compounds, enzymatic activity and zinc concentration in grapevine. Notulae Botanicae Horti Agrobotanici Cluj-Napoca 51(4):13377–13377. https://doi.org/10.15835/nbha51413377
49. Gutiérrez-Miceli FA, Oliva-Llaven MÁ, Luján-Hidalgo MC, Velázquez-Gamboa MC, González-Mendoza D, Sánchez-Roque Y (2021) Zinc oxide phytonanoparticles' effects on yield and mineral contents in fruits of tomato (*Solanum lycopersicum* L. cv. Cherry) under field conditions. Sci World J 5561930. https://doi.org/10.1155/2021/5561930
50. Halawani RF, AbdElgawad H, Aloufi FA, Balkhyour MA, Zrig A, Hassan AH (2023) Synergistic effect of carbon nanoparticles with mild salinity for improving chemical composition and antioxidant activities of radish sprouts. Front Plant Sci 14:1158031. https://doi.org/10.3389/fpls.2023.1158031
51. Hassanpouraghdam MB, Abdollahfam F, Fathi S, Mehrabani LV (2023) Foliar application of iron-oxide nanoparticles coated with citral and chitosan enhances the growth and physiological responses of spearmint exposed to salinity stress. Russ J Plant Physiol 70(6):130. https://doi.org/10.1134/S102144372360109X
52. Hatami M, Naghdi-Badi H, Ghorbanpour M (2019) Nano-elicitation of secondary pharmaceutical metabolites in plant cells: a review. J Med Plants 18:6–36. https://doi.org/10.29252/jmp.3.71.6
53. Hegazi GA, Ibrahim WM, Hendawy MH, Salem HM, Ghareb HE (2020) Improving α-tocopherol accumulation in *Argania spinosa* suspension cultures by precursor and nanoparticles feeding. Plant Arch 20:2431–2437
54. Hernández-Hernández H, Quiterio-Gutiérrez T, Cadenas-Pliego G, Ortega-Ortiz H, Hernández-Fuentes AD, Cabrera de la Fuente M et al (2019) Impact of selenium and copper nanoparticles on yield, antioxidant system, and fruit quality of tomato plants. Plants 8(10):355. https://doi.org/10.3390/plants8100355
55. Husen A Iqbal M (2019) Nanomaterials and plant potential: an overview. In: Nanomaterials and plant potential. Springer, Cham, Switz. https://doi.org/10.1007/978-3-030-05569-1_1
56. Hussein MH, El-Ashry SM, El-Faham SY, Sabbour MM, El-Dok S (2019) Wheat plant dry matter and grains nutrients status and its responses to nanofertilizer under salinity condition. Plant Arch 19(2):2053–2063
57. Husted S, Minutello F, Pinna A, Le Tougaard S, Møs P, Kopittke PM (2023) What is missing to advance foliar fertilization using nanotechnology? Trends Plant Sci 28(1):90–105. https://doi.org/10.1016/j.tplants.2022.08.017
58. Jalali M, Ghanati F, Modarres-Sanavi AM, Khoshgoftarmanesh AH (2017) Physiological effects of repeated foliar application of magnetite nanoparticles on maize plants. J Agron Crop Sci 203(6):593–602. https://doi.org/10.1111/jac.12208
59. Jiang M, Dai S, Wang B, Xie Z, Li J, Wang L et al (2021) Gold nanoparticles synthesized using melatonin suppress cadmium uptake and alleviate its toxicity in rice. Environ Sci Nano 8(4):1042–1056. https://doi.org/10.1039/D0EN01172J
60. Kanjana D (2020) Foliar application of magnesium oxide nanoparticles on nutrient element concentrations, growth, physiological, and yield parameters of cotton. J Plant Nutr 43(20):3035–3049. https://doi.org/10.1080/01904167.2020.1799001
61. Karakaş Ö (2020) Effect of silver nanoparticles on production of indole alkaloids in Isatis constricta. Iran J Sci Technol Trans A Sci 44(3):621–627. https://doi.org/10.1007/s40995-020-00878-4
62. Khan AK, Kousar S, Tungmunnithum D, Hano C, Abbasi BH, Anjum S (2021) Nano-elicitation as an effective and emerging strategy for in vitro production of industrially important flavonoids. Appl Sci 11(4):1694. https://doi.org/10.3390/app11041694
63. Khan MN, Mobin M, Abbas ZK, AlMutairi KA, Siddiqui ZH (2017) Role of nanomaterials in plants under challenging environments. Plant Physiol Biochem 110:194–209. https://doi.org/10.1016/j.plaphy.2016.05.038
64. Khan I, Awan SA, Rizwan M, Hassan ZU, Akram MA, Tariq R et al (2022) Nanoparticle's uptake and translocation mechanisms in plants via seed priming, foliar treatment, and root

exposure: a review. Environ Sci Pollut Res 29:89823–89833. https://doi.org/10.1007/s11356-022-23945-2
65. Kibbey TC, Strevett KA (2019) The effect of nanoparticles on soil and rhizosphere bacteria and plant growth in lettuce seedlings. Chemosphere 221:703–707. https://doi.org/10.1016/j.chemosphere.2019.01.091
66. Kirkby EA (2012) Introduction, definition, and classification of nutrients. In: Marschner's mineral nutrition of plants. Academic, Oxford, UK. https://doi.org/10.1016/C2009-0-63043-9
67. Kranjc E, Drobne D (2019) Nanomaterials in plants: a review of hazard and applications in the agri-food sector. Nanomaterials 9(8):1094. https://doi.org/10.3390/nano9081094
68. Kruszka D, Sawikowska A, Selvakesavan RK, Krajewski P, Kachlicki P, Franklin G (2020) Silver nanoparticles affect phenolic and phytoalexin composition of *Arabidopsis thaliana*. Sci Total Environ 716:135361. https://doi.org/10.1016/j.scitotenv.2019.135361
69. Kumar A, Choudhary A, Kaur H, Mehta S, Husen A (2021) Metal-based nanoparticles, sensors, and their multifaceted application in food packaging. J Nanobiotechnol 19(1):256. https://doi.org/10.1186/s12951-021-00996-0
70. Kumar A, Dhawan A (2013) Genotoxic and carcinogenic potential of engineered nanoparticles: an update. Arch Toxicol 87:1883–1900. https://doi.org/10.1007/s00204-013-1128-z
71. Kumar N, Samota SR, Venkatesh K, Tripathi SC (2023) Global trends in use of nano-fertilizers for crop production: advantages and constraints—a review. Soil Tillage Res 228:105645. https://doi.org/10.1016/j.still.2023.105645
72. León-Silva S, Fernández-Luqueño F, López-Valdez F (2016) Silver nanoparticles (AgNP) in the environment: a review of potential risks on human and environmental health. Water Air Soil Pollut 227(9):306. https://doi.org/10.1007/s11270-016-3022-9
73. Liu RH (2013) Dietary bioactive compounds and their health implications. J Food Sci 78(s1):A18–A25. https://doi.org/10.1111/1750-3841.12101
74. Lo'ay AA, EL-Ezz SFA, Awadeen AA (2021) Effect of different foliar potassium fertilization forms on vegetative growth, yield, and fruit quality of kaki trees grown in sandy soil. Scientia Horticulturae 288:110420. https://doi.org/10.1016/j.scienta.2021.110420
75. Lombi E, Nowack B, Baun A, McGrath SP (2012) Evidence for effects of manufactured nanomaterials on crops is inconclusive. Proc Natl Acad Sci 109(49):E3336–E3336. https://doi.org/10.1073/pnas.1214934109
76. Ma Y, Xie C, He X, Zhang B, Yang J, Sun M et al (2020) Effects of ceria nanoparticles and $CeCl_3$ on plant growth, biological and physiological parameters, and nutritional value of soil grown common bean (*Phaseolus vulgaris*). Small 16(21):1907435. https://doi.org/10.1002/smll.201907435
77. Mahapatra DM, Satapathy KC, Panda B (2022) Biofertilizers and nanofertilizers for sustainable agriculture: phycoprospects and challenges. Sci Total Environ 803:149990. https://doi.org/10.1016/j.scitotenv.2021.149990
78. Martínez-Damián MT, Cano-Hernández R, Moreno-Pérez EDC, Sánchez-del Castillo F, Cruz-Álvarez O (2019) Effect of preharvest growth bioregulators on physicochemical quality of saladette tomato. Revista Chapingo Serie Horticultura 25(1):29–43. https://doi.org/10.5154/r.rchsh.2018.06.013
79. Marzouk NM, Abd-Alrahman HA, EL-Tanahy AMM, Mahmoud SH (2019) Impact of foliar spraying of nano micronutrient fertilizers on the growth, yield, physical quality, and nutritional value of two snap bean cultivars in sandy soils. Bull Natl Res Centre 43(1):1–9. https://doi.org/10.1186/s42269-019-0127-5
80. Mazhar MW, Ishtiaq M, Maqbool M, Atiq Hussain S, Casini R, Abd-ElGawad AM, Elansary HO (2023) Seed nano-priming with calcium oxide maintains the redox state by boosting the antioxidant defense system in water-stressed carom (*Trachyspermum ammi* L.) plants to confer drought tolerance. Nanomaterials 13(9):1453. https://doi.org/10.3390/nano13091453
81. Miransari M (2011) Soil microbes and plant fertilization. Appl Microbiol Biotechnol 92:875–885. https://doi.org/10.1007/s00253-011-3521-y
82. Mishra AK, Sharma K, Misra RS (2012) Elicitor recognition, signal transduction and induced resistance in plants. J Plant Interact 7(2):95–120. https://doi.org/10.1080/17429145.2011.597517

83. Montaño-Herrera A, Santiago-Saenz YO, López-Palestina CU, Cadenas-Pliego G, Pinedo-Guerrero ZH, Hernández-Fuentes AD, Pinedo-Espinoza JM (2022) Effects of edaphic fertilization and foliar application of se and Zn nanoparticles on yield and bioactive compounds in *Malus domestica* L. Horticulturae 8(6):542. https://doi.org/10.3390/horticulturae8060542
84. Morales-Díaz AB, Ortega-Ortiz H, Juárez-Maldonado A, Cadenas-Pliego G, González-Morales S, Benavides-Mendoza A (2017) Application of nanoelements in plant nutrition and its impact in ecosystems—review. Adv Nat Sci Nanosci Nanotechnol 8(1):013001. https://doi.org/10.1088/2043-6254/8/1/013001
85. Mustafa N, Raja NI, Ilyas N, Ikram M, Mashwani ZUR, Ehsan M (2021) Foliar applications of plant-based titanium dioxide nanoparticles to improve agronomic and physiological attributes of wheat (*Triticum aestivum* L.) plants under salinity stress. Green Process Synth 10(1):246–257. https://doi.org/10.1515/gps-2021-0025
86. Naaz S, Sachdev S, Husain R, Pandey V, Ansari MI (2023) Nanomaterials and nanocomposites: significant uses in plant performance, production, and toxicity response. In: Nanomaterials and nanocomposites exposures to plants: response, interaction, phytotoxicity and defense mechanisms. Springer Nature, Singapore. https://doi.org/10.1007/978-981-99-2419-6_1
87. Ndaba B, Roopnarain A, Haripriya RAMA, Maaza M (2022) Biosynthesized metallic nanoparticles as fertilizers: an emerging precision agriculture strategy. J Integr Agric 21(5):1225–1242. https://doi.org/10.1016/S2095-3119(21)63751-6
88. Neube B, Mupangwa W, French A (2018) Precision agriculture and food security. In: Systems analysis approach for complex global challenges. Springer International Publishing AG, Cham, Switzerland. https://doi.org/10.1016/B978-0-12-815870-8.00009-7
89. Ojeda-Barrios DL, Cruz-Alvarez O, Sánchez-Chavez E, Ciscomani-Larios JP (2023) Effect of foliar application of zinc on annual productivity, foliar nutrients, bioactive compounds, and oxidative metabolism in pecan. Folia Hortic 35(1):179–192. https://doi.org/10.2478/fhort-2023-0014
90. Padhan SR, Kar I, Mohanty A, Panigrahi KK (2023) Nanofertilizers: types, synthesis, methods, and mechanisms. In: Nanofertilizers for sustainable agroecosystems: recent advances and future trends. Springer Nature, Switzerland. https://doi.org/10.1007/978-3-031-41329-2_3
91. Parola-Contreras I, Tovar-Perez EG, Rojas-Molina A, Luna-Vázquez FJ, Torres-Pacheco I, Ocampo-Velázquez RV, Guevara-González RG (2020) Changes in affinin contents in *Heliopsis longipes* (chilcuague) after a controlled elicitation strategy under greenhouse conditions. Ind Crops Prod 148:112314. https://doi.org/10.1016/j.indcrop.2020.112314
92. Pérez-Labrada F, López-Vargas ER, Ortega-Ortiz H, Cadenas-Pliego G, Benavides-Mendoza A, Juárez-Maldonado A (2019) Responses of tomato plants under saline stress to foliar application of copper nanoparticles. Plants 8(6):151. https://doi.org/10.3390/plants8060151
93. Raliya R, Saharan V, Dimkpa C, Biswas P (2018) Nanofertilizer for precision and sustainable agriculture: current state and future perspectives. J Agric Food Chem 66(26):6487–6503. https://doi.org/10.1021/acs.jafc.7b02178
94. Ramírez-Gil JG, Lopera AA, García C (2023) Calcium phosphate nanoparticles improve growth parameters and mitigate stress associated with climatic variability in avocado fruit Heliyon 9:e18658. https://doi.org/10.1016/j.heliyon.2023.e18658
95. Ramírez-Rodríguez SC, Ortega-Ortiz H, González-Morales S, Preciado-Rangel P (2023) Chitosan nanoparticles improve yield, enzymatic activity, and bioactive compounds in tomato fruits. Terra Latinoamericana 41(1–11):e1686. https://doi.org/10.28940/terra.v41i0.1686
96. Rao S, Shekhawat GS (2016) Phytotoxicity and oxidative stress perspective of two selected nanoparticles in *Brassica juncea*. 3 Biotech 6(2):244. https://doi.org/10.1007/s13205-016-0550-3
97. Rivera-Gutiérrez RG, Preciado-Rangel P, Fortis-Hernández M, Betancourt-Galindo R, Yescas-Coronado P, Orozco-Vidal JA (2021) Zinc oxide nanoparticles and their effect on melon yield and quality. Revista Mexicana de Ciencias Agrícolas 12(5):791–803. https://doi.org/10.29312/remexca.v12i5.2987

98. Rivero-Montejo SDJ, Vargas-Hernández M, Torres-Pacheco I (2021) Nanoparticles as novel elicitors to improve bioactive compounds in plants. Agriculture 11(2):134. https://doi.org/10.3390/agriculture11020134
99. Rodríguez-González V, Díaz-Cervantes E (2024) Potencial de los nanomateriales en la agricultura: retos y oportunidades. Mundo Nano. Revista Interdisciplinaria en Nanociencias y Nanotecnología 17(32):1e–20e
100. Rossi L, Fedenia LN, Sharifan H, Ma X, Lombardini L (2019) Effects of foliar application of zinc sulfate and zinc nanoparticles in coffee (*Coffea arabica* L.) plants. Plant Physiol Biochem 135:160–166. https://doi.org/10.1016/j.plaphy.2018.12.005
101. Salama DM, Osman SA, Shaaban EA, Abd Elwahed MS, Abd El-Aziz ME (2023) Effect of foliar application of phosphorus nanoparticles on the performance and sustainable agriculture of sweet corn. Plant Physiol Biochem 203:108058. https://doi.org/10.1016/j.plaphy.2023.108058
102. Salcido-Martínez A, Sanchez E, Licon-Trillo LP, Perez-Alvarez S, Palacio-Marquez A, Amaya-Olivas NI, Preciado-Rangel P (2020) Impact of the foliar application of magnesium nanofertilizer on physiological and biochemical parameters and yield in green beans. Notulae Botanicae Horti Agrobotanici Cluj-Napoca 48(4):2167–2181. https://doi.org/10.15835/nbha48412090
103. Sangwan NS, Jadaun JS, Tripathi S, Mishra B, Narnoliya LK, Sangwan RS (2018) Plant metabolic engineering. In: Omics technologies and bio-engineering: towards improving quality of life. Academic, Oxford, UK. https://doi.org/10.1016/B978-0-12-815870-8.00009-7
104. Shahsavari N, Jais HM, Rad AHS (2014) Effect of zeolite and zinc on the biochemical characteristics of canola upon drought stress. Sains Malaysiana 43(10):1549–1555
105. Sharifi-Rad R, Esmaeilzadeh Bahabadi S, Samzadeh-Kermani A, Gholami M (2020) The effect of non-biological elicitors on physiological and biochemical properties of medicinal plant *Momordica charantia* L. Iran J Sci Technol Trans A: Sci 44:1315–1326. https://doi.org/10.1007/s40995-020-00939-8
106. Sharma P, Bhatt D, Zaidi MGH, Saradhi PP, Khanna PK, Arora S (2012) Silver nanoparticle-mediated enhancement in growth and antioxidant status of *Brassica juncea*. Appl Biochem Biotechnol 167:2225–2233. https://doi.org/10.1007/s12010-012-9759-8
107. Sheoran P, Goel S, Boora R, Kumari S, Yashveer S, Grewal S (2021) Biogenic synthesis of potassium nanoparticles and their evaluation as a growth promoter in wheat. Plant Gene 27:100310. https://doi.org/10.1016/j.plgene.2021.100310
108. Sigala-Aguilar NA, González-Fuentes JA, Valdez-Aguilar LA, López-Pérez MG, Medrano-Macias J, Benavides-Mendoza A, González-Morales S (2023) Effect of elicitors and biostimulants on the content of bioactive compounds in raspberry fruits. Hortic Sci (Prague) 50(2):101–111. https://doi.org/10.17221/98/2021-HORTSCI
109. Singh OS, Pant NC, Laishram L, Tewari M, Dhoundiyal R, Joshi K, Pandey CS (2018) Effect of CuO nanoparticles on polyphenols content and antioxidant activity in Ashwagandha (*Withania somnifera* L. Dunal). J Pharmacognosy Phytochem 7(2):3433–3439
110. Singh Y, Kumar U, Panigrahi S, Balyan P, Mehla S, Sihag P et al (2023) Nanoparticles as novel elicitors in plant tissue culture applications: Current status and future outlook. Plant Physiol Biochem 108004. https://doi.org/10.1016/j.plaphy.2023.108004
111. Sturikova H, Krystofova O, Huska D, Adam V (2018) Zinc, zinc nanoparticles and plants. J Hazard Mater 349:101–110. https://doi.org/10.1016/j.jhazmat.2018.01.040
112. Sutulienė R, Brazaitytė A, Małek S, Jasik M, Samuolienė G (2023) Response of oxidative stress and antioxidant system in pea plants exposed to drought and boron nanoparticles. Antioxidants 12(2):528. https://doi.org/10.3390/antiox12020528
113. Sutulienė R, Brazaitytė A, Małek S, Jasik M, Samuolienė G (2023) Biochemical responses of pea plants to drought stress and in the presence of molybdenum trioxide nanoparticles. Plant Soil 492(1):381–397. https://doi.org/10.1007/s11104-023-06179-0
114. Tejada-Alvarado JJ, Meléndez-Mori JB, Ayala-Tocto RY, Goñas M, Oliva M (2023) Influence of silver nanoparticles on photosynthetic pigment content and mineral uptake in pineapple seedlings grown in vitro under aluminum stress. Agronomy 13(5):1186. https://doi.org/10.3390/agronomy13051186

115. Tiwari M, Sharma NC, Fleischmann P, Burbage J, Venkatachalam P, Sahi SV (2017) Nanotitania exposure causes alterations in physiological, nutritional and stress responses in tomato (*Solanum lycopersicum*). Front Plant Sci 8:633. https://doi.org/10.3389/fpls.2017.00633
116. Vargas-Hernández M, Macias-Bobadilla I, Guevara-González RG, Rico-García E, Ocampo-Velázquez RV, Ávila-Juárez L, Torres-Pacheco I (2020) Nanoparticles as potential antivirals in agriculture. Agriculture 10(10):444. https://doi.org/10.3390/agriculture10100444
117. Verma KK, Song XP, Joshi A, Rajput VD, Singh M, Sharma A et al (2022) Nanofertilizer possibilities for healthy soil, water, and food in future: an overview. Front Plant Sci 13:865048. https://doi.org/10.3389/fpls.2022.865048
118. Wang P, Lombi E, Zhao FJ, Kopittke PM (2016) Nanotechnology: a new opportunity in plant sciences. Trends Plant Sci 21(8):699–712. https://doi.org/10.1016/j.tplants.2016.04.005
119. Wang Y, Chang CH, Ji Z, Bouchard DC, Nisbet RM, Schimel JP et al (2017) Agglomeration determines effects of carbonaceous nanomaterials on soybean nodulation, dinitrogen fixation potential, and growth in soil. ACS Nano 11(6):5753–5765. https://doi.org/10.1021/acsnano.7b01337
120. Wu J, Wang G, Vijver MG, Bosker T, Peijnenburg WJ (2020) Foliar versus root exposure of AgNPs to lettuce: phytotoxicity, antioxidant responses and internal translocation. Environ Pollut 261:114117. https://doi.org/10.1016/j.envpol.2020.114117
121. Xuan L, Ju Z, Skonieczna M, Zhou PK, Huang R (2023) Nanoparticles-induced potential toxicity on human health: applications, toxicity mechanisms, and evaluation models. MedCommun 4(4):e327. https://doi.org/10.1002/mco2.327
122. Yadav A, Yadav K, Abd-Elsalam KA (2023) Nanofertilizers: types, delivery and advantages in agricultural sustainability. Agrochemicals 2(2):296–336. https://doi.org/10.3390/agrochemicals2020019
123. Zahedi SM, Abolhassani M, Hadian-Deljou M, Feyzi H, Akbari A, Rasouli F et al (2023) Proline-functionalized graphene oxide nanoparticles (GO-pro NPs): a new engineered nanoparticle to ameliorate salinity stress on grape (*Vitis vinifera* L. cv sultana). Plant Stress 7:100128
124. Zarinkoob A, Esmaeilzadeh Bahabadi S, Rahdar A, Hasanein P, Sharifan H (2021) Ce–Mn ferrite nanocomposite promoted the photosynthesis, fortification of total yield, and elongation of wheat (*Triticum aestivum* L.). Environ Monit Assess 193:800. https://doi.org/10.1007/s10661-021-09506-z
125. Zhai X, Jia M, Chen L, Zheng CJ, Rahman K, Han T, Qin LP (2017) The regulatory mechanism of fungal elicitor-induced secondary metabolite biosynthesis in medical plants. Crit Rev Microbiol 43(2):238–261. https://doi.org/10.1080/1040841X.2016.1201041
126. Zhaohui L, Yingpeng Z, Yan L (2013) Zeolite as slow release fertilizer on spinach yields and quality in a greenhouse test. J Plant Nutr 36(10):1496–1505. https://doi.org/10.1080/01904167.2013.790429
127. Zhou Y, Zou Z, Wang M, Wang Y, Li J, Qiu L et al (2023) Biochar and nano-ferric oxide synergistically alleviate cadmium toxicity of muskmelon. Environ Sci Pollut Res 30(20):57945–57959. https://doi.org/10.1007/s11356-023-26369-8
128. Zulfiqar F, Navarro M, Ashraf M, Akram NA, Munné-Bosch S (2019) Nanofertilizer use for sustainable agriculture: advantages and limitations. Plant Sci 289:110270. https://doi.org/10.1016/j.plantsci.2019.110270

Use of Nanomaterials for the in Vitro Production of Plants and Plant Metabolites

César R. Sarabia-Castillo, Perla C. Meléndez-González, René Juárez-Altamirano, and Juan C. Carrillo-Rodríguez

Abstract The application of nanomaterials in plant biotechnology has turned out to be an innovative and promising idea to develop efficient methods to improve plant growth and increase the production of bioactive compounds such as alkaloids, flavonoids, phenolic acids, glycosides, among others, which are widely studied due to their high value in the food, cosmetic and pharmacological industries. In this sense, it has been seen that some nanomaterials, such as Cu, Ag, TiO_2, ZnO nanoparticles and carbon-based materials such as carbon nanotubes, to mention a few, promote organogenesis and callus production in in vitro cultures; likewise, it has been reported that the use of nanomaterials induces important somaclonal changes, such as changes in metabolism, conferring improvements in survival and reproduction, and allowing the adaptation of plants to different types of biotic and abiotic stresses. On the other hand, it has been found that the application of nanomaterials as abiotic elicitors can increase the production of secondary metabolites in in vitro cultures, improving their availability and specificity. Based on the available studies, the framework of this chapter summarizes the use and impact of some nanomaterials on plant reproduction in vitro, the somaclonal variations that may occur, as well as their influence on the production of secondary metabolites, and their ability to adapt to different types of stresses.

Keywords Bioactive compounds · Nanoelicitor · Somaclonal variation · Tissue culture

C. R. Sarabia-Castillo (✉) · R. Juárez-Altamirano
Sustainability of Natural Resources and Energy Program, Cinvestav-Saltillo, Parque Industrial Saltillo-Ramos Arizpe, Ramos Arizpe, Coahuila, Mexico
e-mail: ing.cesarsarabia@gmail.com; cesar.sarabia@cinvestav.mx

P. C. Meléndez-González
(UANL) Facultad de Ciencias Químicas, Universidad Autónoma de Nuevo León, San Nicolas de los Garza, México

J. C. Carrillo-Rodríguez
Centro de Investigación en Química Aplicada, Saltillo, Coahuila, México

© The Author(s), under exclusive license to Springer Nature Singapore Pte Ltd. 2025
A. Juárez-Maldonado et al. (eds.), *Plant Biostimulation with Nanomaterials*, Smart Nanomaterials Technology, https://doi.org/10.1007/978-981-96-4648-7_5

1 Introducction

Nanomaterials (NMs) are structures that have at least one of their dimensions in the nanometer scale, i.e., they comprise sizes from 1 to 100 nm [46, 71], This characteristic makes nanomaterials acquire novel and unique physicochemical properties different from those of their bulk counterpart, which makes them ideal for innovative applications in various fields of study such as plant biotechnology, such as plant tissue culture [46, 65].

In this sense, it has been reported that the use of NMs in plant tissue culture can improve seed germination rates, enhance plant growth, development, and yield, and improve the production of bioactive compounds of great industrial, pharmacological, and cosmetic importance, among others [37].

Therefore, plant tissue cultures have the advantage of having the ideal conditions for the evaluation of the effects of NMs on plants, since they have an environment free of contamination and the use of stable culture media, which allows that external factors do not interfere in the plant-NMs interaction, resulting in a more accurate interpretation of the direct effects of NMs on the plants under study [84].

For this reason, in this review we analyzed detailed reports on the effects of different NMs applied to in vitro culture production with emphasis on the effects on callus production and induction of organogenesis, the induction of somaclonal variations, the production of bioactive compounds and the usefulness of NMs as enhancers for the tolerance of different types of stress.

2 Effect of Nanomaterials on Callus Induction, and Organogenesis

The plant sciences are focused on tissue culture comprising protection, mass spread, heredity control, creation of bioactive mixes, and plant improvement aimed at enhancing the production of crops [79, 86].

The application of nanotechnology in agriculture and horticulture crops has developed as a strategy to avoid biotic/abiotic stress that improves crop production. Most plants are sensitive to abiotic stress, threatening sustainable agriculture production [1]. Likewise, nanotechnology in plant tissue culture has several reports, and nanotechnology could be classified as nutrients, disinfectants, or both [33, 76, 85].

Nanomaterials are materials of at least one dimension less than 100 nm with distinct physicochemical properties [46]. These properties benefit efficient interaction with plants due to their contribution to optimizing agricultural practices [11]. The plants need micronutrients (e.g. Fe, Zn, Mn, Ni, Mo, B, and Na) for good biological process regulation [29]. These micronutrients can be utilized as nanoparticles (NPs) in their metal or metal oxide form to regulate diverse mechanisms in response to abiotic stress [6].

NPs can affect the tissue culture (e.g. callus induction, morphogenesis, embryogenesis, organogenesis, conservation of germplasm, mass propagation, genetic transformation, and the production of free pathogen plants and secondary metabolites) [77]. Several studies have evaluated the influence of NPs on tissue culture, especially on callus induction and organogenesis, by incorporating different amounts of NPs into the MS basal medium containing explants [50, 82].

Callus induction is forming an undifferentiated cluster of cells from an explant. It can be applied in the plant's rejuvenation and synthesizing metabolites of interest [62, 78]. Ebadollahi et al. [21], evaluated the effect of TiO_2/perlites NPs on growth characteristics and secondary metabolite profiles on explant sources of *Hypericum perforatum* L. callus cultures, indicating a limit of 200 mg L^{-1} no adverse impacts on the growth of *H. perforatum* L. calli. However, the pants obtained in vitro were grown to produce more volatile compounds than those obtained from field developed.

Moreover, Malik et al. [52] investigated the impact of Cu and Ag NPs on callus induction in wheat. The study revealed that the combined use of Cu and Ag (0.015 and 4 mg L^{-1}, respectively) significantly increased callus formation on mature wheat embryo explants. Further, it indicated a possible negative effect on the application of these NPs due to their antimicrobial activity.

Besides, Khan et al. [42], examined the outcome of graphene oxide and carbon nanotubes in callus cultures of *Fagonia indica* Burm.f. in terms of its regeneration, biomass accumulation, and phytochemical content. The study revealed that reduced graphene oxide has better results than carbon nanotubes for inducing a positive effect in plant regeneration, biomass accumulation, and synthesis of secondary metabolites. In contrast, the application of ZnO NPs in the callus of *Nicotiana tabacum* L. had a dose-dependent that induced growth or oxidative stress in the callus cell [55].

Fouda et al. [27], evaluated the application of different NPs (ZnO, SiO_2, and Fe_3O_4) in the development of the callus of *Salvadora persica* L. to induce bioactive metabolites (Benzyl isothiocyanate, BITC). They assessed the cellular and genetic responses. In general, the ZnO, SiO_2 and Fe_3O_4 positively affected the formation of callus and BITC. However, they indicate a genetic variation and bioaccumulation of the NPs that can have side effects.

Organogenesis is the regenerative process in which different organs are developed and created in a plant. The organogenesis could divide into direct organogenesis (it is made from the meristematic regions) and indirect organogenesis (it is produced from the callus phase at the early phase) [14, 49].

The optimization of nutrient medium improved organogenesis; therefore, the utilization of NPs is beneficial for developing explant. The low concentration of Ag NPs in *Campomanesia rufa* (Berg) Niedenzu did not affect the in vitro multiplication. The shoots did not observe a biochemical, morphological, or anatomical change. However, in upper concentrations of 15 mg L^{-1}, there was a reduction in the shot number, as reported by Timoteo et al. [80].

Besides, Krupa-Makiewicz et al. [45] evaluated the effect of different concentrations of Ag NPs in infection limitations, morphological features, and chemical composition at young barley seedlings under in vitro conditions. The morphological characteristics depend on the concentration of Ag NPs,the positive effect is an increased amount in young barley leaves and a high amount of K, Ca, vitamin C, and E in young barley seedlings.

CuO and MnO NPs were used for biomass and bioactive compounds in *Ocimum basilicum* L. CuO increased biomass proliferation and phytochemicals contents. In the in vitro experiments, CuO and MnO NPs demonstrated increased antioxidant activities. They were effective elicitors for critical metabolite accumulation in callus cultures, as reported by Nazir et al. [61].

Chutipaijit and Sutjaritvorakul [17] study the impact of charcoal and nanocarbon on callus induction and plant regeneration in *Oryza sativa* L. The experiment showed a high percentage of green spots and plant regeneration and a high ratio of seedlings to the number of regenerated calli (RSR). Therefore, charcoal and nanocarbon enhanced the callus induction and plant regeneration. Table 1 shows the effect of different nanoparticles on callus induction and organogenesis.

Another application of NPs is their use as antifungal and antibacterial in vitro tissue cultures. For example, Devasia et al. [18] explained the effect of ZnO NPs in vitro culture leas explants of *Coffea arabica* L. They described that ZnO NPs decrease contamination due to their antimicrobial properties. Additionally, it did not show an inhibitory effect on the growth of explants.

Alfarraj et al. [7] have explored the use of Ag NPs to prevent the contamination of the tissue culture and induce callus development. They discovered that 40 mg L^{-1} of Ag NPs effectively reduces microbial contamination and improves callus growth while maintaining its genetic stability.

The use of nanoparticles to generate a positive effect on callus induction and organogenesis has been established in different reports. The difference is the type, morphology, size, and quantity of nanoparticles used. Furthermore, using the same nanoparticle for two types of cellular tissue can generate different effects. Therefore,

Table 1 The influence of diverse nanoparticles on various plant species

Plant	NPs	Size Nm	Concentration mg L^{-1}	Effect	References
Triticum aestivum L	Ag	5–20	2–8	High values of callus induction and embryogenic callus induction	[83]
× *Triticosecale Wittmack*	CaO	66–145	0.5–1.5	Suppress the adverse effect of salt stress on the callus and adjust the micronutrient content	[93]
Paulownia sp.	TiO_2	<25	60	High effect on callus formation and somatic embryogenesis	[19]
Phoenix dactylifera L	ZnO	ND	0–150	Highest shoots content of endogenous	[10]
Momordica charantia L	Se	ND	0–50	High efficiency to stimulate growth and organogenesis	[13]

ND = Not determined

the impact of each nanoparticle on callus induction and organogenesis must be individually evaluated to create the best conditions in which stress is not caused and the best results are produced.

3 Use of Nanomaterials in Crop Improvement Through the Induction of Somaclonal Variations

Currently, over 60% of the world's population relies on agriculture as a means of livelihood, and it is one of the primary activities to meet the demand for food [5]. By 2050, the global population is projected to reach approximately 9 billion people. If consumption and production practices do not change, it is estimated that a 100% increase in agricultural production will be required [64]. However, this sector faces various challenges, such as climate change, which limits crop yields due to nutrient deficiencies, water deficits, and rising temperatures. Additionally, the proliferation of phytopathogens and infectious crop diseases is progressively spreading to other regions of the planet. These challenges pose a threat to food security in several parts of the world and adversely impact the global economy.

Therefore, in recent decades, efforts have been made to find solutions to these problems, ranging from modifying agricultural technologies in search of sustainable alternatives to the quest for improving crop characteristics. In this regard, plant biotechnology has played an essential role in the production of enhanced crops. According to the Food and Agriculture Organization (FAO) database, biotechnologies focus on tissue culture, molecular markers, diagnostic techniques, and microbial products [23].

Plant tissue culture is a technique that manipulates plant organs and cells under sterile conditions, cultivated in a growth medium with regulated temperature, light, and humidity conditions. This technique is used for the mass propagation of plant cells, the production of genetically modified tissues resistant to diseases, and the enhancement of secondary metabolite production [25]. Tissue culture techniques can also be employed to produce identical plants through clonal propagation, known as clones. The production of clones is considered an advantage in maintaining plant characteristics,however, plants grown in vitro exhibit genetic variations due to changes in chromosomes. These genetic changes present in plants produced by tissue culture is known as somaclonal variations [66].

Somaclonal variation can occur in two forms: genetic (heritable) variations and epigenetic variations. Genetic changes represent the expression of pre-existing variation in the original material or changes produced by genetically controlled mechanisms. These genetic changes occur at the chromosomal level, often involving alterations in the number of chromosomes or at the DNA sequence level. In contrast, epigenetic variations are temporary, reversible changes that are not heritable. Genetic variations are desirable due to their stability, as they can persist after repeated asexual propagation or a sexual cycle. Therefore, these variations can be maintained in plants even after planting them in the field [53, 66].

Initially, somaclonal variations posed a challenge because plant tissue culture aimed at micropropagation, the establishment of commercial plantations through standardized procedures for economically important species. These variations hindered the strict maintenance of the characteristics of selected individuals for the initiation of in vitro cultures. Subsequently, another perspective emerged, considering somaclonal variability as a source of genetic diversity that researchers can leverage for generating new plants with superior traits.

This is when somaclonal variation becomes a valuable tool for agriculture, as it has led to the production of crops with variants holding significant agronomic value. These variants include improvements in yield, resistance to diseases, tolerance to water stress, frost resistance, resistance to salinity conditions, and resistance to pathogens. Additionally, biochemical variants can be generated for the study of metabolic pathways. Furthermore, it is considered a relatively simpler technique compared to others, such as recombinant DNA technology [67].

Currently, in the field of biotechnology, plant tissue culture is one of the most promising areas, and nanotechnology has been integrated into these techniques. It has been discovered that nanoparticles can be used with positive effects in plant tissue culture. The addition of different nanoparticles to plant tissue culture media results in the reduction of microbial contamination, increased production of secondary metabolites, improved morphogenetic potential, and the potential induction of somaclonal variations [77]. However, phytotoxicity has also been reported due to the use of nanoparticles, especially at high levels. Although it is acknowledged that phytotoxicity may be attributed to various factors such as nanoparticle type, chemical composition, size, stability, as well as the composition of the culture medium, plant type, application method, etc. [57].

After exposure to plant tissue cultures, nanoparticles penetrate and enter the cell wall and plasma membrane. Subsequently, the nanoparticles translocate to adjacent cells through the apoplast or plasmodesmata. This results in the influence on mitosis, the stabilization of DNA, and the modification of plant proteins (Fig. 1) [77].

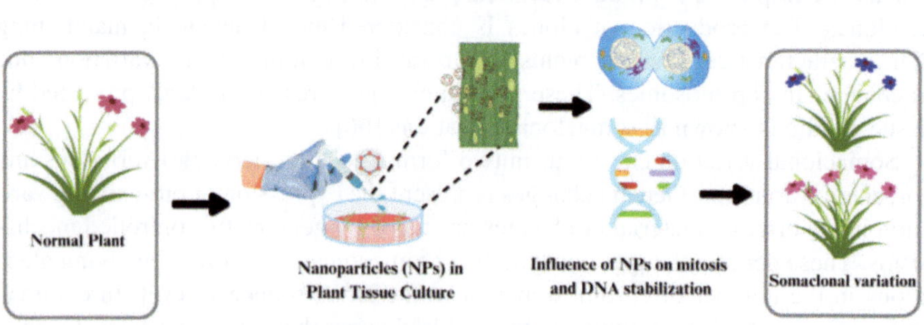

Fig. 1 Nanoparticles for the induction of somaclonal variation

Various nanoparticles have been employed in this field of research, such as carbon nanotubes and metal-based nanomaterials like gold (Au), iron (Fe), silver (Ag), zinc oxides (ZnO), titanium oxides (TiO_2), etc. Depending on the nanomaterial, they exhibit different properties. For example, the addition of silver nanoparticles has been reported to induce variations in the morphology, protein content, anatomy, and DNA of Solanum nigrum [22]. Other plants treated with zinc and zinc oxide nanoparticles showed somaclonal variations that accumulated more proline, chlorophyll, and antioxidant enzyme activity, in addition to exhibiting higher dry weight accumulation [70]. On the other hand, crops treated with gold nanoparticles showed an increase in tetraploids, a genetic mutation in plants that confers improvement in survival and reproduction [44]. Similar results have been reported in plant tissue cultures treated with carbon nanoparticles [68].

Nanoparticle-induced somaclonal variations also exhibit phenotypic changes of value or with useful characteristics, such as plant size, flower color, leaf variety, and fruit maturation. These phenotypic changes result from epigenetic alterations, which, as mentioned earlier, can be reversible. For instance, in pineapple cultivation, variations with less spiny leaves were identified, facilitating the handling of this fruit in the field. In chrysanthemum, changes in the size and number of flowers were observed, while in tobacco cultivation, variations that could enhance productivity, such as increased leaf width, length, and number, were reported [26].

Despite the agronomic advantages provided by nanoparticle-induced somaclonal variations, it is also true that not all variations are desirable, and some may even be detrimental to crops. Firstly, the origin of somaclonal variations is not always clear and can differ among plants. For example, variations could be due to chromosomal rearrangements, DNA replication disturbance, somatic recombination, nucleotide alteration through methylation, or exchange of sister chromatids. The variations can then be unpredictable and, at times, uncontrollable. Many genetic changes are neither heritable nor stable, occurring with variable frequencies, and may not always have any agricultural utility. An example is the occurrence of dwarfism events, which in some crops represents a loss of productivity [20].

Similarly, it is still necessary to conduct more extensive and rigorous research to determine the impact that nanoparticles can have on variations, assess a wide variety of nanoparticles, and understand the associated mechanism of action in somaclonal variations. Additionally, more information is needed about the potential adverse effects of nanoparticles on plant cell and tissue cultures. Limited data exist on the toxicity of nanoparticles, and the mechanisms of absorption and subsequent fate in the food chain remain a mystery. Understanding this could help address potential environmental issues and avoid future challenges, all with the aim of developing sustainable applications of nanomaterials in plant biotechnology for crop improvement.

4 Nanomaterials as Inducers in the in Vitro Production of Secondary Metabolites

Plant secondary metabolites are organic compounds that are biosynthesized through the secondary metabolism of plants, which is of great relevance for their survival as it helps them in functions such as protection against biotic and abiotic factors (herbivores and pathogenic microorganisms, UV radiation, drought, etc.), competition, or in signaling to carry out symbiotic interactions of plants with microorganisms [74].

On the other hand, in addition to the benefits that secondary metabolites confer on plants, many of them are of economic importance, with applications in the pharmaceutical industry, in the production of colorants, pigments, flavorings and fragrances, in the production of agricultural inputs, or as additives in the food industry. However, the production of secondary metabolites naturally is not enough to satisfy the demand for these organic compounds for their application in different industrial sectors.

One of these alternatives is the application of nanotechnology using different nanomaterials (NMs) in vitro cultures that promote the accumulation of specific secondary metabolites of various plant species (Fig. 2). This is because (*i*) NMs can provoke abiotic stress in plants, which in turn, activates the antioxidant mechanism of plants and consequently increases the concentration of secondary metabolites [47], and (*ii*) obtaining in vitro cultures allows increasing the production of secondary metabolites since they have the capacity to increase yields under controlled growth conditions and/or genetic manipulation [54].

In relation to the above, previous studies suggest that the application of NPs in vitro cultures significantly improves the production of secondary metabolites of commercial, medicinal, or industrial interest. Maqbool et al. [54] evaluated the effects of Ag NPs as inducing agents in the production of secondary metabolites in *Aerva sanguinolenta* (L.) callus grown in vitro, finding that Ag NPs increased the total

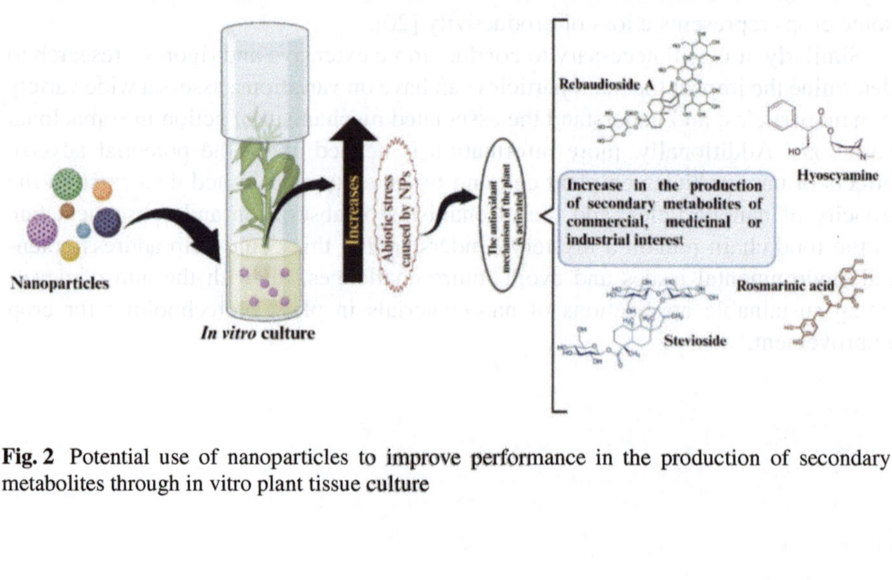

Fig. 2 Potential use of nanoparticles to improve performance in the production of secondary metabolites through in vitro plant tissue culture

phenolic compound (TFC) content, total flavonoid content (TFC), aervin and some antioxidant metabolites, similarly, Ali et al. [8] reported that with the application of Ag NPs it was possible to increase the production of pharmacologically important secondary metabolites (TFC, TFC, phenylalanine ammonialyase activity and antioxidant activity) in callus of *Caralluma tuberculata* N.E.Br., that could be used in drug development and discovery.

On the other hand, Mosavat et al. [59] found that the application of ZnO NPs at a concentration of 150 mg L^{-1}, improved the content of thymol and carvacrol, two secondary metabolites of pharmacological, food and cosmetic importance in vitro cultures of callus of *Thymus kotschyanus* Boiss. and Hohen and *Thymus daenensis* Celak, which represents an alternative in the production and availability of these metabolites. In turn, Shehzad et al. [75] found that adding ZnO NPs (0.15 mg L^{-1}) to in vitro callus cultures of *Silybum marianum* L. and treated under different light conditions improved the content of isomeric flavonolignans such as silydianin, silyristin, silybin disestereoisomers and isosilybin disestereoisomers, which compose silymarin, being present in higher amounts when the callus were grown in complete darkness in a medium enriched with ZnO NPs.

Ahmad et al. [4], on the other hand, evidenced the increase in the production of steviol glycosides such as rebaudioside A and stevioside in *Stevia rebaudiana* Bertoni regenerants grown in vitro when ZnO and CuO NPs were used at concentrations of 2 and 20 mg L^{-1} respectively, thus suggesting that the use of nanostructured metal oxides could improve the production of bioactive compounds in vitro cultures for their application in the nutraceutical industry. Regarding the use of CuO NPs, Chung et al. [16] evaluated the accumulation of secondary metabolites of pharmacological interest in cell suspension cultures of *Gymnema sylvestre* (Retz.) R. Br. was treated with CuO NPs and found that when treated with a dose of NPs 3 mg L^{-1}, higher amounts of gymnemic acids and phenolic compounds, important bioactive compounds used in the treatment of diabetes mellitus were obtained.

As well as the above mentioned NMs, there are other NMs that have been studied as inducers of secondary metabolism. In a study using transformed hairy root cultures of *Hyoscyamus reticulatus* L. and *Hyoscyamu pusillus* L. and treated with SiO_2 NPs, an increase in the production of tropane alkaloids was observed, mainly hyoscyamine and scopolamine, compounds of pharmacological importance due to their antispasmodic and anticholinergic properties [34]. Fe_2O_3 NPs have been evaluated as inducers in the production of medically important active agents such as rosmarinic acid, the main active agent presents in *Dracocephalum kotschyi* Bioss, by in vitro assays of transformed hairy roots treated with Fe_2O_3 NPs, resulting in a significant increase in rosmarinic acid content after 24 h of exposure. Similarly, TiO_2 NPs were tested on transformed hairy roots of *D. kotschyi* Bioss, resulting in an increased production of rosmarinic acid and methoxylated flavonoids such as xanthomicrol, which has anticarcinogenic properties [63].

Of the carbon-based nanomaterials, it has been reported that in vitro cultures of *Satureja khuzestanica* Jamzad treated with multi-walled carbon nanotubes

(MWCNT) increased the content of rosmarinic acid and caffeic acid [28]. Likewise, the capacity of MWCNTs and graphene to stimulate the synthesis of pharmaceutically important alkaloids in vitro cultures of *Catharanthus roseus* (L.) G. Don has been evaluated. In this sense, it was determined that both nanomaterials significantly increased the content of vinblastine and vincristine, alkaloids beneficial for the treatment of some types of cancer and diabetes [56].

Therefore, and due to the growing demand for secondary metabolites with medical, cosmetic, industrial applications, etc. it is necessary to explore and apply new plant secondary metabolism inducers. In this sense, the use of nanomaterials promises to be a promising alternative to improve the production of secondary metabolites in terms of production quantity and specificity, thus overcoming the existing productivity obstacles.

5 Nanomaterials to Improve Tolerance Against Biotic and Abiotic Stress

The decrease in cultivable soil due to global population growth, in addition to pollution and climate change, creates the need to develop techniques to optimize crop yields and production. In vitro propagation of food-important plants is one of the sustainable alternatives to other propagation methods, one of the marked advantages being the rapid availability of food throughout the year, independent of the season. However, one of the disadvantages would be large-scale in vitro regenerated plants and aseptic conditions [31]. The technique of in vitro plant propagation is an efficient technique for biotechnological studies, from which the behavior of plants under different important factors such as oxidative stress due to the low tolerance of plants to biotic or abiotic stress is studied. These conditions such as excessive use of pesticides, drought, salinity, temperature, flooding, toxic metals, high light, and UV radiation. One of the alternatives available is the responsible use of nanomaterials, which, thanks to their unique physical and chemical properties, promise to improve productivity and optimize resistance, in addition to minimizing negative impacts [73, 92].

Different nanomaterials such as metal nanoparticles (NPs), metal oxides, and polymeric nanoparticles have been studied as nanofertilizers or biostimulants and have shown the ability to reduce and control biotic stress in plants, as these have exceptional antimicrobial properties [81] SiO_2 NPs, when combined with specific plant extracts, are shown to enhance insecticidal activity by increasing antifeedant efficacy in insects, reaching up to 90% mortality [89]. Therefore, SiO_2 NPs are used in nanofungicides [94], nanofertilizers [72], and nanopesticides [15] to reduce the abiotic stress on plants.

In situations of abiotic stress, plants tend to adapt, seeking to mitigate the effects of stress through molecular alterations. These alterations translate into changes in morphology, physiological and biochemical characteristics [9, 32]. In vitro cultures are a useful tool to improve biotic and abiotic stress tolerance in crops, using agents such as polyethylene glycol, salicylic acid, mannitol, sucrose for drought tolerance and sodium chloride for salt stress in *O. sativa* L. cultivars [69]. On the other hand, the use of NPs provides plants with an improvement in the management of abiotic stress due to the impact these have on them. One of the important factors depending on this impact is the concentration of the NPs as it plays an important role in plant tolerance, the plants respond differently to specific concentrations [41].

The positive impact of various types of nanoparticulate metal/metalloids on diverse plant species has been observed under different abiotic stress conditions which are explained below:

5.1 Temperature

Extreme temperature changes due to climate change put plants at risk under this type of stress. Heat stress and cold stress generate an imbalance in the cellular homeostasis of the plant, for example, heat stress has consequences such as chlorophyll degradation, disruption of the photosystem and electron flow, as well as carbon fixation, processes that significantly decrease the photosynthetic mechanism [36, 88]. It also causes the generation of reactive oxygen species (ROS), such as heat shock proteins (HSP) which help to maintain stability within this type of stress, therefore nanoparticles have been used to contribute to the biosynthesis of this type of protein, such as carbon nanotubes and CeO_2 NPs [40].

The use of Ag NPs has been effective against thermal stress, promoting plant growth. Iqbal et al. [38], report the application of Ag NPs in *Triticum aestivum* L. obtaining favorable results in morphological growth in root length, shoot length, root number, plant fresh weight and plant dry weight of the plant under thermal stress. Kohan-Baghkheirati and Geisler-Lee [43], studied AgNPs in *Arabidopsis thaliana* (L.), Heynh. by genomic analysis and obtained a regulation of genes responsible for ROS-involving activities. On the other hand, under cold stress, the interaction of SiO_2 NPs on *Agropyron Elogatum* L. has been studied, indicating an 86.3% increase in germination, and seedling growth using a concentration of 60 mg L^{-1} [12].

5.2 Salinity

Salt stress is one of the most notable problems in the growth and productivity of plants. The increase of Na^+ and K^+ ions induce several alterations including the decrease of the soil osmotic potential, and the ionic increase causing toxicity. To reduce the excess of accumulated Na^+ ions in tomato plants, a treatment with chitosan and Cu NPs, which induce tolerance to salt stress, was carried out, increasing the quality of tomato fruit because of the activation of defense enzymes [35]. The role of biosynthesized Ag NPs was investigated in germinated grains of *T. aestivum L.*

under salt stress, obtaining a stimulation in growth parameters and photosynthetic efficiency, as well as the balance of phytohormones [3].

Another effective alternative to improve salt tolerance in *Lupinus termis* Forsk. plants are the use of ZnO NPs at a concentration of 60 mg L^{-1}, this treatment stimulates the growth of stressed plants by enhancing levels of photosynthetic pigments (total chlorophylls and carotenoids) and organic solutes (sugar, proteins, and amino acids); as well as enzyme activities [2].

5.3 Drought

This type of stress in arid and semiarid regions generally, causes loss of turgidity in plants, including irreversible damage to cell metabolism, protoplasm, membranes, and the protein system, which is reflected in the decrease in leaf size and in the accumulation of ions that can become toxic. Another negative effect is the condition of hypoxia in the roots, reducing oxygen levels, which affects gas exchange [58]. Treatments with different NPs have been reported to impart drought stress tolerance in plants. According to Mahmoud et al. [51], the application of SiO_2 nanoparticles on *Musa acuminata* Colla. crops improve growth and chlorophyll content, in addition to balancing ion content. In *T. aestivum* L. plants under drought stress conditions a mixture of nitric oxide and TiO_2 nanoparticles tends to protect photosynthetic development in the seed, reducing oxidative damage and promoting transpiration [24].

5.4 Toxic Metals

Phytoremediation is a sustainable technique to remove toxic agents from hazardous waste in the environment using plants. To mitigate oxidative stress caused in plant species exposed to heavy metal toxicity, the sustainable use of NPs that promote plant growth under toxic metal stress conditions has been expanded. *Vigna radiata L.* plants exposed to As toxicity revealed hindered growth rates and membrane integrity, which was positively promoted by applying TiO_2 NPs treatments, decreasing ROS-related mechanisms [39]. Another of the most known toxic metals is Hg, SiO_2 NPs in *Glycine max* [L] Merr. plant treatments under this type of abiotic factor, were studied to know the distribution of Hg, chlorophyll levels and antioxidant enzymes, resulting in a decrease of Hg concentration, improving the increase of chlorophyll, and the antioxidant enzymes [48].

5.5 Flooding

Crop plants exposed to flood stress exhibit growth retardation, the use of Ag nanoparticles as well as treatments with Al_2O_3 nanoparticles stimulate enzyme up-regulation and enhance growth induced by biochemical mechanisms in *G. max* [L] Merr. plants under this type of stress [30, 60].

Others abiotic stresses such as UV light, high-intensity light, and excess nutrients, also tend to cause the generation of ROS altering the growth cycle and development of crop plants. Wu et al. [91], demonstrated that in *A. thaliana* (L.), Heynh plants under excess light for a certain time, have negative effects on photosynthetic development, for this, these applied treatments of CeO_2 NPs that are transported to the chloroplast obtaining a reduction of ROS in the leaf of 52%.

The reaction mechanisms of enhanced in vitro plants' tolerance to abiotic stress with nanoparticles and metalloids are still unknown, for which genomic or transgenomic studies and transcriptome analysis are needed. One of the mechanisms proposed to explain the functioning of nanoparticles during abiotic stress is the mechanism of antioxidant stress, it is important to remember that oxidative species are generated during plant stress causing a decrease in properties and characteristics [87]. Figure 3 shows two pathways that can occur with NPs treatments, the first pathway is to act as catalysts for oxidative species such as H_2O_2 transforming them into plant nutrients. Second pathway shows how the NPs have photosynthetic efficiency by reducing oxidative stress and/or increasing stress tolerance.

The responsible application of nanomaterials emerges as an essential element in in vitro cultures. This is due to their properties to neutralize the detrimental effects of biotic and abiotic stresses on plants. In addition, these demonstrate the ability to regulate photosynthetic efficiency and detoxify reactive oxygen species, significantly improving plant germination and growth.

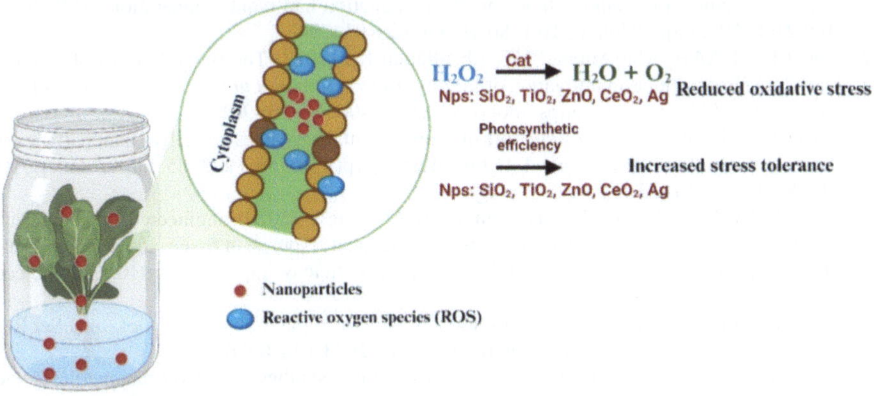

Fig. 3 Antioxidative mechanism of nanoparticles in vitro plants

6 Conclusions

The implementation of nanotechnology in plant biotechnology has generated positive effects, mainly on the production of in vitro cultures and as inducers in the production of secondary metabolites, however, these effects are variable and this is mainly due to the type of nanomaterial, its chemical composition, its shape, size, etc., without forgetting that the type of tissue in which the NMs are evaluated also influences.

However, despite the variant effects that are presented both in the in vitro reproduction of plant tissue and in the production of bioactive compounds, the use of NMs promises to be an alternative for improvement in the production of callus and explants and also as a strategy to induce the synthesis of secondary metabolites in terms of specificity and higher productivity.

On the other hand, it is necessary to evaluate NMs in terms of toxicity, since the existing data are limited, and it is also necessary to evaluate the absorption mechanisms and the final destination of NMs, since they are released in large quantities on a daily basis; this could help to address the possible impacts on the environment and their interactions with other organisms, thus avoiding future challenges.

Acknowledgements The authors acknowledge the Mexican National Council for Humanities, Science and Technology (CONAHCYT) for the postdoctoral fellowship awarded to PCMG and the Ph.D. fellowship awarded to RJA, the Centro de Investigación en Química Aplicada, and the Sustainability of Natural Resources and Energy Program from Cinvestav-Saltillo

References

1. Aazami MA, Rasouli F, Ebrahimzadeh A (2021) Oxidative damage, antioxidant mechanism and gene expression in tomato responding to salinity stress under in vitro conditions and application of iron and zinc oxide nanoparticles on callus induction and plant regeneration. BMC Plant Biol 21(1):597. https://doi.org/10.1186/s12870-021-03379-7
2. Abdel Latef AAH, Abu Alhmad MF, Abdelfattah KE (2017) The possible roles of priming with ZnO nanoparticles in mitigation of salinity stress in lupine (*Lupinus termis*) plants. J Plant Growth Regul 36(1):60–70. https://doi.org/10.1007/s00344-016-9618-x
3. Abou-Zeid H, Ismail G (2018) The role of priming with biosynthesized silver nanoparticles in the response of *Triticum aestivum* L to salt stress. Egypt J Bot 58(1):73–85. https://doi.org/10.21608/ejbo.2017.1873.1128
4. Ahmad MA, Javed R, Adeel M, Rizwan M, Ao Q, Yang Y (2020) Engineered ZnO and CuO nanoparticles ameliorate morphological and biochemical response in tissue culture regenerants of candyleaf (*Stevia rebaudiana*). Molecules 25(6), Article 6. https://doi.org/10.3390/molecules25061356
5. Ahmed MH (2023) Nanoparticles NPs stresses as in vitro plant innovative tool. Int J Sci Res Arch 10(2), Article 2. https://doi.org/10.30574/ijsra.2023.10.2.0976
6. Alabdallah NM, Hasan MdM (2021) Plant-based green synthesis of silver nanoparticles and its effective role in abiotic stress tolerance in crop plants. Saudi J Biol Sci 28(10):5631–5639. https://doi.org/10.1016/j.sjbs.2021.05.081
7. Alfarraj NS, Tarroum M, Al-Qurainy F, Nadeem M, Khan S, Salih AM, Shaikhaldein HO, Al-Hashimi A, Alansi S, Perveen K (2023) Biosynthesis of silver nanoparticles and exploring

their potential of reducing the contamination of the in vitro culture media and inducing the callus growth of *Rumex nervosus* Explants. Molecules 28(9), Article 9. https://doi.org/10.3390/molecules28093666
8. Ali A, Mohammad S, Khan MA, Raja NI, Arif M, Kamil A, Mashwani Z-R (2019) Silver nanoparticles elicited in vitro callus cultures for accumulation of biomass and secondary metabolites in *Caralluma tuberculata*. Artif Cells Nanomed Biotechnol 47(1):715–724. https://doi.org/10.1080/21691401.2019.1577884
9. Al-Khayri JM, Rashmi R, Surya Ulhas R, Sudheer WN, Banadka A, Nagella P, Aldaej MI, Rezk AA-S, Shehata WF, Almaghasla MI (2023) The role of nanoparticles in response of plants to abiotic stress at physiological, biochemical, and molecular levels. Plants 12(2), Article 2. https://doi.org/10.3390/plants12020292
10. Al-Mayahi AMW (2021) The effect of humic acid (HA) and zinc oxide nanoparticles (ZnO-NPS) on in vitro regeneration of date palm (*Phoenix dactylifera* L.) cv. Quntar. Plant Cell, Tissue, Organ Culture (PCTOC) 145(2):445–456. https://doi.org/10.1007/s11240-021-02020-7
11. Altabbaa S, Mann NA, Chauhan N, Utkarsh K, Thakur N, Mahmoud GA-E (2023) Era connecting nanotechnology with agricultural sustainability: issues and challenges. Nanotechnol Environ Eng 8(2):481–498. https://doi.org/10.1007/s41204-022-00289-3
12. Azimi R, Borzelabad MJ, Feizi H, Azimi A (2014) Interaction of SiO nanoparticles with seed prechilling on germination and early seedling growth of tall wheatgrass (*Agropyron Elongatum* L.). Polish J Chem Technol 16(3):25–29
13. Behbahani SR, Iranbakhsh A, Ebadi M, Majd A, Ardebili ZO (2020) Red elemental selenium nanoparticles mediated substantial variations in growth, tissue differentiation, metabolism, gene transcription, epigenetic cytosine DNA methylation, and callogenesis in bittermelon (*Momordica charantia*); an in vitro experiment. PLoS ONE 15(7):e0235556. https://doi.org/10.1371/journal.pone.0235556
14. Bidabadi SS, Jain SM (2020) Cellular, molecular, and physiological aspects of in vitro plant regeneration. Plants 9(6), Article 6. https://doi.org/10.3390/plants9060702
15. Bueno V, Gao X, Abdul Rahim A, Wang P, Bayen S, Ghoshal S (2022) Uptake and translocation of a silica nanocarrier and an encapsulated organic pesticide following foliar application in tomato plants. Environ Sci Technol 56(10):6722–6732. https://doi.org/10.1021/acs.est.1c08185
16. Chung I-M, Rajakumar G, Subramanian U, Venkidasamy B, Thiruvengadam M (2019) Impact of copper oxide nanoparticles on enhancement of bioactive compounds using cell suspension cultures of *Gymnema sylvestre* (Retz.) R. Br. Appl Sci 9(10), Article 10. https://doi.org/10.3390/app9102165
17. Chutipaijit S, Sutjaritvorakul T (2018) Application of activated charcoal and nanocarbon to callus induction and plant regeneration in aromatic rice (*Oryza sativa* L.). Chem Speciat Bioavailab 30(1):1–8. https://doi.org/10.1080/09542299.2017.1418184
18. Devasia J, Muniswamy B, Mishra MK (2020) Investigation of ZnO nanoparticles on in vitro cultures of coffee (*Coffea Arabica* L.). Int J Nanosci Nanotechnol 16(4):271–277
19. Dumani Y, Mortazavian SMM, Izadi-Darbandi A, Ramshini H, Amini F (2022) Titanium dioxide nanoparticles affect somatic embryo initiation, development, and biochemical composition in *Paulownia* sp. seedlings. Ind Crops Prod 176:114398. https://doi.org/10.1016/j.indcrop.2021.114398
20. Duta-Cornescu G, Constantin N, Pojoga D-M, Nicuta D, Simon-Gruita A (2023) Somaclonal variation—advantage or disadvantage in micropropagation of the medicinal plants. Int J Mole Sci 24(1), Article 1. https://doi.org/10.3390/ijms24010838
21. Ebadollahi R, Jafarirad S, Kosari-Nasab M, Mahjouri S (2019) Effect of explant source, perlite nanoparticles and TiO_2/perlite nanocomposites on phytochemical composition of metabolites in callus cultures of Hypericum perforatum. Sci Rep 9(1), Article 1. https://doi.org/10.1038/s41598-019-49504-3
22. Ewais EA, Desouky SA, Elshazly EH (2015) Evaluation of callus responses of *Solanum nigrum* L. exposed to biologically synthesized silver nanoparticles. Nanosci Nanotechnol 5(3):45–56

23. FAO (2017) Status of research and application of crop biotechnologies in developing countries
24. Faraji J, Sepehri A (2020) Exogenous nitric oxide improves the protective effects of TiO_2 nanoparticles on growth, antioxidant system, and photosynthetic performance of wheat seedlings under drought stress. J Soil Sci Plant Nutr 20(2):703–714. https://doi.org/10.1007/s42729-019-00158-0
25. Feizi S (2023) Role of nanomaterials in plant cell and tissue culture. In: Al-Khayri JM, Alnaddaf LM, Jain SM (eds) Nanomaterial interactions with plant cellular mechanisms and macromolecules and agricultural implications. Springer International Publishing, pp 359–397. https://doi.org/10.1007/978-3-031-20878-2_14
26. Ferreira MD, Rocha AD, Nascimento FD, Oliveira WD, Soares JM, Rebouças TA, Morais Lino LS, Haddad F, Ferreira CF, Santos-Serejo JA, Fernández JS, Amorim EP (2023) The role of somaclonal variation in plant genetic improvement: a systematic review. Agronomy 13(3), Article 3. https://doi.org/10.3390/agronomy13030730
27. Fouda MS, Hendawey MH, Hegazi GA, Sharada HM, El-Arabi NI, Attia ME, Soliman ERS (2021) Nanoparticles induce genetic, biochemical, and ultrastructure variations in *Salvadora persica* callus. J Genetic Eng Biotechnol 19(1):27. https://doi.org/10.1186/s43141-021-00124-3
28. Ghorbanpour M, Hadian J (2015) Multi-walled carbon nanotubes stimulate callus induction, secondary metabolites biosynthesis and antioxidant capacity in medicinal plant Satureja khuzestanica grown in vitro. Carbon 94:749–759. https://doi.org/10.1016/j.carbon.2015.07.056
29. Gui J-Y, Rao S, Huang X, Liu X, Cheng S, Xu F (2022) Interaction between selenium and essential micronutrient elements in plants: a systematic review. Sci Total Environ 853:158673. https://doi.org/10.1016/j.scitotenv.2022.158673
30. Hashimoto T, Mustafa G, Nishiuchi T, Komatsu S (2020) Comparative analysis of the effect of inorganic and organic chemicals with silver nanoparticles on soybean under flooding stress. Int J Mole Sci 21(4), Article 4. https://doi.org/10.3390/ijms21041300
31. Hasnain A, Naqvi SAH, Ayesha SI, Khalid F, Ellahi M, Iqbal S, Hassan MZ, Abbas A, Adamski R, Markowska D, Baazeem A, Mustafa G, Moustafa M, Hasan ME, Abdelhamid MMA (2022) Plants in vitro propagation with its applications in food, pharmaceuticals and cosmetic industries; current scenario and future approaches. Front Plant Sci 13. https://doi.org/10.3389/fpls.2022.1009395
32. Hayat F, Khanum F, Li J, Iqbal S, Khan U, Javed HU, Razzaq MK, Altaf MA, Peng Y, Ma X, Li C, Tu P, Chen J (2023) Nanoparticles and their potential role in plant adaptation to abiotic stress in horticultural crops: a review. Sci Hortic 321:112285. https://doi.org/10.1016/j.scienta.2023.112285
33. He X, Deng H, Hwang H (2019) The current application of nanotechnology in food and agriculture. J Food Drug Anal 27(1):1–21. https://doi.org/10.1016/j.jfda.2018.12.002
34. Hedayati A, Hosseini B, Palazon J, Maleki R (2020) Improved tropane alkaloid production and changes in gene expression in hairy root cultures of two Hyoscyamus species elicited by silicon dioxide nanoparticles. Plant Physiol Biochem 155:416–428. https://doi.org/10.1016/j.plaphy.2020.07.029
35. Hernández-Hernández H, González-Morales S, Benavides-Mendoza A, Ortega-Ortiz H, Cadenas-Pliego G, Juárez-Maldonado A (2018) Effects of chitosan–PVA and Cu nanoparticles on the growth and antioxidant capacity of tomato under saline stress. Molecules 23(1):178. https://doi.org/10.3390/molecules23010178
36. Hussain S, Ulhassan Z, Brestic M, Zivcak M, Zhou W, Allakhverdiev SI, Yang X, Safdar ME, Yang W, Liu W (2021) Photosynthesis research under climate change. Photosynth Res 150(1):5–19. https://doi.org/10.1007/s11120-021-00861-z
37. Hwan Kim D, Gopal J, Sivanesan I (2017) Nanomaterials in plant tissue culture: the disclosed and undisclosed. RSC Adv 7(58):36492–36505. https://doi.org/10.1039/C7RA07025J
38. Iqbal M, Raja NI, Mashwani Z-U-R, Hussain M, Ejaz M, Yasmeen F (2019) Effect of silver nanoparticles on growth of wheat under heat stress. Iran J Sci Technol Trans A Sci 43(2):387–395. https://doi.org/10.1007/s40995-017-0417-4

39. Katiyar P, Yadu B, Korram J, Satnami ML, Kumar M, Keshavkant S (2020) Titanium nanoparticles attenuates arsenic toxicity by up-regulating expressions of defensive genes in *Vigna radiata* L. J Environ Sci 92:18–27. https://doi.org/10.1016/j.jes.2020.02.013
40. Khalid MF, Iqbal Khan R, Jawaid MZ, Shafqat W, Hussain S, Ahmed T, Rizwan M, Ercisli S, Pop OL, Alina Marc R (2022) Nanoparticles: the plant saviour under abiotic stresses. Nanomaterials 12(21), Article 21. https://doi.org/10.3390/nano12213915
41. Khan MA, Khan T, Mashwani Z-R, Riaz MS, Ullah N, Ali H, Nadhman A (2019) Chapter two - Plant cell nanomaterials interaction: Growth, physiology and secondary metabolism. In: Verma SK, Das AK (eds) Comprehensive analytical chemistry, vol 84. Elsevier, pp 23–54. https://doi.org/10.1016/bs.coac.2019.04.005
42. Khan S, Khan T, Karim S, Zahoor M, Jan T, Khan MA, Nadhman A (2023) Efficient regeneration of shoots and roots in graphene oxide and carbon nanotubes mediated callus cultures: a qualitative and quantitative study. Ind Crops Prod 204:117262. https://doi.org/10.1016/j.indcrop.2023.117262
43. Kohan-Baghkheirati E, Geisler-Lee J (2015) Gene expression, protein function and pathways of *Arabidopsis thaliana* responding to silver nanoparticles in comparison to silver ions, cold, salt, drought, and heat. Nanomaterials 5(2), Article 2. https://doi.org/10.3390/nano5020436
44. Kokina I, Mickeviča I, Jermaļonoka M, Bankovska L, Gerbreders V, Ogurcovs A, Jahundoviča I (2017) Case study of somaclonal variation in resistance genes *mlo* and *Pme3* in flaxseed (*Linum usitatissimum* L.) induced by nanoparticles. Int J Genomics e1676874. https://doi.org/10.1155/2017/1676874
45. Krupa-Małkiewicz M, Oszmiański J, Lachowicz S, Szczepanek M, Jaśkiewicz B, Pachnowska K, Ochmian I (2019) Effect of nanosilver (nAg) on disinfection, growth, and chemical composition of young barley leaves under in vitro conditions. J Integr Agric 18(8):1871–1881. https://doi.org/10.1016/S2095-3119(18)62146-X
46. Kumar A, Gupta K, Dixit S, Mishra K, Srivastava S (2019) A review on positive and negative impacts of nanotechnology in agriculture. Int J Environ Sci Technol 16(4):2175–2184. https://doi.org/10.1007/s13762-018-2119-7
47. Lala S (2021) Nanoparticles as elicitors and harvesters of economically important secondary metabolites in higher plants: a review. IET Nanobiotechnol 15(1):28–57. https://doi.org/10.1049/nbt2.12005
48. Li Y, Zhu N, Liang X, Bai X, Zheng L, Zhao J, Li Y, Zhang Z, Gao Y (2020) Silica nanoparticles alleviate mercury toxicity via immobilization and inactivation of Hg (II) in soybean (*Glycine max*). Environ Sci Nano 7(6):1807–1817. https://doi.org/10.1039/D0EN00091D
49. Long Y, Yang Y, Pan G, Shen Y (2022) New insights into tissue culture plant-regeneration mechanisms. Front Plant Sci 13. https://doi.org/10.3389/fpls.2022.926752
50. Mahajan S, Kadam J, Dhawal P, Barve S, Kakodkar S (2022) Application of silver nanoparticles in in-vitro plant growth and metabolite production: revisiting its scope and feasibility. Plant Cell, Tissue Organ Culture (PCTOC) 150(1):15–39. https://doi.org/10.1007/s11240-022-02249-w
51. Mahmoud LM, Dutt M, Shalan AM, El-Kady ME, El-Boray MS, Shabana YM, Grosser JW (2020) Silicon nanoparticles mitigate oxidative stress of in vitro-derived banana (*Musa acuminata* 'Grand Nain') under simulated water deficit or salinity stress. S Afr J Bot 132:155–163. https://doi.org/10.1016/j.sajb.2020.04.027
52. Malik WA, Mahmood I, Razzaq A, Afzal M, Shah GA, Iqbal A, Zain M, Ditta A, Asad SA, Ahmad I, Mangi N, Ye W (2021) Exploring potential of copper and silver nano particles to establish efficient callogenesis and regeneration system for wheat (*Triticum aestivum* L.). GM Crops Food 12(1):564–585. https://doi.org/10.1080/21645698.2021.1917975
53. Manchanda P, Kaur A, Gosal SS (2018) Somaclonal variation for sugarcane improvement. In: Gosal SS, Wani SH (eds) Biotechnologies of crop improvement, Volume 1: cellular approaches. Springer International Publishing, pp 299–326. https://doi.org/10.1007/978-3-319-78283-6_9
54. Maqbool M, Ishtiaq M, Mazhar MW, Casini R, Mahmoud EA, Elansary HO (2023) Enhancing bioactive metabolite production in *Aerva sanguinolenta* callus cultures through silver nanoparticle and salicylic acid elicitation. Sustainability 15(13), Article 13. https://doi.org/10.3390/su151310395

55. Mazaheri-Tirani M, Dayani S (2020) In vitro effect of zinc oxide nanoparticles on *Nicotiana tabacum* callus compared to ZnO micro particles and zinc sulfate ($ZnSO_4$). Plant Cell, Tissue Organ Culture (PCTOC) 140(2):279–289. https://doi.org/10.1007/s11240-019-01725-0
56. McGehee DL, Alimohammadi M, Khodakovskaya MV (2019) Carbon-based nanomaterials as stimulators of production of pharmaceutically active alkaloids in cell culture of *Catharanthus roseus*. Nanotechnology 30(27):275102. https://doi.org/10.1088/1361-6528/ab1286
57. Mishra D, Tripathi S, Pant S, Chaturvedi P (2022) Nanomaterials and plant tissue culture: developmental path and contradictory facts. In: Gupta S, Chaturvedi P (eds) Commercial scale tissue culture for horticulture and plantation crops. Springer Nature, pp 153–172. https://doi.org/10.1007/978-981-19-0055-6_7
58. Mohapatra B, Chamoli S, Salvi P, Saxena SC (2023) Fostering nanoscience's strategies: a new frontier in sustainable crop improvement for abiotic stress tolerance. Plant Nano Biol 3:100026. https://doi.org/10.1016/j.plana.2023.100026
59. Mosavat N, Golkar P, Yousefifard M, Javed R (2019) Modulation of callus growth and secondary metabolites in different Thymus species and Zataria multiflora micropropagated under ZnO nanoparticles stress. Biotechnol Appl Biochem 66(3):316–322. https://doi.org/10.1002/bab.1727
60. Mustafa G, Komatsu S (2016) Insights into the response of soybean mitochondrial proteins to various sizes of aluminum oxide nanoparticles under flooding stress. J Proteome Res 15(12):4464–4475. https://doi.org/10.1021/acs.jproteome.6b00572
61. Nazir S, Jan H, Zaman G, Khan T, Ashraf H, Meer B, Zia M, Drouet S, Hano C, Abbasi BH (2021) Copper oxide (CuO) and manganese oxide (MnO) nanoparticles induced biomass accumulation, antioxidants biosynthesis and abiotic elicitation of bioactive compounds in callus cultures of *Ocimum basilicum* (Thai basil). Artif Cells Nanomed Biotechnol 49(1):625–633. https://doi.org/10.1080/21691401.2021.1984935
62. Noor W, Lone R, Kamili AN, Husaini AM (2022) Callus induction and regeneration in high-altitude Himalayan rice genotype SR4 via seed explant. Biotechnol Rep 36:e00762. https://doi.org/10.1016/j.btre.2022.e00762
63. Nourozi E, Hosseini B, Maleki R, Abdollahi Mandoulakani B (2021) Inductive effect of titanium dioxide nanoparticles on the anticancer compounds production and expression of rosmarinic acid biosynthesis genes in *Dracocephalum kotschyi* transformed roots. Plant Physiol Biochem 167:934–945. https://doi.org/10.1016/j.plaphy.2021.09.017
64. ONU (2013) World population prospects: 2012 revision. United Nations Department of Economic Social Affairs, New York, NY, USA
65. Pathak A, Haq S, Meena N, Dwivedi P, Kothari SL, Kachhwaha S (2023) Multifaceted role of nanomaterials in modulating in vitro seed germination, plant morphogenesis, metabolism and genetic engineering. Plants 12(17), Article 17. https://doi.org/10.3390/plants12173126
66. Rajan RP, Singh G (2021) A review on application of somaclonal variation in important horticulture crops. Plant Cell Biotechnol Mole Biol 161–175
67. Ranghoo-Sanmukhiya VM (2021) Somaclonal variation and methods used for its detection. In: Siddique I (ed) Propagation and genetic manipulation of plants. Springer, pp 1–18. https://doi.org/10.1007/978-981-15-7736-9_1
68. Safdar M, Kim W, Park S, Gwon Y, Kim Y-O, Kim J (2022) Engineering plants with carbon nanotubes: a sustainable agriculture approach. J Nanobiotechnol 20(1):275. https://doi.org/10.1186/s12951-022-01483-w
69. Sahu M, Maurya S, Jha Z (2023) In vitro selection for drought and salt stress tolerance in rice: An overview. Plant Physiol Rep 28(1):8–33. https://doi.org/10.1007/s40502-023-00713-z
70. Salih AM, Al-Qurainy F, Khan S, Tarroum M, Nadeem M, Shaikhaldein HO, Gaafar A-RZ, Alfarraj NS (2021) Biosynthesis of zinc oxide nanoparticles using *Phoenix dactylifera* and their effect on biomass and phytochemical compounds in *Juniperus procera*. Sci Rep 11(1), Article 1. https://doi.org/10.1038/s41598-021-98607-3
71. Sarabia-Castillo CR, Pérez-Moreno AY, Fernández-Luqueño F (2023) Metal oxide nanoparticles (Tio_2, Zno, and Fe_2O_3) change the functional groups, but not the plant tissue content of common bean plants grown in a greenhouse. Polish J Environ Stud 32(2):1343–1352. https://doi.org/10.15244/pjoes/156777

72. Sarkar MM, Mathur P, Mitsui T, Roy S (2022) A review on functionalized silica nanoparticle amendment on plant growth and development under stress. Plant Growth Regul 98(3):421–437. https://doi.org/10.1007/s10725-022-00891-0
73. Sarraf M, Vishwakarma K, Kumar V, Arif N, Das S, Johnson R, Janeeshma E, Puthur JT, Aliniaeifard S, Chauhan DK, Fujita M, Hasanuzzaman M (2022) Metal/metalloid-based nanomaterials for plant abiotic stress tolerance: an overview of the mechanisms. Plants 11(3), Article 3. https://doi.org/10.3390/plants11030316
74. Selvakesavan RK, Kruszka D, Shakya P, Mondal D, Franklin G (2023) Impact of nanomaterials on plant secondary metabolism. In: Al-Khayri JM, Alnaddaf LM, Jain SM (eds) Nanomaterial interactions with plant cellular mechanisms and macromolecules and agricultural implications. Springer International Publishing, pp 133–170. https://doi.org/10.1007/978-3-031-20878-2_6
75. Shehzad MA, Khan MA, Ali A, Mohammad S, Noureldeen A, Darwish H, Ali A, Ahmad A, Khan T, Khan RS (2021) Interactive effects of zinc oxide nano particles and different light regimes on growth and silymarin biosynthesis in callus cultures of *Silybum marianum* L. Artif Cells Nanomed Biotechnol 49(1):523–535. https://doi.org/10.1080/21691401.2021.1946069
76. Shinde S, Paralikar P, Ingle AP, Rai M (2020) Promotion of seed germination and seedling growth of Zea mays by magnesium hydroxide nanoparticles synthesized by the filtrate from *Aspergillus niger*. Arab J Chem 13(1):3172–3182. https://doi.org/10.1016/j.arabjc.2018.10.001
77. Singh Y, Kumar U, Panigrahi S, Balyan P, Mehla S, Sihag P, Sagwal V, Singh KP, White JC, Dhankher OP (2023) Nanoparticles as novel elicitors in plant tissue culture applications: current status and future outlook. Plant Physiol Biochem 203:108004. https://doi.org/10.1016/j.plaphy.2023.108004
78. Syeed R, Mujib A, Malik MQ, Mamgain J, Ejaz B, Gulzar B, Zafar N (2021) Mass propagation through direct and indirect organogenesis in three species of genus Zephyranthes and ploidy assessment of regenerants through flow cytometry. Mol Biol Rep 48(1):513–526. https://doi.org/10.1007/s11033-020-06083-1
79. Tariq A, Ilyas S, Naz S (2020) Nanotechnology and plant tissue culture. In: Javad S (ed) Nanoagronomy. Springer International Publishing, pp 23–35. https://doi.org/10.1007/978-3-030-41275-3_2
80. Timoteo C de O, Paiva R, dos Reis MV, Claro PIC, da Silva DPC, Marconcini JM, de Oliveira JE (2019) Silver nanoparticles in the micropropagation of Campomanesia rufa (O. Berg) Nied. Plant Cell Tissue Organ Culture (PCTOC) 137(2):359–368. https://doi.org/10.1007/s11240-019-01576-9
81. Tortella G, Rubilar O, Pieretti JC, Fincheira P, de Melo Santana B, Fernández-Baldo MA, Benavides-Mendoza A, Seabra AB (2023) Nanoparticles as a promising strategy to mitigate biotic stress in agriculture. Antibiotics 12(2), Article 2. https://doi.org/10.3390/antibiotics12020338
82. Tung HT, Bao HG, Cuong DM, Ngan HTM, Hien VT, Luan VQ, Vinh BVT, Phuong HTN, Nam NB, Trieu LN, Truong NK, Hoang PND, Nhut DT (2021) Silver nanoparticles as the sterilant in large-scale micropropagation of chrysanthemum. In Vitro Cellular Dev Biol Plant 57(6):897–906. https://doi.org/10.1007/s11627-021-10163-7
83. Türkoğlu A, Haliloğlu K, Demirel F, Aydin M, Çiçek S, Yiğider E, Demirel S, Piekutowska M, Szulc P, Niedbała G (2023) Machine learning analysis of the impact of silver nitrate and silver nanoparticles on wheat (*Triticum aestivum* L.): callus induction, plant regeneration, and DNA methylation. Plants 12(24), Article 24. https://doi.org/10.3390/plants12244151
84. Tymoszuk A (2021) Silver nanoparticles effects on *in vitro* germination, growth, and biochemical activity of tomato, radish, and kale seedlings. Materials 14(18), Article 18. https://doi.org/10.3390/ma14185340
85. Tymoszuk A, Miler N (2019) Silver and gold nanoparticles impact on in vitro adventitious organogenesis in chrysanthemum, gerbera and Cape Primrose. Sci Hortic 257:108766. https://doi.org/10.1016/j.scienta.2019.108766
86. van Dijk M, Morley T, Rau ML, Saghai Y (2021) A meta-analysis of projected global food demand and population at risk of hunger for the period 2010–2050. Nature Food 2(7), Article 7. https://doi.org/10.1038/s43016-021-00322-9

87. Villalobos-López MA, Arroyo-Becerra A, Quintero-Jiménez A, Iturriaga G (2022) Biotechnological advances to improve abiotic stress tolerance in crops. Int J Mole Sci 23(19), Article 19. https://doi.org/10.3390/ijms231912053
88. Vinay JU, Kurjogi M, Uppar A, Vanti G (2023) Chapter 10—Engineered nanomaterials in crop plants temperature and or heat stress management. In: Husen A (ed) Engineered nanomaterials for sustainable agricultural production, soil improvement and stress management. Academic, pp 227–239. https://doi.org/10.1016/B978-0-323-91933-3.00006-4
89. Wang L, Ning C, Pan T, Cai K (2022) Role of silica nanoparticles in abiotic and biotic stress tolerance in plants: a review. Int J Mole Sci 23(4), Article 4. https://doi.org/10.3390/ijms23041947
90. World Population Prospects, the 2012 Revision (2013) [Report]. https://www.un.org/en/development/desa/publications/world-population-prospects-the-2012-revision.html
91. Wu H, Tito N, Giraldo JP (2017) Anionic cerium oxide nanoparticles protect plant photosynthesis from abiotic stress by scavenging reactive oxygen species. ACS Nano 11(11):11283–11297. https://doi.org/10.1021/acsnano.7b05723
92. Yasmeen F (2023) Role of Nanomaterials in improving crop tolerance to abiotic stress. In: Al-Khayri JM, Alnaddaf LM, Jain SM (eds) Nanomaterial interactions with plant cellular mechanisms and macromolecules and agricultural implications. Springer International Publishing, pp 423–453. https://doi.org/10.1007/978-3-031-20878-2_16
93. Yazıcılar B, Böke F, Alaylı A, Nadaroglu H, Gedikli S, Bezirganoglu I (2021) In vitro effects of CaO nanoparticles on Triticale callus exposed to short and long-term salt stress. Plant Cell Rep 40(1):29–42. https://doi.org/10.1007/s00299-020-02613-0
94. Zhao P, Wang C, Zhang S, Zheng L, Li F, Cao C, Cao L, Huang Q (2022) Fungicide-loaded mesoporous silica nanoparticles promote rice seedling growth by regulating amino acid metabolic pathways. J Hazard Mater 425:127892. https://doi.org/10.1016/j.jhazmat.2021.127892

Physiological and Morphological Responses of Crops When Applying Nanostimulant

Nayelli Azucena Sigala-Aguilar, René Juárez-Altamirano, Dulce Flores-Rentería, Andrés Torres-Gómez, and Fabián Fernández-Luqueño

Abstract Nanostimulants play a crucial role in improving plant productivity and promoting agricultural sustainability. The inadequate and excessive use of agrochemicals and fertilizers on soils to increase crop yields has raised alarms regarding shape sustainability. This chapter highlights different cutting-edge technologies with the potential to produce affordable and innocuous food. Therefore, plant-associated soil microorganisms should be considered when new technologies seek better productivity. Microbial communities and plant cells use different uptaken mechanisms and compartmentalization of nanomaterials (NMs), which could change the bioavailability and effect of these NMs on the targeted crop and ecosystems. It has to be considered that when new benefits are expected, unfavorable effects could also be triggered. Therefore, collateral effects should be considered in long-term experiments under land conditions when nanosized stimulants are used.

Keywords Agriculture using nanotechnology · Crop nutrients · Food production · Soil microorganisms · Sustainable agriculture

1 Introduction

In recent years, new technologies have been used to increase crop yields and produce affordable and innocuous food with minimal environmental damage. In this sense, the use of cutting-edge technologies and procedures never reported before has emerged

N. A. Sigala-Aguilar · R. Juárez-Altamirano · D. Flores-Rentería · A. Torres-Gómez · F. Fernández-Luqueño (✉)
Center for Research and Advanced Studies of the National Polytechnic Institute (Cinvestav-Saltillo), Sustainability of Natural Resources and Energy Programs, Ramos Arizpe, Coahuila de Zaragoza, Mexico
e-mail: fabian.fernandez@cinvestav.edu.mx

D. Flores-Rentería
CONACYT-Center for Research and Advanced Studies of the National Polytechnic Institute (Cinvestav-Saltillo), Sustainability of Natural Resources and Energy Programs, Ramos Arizpe, Coahuila de Zaragoza, Mexico

© The Author(s), under exclusive license to Springer Nature Singapore Pte Ltd. 2025
A. Juárez-Maldonado et al. (eds.), *Plant Biostimulation with Nanomaterials*, Smart Nanomaterials Technology, https://doi.org/10.1007/978-981-96-4648-7_6

with diverse nanodevices and methodologies, but also with doubts and concerns regarding the collateral effects of these nanosized materials on soil and human health.

Fortunately, several long-term experiments have been reported with outstanding results, but additional studies are required. Meanwhile, optimism is involved regarding using nanosized materials in agriculture and environmental areas. Different studies have reported evidence regarding the benefits of nanostimulants in different crops [18, 48, 62]. Nanostimulants plants positively regulate the expression of genes and enzymes involved in plants' physiological or metabolic pathways. This cascade triggering is involved in crop growth and development.

The benefits of nanosized materials in agriculture have also been frequently studied. However, the spread of their use worldwide still requires solving some questions regarding nanotechnologies' harmlessness, focusing on soil organisms and other properties regarding the quality and health of the soil. Besides, the effect of nanotechnology on the abundance and diversity of organisms, i.e., biodiversity, has also caught the attention of specialists and governments [5, 23]. Still, no regulations regarding using nanotechnology devices in agriculture and the environment have been stated worldwide.

This chapter aims to discuss the potential of nanostimulants in the physiological and morphological responses of crops to use them to improve the quality and quantity of food produced in the agricultural sector without jeopardizing human and soil health.

2 Differential Effects of Traditional Fertilizers and Nanofertilizers on Plant-Associated Soil Microbes

With the constant and rapid increase in the human population, the demand for agricultural products will continue to rise. The growing need for food and the scarcity of agricultural land will require doubling crop yields, and this pressure on fields necessitates introducing changes to the agricultural system to achieve sustainability. To promote sustainability in agricultural practices, special attention must be paid to the plant rhizosphere, the soil zone closest to the root system supporting crop production [4]. Evaluating specific soil components can help ensure the sustainability of agricultural production. Soil biota components, such as microbial community, abundance, diversity, activity, and stability, are essential soil quality indicators [65].

The soil microbial communities contain groups of organisms with either positive or negative effects on plant productivity. On the positive side, there are hundreds of microbial species, such as *Rhizobium hainanense*, *Parasponia andersonii*, *Bacillus subtilis*, and *Pseudomonas*, as well as some fungal species like *Aspergillus quadrilineatus*, *Aspergillus rugulosus*, *Penicillium corylophilum*, *Cladosporium tenuissimum*, among others. On the negative side, soils may harbor groups of pathogenic microorganisms such as those members of oomycetes, bacteria, and nematodes that can accumulate in agroecosystems, reducing production and requiring strategies for pest and disease management [67].

The soil microbiota plays a crucial role in accelerating the decomposition rate of plant residues to form nutrients easily absorbed by plants for their growth and development. Soil microorganisms such as rhizobial bacteria and mycorrhizal fungi aid nutrient acquisition from insoluble elements. For example, they transform nitrogen into organic and inorganic forms, making it easier for plants to uptook. Similarly, they play a significant role in phosphorus solubilization and act as biocontrol agents in agricultural ecosystems [12]. The soil microorganism populations like arbuscular mycorrhizal fungi (AMF) and beneficial nematodes are highly correlated with crop yield, soil water retention, and nutrient cycling. They play critical roles in plant health by protecting them from various diseases, aiding soil aggregation, and facilitating nutrient transfer [65]. The interaction among microbes, plants, and soil is a vital process that often depends on or is affected by agricultural practices. In many cases, farmers employ techniques such as the use of chemical, organic, or nanosized fertilizers that can have an impact on microorganisms and ecosystems.

Crops require 17 chemical elements named essential nutrients for their growth and development. These essential nutrients are divided into macro and micronutrients. Macronutrients include carbon (C), hydrogen (H), oxygen (O), nitrogen (N), phosphorus (P), potassium (K), calcium (Ca), magnesium (Mg), and sulfur (S). Micronutrients are chlorine (Cl), Boron (B), iron (Fe), manganese (Mn), Copper (Cu), zinc (Zn), nickel (Ni), and Molybdenum (Mo). Besides, at least ten chemical elements have been named beneficial elements for crops, such as sodium (Na), selenium (Se), cobalt (Co), aluminum (Al), silicon (Si), cerium (Ce), iodine (I), lanthanum (La), titanium (Ti), and vanadium (V) [35]. It has to be stated that while beneficial elements are not necessary for all plant species, they influence the absorption, movement, and usage of essential elements. They also aid in synthesizing crucial metabolites and may counteract the harmful effects of some toxic elements or anti-metabolites. However, crop plants often experience nutrient imbalances that negatively impact various metabolic processes.

Fertilizers are any organic or inorganic material of natural or synthetic origin added to fields to supply one or more essential nutrients [2]. Since the early days of agriculture, farmers have used various sources to fertilize soils. In ancient times, reliance was primarily on organic fertilizers such as plant residues, manure, and compost. However, these had the limitation of releasing nutrients into the soil slowly. After several decades, fertilization technologies and crop management have been improved through new and never-before-seen technologies such as nanosized fertilizers named nanofertilizers.

During the past century, there was a rapid increase in the demand for agricultural products, leading to the need to overcome the limitation of organic fertilizers. This led to the introduction of synthetic or conventional fertilizers characterized by one of their main advantages: the exact composition and concentration of nutrients are known and readily released at the time of application, playing an indispensable role in meeting human needs regarding the production of food and raw materials.

Although these advances have helped increase crop production, it has also been discovered that fertilizers bring various adverse environmental effects. For instance, the biogeochemical cycle is often affected by their application. Besides, if the fertilizer is not applied correctly, it can decrease crop productivity. Conversely, if applied excessively, it may remain available in the soil. Plants are estimated not to utilize more than 50% of the chemical fertilizers added, eventually leaching into nearby water streams over time [14]. The excess residues from fertilization lead to the gradual degradation of the soil, altering its salinity and pH levels. Some fertilizers, such as urea and ammonium salt, serve as sources of ammonia, and their migration to nearby ecosystems disrupts vegetation. Acid rain can occur when a portion of the ammonia converts to nitric acid and combines with sulfuric acid. This acid rain causes soil erosion and negatively impacts vegetation [2].

On the other hand, it has been studied that standard fertilization practices with a long-term unbalanced application could affect microbial diversity and abundance. The impact of fertilizers on soil microorganisms can be either beneficial or detrimental depending on the duration, quantity, quality, and method of fertilizer application [56]. For example, Nitrogen-based fertilizers have direct and indirect effects on soil microbial diversity by influencing the physicochemical properties of the soil. These fertilizers acidify the soil, which benefits fungi but negatively influences bacterial abundance. Moreover, the chemical reactions with these fertilizers produce ammonia and carbon dioxide, creating an inhospitable environment that can harm seedlings, seeds, and soil-dwelling organisms [42]. Similar problems have been identified when phosphorus is added. The application of phosphate fertilizers generally leads to a reduction in mycorrhizal fungi. Still, the intensity of this effect depends on the fungal species present in the soil and the amount of available phosphorus [15].

Excessive potassium-based fertilizer can result in calcium deficiencies in plants and the loss of soil structure, leading to reduced soil aeration. Consequently, this diminishes root respiration capacity and promotes the generation of harmful compounds in plants. The reduction in soil aeration and calcium deficiency then leads to a decrease in the number of soil microbes and a subsequent reduction in the supply of organic matter [56]. At the same time, fertilization increases aboveground plant biomass, causing higher transpiration rates, which can decrease water available for soil organisms. It has also been discovered that fertilization may be associated with increased soil pathogens, necessitating biocides and disease-resistant plant varieties to counteract these effects [67]. The increase in soil pathogens can inhibit the positive effects of other organisms and impact soil microbial biodiversity. It has to be shown that, in general, fertilizers enhance microbial catabolic functions. Still, they also bring changes in microbial communities, which can significantly affect soil fertility and productivity.

In recent years, special attention has been given to nanotechnology in pursuing alternatives for more sustainable agricultural practices to restore soil health affected by conventional fertilizers. Nanomaterials combined with traditional methods can have innovative applications in agriculture, ensuring global food security. Using nanostructured systems aims to achieve precision agriculture by supplying the soil with less fertilizer, improving application efficiency, and reducing transportation

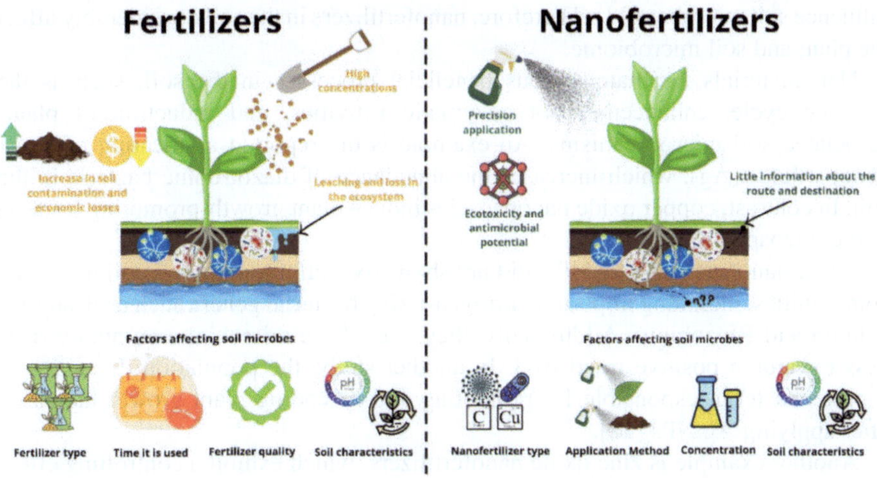

Fig. 1 The main differences between using traditional or nanosized fertilizers for growing crops

costs. This, in turn, would decrease nutrient loss through leaching or runoff and reduce degradation and volatility, thereby enhancing nutrient quality soil fertility and promoting long-term crop productivity [54] (Fig. 1).

To be more precise, nanofertilizers are nutrients encapsulated or coated within nanomaterials to allow controlled release and subsequent slow diffusion into the soil [9]. In agriculture, the use of natural mineral nanoparticles such as zeolite has been leveraged for some years. However, they can now be synthesized by designing them with various physical and chemical properties to meet different needs [28]. Nanofertilizers have been classified into three main categories: (i) nanomaterials used as micronutrients based on zinc, copper, manganese, and iron, which have received particular attention in recent years, (ii) macronutrients of N, P, and K, and (iii) carriers of macronutrients such as nano clays, mesoporous silica, carbon-based nanomaterials, polymeric nanoparticles, etc. [54]. Their application method is also essential, and various methods have been adopted, including soil mixing, foliar application, and seed preparation methods. The choice of nanofertilizer type and its application method depends on the nutritional requirements for plant growth and development [51].

The potential of nanofertilizers in agriculture and their positive effects on crop development are well-known. However, there are knowledge gaps that need to be addressed. The impact of nanofertilizers on beneficial soil microbes is gradually emerging and not yet well understood. Understanding the effects of nanofertilizers on soil microbial communities is crucial as it will largely determine the long-term effects on the plant-associated environment and, consequently, the success of nanofertilizers.

When nanofertilizers are applied, they accumulate in plants over extended periods. Subsequently, plants secrete various exudates into their rhizosphere. In the presence of nanomaterials, these plant exudates and extracellular substances significantly influence soil microbes [28]. Therefore, nanofertilizers in the soil considerably affect the plant and soil microbiome.

Nanomaterials facilitate various beneficial processes in the soil, such as the nitrogen cycle, enhancement of enzymatic activities, and induction of plant-associated soil microorganisms. An example is the reported application of silver nanoparticles (Ag), which increases the abundance of diazotrophic bacteria in the soil. In contrast, copper oxide nanoparticles induce plant growth-promoting bacteria in the rhizosphere [61].

Silver nanoparticles (Ag NPs) did not show toxic effects on native soil microbial communities, including important nitrogen-fixing bacterial genera such as Bradyrhizobium and Rhizobium. Additionally, they altered the microbial community richness towards a positive trend [61]. In another study, the population dynamics of α-proteobacteria responsible for promoting and defending plant growth increased after applying Ag NPs [22].

Another example is zinc oxide nanofertilizers, which exhibit a controlling effect against invasive plant pathogens. Furthermore, metal and oxide nanomaterials are effective against significant phytopathogens such as *Phytophthora, Fusarium, Colletotrichum, Alternaria*, etc. [28].

Despite these advantages, it is also known that nanomaterials have specific ecotoxicological implications. Nanoparticles have been extensively studied for their antimicrobial potential; consequently, nanofertilizers would inevitably adversely affect microbial communities that are beneficial for soil and plants [1].

Soil microbial communities respond differently to the application of various types of nanofertilizers. For example, the direct application of ZnO to the soil at concentrations greater than 2 mg/g, with the highest concentration, showed a toxic effect on the bacterial community. Additionally, these nanoparticles generate reactive oxygen species that could harm microbes [71]. On the other hand, the application of CeO_2 nanoparticles decreased rhizospheric microbial communities [11]. Regarding the application of Cu nanofertilizers in the soil, not only did they inhibit pathogenic microbes, but they also disrupted the diversity of beneficial microbes [43].

Introducing nanofertilizers into the soil ecosystem could impact the microbial community. However, this impact is variable and depends on the nanofertilizer type, nanoparticle size, concentration, application period, and soil properties like pH, ionic strength, organic matter content, etc. [13]. Therefore, the interplay of all these factors will determine the successful future application of nanofertilizers in agriculture. There is still limited information on the route and ultimate fate of introduced nanoparticles in the soil.

Apparent differences emerge when contrasting the use of conventional fertilizers and nanofertilizers. Despite decades of aiding agriculture, conventional fertilizers have low efficiency, leading to environmental losses and incomplete plant utilization. It has caused environmental pollution and economic losses. In contrast, nanofertilizers, when used in smaller quantities, exhibit higher efficiency and are better utilized by crops. Regarding their effects on plant-associated soil microbes, both resources have positive and negative effects. However, nanofertilizers, while still being studied to improve their application, for example, by determining concentrations that do not affect microbial communities, represent an advancement. Nanofertilizers can thus positively contribute to the agricultural sector by enhancing crop production, ensuring economic stability, and promoting ecological sustainability.

3 Uptake Mechanisms and Compartmentalization of ENMs in Microbial Cells

In the soil ecosystem, ENMs from various sources, such as nanofertilizers, undergo transformation, retention, and degradation processes through physical, chemical, and biological action [28]. Depending on the physical, chemical, and biological characteristics of the soil, as well as the type of ENMs and properties, the ENMs can be homogeneously or heterogeneously agglomerated with particles of the exact nature and organic compounds present in the soil or the soluble ENMs are ionized in the soil solution [63]. When ENMs come into direct contact with the soil ecosystem, ENMs adhere molecules to their structure through intramolecular forces. The layer of molecules attached to the surface of the ENMs is known as the corona, and the corona's composition is a function of the affinity between the interfacial charges of the ENMs and the compounds present [19]. Furthermore, the size and composition of the corona are important since the interfacial charges of the corona interfere with the absorption and reactivity of ENMs in microbial cells [52]. The negative charges of the membrane components of microorganisms adhere to cationic ENMs through attraction between the charges [53]. Therefore, the physicochemical properties of ENMs, such as nature, size, Z potential, functionalization, morphology, and solubility, are some of the factors that interfere with the absorption and reactivity of ENMs in microbial cells [1, 6].

The mechanisms of ENMs absorption in microbial cells are still under investigation. However, the evidence suggests that ENMs can be absorbed and degraded by microbiological action [63]. Endocytosis is a biological absorption mechanism that cells use to internalize extracellular particles. In endocytosis, microbial cells can selectively or not capture some ENMs through the cell membrane [50]. In absorption, the cell membrane receptors of microorganisms perceive the presence of ENMs,this perception causes the formation of endocytic vesicles, which surround the ENMs, internalizing ENMs in the nucleus of the microbial cell (Fig. 2a) [32].

Pinocytosis is another process of internalizing ENMs in cells; in this process, the cells capture fluids, along with nanoparticles or dissolved particles. However, the mechanisms of ENM absorption by endocytosis are mostly related to and studied in eukaryotic cells of higher organisms such as plants and mammals [32]. In microbial cells, pinocytosis-like uptake mechanisms may be how soil microorganisms could absorb ENM ions dissolved in the soil solution, such as iron (Fe) and zinc (Zn) nanoparticles. Soil microorganisms such as fungi and bacteria absorb extracellular

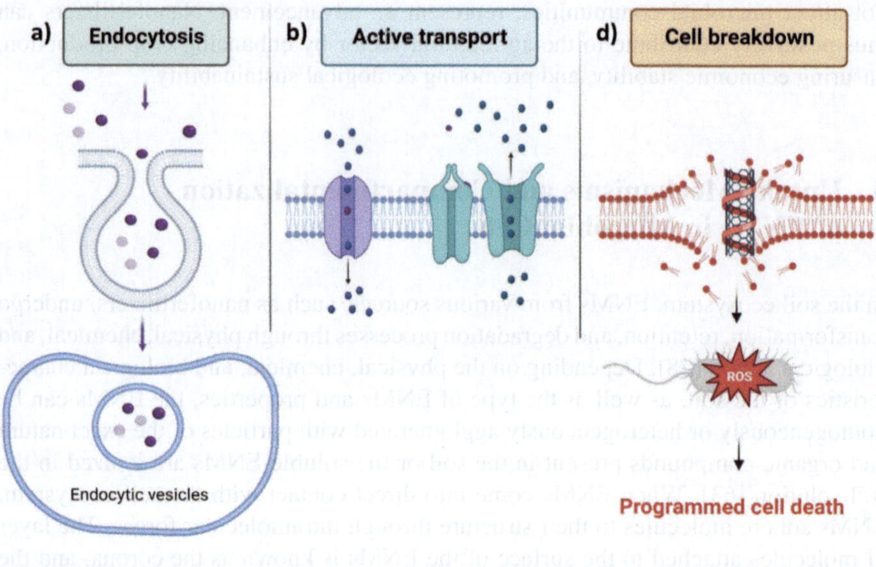

Fig. 2 Uptake mechanisms and compartmentalization of ENMs in microbial cells. (**a**) capture of ENMs by endocytosis; (**b**) passive absorption of ENMs in microbial cells by proton gradient; (**c**) ENMs entry into microbial cells by creating pores

particles used for energy and nutrition. Additionally, microbial cells release extracellular enzymes to decompose complex compounds present in the soil, such as ENMs [41].

On the other hand, the absorption process by endocytosis involves physicochemical events, such as the interfacial interaction between the surface charges of ENMs and compounds of the cell membrane, as well as the deformation of the membrane for the formation of vesicles [50]. These events can trigger a series of intracellular and microbial community responses, ranging from stimulation to toxicity [28]. The responses will depend mainly on the type of ENMs, physicochemical properties of ENMs, concentration, exposure time, and physicochemical and biological properties of the environment where ENMs are dispersed [68]. For example, a study reported that soils incubated for 30 days with carbon-based nanomaterials increased the composition of microbial populations of the Alpha class (belonging to the *Phylum Proteobacteria*),however, after 360 days of incubation, the carbon nanotubes (CNTs) altered the Beta diversity. (also belonging to the *Phylum Proteobacteria*) of the soil microbial community [69].

Depending on the type and size, ENMs can be absorbed into microbial cells by active transport or by creating new pores. ENMs can even translocate into the membrane and periplasm of microbial cells [6]. Soil microorganisms use active transport systems through proton pumps, creating a proton gradient that drives ions into the cell (Fig. 2b) [53]. In this sense, absorption by active transport is probably another capture mechanism by which microbial cells absorb ions from water-soluble ENMs.

Regardless of the absorption mechanism, the perception and absorption of ENMs in microbial cells cause physicochemical changes in the lipid and protein composition of the cell membrane of microorganisms [1]. At the same time, the perception and absorption of ENMs in microbial cells cause oxidative stress in microbial cells, inducing the production of reactive oxygen species (ROS). ROS activates redox pathways, which can oxidize subcellular molecules such as DNA, lipids, and proteins [53]. The expansion of ROS plays a key role in intracellular signaling that leads to the activation of programmed cell death, causing membrane rupture and the release of cytoplasmic material [3].

Due to their shape and hydrophobicity, some ENMs, such as CNTs, can cause physicochemical alterations in microbial cells by penetrating the cell wall of microorganisms through the creation of new pores (Fig. 2c) [52, 53]. Nevertheless, electron and transmission microscopy images showed that the absorption of ENM depended mainly on the physicochemical properties of the ENM and microbial species. In one study, 0.5 mg ml^{-1} of multiwalled carbon nanotubes (MWCNTs) were not taken up by *Azotobacter chroococcum* cells,the MWCNTs remained on the cell surface. This same study reported that applying MWCNTs caused oxidative stress in nitrogen-fixing bacteria without exceeding cellular homeostasis; therefore, no damage was observed in the activity of the nitrogenase enzyme and nitrogen fixation in the soil [55]. In another, titanium oxide (TiO_2) nanoparticles were accumulated in the mycelium of *Aspergillus niger* fungi [60]. A study reported that silver (Ag) nanoparticles accumulated in the cell wall of the phytopathological fungus *A. niger*, accelerating the production of ROS, which inhibited the growth of the hyphae and mycelium of the fungus [52].

The release of ENMs into soil from various sources is a continuous process in ecosystems. Therefore, it is crucial to investigate and understand the interfacial interactions in the soil ecosystem. As mentioned above, the mechanisms by which soil microorganisms interact and absorb ENMs is an area of science still under investigation. However, soil microorganisms are in multicellular colonies in the soil. Therefore, it is of utmost importance to evaluate the impact of ENMs on soil microorganisms in more real conditions since the disproportionate and unregulated use of ENMs for agricultural purposes could directly affect biogeochemical processes and, thus, ecosystem services.

4 Effects of ENMs on the Role of Microbes in Soil Health

Soil is a complex ecosystem providing vast services and has been the foundation of terrestrial and agricultural ecosystems. The ability of soil to operate and offer ecosystem services is characterized by soil health [38]. Soil health is the soil's continued capacity to function as a vital living ecosystem that sustains plants, animals, and humans. Soil health connects agricultural management practices and soil science to policy, stakeholder needs, and sustainable supply chain management [29, 38].

The four principal ecosystem services soils provide sustainable plant production, water-quality control, human health advancement, and climate-change mitigation. Soil microorganisms directly influence soil ecosystem processes, especially the decomposition of soil organic matter and the cycling of nutrients. Soil microorganisms generate polysaccharides that function as binding agents in soil, contributing to the maintenance of soil structure, enhanced aeration, increased water retention, and the prevention of soil crusting and compaction. Consequently, soil microbes facilitate optimal plant production, so it is essential to manage microbial abundance, activity, community composition, and specific functions [38]. Strategies to mitigate climate change, such as carbon sequestration in the soil as organic matter, can be realized by promoting aggregation and controlling microbial mineralization to carbon dioxide, ultimately fostering plant growth [36]. Soil microbes are used as indicators of soil health, providing a comprehensive measure due to their rapid responses to changes in the surrounding environment and rapid adaptation to environmental stress conditions [59].

The utilization of Engineered Nanomaterials (ENMs) across diverse industries is experiencing a rapid surge. Consequently, an escalating concentration of these nanomaterials is permeating different environments, including soil. The intentional or unintentional release of these ENMs into various environments occurs during their preparation, utilization, and disposal processes [44]. Current estimates suggest that thousands of commercially available nanomaterial-based products are in the market, which is anticipated to grow [25]. The most prevalent ENMs, considering their element compositions, can be categorized as inorganic (e.g., metallic or metal oxides such as Au, Ag, Al_2O_3, ZnO, TiO_2, CuO, FeO_2, SiO_2, among others), carbon-based nanomaterials (e.g., carbon nanotubes, graphene, fullerenes, among others), organic (e.g., linear, branched, or coordination polymers), nonmetallic (e.g., B, N, P and S), and hybrid nanomaterial (e.g., metal–organic or nanocomposites) [72].

Numerous ENMs are widely acknowledged for their beneficial use as additives in agriculture, including applications such as nano-fertilizers, nano-pesticides, and nano-nutrition, aimed at enhancing crop yields. Nevertheless, the excessive migration of these ENMs from contaminated soils and water during irrigation poses various adverse effects on plant growth. These effects manifest as delays in germination and reduced production of biomass, roots, and leaves, ultimately impacting crop quality through plant toxicity [34]. It is crucial to comprehend the dynamics and transformation mechanisms of ENMs within natural ecosystems to precisely evaluate the consequences of their secure utilization for enhancing the sustainable development of the agricultural sector.

4.1 *Release Routes and Fate of Engineered Nanomaterials (ENMs) in Soil*

While there is consensus that soil serves as a significant sink for ENMs [17, 25, 31, 33, 37], the ecological consequences of their accumulation are still subject to intense research. It has been calculated that around 60–90% of ENMs are directed to landfill sites, where they interact with soil. Additionally, a significant proportion of ENMs, ranging from 10 to 25%, is directly disposed of into the soil. The most predominant ENMs released into diverse environments include SiO_2, TiO_2, ZnO, FeO_2, and Al_2O_3 [31]. ENMs such as TiO_2 also enter the soil through irrigation with sewage treatment effluents that contain these nanomaterials. A study indicates that wastewater irrigation, likely containing ENMs, significantly raises nanomaterial concentrations in soil. For instance, approximately 90 μg of TiO_2 per kilogram of soil is detected annually [21]. The same study calculates that approximately 0.3–1.3 μg/kg of TiO_2 reaches European soil each year. In comparison, around 0.05 and 0.09 μg of ZnO per kilogram of soil is deposited in the soil in the USA and Europe, respectively.

Nanopesticides, nanofertilizers, hydroponic solutions, and seed treatments have various applications in soil and agro-environments, potentially introducing new nanoparticle (NP) release pathways. For example, Cu-ENM is speculated to result in 95% of the metal eventually ending up in aquatic sediments and soils at concentrations of 500 μg/kg [30]. Furthermore, the bioavailability of NPs, to obtain the benefits they were added in the agricultural context, primarily depends on their transformations in the soil, with dissolved forms presenting higher environmental risk and bioavailability [37]. Therefore, accurately and promptly assessing NP bioavailability in soil is a critical concern.

Upon release into the environment, the distribution and behavior of ENMs are influenced by both their inherent properties (particle size, surface coating, surface area, morphology, zeta potential, colloidal stability) and the prevailing environmental conditions (temperature, soil texture, pH, porosity, composition, organic and inorganic colloids, ionic strength, multiple dynamic interactions) [17]. In this sense, ENMs can acquire a coating of organic substances, such as humic or fulvic acids found in the soil environment, causing an increase in particle size. This soil-derived coating induces a shift in the zeta potential of the particles from positive to negative. Sometimes, clay particles, or the soil organic matter, dissolve in an ionic form, leading to heteroaggregation. Consequently, this process reduces the available surface area for chemical interactions [33].

4.2 *Toxicity and Ecological Implications of ENMs on Soil*

The toxicity and ecological consequences of ENMs on soil microorganisms have been the subject of several studies. ENMs can impact the soil microbiome, including microorganisms and plants, and their toxicity is based on various physicochemical properties such as size, morphology, and chemical characteristics [31]. While the precise mechanisms of toxicity of ENMs on microbial communities remain incompletely understood, potential mechanisms include cell membrane damage, protein oxidation, genotoxicity, interaction with the respiratory chain and local proton depletion, and production of reactive oxygen species (ROS) or apoptosis [26, 45].

Some ENMs have been shown to reduce soil microbial communities, and their impact on ecologically relevant microorganisms has been a research focus [26]. The rate of bacterial growth and the capability of bacteria to produce extracellular polysaccharides and their enzyme activity can be affected by ENMs [31]. For instance, soil from pasturelands added with relatively low concentration (50 mg kg^{-1} soil) of silver nanoparticles (AgNPs) has a significant negative impact on total soil dehydrogenase and urease activity, also leading to substantial shifts in bacterial, archaeal, and fungal community composition. However, due to the current limited understanding of the ecological roles of bacterial and fungal communities affected by AgNPs, evaluating the significance of their decline in AgNP-amended soil is still challenging [49]. Additionally, soil samples treated with AgNPs negatively impact the enzymatic activities of crucial enzymes such as acid phosphatase, β-glucosaminidase, β-glucosidase, and arylsulfatase. These enzymes are critical in carbon and other nutrient cycling [16].

Similarly, single-walled carbon nanotubes (SWs) negatively impact soil bacterial abundance, diversity, and composition [69]. However, many phyla demonstrated resilience and returned to their initial relative abundance within 56 days, indicating a high resilience of the soil bacterial community. Contrarily, MWCNTs shower less impact in the bacterial community, with a differential impact depending on the MWCNTs outer diameters (10, 30, and 50 nm). These differences likely stem from the materials' unique physical structure and surface composition, influencing their soil bioavailability [69]. In this sense, the natural organic matter in the environment and the sorption of ENMs to biosolids may influence their properties and bioavailability [26].

Furthermore, the impact of ENMs on plant growth is not solely determined by toxicity or direct uptake but is mediated by their interaction with microorganisms and soil physicochemical properties. Plants generally establish complex symbiotic relationships with microbes in the rhizosphere, including nitrogen-fixing and beneficial bacteria. The indirect effects of ENM on these species, as highlighted by Gardea-Torresdey et al. [20], can disrupt microbial populations, potentially leading to nitrogen starvation and resulting in stunted plant growth. The diverse factors influencing the destiny and behavior of ENMs could account for the contradictory results in the literature, including variations in the plant species, geographical regions, and the measurement of endpoints.

While there is a growing body of research on the toxicity of ENMs to soil microorganisms, there are still knowledge gaps in understanding nanotoxicity to bacteria, including the role of physicochemical properties of ENMs in toxicity responses [40]. Proper assessment of ENMs' accumulation in the environment, particularly in the soil and plants, and their threat to the microbiomes is considered highly desirable [31].

Summarizing, soil health is integral to the functioning of terrestrial and agricultural ecosystems, connecting agricultural management practices, soil science, and sustainable supply chain management. The growing use of ENMs across industries poses challenges to soil ecosystems, specifically for soil microorganisms. The beneficial use of ENMs in agriculture, such as nanopesticides and nanofertilizers, has mainly been recognized, but their excessive migration into soils can negatively impact plant growth. Understanding the fate and behavior of ENMs in the soil is crucial for assessing their environmental impact accurately. While research on the toxicity of ENMs to soil microorganisms is expanding, there are still knowledge gaps, particularly regarding the role of ENMs' physicochemical properties in the responses of the soil microbial communities and the effect on the bioavailability of the ENMs. A thorough assessment of ENMs' accumulation in the environment, especially in soil and plants, and their implications for microbiomes is essential for addressing potential environmental risks.

5 Approaches for Minimizing the Adverse Effects of ENMs on Plant-Associated Soil Microbes

Sustainability and innovation are fundamental to any production system for providing food to the population; nanotechnology in biofertilizer systems can be a promising technology; however, it requires extensive field testing because the toxicity to the soil microbe's communities is not verified. Nanoencapsulation can be a way to protect microorganisms and slow-release nutrients. Several studies demonstrate that this combination of biofertilizers with nanotechnologies can improve and increment soil microbial populations and not harm them [58].

Biostimulation of an organism is a modification in the metabolic process that results in better use or efficient function, more significant growth, and yield. The stimulation depends on the organism's structure, the interaction between the external impulses, and the stimulant source. ENMs (engineering nanomaterials) can modify the composition and improve food crops' quality and stress tolerance. For example, some NPs, like Cu sources, become very toxic but can benefit crops in small amounts [27]. When the NMs are manufactured via biosynthesis or eco-friendly methods, they could benefit soil microorganisms, help avoid harmful chemicals, and help create a healthy environment. Studies demonstrate that these biogenic methods are better than traditional ones [47].

Excessive application of metal-containing agents leads to their accumulation in the soil, which affects non-target soil organisms too; for example, Cu Cu-containing NMs severely affected non-target soil communities in gram-positive ones. This toxicity occurs when the NMs are applied in the top horizon, depending on the soil type and the major or minor affectation. Metal and metal-oxide NMs can affect the non-target soil microbial community's structural and functional microbial diversity [5].

ENMs and nanotechnology have been used for many remediation techniques and innovative strategies in pollutant removal and resource reduction. The anthropogenic introduction of these materials has not been regulated. The consequences on soil microorganisms influence growth and distort metabolism, showing adverse effects on some macro-organisms like snails, earthworms, larvae, and insects. These materials may become the food chain for humans. Studies showed cellular and genetic impact, biodiversity, enzyme activity, and community structure. CNTs help inhibit the mineralization and mobilization of pollutants, reducing the availability of PAHs to plant and soil microorganisms [57].

CNTs can be degraded by soil microbes in the presence of carbon sources via co-metabolism; however, microbes in natural soils need to evolve and adapt to the manufactured carbon materials. The significant amount of these materials used in several industries becomes an issue to the soil because it acts like a sink of these materials, and in agriculture, too, MWCNTs can form C = O and COOH groups on the outer walls and increase the oxygen species [10]. In others, studies related to microorganisms and biofilms on surfaces demonstrate that MWCNTs affect the metabolic activities of bacteria. The most damaging effect of MWCNTs was on Gram-negative bacteria [46].

Combining graphene oxide (GO) with nano sulfur benefits some soil microorganisms; a tremendous positive effect on soil respiration was observed when GO was used alone and in combination with high nano sulfur (0.5 g Kg^{-1}). However, with a lower concentration (0.05 g Kg^{-1}), some soil respiration was negatively affected. Elemental Nanos would probably control the GO-mediated effect of organic carbon oxidation [24].

Many ENMs can be transformed during wastewater treatment and ultimately enter terrestrial ecosystems. A novel method to determine how the soil will be affected by exposure to these materials is the measure of O_2 consumption by microbes. Studies showed that metal NMs-induced species-specific respiration responses directly affect soil microorganisms [39].

The source that ENMs enter the environment, size, coexistence with pollutants, structure, and surface modification may cause changes in the structure and function of microbial communities in soil. Nano pesticides and nano fertilizers are widely used in agricultural products. Soil microorganisms participate in almost all biochemical reactions, and they are related to soil health and quality; the effects of these ENMs over them depend on doses, particulate state, ion release, and shape. According to Sun and Collaborates [64], TiO_2 NMs affect biomass in clay soil,ZnO NMs inhibit some *Azotobacter* bacteria and reduce the abundance of functional bacteria; Non-metal NMs lowered soil enzyme activity in the short term, CNTs affect the structure and function of soil microorganisms.

Molybdenum (Mo) NMs cause serious toxic effects on plants; however, they help microorganisms fix N_2; this element source NMs affects the principal structure of the microbial community that colonizes roots in legumes and free life in the rhizosphere [73]. Mo, combined with sulfur in MoS_2 nanosheets via the soil route, perturbed soil microorganisms' nutrient uptake and toxicity. Mo ions increase with time in soil, and the dissolution is even greater than water dissolution in soil. These NMs change the structure of soil microbial communities at the phylum level after high concentrations, decreasing soil fertility [7].

Toxic effects depend on the physicochemical properties of ENMs themselves, the exposure concentration, and duration in the soil. For example, studies with metal NMs have poisonous effects on soil enzyme activity and decreased soil microbial abundance, diversity, and biomass; metal and non-metal induced the generation of reactive oxygen species to cause membrane damage and harm microorganisms. Figure 3 shows the possible membrane damage of some microorganisms through the cell membrane [70].

Some NMs from Ag sources are toxic to bacteria, attributed to Ag's interaction with thiol groups of membrane-associated proteins, including respiratory enzymes. The interactions alter the structure of the membrane, destabilize the proton gradient, and promote uncoupling of the respiratory chain from oxidative phosphorylation. Some microorganisms can convert toxic heavy metals into non-toxic species along with the precipitation of metal clusters in dimensions of nano range and defined shape [66].

Another of the most common NMs is from Zn sources. Nanosized fertilizer is more efficient than bulk-size fertilizer. Still, rhizosphere microorganisms decrease the abundance of some bacteria more than bulk ones, and the nano Zn in high doses can impact the structure and function of rhizobacteria. Enhanced carbohydrate metabolism promotes the turnover rate of available carbon by minerals [8].

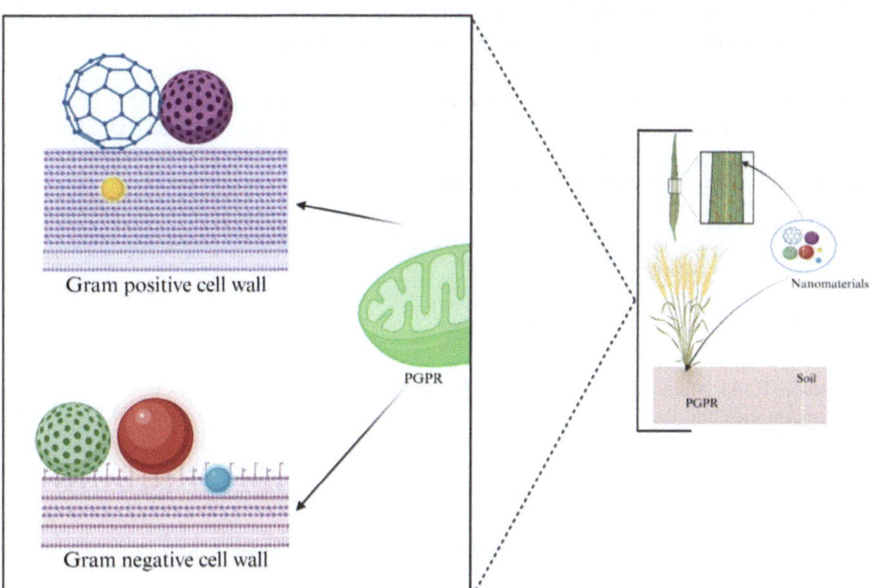

Fig. 3 ENMs can damage the microorganism's cell membrane, depending on the material's shape, structure, and size. The difference in cell membranes can be a factor of damage

The use of nanotechnology in biofertilizer systems has the potential to enhance sustainability and innovation in food production. However, it is crucial to conduct extensive field testing to determine the potential toxicity of nanomaterials to soil microbe communities. Although developing eco-friendly nanomaterial synthesis methods shows promise in benefiting soil microorganisms and creating a healthier environment, humans' unregulated introduction of nanomaterials can be risky. Thorough investigations are necessary to determine the long-term implications of nanosized fertilizers, especially those derived from metal and metal oxide sources.

6 Conclusions

Using nanosized materials in agriculture has several advantages when growing crops in different soil conditions and environments. Nanostimulants improve crop yields and physiological and morphological characteristics when crops are amended with them. Besides, nanostimulants could preserve the diversity and abundance of soil microbial communities. Despite the above, additional studies are required to guarantee the environmental equilibrium when external nanomaterials are added to the agricultural environment. Applying nanotechnological products in soils or plants to increase productivity could have collateral and adverse effects. Therefore, holistic studies for extended periods should be required before accepting that nanosized products are innocuous. Technicians, scientists, and decision-makers should also pay special attention to local or international regulations.

Acknowledgements This research was founded by the projects' Ciencia Básica SEP-CONACyT-151881', 'FONCYT-COAHUILA COAH-2019-C13-C006', and 'FONCYT-COAHUILA COAH-2021-C15-C095', by the Sustainability of Natural Resources and Energy Programs (Cinvestav-Saltillo), and by Cinvestav Irapuato and Cinvestav Zacatenco.

References

1. Ameen F, Alsamhary K, Alabdullatif JA, ALNadhari S (2021) A review on metal-based nanoparticles and their toxicity to beneficial soil bacteria and fungi. Ecotoxicol Environ Saf 213:112027
2. Ashitha A, Rakhimol K, Mathew J (2021) Fate of the conventional fertilizers in environment. In: Controlled release fertilizers for sustainable agricultura. Elsevier, pp 25–39
3. Bapat MS, Singh H, Shukla SK, Singh PP, Vo DVN, Yadav A, Goyal A, Sharma A, Kumar D (2022) Evaluating green silver nanoparticles as prospective biopesticides: an environmental standpoint. Chemosphere 286:13176
4. Berendsen R, Pieterse CM, Bakker PA (2012) The rhizosphere microbiome and plant health. Trends Plant Sci 17(8):478–486

5. Borymski S, Markowicz A, Nowak A, Matus K, Dulski M, Sułowicz S (2023) Copper-oxide nanoparticles exert persistent changes in the structural and functional microbial diversity: a 60-day mesocosm study of zinc-oxide and copper-oxide nanoparticles in the soil-microorganism-nanoparticle system. Micro Biol Res 127395
6. Brandelli A (2020) The interaction of nanostructured antimicrobials with biological systems: cellular uptake, trafficking and potential toxicity. Food Sci Hum Wellness 9:8–20
7. Chen C, Zhao Y-Y, Wang D, Ren Y-H, Liu H-L, Tian Y et al (2023) Effects of nanoscale zinc oxide treatment on growth, rhizosphere microbiota, and metabolism of Aconitum carmichaelii. Peer J Hubs 11:e16177
8. Chen S, Kang Z, Peralta-Videa JR, Zhao L (2023) Environmental implication of MoS_2 nanosheets: effects on maize plant growth and soil microorganisms. Sci Total Environ 860:160362
9. Chhipa H (2017) Nanofertilizers and nanopesticides for agriculture. Environ Chem Lett 15:15–22
10. Chouhan RS, Qureshi A, Yagci B, Gülgün MA, Ozguz V, Niazi JH (2016) Biotransformation of multiwalled carbon nanotubes mediated by nanomaterial resistant soil bacteria. Chem Eng J 298:1–9
11. Dai Y, Chen F, Yue L, Li T, Jiang Z, Xu Z, Wang Z, Xing B (2020) Uptake, transport, and transformation of CeO2 nanoparticles by strawberry and their impact on the rhizosphere bacterial community. ACS Sustain Chem Eng 8(12):4792–4800
12. Das P, Singh KR, Nagpure G, Mansoori A, Singh RP, Ghazi IA, Kumar A, Singh J (2022) Plant-soil-microbes: a tripartite interaction for nutrient acquisition and better plant growth for sustainable agricultural practices. Environ Res 214:113821
13. Dimkpa CO (2018) Soil properties influence the response of terrestrial plants to metallic nanoparticles exposure. Curr Opin Environ Sci Health 6:1–8
14. Dincă LC, Grenni P, Onet C, Onet A (2022) Fertilization and soil microbial community: a review. Appl Sci 12(3):1198
15. Du-Preez C, Van Huyssteen C, Kotzé E, Van Tol J (2020) Ecosystem services in sustainable food systems: operational definition, concepts, and applications. In: The role of ecosystem services in sustainable food systems. Elsevier, pp 17–42
16. Eivazi F, Afrasiabi Z, Jose E (2018) Effects of silver nanoparticles on the activities of soil enzymes involved in carbon and nutrient cycling. Pedosphere 28:209–214
17. El-Kady MM, Ansari I, Arora C, Rai N, Soni S, Verma DK, Singh P, Mahmoud AED (2023) Nanomaterials: a comprehensive review of applications, toxicity, impact, and fate to environment. J Mol Liq 370:121046
18. Fan GJ, Xiao QL, Xia YL, Feng H, Ma XZ, Cai L, Sun XC (2024) Antimicrobial mechanisms of ZnO nanoparticles to phytopathogen Pseudomonas syringae: damage of cell envelope, suppression of metabolism, biofilm and motility, and stimulation of stomatal immunity on host plant. Pest Biochem Physiol 194:105455
19. Francia V, Yang K, Deville S, Reker-Smit C, Nelissen I, Salvati A (2019) Corona composition can affect the mechanisms cells use to internalize nanoparticles. ACS Nano 13:11107–11121
20. Gardea-Torresdey JL, Rico CM, White JC (2014) Trophic transfer, transformation, and impact of engineered nanomaterials in terrestrial environments. Environ Sci Technol 48:2526–2540
21. Gottschalk F, Sonderer T, Scholz RW, Nowack B (2009) Modeled environmental concentrations of engineered nanomaterials (TiO_2, ZnO, Ag, CNT, Fullerenes) for different regions. Environ Sci Technol 43:9216–9222
22. Grün AL, Straskraba S, Schulz S, Schloter M, Emmerling C (2018) Long-term effects of environmentally relevant concentrations of silver nanoparticles on microbial biomass, enzyme activity, and functional genes involved in the nitrogen cycle of loamy soil. J Environ Sci 69:12–22
23. Guo J, Fu Q, Tang MS, Bai JR, Liu RY, Zhang HY, Siddique KHM, Mao H (2024) Fulvic acid modified ZnO nanoparticles improve nanoparticle stability, mung bean growth, grain zinc content, and soil biodiversity. Sci Total Environ 913:169840

24. Hammerschmiedt T, Holatko J, Zelinka R, Kintl A, Skarpa P, Bytesnikova Z et al (2023) The combined effect of graphene oxide and elemental nano-sulfur on soil biological properties and lettuce plant biomass. Front Plant Sci 14:1057133
25. Hansen SF, Hansen OFH, Nielsen MB (2020) Advances and challenges towards consumerization of nanomaterials. Nat Nanotechnol 15:964–965
26. Hegde K, Brar SK, Verma M, Surampalli RY (2016) Current understandings of toxicity, risks and regulations of engineered nanoparticles with respect to environmental microorganisms. Nanotechnol Environ Eng 1:5
27. Juárez-Maldonado A, Ortega-Ortíz H, Morales-Díaz AB, González-Morales S, Morelos-Moreno Á, Cabrera-De la Fuente M et al (2019) Nanoparticles and nanomaterials as plant biostimulants. Int J Mol Sci 20(1):162
28. Kalwani M, Chakdar H, Srivastava A, Pabbi S, Shukla P (2022) Effects of nanofertilizers on soil and plant-associated microbial communities: emerging trends and perspectives. Chemosphere 287:132107
29. Karlen DL, Veum KS, Sudduth KA, Obrycki JF, Nunes MR (2019) Soil health assessment: past accomplishments, current activities, and future opportunities. Soil Till Res 195:104365
30. Keller AA, Adeleye AS, Conway JR, Garner KL, Zhao L, Cherr GN, Hong J, Gardea-Torresdey JL, Godwin HA, Hanna S, Ji Z, Kaweeteerawat C, Lin S, Lenihan HS, Miller RJ, Nel AE, Peralta-Videa JR, Walker SL, Taylor AA, Torres-Duarte C, Zink JI, Zuverza-Mena N (2017) Comparative environmental fate and toxicity of copper nanomaterials. NanoImpact 7:28–40
31. Khan ST, Adil SF, Shaik MR, Alkhathlan HZ, Khan M, Khan M (2021) Engineered nanomaterials in soil: their impact on soil microbiome and plant health. Plants (Basel) 11
32. Khan S, Mansoor S, Rafi Z, Kumari B, Shoaib A, Saeed M, Alshehri S, Ghoneim MM, Rahamathulla M, Hani U, Shakeel F (2022) A review on nanotechnology: properties, applications, and mechanistic insights of cellular uptake mechanisms. J Mol Liq 348:118008
33. Kokilavani S, Janani B, Balasurya S, Khan SS (2023) Accumulation of engineered nanomaterials in soil, water, and air. In: Fernandez-Luqueno F, Patra JK (eds) Agricultural and environmental nanotechnology: novel technologies and their ecological impact. Springer Nature Singapore, Singapore, pp 551–582
34. Kumar A, Gupta K, Dixit S, Mishra K, Srivastava S (2019) A review on positive and negative impacts of nanotechnology in agriculture. Int J Environ Sci Technol 16:2175–2184
35. Kumar S, Kumar S, Mohapatra T (2021) Interaction between macro-and micro-nutrients in plants. Front Plant Sci 12:665583
36. Lal R (2004) Soil carbon sequestration impacts on global climate change and food security. Science 304:1623
37. Lead JR, Batley GE, Alvarez PJJ, Croteau M-N, Handy RD, McLaughlin MJ, Judy JD, Schirmer K (2018) Nanomaterials in the environment: behavior, fate, bioavailability, and effects—an updated review. Environ Toxicol Chem 37:2029–2063
38. Lehmann J, Bossio DA, Kögel-Knabner I, Rillig MC (2020) The concept and future prospects of soil health. Nat Rev Earth Environ 1:544–553
39. Lewis RW, Unrine J, Bertsch PM, McNear DH (2017) Silver engineered nanomaterials and ions elicit species-specific O_2 consumption responses in plant growth promoting rhizobacteria. Biointerphases 12(5)
40. Lewis RW, Bertsch PM, McNear DH (2019) Nanotoxicity of engineered nanomaterials (ENMs) to environmentally relevant beneficial soil bacteria—a critical review. Nanotoxicology 13:392–428
41. Lin J, Ma K, Chen H, Chen Z, Xing B (2022) Influence of different types of nanomaterials on soil enzyme activity: a global meta-analysis. Nano Today 42:101345
42. Liu CW, Sung Y, Chen BC, Lai HY (2014) Effects of nitrogen fertilizers on the growth and nitrate content of lettuce (Lactuca sativa L.). Int J Environ Res Public Health 11(4):4427–4440
43. Lofts S, Criel P, Janssen CR, Lock K, McGrath SP, Oorts K, Rooney CP, Smolders E, Spurgeon DJ, Svendsen C (2013) Modelling the effects of copper on soil organisms and processes using the free ion approach: towards a multi-species toxicity model. Environ Pollut 178:244–253

44. Loureiro S, Tourinho PS, Cornelis G, Van Den Brink NW, Díez-Ortiz M, Vázquez-Campos S, Pomar-Portillo V, Svendsen C, Van Gestel CA, Duarte AC (2018) Soil pollution. Elsevier Cambridge, MA, USA
45. Maksimova YG (2019) Microorganisms and carbon nanotubes: interaction and applications (review). Appl Biochem Microbiol 55:1–12
46. Maksimova Y, Bykova Y, Maksimov A (2022) Functionalization of multiwalled carbon nanotubes changes their antibiofilm and probiofilm effects on environmental bacteria. Microorganisms 10(8):1627
47. Malini S, Raj K, Madhumathy S, El-Hady KM, Islam S, Dutta M (2022) Bioinspired advances in nanomaterials for sustainable agriculture. J Nanomater
48. Masood S, Khan NR, Fayyaz M, Qayum M, Khatoon A, Jamil M (2024) Synthesis of plant-derived smoke-mediated silver nanoparticles and its stimulatory effects on maize growth under wastewater stress. Arab J Sci Eng Early Access. https://doi.org/10.1007/s13369-024-09049-8
49. McGee CF, Storey S, Clipson N, Doyle E (2017) Soil microbial community responses to contamination with silver, aluminium oxide and silicon dioxide nanoparticles. Ecotoxicology 26:449–458
50. Means N, Elechalawar CK, Chen WR, Bhattacharya R, Mukherjee P (2022) Revealing macropinocytosis using nanoparticles. Mol Aspects Med 83:100993
51. Moaveni P, Kiapour H, Sani B, Rajabzadeh F, Mozafari H (2020) Changes in some physiological traits and mucilage yield of sour tea (Hibiscus Sabdariffa L.) under foliar application of magnesium and iron oxide nanoparticles. Iran J Plant Physiol 10(4):3333–3341
52. Molina-Hernández JB, Scroccarello A, Della Pelle F, De Flaviis R, Compagnone D, Del Carlo M, Paparella A, Chaves López C (2022) Synergistic antifungal activity of catechin and silver nanoparticles on Aspergillus niger isolated from coffee seeds. Lwt 169:113990
53. Nisar P, Ali N, Rahman L, Ali M, Shinwari ZK (2019) Antimicrobial activities of biologically synthesized metal nanoparticles: an insight into the mechanism of action. J Biol Inorg Chem 24:929–941
54. Nongbet A, Mishra AK, Mohanta YK, Mahanta S, Ray MK, Khan M, Baek K-H, Chakrabartty I (2022) Nanofertilizers: a smart and sustainable attribute to modern agriculture. Plants 11(19):2587
55. Ouyang B, Yilihamu A, Liu D, Ouyang P, Zhang D, Wu X, Yang ST (2021) Toxicity and environmental impact of multiwalled carbon nanotubes to nitrogen-fixing bacterium Azotobacter chroococcum. J Environ Chem Eng 9:105291
56. Pahalvi HN, Rafiya L, Rashid S, Nisar B, Kamili AN (2021) Chemical fertilizers and their impact on soil health. microbiota and biofertilizers, vol 2: ecofriendly tools for reclamation of degraded soil environs. Cham, Springer International Publishing, pp 1–20
57. Qian Y, Qin C, Chen M, Lin S (2020) Nanotechnology in soil remediation—applications versus implications. Ecotoxicol Environ Saf 201:110815
58. Rai PK, Rai A, Sharma NK, Singh T, Kumar Y (2023) Limitations of biofertilizers and their revitalization through nanotechnology. J Clean Prod 138194
59. Sarkar S, Kumar R, Kumar A, Kumar U, Singh DK, Mondal S, Kumawat N, Singh AK, Raman RK, Sundaram PK, Gupta AK, Yadav SK (2022) Role of soil microbes to assess soil health. In: Giri B, Kapoor R, Wu Q-S, Varma A (eds) Structure and functions of pedosphere. Springer Nature Singapore, Singapore, pp 339–363
60. Šebesta M, Nemček L, Urík M, Kolenčík M, Bujdoš M, Hagarová I, Matúš P (2020) Distribution of TiO_2 nanoparticles in acidic and alkaline soil and their accumulation by aspergillus Niger. Agronomy 10:1833
61. Shah V, Collins D, Walker VK, Shah S (2014) The impact of engineered cobalt, iron, nickel and silver nanoparticles on soil bacterial diversity under field conditions. Environ Res Lett 9(2):024001
62. Sigala Aguilar NA, López MG, Fernández-Luqueño F (2024) Carbon-based nanomaterials as inducers of biocompounds in plants: potential risks and perspectives. Plant Physiol Biochem 212:108753

63. Sun W, Dou F, Li C, Ma X, Ma LQ (2021) Impacts of metallic nanoparticles and transformed products on soil health. Crit Rev Environ Sci Technol 51:973–1002
64. Sun C, Hu K, Mu D, Wang Z, Yu X (2022) The widespread use of nanomaterials: the effects on the function and diversity of environmental microbial communities. Microorganisms 10(10):2080
65. Tahat MM, Alananbeh KM, Othman YA, Leskovar DI (2020) Soil health and sustainable agriculture. Sustainability 12(12):4859
66. Tauseef A, Hisam F, Hussain T, Caruso A, Hussain K, Châtel A et al (2023) Nanomicrobiology: emerging trends in microbial synthesis of nanomaterials and their applications. J Clu Sci 34(2):639–664
67. Usero FM, Armas C, Morillo JA, Gallardo M, Thompson RB, Pugnaire FI (2021) Effects of soil microbial communities associated to different soil fertilization practices on tomato growth in intensive greenhouse agriculture. Appl Soil Ecol 162:103896
68. Vineeth Kumar CM, Karthick V, Kumar VG, Inbakandan D, Rene ER, Suganya KSU, Embrandiri A, Dhas TS, Ravi M, Sowmiya P (2022) The impact of engineered nanomaterials on the environment: release mechanism, toxicity, transformation, and remediation. Environ Res 212:113202
69. Wu F, You Y, Zhang X, Zhang H, Chen W, Yang Y, Werner D, Tao S, Wang X (2019) Effects of various carbon nanotubes on soil bacterial community composition and structure. Environ Sci Technol 53:5707–5716
70. Xu Z, Long X, Jia Y, Zhao D, Pan X (2022) Occurrence, transport, and toxicity of nanomaterials in soil ecosystems: a review. Env Chem L 20(6):3943–3969
71. You T, Liu D, Chen J, Yang Z, Dou R, Gao X, Wang L (2018) Effects of metal oxide nanoparticles on soil enzyme activities and bacterial communities in two different soil types. J Soils Sediments 18:211–221
72. Younis SA, Kim K-H, Shaheen SM, Antoniadis V, Tsang YF, Rinklebe J, Deep A, Brown RJC (2021) Advancements of nanotechnologies in crop promotion and soil fertility: benefits, life cycle assessment, and legislation policies. Renew Sustain Energy Rev 152:111686
73. Zhou Y, Ma J, Yang J, Lv Z, Song Z, Han H (2023) Soybean rhizosphere microorganisms alleviate Mo nanomaterials induced stress by improving soil microbial community structure. Chemosphere 310:136784

Induction of Tolerance to Biotic Stress by Applying Nanostimulats to Crops

R. Paola Fincheira, Adalberto Benavides-Mendoza, Olga Rubilar, Martín Fernández-Baldo, and Gonzalo Tortella Fuentes

Abstract The intensifying pressures on agriculture, driven by a growing global population and climate change, demand innovative strategies to ensure food security. Biotic stressors, including insects, fungi, bacteria, and weeds, pose significant challenges to crop productivity. This chapter explores the application of nanotechnology as a transformative tool for mitigating biotic stresses in agriculture. Nanoparticles (NPs) offer unique properties, such as enhanced reactivity, controlled release, and targeted delivery, improving pest control efficacy while minimizing environmental impacts. Metal-based, polymeric, and lipid nanoparticles have demonstrated remarkable antimicrobial, antifungal, and pest management capabilities. Furthermore, NPs can activate plant defense mechanisms, including systemic acquired resistance, by modulating phytohormones and reactive oxygen species pathways. Advanced nanoformulations provide prolonged protection and reduce agrochemical dependency, addressing pesticide resistance and environmental contamination. This chapter highlights groundbreaking research and practical applications of NPs in biotic stress management, emphasizing their role in sustainable agriculture. Integrating nanotechnology into crop protection enhances productivity, minimizes ecological risks, and supports the transition to resilient agricultural practices amid global environmental challenges.

Keywords Nanotechnology · Agriculture · Lipid nanoparticles · Abiotic stress

R. Paola Fincheira (✉) · O. Rubilar · G. Tortella Fuentes
Facultad de Ingeniería y Ciencias, Centro de Excelencia en Investigación Biotecnológica Aplicada Al Medio Ambiente (CIBAMA), Universidad de La Frontera, Temuco, Chile
e-mail: paola.fincheira@ufrontera.cl

Department of Chemical Engineering. Faculty of Engineering and Sciences, Universidad de La Frontera, Temuco, Chile

A. Benavides-Mendoza
Horticulture Department, Agronomy Division, Universidad Autónoma Agraria Antonio Narro, Saltillo, México

M. Fernández-Baldo
Instituto de Química de San Luis INQUISAL, CONICET San Luis, Universidad de San Luis, San Luis, Argentina

© The Author(s), under exclusive license to Springer Nature Singapore Pte Ltd. 2025
A. Juárez-Maldonado et al. (eds.), *Plant Biostimulation with Nanomaterials*, Smart Nanomaterials Technology, https://doi.org/10.1007/978-981-96-4648-7_7

1 Introduction

The pressure on agriculture to produce more food is intensifying due to the projected increase in world population. The agricultural sciences have presented important challenges in recent years to overcome the effects of the huge climate change crisis, decreased soil fertility, toxic residues, and contaminants present in the soil [78]. In this context, farmers' agrochemical products have been implemented to improve the efficiency and productivity of food crops. Nevertheless, global agriculture and the balance of crop yield are strongly dependent on various stress factors [37]. In particular, biotic stresses stand out as a formidable adversary, constantly threatening the world's food security. Biotic stresses in agriculture are primarily attributed to living organisms, including insects, fungi, bacteria, nematodes, and weeds, which adversely impact crop growth, development, and productivity. The understanding and effectively managing biotic stress becomes imperative for ensuring sustainable crop production (Fig. 1).

Commercial pesticides are the principal product used for pest control management due to their low cost and efficiency [81]. The data estimated that pesticides protect at least around one-third of crops worldwide. According to Pathak et al. [68], global pesticide consumption reached approximately 4.19 million tons in 2019, and the main consuming countries are China, the United States, Brazil, and Argentina. Pesticides are generally classified according to their chemical structure: organochlorine, organophosphate, chlorophenols, carbamate, carboxylic acid, and triazine. Great losses in crop yield are reported worldwide due to pest attacks, which reach from 20 to 40% annually [71]. Worldwide crop damage is caused by around 9,000 mites and insects, 8,000 weeds, and 50,000 plant pathogen species. The most consumed pesticides in agriculture are herbicides (47.5%), insecticides (29.5%), and fungicides (17.5%).

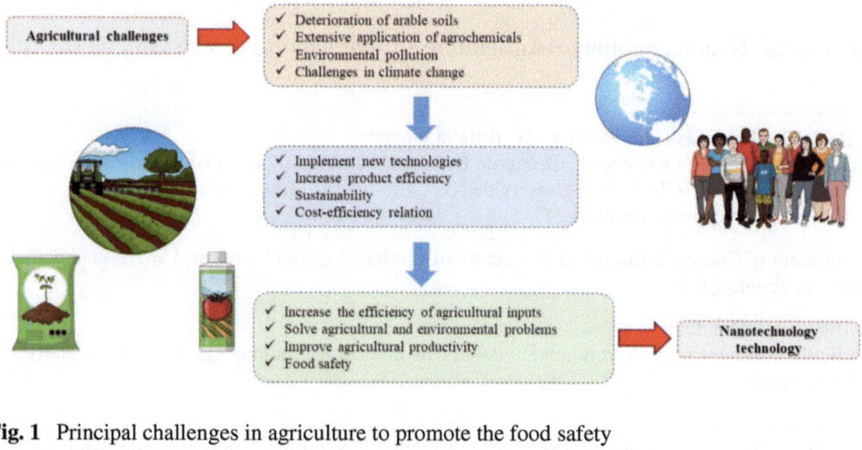

Fig. 1 Principal challenges in agriculture to promote the food safety

The intensive and excessive application of commercial pesticides to pest control has adversely affected human health and the ecosystem [68]. In addition, developing resistant organisms to pesticides has produced an additional serious problem in agriculture. Otherwise, the accumulation of pesticides in aquatic organisms and sediments represents a great toxicity risk to the water sources and ecosystem. In this context, nanotechnology emerges as an important strategy to alleviate the damage caused by pests. Nanoparticles (NPs) refer to particles or assemblies of at least one dimension ranging from 1 to 100 nm [54]. Moreover, NPs used as nanocarriers can reach 1,000 nm in three dimensions. The size-dependent properties of NPs led to unique physical, chemical, and biological characteristics. Moreover, NPs can enhance reactivity, increase surface area, and modify optical properties compared to their bulk counterparts.

Polymeric, metal, and metal-oxide NPs have demonstrated efficient activity to control the management of fungi, weeds, insects, bacteria, and nematodes in agriculture. For example, copper and silver NPs have been widely reported for their antimicrobial activity, and silica NPs for their role in inducing plant resistance. Furthermore, polymeric NPs have attracted special interest due to their biocompatibility, biodegradability, and controlled release of the bioactive compound to pest management control. On the other hand, applying NPs to plants can elicit specific responses to induce defense mechanisms against fungi, insects, and bacteria, among others. Therefore, this chapter considers the current and relevant information about findings related to using NPs to control agriculture pests and future guidelines.

2 The Importance of Nanotechnology in Mitigating Biotic Stresses

The manipulation of particles at the nanoscale range holds immense promise in mitigating biotic stresses that influence agriculture yield [35]. Nanotechnology can play a pivotal role in reducing pests' adverse impact, offering innovative solutions to enhance crop yield and protect plants [33]. The small size of NPs allows it to penetrate plant and pathogen tissues more effectively, enabling targeted delivery of pesticides. Besides, the activities of NPs in specific cellular locations where pests or pathogens are most active improve the efficacy. This targeted approach minimizes environmental impact by reducing the chemicals needed and maximizing their efficiency. Nanotechnology provides a wide variety of tools to mitigate biotic stresses, i.e., nanoencapsulation of pesticides allows for controlled release, ensuring prolonged and sustained activity factors [37]. Furthermore, NPs exhibit inherent antimicrobial properties, directly combating pathogens and pests due to their capacity to disrupt pathogens' cellular structures and reduce their growth [53]. These functionalities of NPs, both as carriers for agrochemicals and as direct antimicrobial agents, amplify their potential application to crop protection to crops against a spectrum of biotic stresses (Table 1).

Table 1 Principal nanoparticles used to mitigate biotic stresses

Type of nanoparticle	Function	Reference
Silver NPs	Exhibit significant antimicrobial properties, inhibiting the growth of bacteria and fungi	Bruna et al. [15]
Copper NPs	Antimicrobial properties focus on fungicides. They can be effective in controlling fungal diseases in crops	Rojas et al. [80]
Polymeric NPs	These NPs can encapsulate the sustainable release of pesticides, reducing the need for frequent applications	Vinzant et al. [98]
Carbon nanotubes and graphene NPs	NPs have been explored for their potential as carriers for delivering pesticides, enhancing some agrochemicals' solubility and stability, and enabling targeted delivery	Majeed et al. [64]
Lipid-NPs	Lipid-NPs can encapsulate pesticides, improving their bioavailability and enabling controlled release, minimizing environmental impact	Ganilho et al. [32]
Nanoemulsions	It provides a stable formulation for incorporating and delivering pesticides. Nanoemulsions enhance the efficacy of the active ingredients and improve the adhesion	Faikah Mustafa and Zobir Hussein [28]
Silica-NPs	Silica NPs can be engineered to carry and release pesticides in a controlled manner. Their porous structure allows an efficient loading and release of active ingredients	Thabet et al. [93]
Chitosan-NPs	Chitosan NPs have antifungal and antibacterial properties and are used to encapsulate active compounds to protect crops against diseases caused by various pathogens	Amin Mousavi et al. [8]

This increases the efficacy of the pesticides and reduces the frequency of application, decreasing the risk of environmental pollution. Moreover, NPs can be engineered to have specific properties that enhance their interaction with pests, making them more lethal. This precision in targeting harmful organisms is a groundbreaking advancement in sustainable agriculture. In this sense, NPs are a valuable tool to

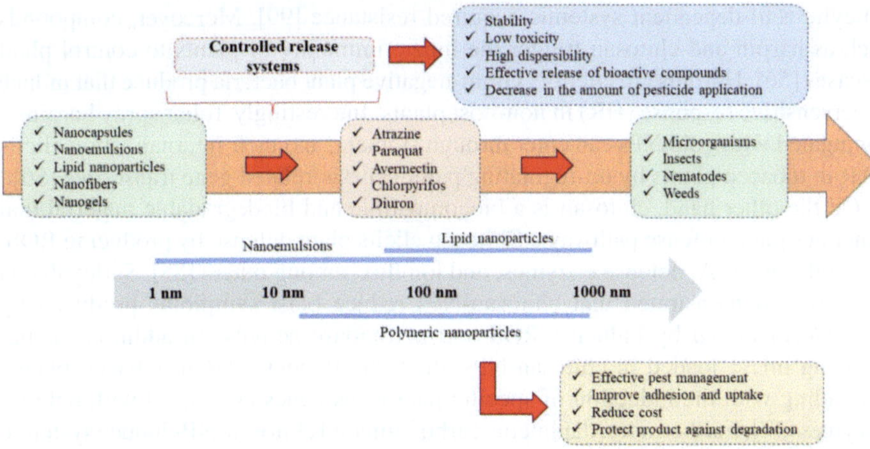

Fig. 2 The characteristics of nanoparticles to formulate controlled release systems of commercial or natural compounds used as a pesticide

overcome challenges derived from pest management in agriculture, with a significant contribution to the well-being of ecosystems, human health, and sustainable development (Fig. 2).

3 Nanoparticle Application to Induce Plant Tolerance

Applying chemical products has demonstrated an effective action in inducing plant defense response, but their irrational and extensive use produces phytopathogen resistance and damage to the ecosystem. In this sense, environmentally sustainable plant defense elicitors with long-lasting resistance and spread spectrum have emerged as important tools to plant disease control [100]. Plants are constantly exposed to biotic stress, where the defense is activated once pathogens are detected [44]. The biotic stresses plants trigger the activation of regulatory or transcriptional systems to generate a response [44]. Salicylic acid (SA) and jasmonic acid (JA) are phytohormones that modulate signaling networks to activate defense response to biotic stresses. SA is involved in the modulation of immune signals, amplification, and suppression. Exogenous application of SA is attractive to induce systemic acquired resistance (SAR) to establish plant immunity [96]. The induction of SAR constitutes one of the best alternatives for pest management because it provides a wide spectrum of long-term protection against phytopathogens in plants [24]. Otherwise, reactive oxygen species (ROS) are signal molecules involved in plant defense, where hydrogen peroxide (H_2O_2) mediates the induction of plant defense response, SAR, and expression of pathogen-related (PR) genes [67]. The activation of plant defense response has been associated with the elimination of ROS and the activation of salicylic acid-dependent systemic acquired resistance [99]. Moreover, compounds such as harpin and chitosan trigger the innate immunity of plants to control plant diseases [56]. Harpins are elicitors Gram-negative plant bacteria produce that induce hypersensitive response (HR) in non-host plants. Interestingly, foliar spray harpin$_{Pss}$ conjugated with gold NPs can enter through stomata, and their internalization elicits SAR in tobacco plants by up-regulating pathogenesis-related gene transcripts [56].

On the other hand, chitosan is a biocompatible and biodegradable material that generates plant defense pathways. Chitosan elicits plant defense by producing ROS, biosynthesis of JA, defense enzymes, and ion flux, among others [85]. Sathiyabama et al. [85] demonstrated that chitosan-NPs reduce blast symptoms produced by *Pyricularia grisea* by inducing ROS and peroxidase activity. In addition, harpin at 30 ng mL^{-1} loaded in chitosan-NPs elicit an efficient SAR in tobacco plants, coinciding with the induction of specific pathogens genes associated with defense enzymes [57]. Furthermore, daphnetin-carboxymethyl chitosan-NPs induce systemic resistance in tobacco plants against *Ralstonia solanacearum* by increasing defense enzymes, including catalase, polyphenol oxidase, and superoxide dismutase and the up-expression of PR proteins [100]. Chitosan-fabricated silver-NPs protect against bacterial leaf spots produced by *Xanthomonas campestris* in tomatoes [36]. The expression of genes responsive related to growth, early defense, defense, hypersensitivity, and stress hormones were regulated to decrease the disease triggered by *X. campestris* [36].

Selenium (Se)-NPs have been studied for eliciting plant defense on *Salvia miltiorrhiza*, modulating plant hormones, photosynthesis, and antioxidant response. Se-NPs activate SA and JA pathways, increasing the contents of tanshinones and salvianolic acids to protect plants against aphids [104]. The seed priming in tomatoes with Se-NPs increases the protection against *Phytopthora infestans* by 72.9% [49]. The results showed an accumulation of H_2O_2, lignin, and callose and elevated activity of β-1, 3-glucanase, phenylalanine lyase, superoxide dismutase, and lipoxygenase in primed plants [49]. Besides, the foliar spray of Fe_3O_4-NPs elicits the resistance of *Nicotiana benthamiana* against *Tobacco mosaic virus*, up-regulating SA synthesis and the expression of SA-responsive PR genes [16]. This study suggested that the deposition of Fe_3O_4-NPs induced the accumulation of SA to promote plant defense. Moreover, CuO and Ag-NPs have an efficient activity in inducing defense in *Oryza sativa* to control *Nilaparvata lugens* by up-regulating linalool synthase gene and other genes involved in plant defense response. Silica(Si)-NPs significantly reduced by 20.7–87.9% the severity triggered by *R. solanacearum* in tobacco plants due to the reduction of H_2O_2 and malondialdehyde. In addition, the defense enzymes and the expression of SA-related genes were up-regulated in the leaves of tomatoes exposed to Si-NPs [99]. SA-doped iron-NPs suppressed the disease triggered by *Fusarium oxysporum* f. sp. *niveum* in watermelon by improving antioxidant capacity and primed SAR by the SA signaling pathway [73]. Furthermore, Ag-NPs reduce the damage produced by *Clavibacter michiganensis* subsp *michiganesis* in tomato plants, suggesting a key role in the SAR pathway. Besides, the foliar application of ZnO and CuO-NPs promotes the defense mechanism in *Glycine* max against *Fusarium virguliforme* by enhancing the expression of pathogenesis-related GmPR genes encoding defense proteins [52]. The application of ZnO-NPs and CuO-NPs in *Glycine* max improves the disease resistance against *F. virguliforme* by increasing in 300% the expression of pathogenesis-related GmPR genes encoding defense and antifungal proteins [52].

4 Nanoparticles to Fungi Control

NPs have emerged as an innovative solution for controlling fungi in agriculture, offering a multifaceted approach to combat fungi diseases [12]. Fungal phytopathogens significantly threaten global food security by causing devastating diseases in crops, leading to yield and economic losses. The utilization of NPs in fungi control represents a promising alternative to overcome the limitations of conventional methods due to the enhanced efficacy and sustainability. NPS offers innovative strategies for mitigating infection through different action mechanisms associated with physiological functions [47]. For example, NPs exert their antifungal effects by disrupting fungi cell walls and cell membranes, leading to structural damage and increased permeability, compromising physiological functions. Furthermore, the small size and high surface area of NPs promote the generation of ROS, producing oxidative damage to cellular components, such as proteins, lipids, and DNA. Consequently, NPs can lead to disrupting vital metabolic processes, contributing to the inhibition of fungal growth. On the other hand, NPs can interfere with the signaling pathways and enzymatic activities essential for fungal development. The disruption of key cellular processes by NPs prevents fungi phytopathogens from establishing and spreading infection [59]. Interestingly, functionalized NPs can be designed to specifically target and adhere to structures to enhance selectivity and efficiency and minimize potential impacts on beneficial organisms. Otherwise, the controlled release of fungicidal agents from NPs sustained their antifungal activity over time. The multifaceted actions of NPs show their potential as effective and environmentally sustainable tools for managing fungi infections in agriculture, offering a novel approach to enhance crop protection and yield. For that, the nanoencapsulation of bioactive agents emerges as a strategy to prolong the shelf life and the controlled release [82].

The chitosan has been widely studied for its eco-friendly and sustainable properties. In this sense, chitosan has played an important role in formulating NPs to encapsulate active compounds with antifungal activity [50]. Chitosan-thymol NPs act as an effective fungicide against *Botrytis cinerea* with a significant pH-responsiveness to release thymol under acidic conditions, demonstrated with detached leaf and pot experiments [105]. This study evidenced a protective efficacy of 78.7% in pot experiments with cucumber seedlings. Similarly, the controlled release of garlic essential oils from chitosan-NPs reduces the infection by *Aspergillus flavus* due to changes in ions leakage, ergosterol concentration, alterations in mitochondrial membrane potential, and antioxidant system [91]. The in vitro release profile showed an accumulative release of 50% at hour 70. Similarly, the encapsulation of essential oils of

Nepeta hormozganica and *Nepeta Dschuprensis* into chitosan-NPs inhibited around 100% of the radial growth of *Sclerotinia sclerotiorum*, *B. cinerea*, *Fusarium oxysporium*, *Rhizoctonia solani*, and *Alternaria alternate*. Interestingly, chitosan-NPs promote the control of wilt disease caused by *Fusarium andiyazi* by inhibiting radial growth in 73.81%. The effect was related to the up-regulation of genes associated with β-1,3-glucanase, chitinase, PR1, and PR10 in 1.48, 1.15, 1.15, and 1.41-fold [20]. A significant transcript profile of superoxide dismutase was fold-expressed in 4.5. Moreover, chitosan silver nanocomposites at 0.0005% showed fungicidal activity against the anthracnose pathogen *Colletotrichum truncatum* by inhibiting the conidia germination [38]. Assays evidenced an important reduction of anthracnose in *Capsicum annuum*. Chitosan-NPs loaded SA at 400 ppm, and chitosan-NPs loaded silver at 200 ppm inhibited the conidial germination of *A. alternata* [42]. Moreover, chitosan-coated Ag-NPs had antifungal activity against *F. oxysporum* and *Aspergillus niger* at 41.7 and 208.3 μg mL^{-1} [70]. Other techniques and polymers have been implemented to nanoencapsulate compounds. For example, the encapsulation of polyphenols in β-lactoglobulin-NPs significantly increased the antioxidant activity and improved the antifungal activity against the spore germination and expansion of *Penicillium digitatum* [100]. This study evidenced an efficient antifungal activity of the encapsulated to protect orange fruits against green mold infection decay, maintaining fruit quality. Besides, essential oil-encapsulated lipid nanoemulsions exhibited great antifungal activity to reduce leaf spots of *Neopestalotiopasis rosae* on strawberries, even preventing the disease by 100% [95]. In addition, oregano essential oil-loaded nanosilica had antifungal activity against *Curvularia lunata* found in *Agaricus bisporus*. The results showed a minimum fungicidal concentration at 0.20 and 0.40 mg mL^{-1} of NPs. NPs produced the exudation of proteins and nucleic acids due to the disruption of nuclear and cellular membranes [17].

Moreover, numerous metal-NPs have been synthesized to evaluate their capacity to control fungi phytopathogen. Prominent results of metal NPs have demonstrated their ability to reduce conidia germination and radial mycelia growth of various fungi of economic and agricultural importance [21]. For example, Ag-NPs have been reported to control fungi phytopathogens such as *A. alternata*, *Macrophomina phaseolina*, *Curvularia lunata*, *B. cinerea*, *Rhizoctonia solani*, *P. digitatum*, and *S. sclerotiorum*, among others [66]. Ag-NPs produce an antifungal activity by producing free radicals and ROS, which induces protein denaturation, cell wall damage, and lipid peroxidation [66]. The grey mold disease triggered by *B. cinerea* significantly reduced their infection in tomato plants treated with concentrations from 0.03 to 0.05% of Ag-NPs [11]. In addition, Ag-NPs at 150 μg mL^{-1} showed significant inhibition of *Trichoderma harzianum* MTCC 801 [86]. Similarly, Ag-NPs at 1 mM showed an efficient activity to control *F. oxysporum* with a protection rate of 75% due to the improvement of defense by activating plant immunity and the increasing of polyphenol oxidase and peroxidase [2]. Ag-NPs at 40 mg L^{-1} improve the antioxidant activity, effectively reducing the intensity of spot blotch and the yield losses of wheat *Bipolaris sorokiniana* [13]. Besides, Ag-NPs inhibited the growth of *R. solani* up to 82% in vitro conditions, which can be attributed to the accumulation of ROS and the decrease of mitochondrial membrane potential. The results showed an elevated cytochrome oxidase three and NADPH dehydrogenase expression [19]. The primed rice seedlings with Ag-NPs showed increased peroxidase activity, superoxide dismutase, and phenylalanine ammonia [19]. Moreover, Ag-NPs at 150 and 300 μg L^{-1} completely inhibited the growth of *R. solani* GPB and *Sclerotium rolfsii* [51]. Exposure to Ag-NPs enhances lipid peroxidation, causing toxicity to fungal cells and increasing peroxidase and catalase activities. A significant reduction of growth of *A. alternata* and *S. sclerotiorum* was triggered by Ag-NPs. Ag-NPs at 50 μg mL^{-1} reduced the radial expansion of *Pestalotiopsis magniferae* isolated from the coconut palm, reaching around 50% and 60% inhibition [46].

In addition, other metal-NPs have demonstrated an important antifungal activity. For example, the spray of MgO-NPs reduces the *Alternaria dauci* infection on *Daucus carota*L., inhibiting 60% under greenhouse conditions [3]. TiO_2-NPs decrease 58% of the fungal disease of *Ustilago tritici* in *Triticum aestivum* L [6]. CeO_2-NPs reduce the disease of *Ustilago tritici* in *T. aestivum* L by decreasing disease severity and fungal infection and modifying the antioxidant system [9]. In addition, MnO-NPs evidenced a significant mycelia growth inhibition of *A. niger*, reducing the black rot disease of tomatoes. CuO-NPs had a relevant antifungal activity against *Penicillium expansum*, *R. solani*, and *F. oxysporium* with a dose-dependent response from 125 to 2000 μg mL^{-1} [40]. The foliar treatment of *Glycine* max with ZnO-NPs and CuO-NPs increases in 300% the expression of GmPR genes encoding defense and antifungal proteins [52]. Interestingly, the combined application of Zn-NPs with *Bacillus amyloliquefaciens* produces resistance against the disease severity of *Alternaria solani* in tomato plants [5]. Ag_2O nanostructures inhibited *Macrophomina phaseolina* growth, decreasing the severity of charcoal rot disease in strawberry modulating defense gene *PR-1* [22]. Si-NPs disrupted the cell morphology of *Ralstonia solanacearum*, mitigating their impact on tomato plants [55].

Carboxymethyl cellulose trimetallic magnetic Cu-Zn-Fe-NPs interfere with ergosterol production to inhibit the growth of *F. oxysporium*. Moreover, trimetallic-NPs-NPs down-regulated virulence factors and velvet complex [10]. A significant antifungal activity of Tb-ZnO NPs (200 ppm) was found against *Alternaria brassicae*, which causes the blight disease in *Brassica juncea* [23]. Ag-SeS_2-NPs at 100 μg mL^{-1} had a fungistatic activity against *Phytophthora cactorum* by inhibiting mycelial growth up to 40% [60].

5 Nanoparticles with Insecticide Activity

The impact of NPs on decreasing insect pests in agriculture is a subject of growing interest and research, presenting a range of potential benefits and challenges. NPs with insecticidal properties have shown promise in providing a targeted and effective approach to pest management [74]. One notable impact of NPs is insecticide's enhanced efficiency through improved delivery and controlled release over time.

This controlled release ensures a sustained exposure of pests to the active ingredients, potentially reducing the need for frequent applications and minimizing the environmental impact of conventional insecticides. NPs exhibit physiological actions to reduce insect populations through various mechanisms, presenting innovative approaches to insect pest management [45]. One significant mechanism involves the interaction of NPs with the insect cuticle. The small size of NPs enables them to penetrate the cuticle and other protective layers of insects more effectively, enhancing their direct interaction with vital physiological processes, disrupting cellular functions, and leading to increased mortality among insect pests. NPs, such as those composed of silica or metal oxides, can disrupt the integrity of the insect exoskeleton. The disruption of insect exoskeleton can lead to desiccation, loss of water regulation, and increased vulnerability to environmental stress, compromising physiological functions and survival of insects. In addition, NPs may interfere with insect reproductive processes by disrupting the development of eggs or larvae, leading to reduced pest populations over time. Some studies have demonstrated that NPs can influence the behavior of insects, affecting feeding and mating and reducing the pest damage to crops.

NPs can also be used as carriers for conventional or natural pesticides, allowing targeted delivery of pesticides for insect control. The application of natural insect repellents through NPs improves their efficiency. In addition, the nanoencapsulation of these compounds reduces the application of chemical insecticides, potentiating the implementation of eco-friendly products [58]. For example, the thermoresponsive nanoencapsulation of thiamethoxam-loaded mesoporous silica-NPs increases the mortality of *Nilaparvata lugens* (Hemiptera: Delphacidae), reducing their negligible effects on rice seedling growth [31]. In addition, the results showed that thiamethoxam-loaded mesoporous silica-NPs had bioactivity for 14 days, showing negligible effects on rice seedlings [31]. Furthermore, the nanoform of emamectin benzoate increases its toxicity, decreasing the mean number of lai eggs/female of *Spodoptera littoralis* (Boisduval) (Lepidoptera: Noctuidae) [41]. The application of Ag-NPs with emamectin benzoate nanoform decreased the percentage of pupation and adult emergence in larvae at LC_{50} of 43.6 and 70.3 ppm [41]. A study with the nanoencapsulation of dimethoate in alginate chitosan-NPs showed a slow-release over time with 52% after 48 h, evidencing the reduction of toxicity of insecticides with non-target species. In addition, the smart release of avermectin from metal-organic frameworks had higher insecticidal efficacy in controlling *Spodeptera litura,* and no adverse effects were detected on plant growth and human health, supporting the biosafety toward non-target organisms [62]. Meanwhile, the co-delivery of imidacloprid and *CYP6CY13* double-stranded RNA increases the susceptibility of *Aphis gossypii* to chemical insecticide, enhancing the toxicity in 1.95-fold to conventional imidacloprid and improving the effectiveness in 19.95% in 5 days [63]. A study showed that indoxacarb-loaded hollow mesoporous silica-β-cyclodextrin NPs exhibited an important insecticidal activity in controlling *Spodoptera frugiperda* at concentrations from 5 to 20 mg L^{-1}, reaching mortality percentage from 50 to 100% [101]. Furthermore, the controlled system decreased the toxicity in non-target organisms 5-fold (i.e., zebrafish) [101].

On the other hand, metal-NPs have been evidenced to be an important insecticidal activity. For example, Si-NPs can act as insecticides or nanocarriers to manage pests or improve plant resistance against insects [88]. In addition, Si-NPs coated with carboxymethyl chitosan as a stimuli-response system had high efficiency in the sustainable release of chlorantraniliprole to control *Spodoptera frugiperda* in *Zea mays* plants [29]. A study showed that 5% of Cu-doped ZnO-NPs exhibited a toxic effect on *Spodoptera littoralis*, reducing their impact [1]. Besides, the extract of *Codium adule* mixed with Ag-NPs (LC_{50} 300.37 ppm) increased the mortality from 21.8 to 95.8% of *Musco domestica* [65]. The treatments with Fe-NPs produced the mortality of *Helicoverpa armigera* and *Spodoptera litura* at 91.4 and 92.8% after 72 h [72]. The antifeeding activity increased from 10 to 90% according to the growing concentration of Fe-NPs (25 to 250 mg L^{-1}). Furthermore, Ag/Ag_2O-NPs applied through water spray showed good insecticidal activity against agricultural pests such as mustard and *Tulsi* aphids [89]. Using KI-NPs at 10,000 ppm according to the larval dip method triggers toxic effects against *S. frugiperda*, reaching 98% mortality in five days [77]. Interestingly, the nanohybrid of CeO_2 intercalated into mesoporous organosilica-NPs enhanced the toxicity of nitenpyram, clothianidin, and sulfoxaflor against *Nilaparvata lugens* [103]. The bioassays demonstrated that nanohybrids modulate the detoxification mechanism, down-regulating ROS-dependent P_{450} gene expression to overcome insecticide resistance. In addition, the nanohybrid influences the insecticide resistance in *Aphis gossypii*, *Sogatella furcifera*, and *S. frugiperda* [103]. In addition, the combination of cellulose nanofiber and chitosan-NPs has been used for the controlled release of citronella essential oil to reduce the impact of *S.littoralis* (Boisd.) [43]. The bioassays showed that nanoformulation disrupts the development of *S.littoralis*, reducing pupal weight, adult longevity, and female fecundity. Semi-field results evidenced persistence and larval mortality [43].

Additionally, the NPs used as a controlled release system are highly efficient in controlling insect pests. For example, the essential oil of *Rosmarinus officinalis* incorporated into poly-(ε-caprolactone) polymer had an insecticidal activity (LD_{50} 9.1 g L^{-1}) against *Drosophila suzukii* (Diptera: Drosophilidae) [90]. Peppermint and palmarosa oils included in polyethylene glycol 6000 matrix enhanced the lethal and sub-lethal effects on *Blatella germanica* (Blattaria: Blatellidae) with LC_{50} of 31.43 and 25.41 $\mu g\ cm^{-2}$ at 24 h of exposure [102]. Furthermore, chitosan/carboxymethyl chitosan NPs at 0.2% produce 98% of the mortality of *Solenopsis invicta* (Hymenoptera: Formicidae) due to the inhibition of food intake and growth and development.

6 Nanoparticles with Nematicidal Activity

It has been reported that NPs can impact nematodes due to interactions with cellular and molecular components, leading to disruptions in vital processes for the survival of nematodes. NPs have significantly impacted nematode physiology [26]. For example, Ag-NPs damage the nematode's cuticle and protective outer layer, increasing the permeability and osmoregulation. Also, NPs may induce oxidative stress within nematode cells, leading to the generation of ROS that alter the antioxidant defense mechanisms of nematodes. This oxidative stress can cause damage to lipids, proteins, and DNA, compromising the cellular integrity and function of nematodes. Otherwise, NPs may interfere with key metabolic pathways, disrupting nematode energy production and nutrient uptake. In general, the wide impact of NPs in the physiology of nematodes covers biochemical and metabolic alterations that contribute to their nematicide effects [79].

Cu-NPs at a concentration from 100 to 250 ppm exhibited a nematicide activity on infective second-stage juveniles (J2s) and eggs of *Meloidogyne incognita*, which cause root-knot disease of cucumber [27]. Similarly, Ag-NPs reduced egg masses per root, nematode galls, and egg/egg mass of *M. incognita* by 96%, increasing tomato growth [34]. In addition, controlled release systems have exhibited an important activity against *M. incognita*,for example, nanoemulsions containing extracts of *Artimisia annua* and *Lepidium sativum* were applied in tomato plants infected with *M. incognita*, showing that nanoformulations produced higher mortality compared with original extracts [69]. Biochemical assays showed that the controlled release of extracts induced the accumulation of phenolic compounds and increased protease, chitinase, and polyphenol oxidase activity in tomato plants [69]. Moreover, chitosan-NPs decreased the densities of *M. incognita* from 45 to 66%, and the root gall density was reduced from 10 to 67% due to the modulation in total protein, polyphenol oxidase, and soluble phenol concentrations [83]. The application of ZnO-NPs with oxamyl triggers 98.9% of the mortality of second-stage juveniles2 (J2s) of *M. incognita*,in addition, this treatment showed a significant efficacy to decrease J2s community in 82.7% in soil [87]. The sustainable release of essential oil of *Carlina acaulis*-based nanoemulsion (40–80 μL Kg^{-1}) exhibited a significant reduction of *M. incognita* into tomato-infested roots with EC_{50} of 40.31 μL kg^{-1} [75].

7 Nanoparticles with Herbicidal Activity

The impact of NPs on the physiological system of weeds involves complex interactions with cellular and molecular systems, leading to disruptions in essential processes crucial for growth and development [30]. Some NPs can interact with the lipid bilayer of weed cell membranes, leading to structural changes and alterations in permeability. These alterations compromise the integrity of cell membranes, affecting nutrient uptake and cellular communication. In addition, NPs may induce oxidative stress in weed cells by generating ROS. Elevated levels of ROS can reduce the antioxidant defense mechanisms of weeds, damaging cellular components such as proteins, lipids, and DNA. Also, NPs can interfere with the photosynthetic process in weeds, reducing chlorophyll production and energy metabolism. At the nutritional level, NPs alter weeds' availability and uptake of essential nutrients, disrupting the nutrient balance and affecting vital physiological processes associated with growth and vigor [92]. These effects can be related to reduced root development and decreased nutrient and water uptake. This can compromise the overall health and competitiveness of weed plants.

It notes that the specific physiological impact of NPs on weeds can vary depending on the type of NPs, concentration, and the weed species involved. Additionally, NP-based weed control strategies' environmental and ecological implications need extensive research to ensure their safety and sustainability. In this sense, nanoherbicides can be based on inorganic, organic, or hybrid materials to control the release of ions or organic molecules and provide better weed management [30]. The controlled release of herbicides decreases toxic effects in non-target organisms of the ecosystem [30]. This contributes to ensuring and improving the properties of nanoherbicides on target organisms. Chitosan-NPs loaded with paraquat showed less toxicity than pure compounds, evidencing its advantages to weed control and reducing the environmental impact [39]. Besides, the herbicidal activity of paraquat was preserved with the controlled release from chitosan-NPs [39]. Applying nanoemulsion-loaded glyphosate isopropylamine decreased the deposition on creeping foxglove, slender button weed, and buffalo grass compared to the commercial product Roundup® [48]. The weed control with the nanoemulsion-loaded glyphosate isopropylamine reached between 86.6 and 96.6%. Nanoherbicide formulated with a triple structure and dicamba herbicide as a model using dicamba@mSiO_2/ATA/TiO_2 showed significant efficiency in the controlled release of herbicide, decreasing the degradation of herbicide by photocatalysts process [18]. These results suggested that the nanoherbicide improves dicamba's efficacy and reduces its residues. Furthermore, the encapsulation of atrazine into poly(epsilon-caprolactone) nanocapsules increases the post-emergence control of *Alternanthera tenella* Colla plants by inhibiting the maximum quantum efficiency of photosystem II compared to conventional atrazine after 48 h [94]. Atrazine-loaded polymeric NPs showed an increase of 40% in absorption rate on mustard leaves after 24 h of application. Besides, the nanoencapsulation promotes a higher inhibition of photosystem II activity [92]. Bioassay evidenced that stomatal uptake can be involved in the enhanced activity of atrazine.

8 Nanoparticles to Bacteria Control

NPs can produce physiological alterations to reduce the impact of bacteria in agriculture, offering potential applications in disease control and crop protection. The effects depend on the type of NPs and their interaction with bacterial cells. Certain NPs, such as Ag-NPs, can interact with bacterial cell membranes, compromising the integrity and permeability of cell membranes [4]. Under this context, NPs can play a vital role in the leakage of cellular contents, interfering with essential bacterial functions. Also, NPs may induce oxidative stress in bacteria by generating ROS. Elevated ROS levels can modulate bacterial antioxidant defenses, causing damage to proteins, lipids, and DNA. This oxidative stress can trigger cell death and a reduction in bacterial viability. Moreover, NPs can interfere with the activity of essential enzymes in bacterial cells, affecting metabolic processes critical for bacterial growth and survival [84]. Furthermore, NPs can alter bacterial genetic material's replication and transcription processes, reducing their infection. In this sense, NPs can damage DNA and influence gene expression in bacterial cells. This modulation at the genetic level may impact bacterial virulence factors, stress response mechanisms, and other essential functions. In addition, some studies have evidenced that metal-based NPs, such as Cu or Zn NPs, can release metal ions that exert antibacterial effects. These metal ions may disrupt bacterial cell membranes and interfere with intracellular processes.

It's important to note that using NPs in agriculture for bacterial control requires careful consideration of potential environmental and ecological impacts. Additionally, research must optimize nanoparticle formulations, concentrations, and delivery methods to ensure effectiveness. TiO_2-NPsat 75 and 100 μL have evidenced an efficient activity to reduce bacterial pathogens such as *Xanthomonas phaseoli*, which is attributed to oxidative stress produced by the release of Ti^+ [7]. Furthermore, Ag-NPs at 500 μg mL^{-1} in combination with *Stachys parviflora* extract had an effective activity against *Xanthomonas campestris* [25]. Similarly, Ag-NPs exhibited an important action against *Xanthomonas axonopodis* pv. *punicae* and *Ralstonia solanacearum* with minimum inhibitory concentrations of 6.25 and 12.5 μg mL^{-1}. Ag-NPs promote the release of intracellular reactive oxygen species, DNA damage, and inhibition in the replication of DNA [97].

ZnO-NPs showed a significant activity to control *Pseudomonas syringae* pv. tabaci on tobacco by down-regulating metabolic pathways and functions associated with chemotaxis, biofilm formation, and ABC transporters [29]. Besides, this work evidenced increased plant growth and immunity through global molecular response analysis. Se-NPs at 6.25 μg mL^{-1} exhibited a bacteriostatic action against *Clavibacter michiganesis sepedonicus* due to a decrease of transmembrane activity, inhibited cell respiration, produced changes in the cell wall membrane, and a reduction in dehydrogenase activity [61]. In addition, Ag-NPs protected leaves of *Solanum lycopersicum* against *Clavibacter michiganesis* subsp. *michiganensis*, decreasing the disease severity by 86.2% due to a possible role of the SAR pathway (Mercado [67]. Co_3O_4-NPs at 200 μg mL^{-1} inhibited the growth of *Xanthomonas oryzae* pv. *oryzae* due to the decrease of biofilm formation. Rice plants inoculated with *X. oryzae* decreased leaf disease from 57.2 to 11.0% [76]. Furthermore, plants treated with Co_3O_4-NPs showed increased root length, shoot length, fresh weight, dry weight, and photosynthetic activity [76]. Otherwise, encapsulating essential oils of cinnamon and mustard in mesoporous silica NPs enhanced 10-fold the antimicrobial activity against *Pseudomonas fluorescence*, *Pseudomonas syringae* pv. *Pisi* and *Pectobacterium carotovorum* subsp. *carotovorum* compared to free essential oil [14].

9 Perspectives

The application of NPs in addressing biotic stresses in agriculture holds immense promise and offers a transformative perspective on crop protection. To combat pests and diseases, NPs provide a versatile strategy for precision delivery of various bioactive compounds, including mematicides, fungicides, and elicitors. The small size of NPs allows their efficient penetration into plant tissues, facilitating targeted delivery and enhancing the bioavailability of active ingredients. NPs offer a unique potential to mitigate pests and improve disease management by providing targeted solutions with reduced environmental impact. Engineered NPs can be used to disrupt the physiological processes of pests, offering a novel approach that may mitigate the development of resistance. In addition, incorporating NPs into seed coatings and developing biocompatible formulations further underscores their potential to promote crop health. This targeted approach maximizes the efficacy of pest and disease control and reduces the environmental impact by minimizing the overall quantity of agrochemicals required.

Additionally, NPs can act as carriers for biological control agents, promoting the integration of sustainable pest management practices. The ability of NPs to stimulate plant defense mechanisms further contributes to a holistic strategy against biotic stresses. In this sense, nanotechnology advances and their perspectives in agriculture are optimistic, envisioning a future where these NPs play a pivotal role in developing resilient, sustainable, and environmentally friendly approaches to mitigate the impact of biotic stresses on global food production.

Acknowledgements ANID/FONDECYT 11220070, ANID/FOVI220003

References

1. Abd El-Latef EA, Wahba MN, Mousa S et al (2023) Cu-doped ZnO-nanoparticles as a novel eco-friendly insecticide for controlling *Spodoptera littoralis*. Biocatal Agric Biotechnol 52:102823. https://doi.org/10.1016/j.bcab.2023.102823
2. Abdelaziz AM, Elshaer MA, Abd-Elraheem MA et al (2023) Ziziphus spina-christi extract-stabilized novel silver nanoparticle synthesis for combating *Fusarium oxysporum*-causing pepper wilt disease: in vitro and in vivo studies. Arch Microbiol 205:69. https://doi.org/10.1007/s00203-023-03400-7
3. Ahamad L, Ali Khan A, Khan M et al (2023) Exploring the nano-fungicidal efficacy of green synthesized magnesium oxide nanoparticles (MgO NPs) on the development, physiology, and infection of carrot (*Daucus carota* L.) with Alternaria leaf blight (ALB): molecular docking. J Integr Agric 22:3069–3080. https://doi.org/10.1016/j.jia.2023.02.034
4. Ahmad SA, Sachi Das S, Khatoon A et al (2020) Bactericidal activity of silver nanoparticles: a mechanistic review. Mater Sci Energy Technol 3:756–769. https://doi.org/10.1016/j.mset.2020.09.002
5. Aldayel MF, Alrajeh HS, Ateaf Sallam NM et al (2023) *Bacillus Amyloliquefaciens* IKMM and zinc nanoparticles as biocontrol candidate induce the systemic resistance by producing

antioxidants in tomato plants challenged with early blight pathogen. Gesunde Pflanzen. https://doi.org/10.1007/s10343-023-00942-0
6. Alabdallah NM, Atif Irshad M, Rizwan M et al (2023) Synthesis, characterization and antifungal potential of titanium dioxide nanoparticles against fungal disease (*Ustilago tritici*) of wheat (*Triticum aestivum* L.). Environ Res 228:115852. https://doi.org/10.1016/j.envres.2023.115852
7. Alasmary FA, Kumar Rajaram S, Rajabathar J et al (2022) Titanium dioxide nanoparticles fabrication from Parmotrema austrosinense (Zahlbr.) Hale extracts and its antimicrobial efficacy against plant pathogens. Inorg Chem Commun 145:110007. https://doi.org/10.1016/j.inoche.2022.110007
8. Amin Mousavi S, Ghotaslou R, Kordi S et al (2018) Antibacterial and antifungal effects of chitosan nanoparticles on tissue conditioners of complete dentures. Int J Biol Macromol 118:881–885. https://doi.org/10.1016/j.ijbiomac.2018.06.151
9. Alotaibi MO, Alotaibi NM, Ghoneim AM et al (2023) Effect of green synthesized cerium oxide nanoparticles on fungal disease of wheat plants: a field study. Chemosphere 339:139731. https://doi.org/10.1016/j.chemosphere.2023.139731
10. Alsayd Bouqellah N (2023) In silico and in vitro investigation of the antifungal activity of trimetallic Cu—Zn-magnetic nanoparticles against *Fusarium oxysporum* with stimulation of the tomato plant's drought stress tolerance response. https://doi.org/10.1016/j.micpath.2023.106060.
11. Anum F, Jabeen K, Javad S et al (2023) Management of botrytis grey mold of tomato using bio-fabricated silver nanoparticles. S Afr J Bot 159:642–652. https://doi.org/10.1016/j.sajb.2023.06.019
12. Bashir Kutawa A, Ahmad K, Ali A et al (2021) Trends in nanotechnology and its potentialities to control plant pathogenic fungi: a review. Biology 10:881. https://doi.org/10.3390/biology10090881
13. Bibi S, Raza M, Shahbaz M et al (2023) Biosynthesized silver nanoparticles enhanced wheat resistance to *Bipolaris sorokiniana*. Plant Physiol Biochem 203:108067. https://doi.org/10.1016/j.plaphy.2023.108067
14. Bravo Cadena M, Preston GM, Van der Hoorn RAL et al (2018) Species-specific antimicrobial activity of essential oils and enhancement by encapsulation in mesoporous silica nanoparticles. Ind Crops Prod 122:582-590. https://doi.org/10.1016/j.indcrop.2018.05.081
15. Bruna T, Maldonado-Bravo F, Jara P et al (2022) Silver nanoparticles and their antibacterial applications. Int J Mol Sci 22:7202. https://doi.org/10.3390/ijms22137202
16. Cai L, Cai L, Jia H et al (2020) Foliar exposure of Fe3O4 nanoparticles on *Nicotiana benthamiana*: Evidence for nanoparticles uptake, plant growth promoter and defense response elicitor against plant virus. J Hazard Mater 393:122415. https://doi.org/10.1016/j.jhazmat.2020.122415
17. Cheng M, Yan X, Wang X et al (2023) Fabrication, characterization, and antifungal assessment of oregano essential oil-loaded nano-silica against *Curvularia lunata* in brown rot of *Agaricus bisporus* storage. Food Bioproc Tech 16:2921–2934. https://doi.org/10.1007/s11947-023-03125-x
18. Cheng J, Li M, Zhu RX et al (2023) Timing self-cleaning nanoherbicide: design of triple-structure nanovectors for weed control and pesticide residues treatment. Adv Agro Chem. https://doi.org/10.1016/j.aac.2023.10.002
19. Chintala S, Laishram R, Mondal P et al (2023) Turmeric (*Curcuma longa* L.) rhizome extract mediated silver nanoformulation exhibits enhanced antifungal property against *Rhizoctonia solani* and boosts innate immunity of rice. Ind Crops Prod 206:117616. https://doi.org/10.1016/j.indcrop.2023.117616
20. Chun SC, Chandrasekaran M (2019) Chitosan and chitosan nanoparticles induced expression of pathogenesis-related proteins genes enhances biotic stress tolerance in tomato. Int J Biol Macromol 125:948–954. https://doi.org/10.1016/j.ijbiomac.2018.12.167
21. Cruz-Luna AR, Cruz-Martínez H, Vásquez-López A et al (2021) Metal nanoparticles as novel antifungal agents for sustainable agriculture: current advances and future directions. J Fungi 7:1033. https://doi.org/10.3390/jof7121033

22. Derbalah A, Essa T, Kame SM et al (2021) Silver oxide nanostructures as a new trend to control strawberry charcoal rot induced by *Macrophomina phaseolina*. Pest Manag Sci 78:4638–4648. https://doi.org/10.1002/ps.7084
23. Dhiman S, Varma A, Rao M et al (2023) Deciphering the fertilizing and disease suppression potential of phytofabricated zinc oxide nanoparticles on *Brassica juncea*. Environ Res 231:116276. https://doi.org/10.1016/j.envres.2023.116276
24. Ding LN, Li YT, Wu YZ et al (2022) Plant disease resistance-related signaling pathways: recent progress and future prospects. Int J Mol Sci 23:16200. https://doi.org/10.3390/ijms232416200
25. Dilbar S, Sher H, Ali A et al (2023) Biological synthesis of Ag-nanoparticles using *Stachys parviflora* and its inhibitory potential against *Xanthomonas campestris*. S Afr J Bot 157:409-422. https://doi.org/10.1016/j.sajb.2023.04.034
26. El-Ashry RM, El-Saadony MT, El-Sobki AEA et al (2022) Biological silicon nanoparticles maximize the efficiency of nematicides against biotic stress induced by *Meloidogyne incognita* in eggplant. Saudi J Biol Sci 29:920–932. https://doi.org/10.1016/j.sjbs.2021.10.013
27. El-Ashry RM, Nader MM, Shami A et al (2023) Ecofriendly synthesis and nematicidal application of copper nanoparticles fabricated from *Bacillus subtilis* AM18, against root-knot nematode of cucumber. Eur J Plant Pathol. https://doi.org/10.1007/s10658-023-02727-7
28. Faikah Mustafa I, Zobir Hussein M (2020) Synthesis and technology of nanoemulsion-based pesticide formulation. Nanomaterials 10:1608. https://doi.org/10.3390/nano10081608
29. Fan G, Xiao Q, Li Q et al (2023) Antimicrobial mechanisms of ZnO nanoparticles to phytopathogen *Pseudomonas syringae*: damage of cell envelope, suppression of metabolism, biofilm and motility, and stimulation of stomatal immunity on host plant. Pestic Biochem Physiol 194:105455. https://doi.org/10.1016/j.pestbp.2023.105455
30. Forini MML, Pontes MS, Antunes DR et al (2022) Nano-enabled weed management in agriculture: from strategic design to enhanced herbicidal activity. Plant Nano Biology 1:100008. https://doi.org/10.1016/j.plana.2022.100008
31. Gao Y, Xiao Y, Mao K et al (2020) Thermoresponsive polymer-encapsulated hollow mesoporous silica nanoparticles and their application in insecticide delivery. J Chem Eng 383:123169. https://doi.org/10.1016/j.cej.2019.123169
32. Ganilho C, Bessa da Silva M, Paiva C et al (2022) Environmental safety assessments of lipid nanoparticles loaded with lambda-cyhalothrin. Nanomaterials 12:2576. https://doi.org/10.3390/nano12152576
33. Gari Saritha GN, Anju T, Kumar A (2022) Nanotechnology-big impact: how nanotechnology is changing the future of agriculture? J Agric Food Res 10:100457. https://doi.org/10.1016/j.jafr.2022.100457
34. Ghareeb RY, El-Din Shams El-Din NG, El Maghraby DM et al (2022) Nematicidal activity of seaweed-synthesized silver nanoparticles and extracts against *Meloidogyne incognita* on tomato plants. Sci Rep 12:3841. https://doi.org/10.1038/s41598-022-06600-1
35. Gimenez E, Salinas M, Manzano-Agugliaro F (2018) Worldwide research on plant defense against biotic stresses as improvement for sustainable agriculture. Sustainability 10:391. https://doi.org/10.3390/su10020391
36. Giri VP, Pandey S, Srivastava S et al (2023) Chitosan fabricated biogenic silver nanoparticles (Ch@BSNP) protectively modulate the defense mechanism of tomato during bacterial leaf spot (BLS) disease. Plant Physiol Biochem 197:107637. https://doi.org/10.1016/j.plaphy.2023.03.014
37. González Guzmán M, Cellini F, Fotopoulos V et al (2022) New approaches to improve crop tolerance to biotic and abiotic stresses. Physiol Plant 174:e13547. https://doi.org/10.1111/ppl.13547
38. Gowda S, Sriram S (2023) Green synthesis of chitosan silver nanocomposites and their antifungal activity against *Colletotrichum truncatum* causing anthracnose in chillies. Plant Nano Biol 5:100041. https://doi.org/10.1016/j.plana.2023.100041
39. Grillo R, Pereira AES, Nishisaka CS et al (2014) Chitosan/tripolyphosphate nanoparticles loaded with paraquat herbicide: an environmentally safer alternative for weed control. J Hazard Mater 278:163–171. https://doi.org/10.1016/j.jhazmat.2014.05.079

40. Hasanin M, Hashem AH, Lashin I et al (2023) In vitro improvement and rooting of banana plantlets using antifungal nanocomposite based on myco-synthesized copper oxide nanoparticles and starch. Biomass Convers Biorefin 13:8865–8875. https://doi.org/10.1007/s13399-021-01784-4
41. Hassan Amin H, Salama Ali MS, Abd El-Hamed El-Sheikh TA et al (2022) Hematological and histopathological impacts of nano-emamectin benzoate against the larvae of the cotton leafworm, *Spodoptera littoralis* (Boisduval) (Lepidoptera: Noctuidae) under laboratory conditions. Beni-Suef Univ J Basic Appl Sci 11:133. https://doi.org/10.1186/s43088-022-00315-y
42. Hoang IJ, Thanh TL, Sangpueak R et al (2023) The effect of chitosan nanoparticle formulations for control of leaf spot disease on cassava. Phytoparasitica 51:621–636. https://doi.org/10.1007/s12600-023-01065-y
43. Ibrahim S, Abou-Elseoud WS, Elbehery HH et al (2022) Chitosan-cellulose nanoencapsulation systems for enhancing the insecticidal activity of citronella essential oil against the cotton leafworm *Spodoptera littoralis*. Ind Crops Prod 184:115089. https://doi.org/10.1016/j.indcrop.2022.115089
44. Iqbal Z, Shariq Iqbal M, Hashem A et al (2021) Plant defense responses to biotic stress and its interplay with fluctuating dark/light conditions. Front Plant Sci 12:631810. https://doi.org/10.3389/fpls.2021.631810
45. Jameel M, Schoeb M, Talib Khan M et al (2020) Enhanced insecticidal activity of thiamethoxam by zinc oxide nanoparticles: a novel nanotechnology approach for pest control. ACS Omega 5:1607–1615. https://doi.org/10.1021/acsomega.9b03680
46. Jenish A, Ranjani S, Hemalatha S (2022) *Moringa oleifera* nanoparticles demonstrate antifungal activity against plant pathogenic fungi. Appl Biochem Biotechnol 194:4959–4970. https://doi.org/10.1007/s12010-022-04007-2
47. Jian Y, Chen X, Ahmed T et al (2022) Toxicity and action mechanisms of silver nanoparticles against the mycotoxin-producing fungus *Fusarium graminearum*. J Adv Res 38:1–12. https://doi.org/10.1016/j.jare.2021.09.006
48. Jiang Lim C, Basri M, Omar D et al (2012) Green nanoemulsion-laden glyphosate isopropylamine formulation in suppressing creeping foxglove (*A. gangetica*), slender button weed (*D. ocimifolia*) and buffalo grass (*P. conjugatum*). Pest Manag Sci 69:104–111. https://doi.org/10.1002/ps.3371
49. Joshi SM, De Britto S, Jogaiah S (2021) Myco-engineered selenium nanoparticles elicit resistance against tomato late blight disease by regulating differential expression of cellular, biochemical and defense responsive genes. J Biotech 325:196–206. https://doi.org/10.1016/j.jbiotec.2020.10.023
50. Kalagatur NK, Nirmal Ghosh OS, Sundararaj N et al (2018) Antifungal activity of chitosan nanoparticles encapsulated with *Cymbopogon martinii* essential oil on plant pathogenic fungi *Fusarium graminearum*. Front Pharmacol 9:610. https://doi.org/10.3389/fphar.2018.00610
51. Kanaujiya D, Kumar V, Dwivedi SK et al (2020) Photobiosynthesis of silver nanoparticle using extract of *Aspergillus flavus* CR500: its characterization, antifungal activity and mechanism against *Sclerotium rolfsii* and *Rhizoctonia solani*. J Clust Sci 31:1041–1050. https://doi.org/10.1007/s10876-019-01709-2
52. Karmous I, Vaidya S, Dimkpa S et al (2023) Biologically synthesized zinc and copper oxide nanoparticles using *Cannabis sativa* L. enhance soybean (*Glycine max*) defense against *Fusarium virguliforme*. Pestic Biochem Physiol 194:105486. https://doi.org/10.1016/j.pestbp.2023.105486
53. Khezerlou A, Alizadeh-Sani M, Azizi-Lalabadi M et al (2018) Nanoparticles and their antimicrobial properties against pathogens including bacteria, fungi, parasites and viruses. Microb Pathog 123:505–526. https://doi.org/10.1016/j.micpath.2018.08.008
54. Khan I, Saeed K, Khan I (2019) Nanoparticles: properties, applications and toxicities. Arab J Chem 12:908–931. https://doi.org/10.1016/j.arabjc.2017.05.011
55. Khan RA, Sartaj Alam S, Najeeb S et al (2023) Mitigating Cd and bacterial wilt stress in tomato plants through trico-synthesized silicon nanoparticles and Trichoderma metabolites. Environ Pollut 333:122041. https://doi.org/10.1016/j.envpol.2023.122041

56. Kongala SI, Rani Nadendla S, Mamidala P (2021) Internalization and induction of defense responses in tobacco by harpinPss conjugated gold nanoparticles as a foliar spray. Colloids Interface Sci Commun 43:100438. https://doi.org/10.1016/j.colcom.2021.100438
57. Kongala SI, Mamidala P (2023) Harpin-loaded chitosan nanoparticles induced defense responses in tobacco. Carbohydr Polym Technol Appl 5:100293. https://doi.org/10.1016/j.carpta.2023.100293
58. Kumar Dubey A, Mostafavi E (2023) Phyto-insect repellents: a nanotechnology-based approach of sustainability towards synthetic insect repellents. Curr Opin Green Sustain Chem 41:100827. https://doi.org/10.1016/j.cogsc.2023.100827
59. Kumar Ray M, Kumar Mishra A, Kishore Mohanta Y et al (2023) Nanotechnology as a promising tool against phytopathogens: a futuristic approach to agriculture. Agriculture 13:1856. https://doi.org/10.3390/agriculture13091856
60. Lesnichaya M, Tsivileva O (2023) Arabinogalactan-stabilized selenium sulfide nanoparticles and their fungistatic activity against *Phytophthora cactorum*. J Clust Sci 34:853–864. https://doi.org/10.1007/s10876-022-02264-z
61. Lesnichaya M, Perfileva A, Nozhkina O et al (2022) Synthesis, toxicity evaluation and determination of possible mechanisms of antimicrobial effect of arabinogalactane-capped selenium nanoparticles. J Trace Elem Med Biol 69:126904. https://doi.org/10.1016/j.jtemb.2021.126904
62. Liu J, Xu D, Xu G et al (2023) Smart controlled-release avermectin nanopesticides based on metal—organic frameworks with large pores for enhanced insecticidal efficacy. J Chem Eng 475:146312. https://doi.org/10.1016/j.cej.2023.146312
63. Lv H, Li X, Li J et al (2023) Overcoming resistance in insect pest with a nanoparticle-mediated dsRNA and insecticide co-delivery system. J Chem Eng 475:146239. https://doi.org/10.1016/j.cej.2023.146239
64. Majeed N, Panigrahi KCS, Sukla LB et al (2020) Application of carbon nanomaterials in plant biotechnology. Mater Today Proc 30:340–345. https://doi.org/10.1016/j.matpr.2020.01.618
65. Mahyoub JA (2021) Bioactivity of two marine algae extracts and their synthesized silver nanoparticles as safe controls against *Musca domestica* housefly. Entomol Res 51:323–330. https://doi.org/10.1111/1748-5967.12512
66. Mansoor S, Zahoor I, Rehman Baba T et al (2021) Fabrication of silver nanoparticles against fungal pathogens. Front Nanotechnol 3:679358. https://doi.org/10.3389/fnano.2021.679358
67. Mercado-Meza DY, Guevara-González RG, Esquivel K et al (2023) Green silver nanoparticles display protection against *Clavibacter michiganensis* subsp. *michiganensis* in tomato plants (*Solanum lycopersicum* L.). Plant Stress 10:100256. https://doi.org/10.1016/j.stress.2023.100256
68. Mohan Pathak V, Verma VK, Singh Rawat B et al (2022) Current status of pesticide effects on environment, human health and it's eco-friendly management as bioremediation: a comprehensive review. Front Microbiol 13:962619. https://doi.org/10.3389/fmicb.2022.962619
69. Mohammad AA, Amer HM, El-Sawy SM et al (2022) Nematicidal activity of sweet annie and garden cress nano-formulations and their impact on the vegetative growth and fruit quality of tomato plants. Sci Rep 12:22302. https://doi.org/10.1038/s41598-022-26819-2
70. Mondéjar-López M, López-Jimenez AJ, Ahrazem O et al (2023) Chitosan coated—biogenic silver nanoparticles from wheat residues as green antifungal and nanoprimig in wheat seeds. Int J Biol Macromol 225:964–973. https://doi.org/10.1016/j.ijbiomac.2022.11.159
71. MujtabaShah Z, Naz R, Naz S et al (2023) Incorporation of zinc sulfide nanoparticles, *Acinetobacter pittii* and *Bacillus velezensis* to improve tomato plant growth, biochemical attributes and resistance against *Rhizoctonia solani*. Plant Physiol Biochem 202:107909. https://doi.org/10.1016/j.plaphy.2023.107909
72. Muthusamy R, Ramkumar G, Kumarasamy S et al (2023) Synergism and toxicity of iron nanoparticles derived from *Trigonella foenum-graecum* against pyrethriod treatment in *S. litura* and *H. armigera* (Lepidoptera: Noctuidae). Environ Res 231:116079. https://doi.org/10.1016/j.envres.2023.116079

73. Noman M, Ahmed T, Shahid M et al (2023) Salicylic acid-doped iron nano-biostimulants potentiate defense responses and suppress *Fusarium* wilt in watermelon. J Adv Res. https://doi.org/10.1016/j.jare.2023.06.011
74. Nie D, Li J, Xie Q et al (2023) Nanoparticles: a potential and effective method to control insect-borne diseases. Bioinorg Chem Appl 2023:5898160. https://doi.org/10.1155/2023/5898160
75. Ntalli N, Zochios G, Nikolaou P et al (2023) *Carlina acaulis* essential oil nanoemulsion for managing *Meloidogyne incognita*. Ind Crops Prod 193:116180. https://doi.org/10.1016/j.indcrop.2022.116180
76. Olaitan Ogunyemi S, Xu X, Xu L et al (2023) Cobalt oxide nanoparticles: an effective growth promoter of *Arabidopsis* plants and nano-pesticide against bacterial leaf blight pathogen in rice. Ecotoxicol Environ Saf 257:114935. https://doi.org/10.1016/j.ecoenv.2023.114935
77. Pittarate S, Perumal V, Kannan S et al (2023) Insecticidal efficacy of nanoparticles against *Spodoptera frugiperda* (J.E. Smith) larvae and their impact in the soil. Heliyon 9:e16133. https://doi.org/10.1016/j.heliyon.2023.e16133
78. Popescu GC, Popescu M, Khondker M et al (2022) Agricultural sciences and the environment: reviewing recent technologies and innovations to combat the challenges of climate change, environmental protection, and food security. Agron J 114:1895–1901. https://doi.org/10.1002/agj2.21164
79. Rani K, Devi N, Banakar P et al (2022) Nematicidal potential of green silver nanoparticles synthesized using aqueous root extract of *Glycyrrhiza glabra*. Nanomaterials 12:2966. https://doi.org/10.3390/nano12172966
80. Rojas B, Soto N, Villalba M et al (2021) Antibacterial activity of copper nanoparticles (CuNPs) against a resistant calcium hydroxide multispecies endodontic biofilm. Nanomaterials 1:2254. https://doi.org/10.3390/nano11092254
81. Rossi V, Caffi T, Salotti I et al (2023) Sharing decision-making tools for pest management may foster implementation of integrated pest management. Food Secur 15:1459–1474. https://doi.org/10.1007/s12571-023-01402-3
82. Saberi Riseh R, Hassanisaadi M, Vatankhah M et al (2022) Nano/microencapsulation of plant biocontrol agents by chitosan, alginate, and other important biopolymers as a novel strategy for alleviating plant biotic stresses. Int J Biol Macromol 222:1589–1604. https://doi.org/10.1016/j.ijbiomac.2022.09.278
83. Salah M, Abd El-Aziz MH, El-Sayed Selim R (2022) Physiological and morphological response of tomato plants to nano-chitosan used against bio-stress induced by root-knot nematode (*Meloidogyne incognita*) and *Tobacco mosaic* tobamovirus (TMV). Eur J Plant Pathol 163:799–812. https://doi.org/10.1007/s10658-022-02516-8-
84. Slavin YN, Asnis AJ, Häfeli UO et al (2017) Metal nanoparticles: understanding the mechanisms behind antibacterial activity. J Nanobiotechnol 15:65. https://doi.org/10.1186/s12951-017-0308-z
85. Sathiyabama M, Manikandan A (2016) Chitosan nanoparticle induced defense responses in fingermillet plants against blast disease caused by *Pyricularia grisea* (Cke.) Sacc. Carbohydr Polym 154:241–246. https://doi.org/10.1016/j.carbpol.2016.06.089
86. Saxena J, Ayushi KM (2023) Evaluation of *Sclerotinia sclerotiorum* MTCC 8785 as a biological agent for the synthesis of silver nanoparticles and assessment of their antifungal potential against *Trichoderma harzianum* MTCC 801. Environ Res 216:114752. https://doi.org/10.1016/j.envres.2022.114752
87. Sayed M, Hamouda RA, Elshamy MM (2022) Using biosynthesized zinc oxide nanoparticles as a pesticide to alleviate the toxicity on banana infested with parasitic-nematode. Waste Biomass Valorization 13:405–415. https://doi.org/10.1007/s12649-021-01527-6
88. Saw G, Nagdev P, Jeer M et al (2023) Silica nanoparticles mediated insect pest management. Pestic Biochem Physiol 194:105524. https://doi.org/10.1016/j.pestbp.2023.105524
89. Shinde BH, Shinde PB, Inamdar AK et al (2023) Green synthesis of silver/silver oxide composite nanoparticles using Cuscuta Reflexa plant for the insecticidal applications. Mater Today Proc 92:549–553. https://doi.org/10.1016/j.matpr.2023.03.747

90. Silva Caetano A, das Graças Cardoso M, Haddi K et al (2022) *Rosmarinus officinalis* essential oil incorporated into nanoparticles as an efficient insecticide against *Drosophila suzukii* (Diptera: Drosophilidae). Austral Entomol 61:265–272. https://doi.org/10.1111/aen.12592
91. Sindhu M, Rajkumar V, Alagubrahmam Annapoorani C et al (2023) Nanoencapsulation of garlic essential oil using chitosan nanopolymer and its antifungal and anti-aflatoxin B1 efficacy in vitro and in situ. Int J Biol Macromol 243:125160. https://doi.org/10.1016/j.ijbiomac.2023.125160
92. Takeshita V, Teixeira de Sousa B, Preisler AC et al (2021) Foliar absorption and field herbicidal studies of atrazine-loaded polymeric nanoparticles. J Hazard Mater 418:126350
93. Thabet AF, Boraei HA, Galal OA et al (2021) Silica nanoparticles as pesticide against insects of different feeding types and their non-target attraction of predators. Sci Rep 11:14484. https://doi.org/10.1038/s41598-021-93518-9
94. Teixeira de Sousa B, do Espirito Santo Pereira A, Fernandes Fraceto L et al (2022) Post-emergence herbicidal activity of nanoatrazine against Alternanthera tenella Colla plants compared to other weed species. Heliyon 8:e09902. https://doi.org/10.1016/j.heliyon.2022.e09902
95. Tran TNM, Vu NBD, Nguyen MH (2023) Antifungal activity of essential oil-encapsulated lipid nanoemulsions against *Neopestalotiopsis rosae* causing leaf spot on strawberry. J Plant Dis Prot 130:823–832. https://doi.org/10.1007/s41348-023-00760-6
96. Tripathi D, Singh M, Pandey-Rai S (2022). Crosstalk of nanoparticles and phytohormones regulate plant growth and metabolism under abiotic and biotic stress. Plant Stress 6:100107. https://doi.org/10.1016/j.stress.2022.100107
97. Vanti GL, Kurjogi M, Basavesha KN et al (2020) Synthesis and antibacterial activity of Solanum torvum mediated silver nanoparticle against *Xxanthomonas axonopodis* pv. *punicae* and *Ralstonia solanacearum*. J Biotech 309:20–28. https://doi.org/10.1016/j.jbiotec.2019.12.009
98. Vinzant K, Rashid M, Khodakovskaya MV (2023) Advanced applications of sustainable and biological nano-polymers in agricultural production. Front Plant Sci 13:1081165. https://doi.org/10.3389/fpls.2022.1081165
99. Wang L, Pan T, Gao X et al (2022) Silica nanoparticles activate defense responses by reducing reactive oxygen species under *Ralstonia solanacearum* infection in tomato plants. NanoImpact 28:100418. https://doi.org/10.1016/j.impact.2022.100418
100. Wang Y, Yang L, Zhou X et al (2023) Molecular mechanism of plant elicitor daphnetin-carboxymethyl chitosan nanoparticles against *Ralstonia solanacearum* by activating plant system resistance. Int J Biol Macromol 241:124580. https://doi.org/10.1016/j.ijbiomac.2023.124580
101. Yang L, Kaziem AE, Lin Y et al (2021) Carboxylated β-cyclodextrin anchored hollow mesoporous silica enhances insecticidal activity and reduces the toxicity of indoxacarb. Carbohydr Polym 266:118150. https://doi.org/10.1016/j.carbpol.2021.118150
102. Yeguerman C, Jesser E, Massiris M et al (2020) Insecticidal application of essential oils loaded polymeric nanoparticles to control German cockroach: design, characterization and lethal/sublethal effects. Ecotoxicol Environ Saf 189:110047. https://doi.org/10.1016/j.ecoenv.2019.110047
103. Zeng Q, Yu C, Chang X et al (2022) CeO_2 nanohybrid as a synergist for insecticide resistance management. J Chem Eng 446:137074. https://doi.org/10.1016/j.cej.2022.137074
104. Zhang N, Hu W, Hou R et al (2023) Enhanced fungicidal activity and mechanism of pyraclostrobin nanoparticle with reactive oxygen species responsiveness against *Rhizoctonia solani*. J Clean Prod 421:138494. https://doi.org/10.1016/j.jclepro.2023.138494
105. Zhao S, Zhang Y, Chen L et al (2023) Chitosan-thymol nanoparticle with pH responsiveness as a potential intelligent botanical fungicide against *Botrytis cinerea*. Pestic Biochem Physiol 195:105571. https://doi.org/10.1016/j.pestbp.2023.105571

Induction of Tolerance to Abiotic Stress by Applying Nanostimulants to Crops

Gonzalo Tortella Fuentes, Rodrigo Rodríguez, Paola Duran, Martín Fernández-Baldo, Mauricio Schoebitz, Ricardo Tighe, Joana Pieretti, and Paola Fincheira

Abstract Abiotic stress, including drought, salinity, and extreme temperatures, significantly impacts crop production and yield, establishing a significant challenge to global food security. Nanostimulants are designed to enhance plant resilience and have emerged as a promising approach to mitigate the adverse effects caused by abiotic stresses. Therefore, this chapter is related to how nanostimulants can induce tolerance to abiotic stress in crops, focusing on their role in improving nutrient uptake, stimulation, and antioxidant enzyme activity. By leveraging nanotechnology, specific nanoparticles such as zinc oxide, chitosan, copper, silver, and silica can be precisely applied to crops, enabling targeted responses at the cellular and molecular levels. These nanoparticles interact with crucial signaling molecules and pathways in

The original version of the chapter has been revised. A correction to this chapter can be found at https://doi.org/10.1007/978-981-96-4648-7_17

G. Tortella Fuentes (✉) · P. Fincheira
Centro de Excelencia en Investigación Biotecnológica Aplicada al Medio Ambiente (CIBAMA), Facultad de Ing. Y Ciencias, Universidad de La Frontera, Temuco, Chile
e-mail: gonzalo.tortella@ufrontera.cl

R. Rodríguez · P. Duran
Biocontrol Research Laboratory, Universidad de La Frontera, Temuco, Chile

M. Fernández-Baldo
Facultad de Química, Bioquímica y Farmacia, Instituto de Química de San Luis, INQUISAL (UNSL—CONICET), Universidad Nacional de San Luis, San Luis, Argentina

M. Schoebitz
Departamento de Suelos y Recursos Naturales, Facultad de Agronomía, Universidad de Concepción, Casilla, Chile

Center of Biotechnology, Universidad de Concepción, Barrio Universitario S/N, Concepción, Chile

R. Tighe
Departamento de Ciencias Agropecuarias y Acuícolas, Universidad Católica de Temuco, Temuco, Chile

J. Pieretti
Center for Natural and Human Sciences, Universidade Federal Do ABC, Santo André, Brazil

© The Author(s), under exclusive license to Springer Nature Singapore Pte Ltd. 2025, corrected publication 2025

A. Juárez-Maldonado et al. (eds.), *Plant Biostimulation with Nanomaterials*, Smart Nanomaterials Technology, https://doi.org/10.1007/978-981-96-4648-7_8

stress response, enhancing physiological and biochemical adaptations. Findings from various crops demonstrate that nanostimulants reinforce plant defense mechanisms and promote growth and productivity under stress conditions. As agriculture faces increasing abiotic pressures due to climate change, nanostimulants offer a sustainable and effective tool to improve crop resilience. Future research should continue to optimize nanoparticle formulations and application methods to maximize benefits while ensuring environmental safety.

Keywords Antioxidant enzyme activity · Biostimulation · Crop growth and productivity · Nanotechnology · Physiological and biochemical adaptations

1 Introduction

In modern agriculture, the concept of "Induction of Tolerance to Abiotic Stress by Applying Nanostimulants to Crops" is a pivotal area of research and innovation. This topic addresses crops' most pressing challenges under varying environmental stress conditions, such as drought, salinity, and extreme temperatures.

Abiotic stresses are non-living environmental factors that significantly impact plant health and productivity. These stresses include a wide range of conditions, from water scarcity, which leads to drought, to the excess of salts in the soil, causing salinity issues, and the extremes of hot and cold temperatures [1]. Each of these stresses triggers a unique set of responses in plants, leading to a decline in growth and yield. For instance, drought stress causes a reduction in water availability for plants, leading to wilting, leaf scorch, and, in severe cases, plant death. This stress significantly impairs the plant's ability to perform essential functions like photosynthesis and nutrient uptake, culminating in reduced growth and yield [2].

Conversely, salinity affects plants by disrupting their cells' ionic and osmotic balance [3]. It poses a significant challenge, especially in areas with high soil salinity or where irrigation practices involve saline water. Salinity stress can cause ion toxicity, osmotic stress, and nutrient imbalance in plants, leading to symptoms like leaf burn, stunted growth, and, in extreme cases, plant death [4]. This not only reduces the productivity of crops but also affects the quality of the yield [4, 5].

Temperature extremes (cold or heat) represent another category of abiotic stress. Heat stress can disrupt critical physiological processes in plants, such as enzyme activities and photosynthesis, resulting in symptoms like leaf wilting, sunburn, and reduced growth rates [6]. Conversely, cold stress can cause chilling injuries, affecting membrane fluidity and enzyme activities in plants, leading to wilting, chlorosis, and potentially killing plant tissues [7, 8]. Both heat and cold stresses are detrimental to crop growth and yields, posing a significant challenge to agricultural productivity.

In response to these challenges, using nanostimulants in agriculture has gained considerable attention [9]. Nanostimulants, nanoparticles, or nanomaterials designed to stimulate plant growth and development offer a novel approach to enhancing crop resilience against abiotic stresses [10]. Their unique properties, such as increased solubility, targeted delivery, and controlled release, make them more efficient and effective than conventional stimulants [11]. These nanostimulants can interact with plants at the molecular and cellular levels, potentially enhancing their ability to withstand environmental stressors [12].

The induction of crop tolerance against abiotic stress through nanostimulants involves complex interactions between these nanomaterials and plant systems [13, 14]. Studies suggest nanoparticles can influence plant signaling pathways, gene expression, and metabolic processes, enhancing stress resilience [12, 13, 15]. However, the exact mechanisms of how nanostimulants confer stress tolerance are still being explored, necessitating more comprehensive research.

Moreover, while the potential of nanostimulants in agriculture is significant, it comes with challenges and considerations. The safety, environmental impact, and regulatory aspects of using nanomaterials in agriculture must be thoroughly examined [16, 17]. Developing sustainable and eco-friendly nanomaterials, understanding their long-term effects on soil health and the broader ecosystem, and establishing guidelines for their safe use are critical for the responsible implementation of this technology.

Using nanostimulants to induce tolerance to abiotic stress in crops represents a promising yet complex frontier in agricultural science. This interdisciplinary field, straddling nanotechnology, plant science, and environmental science, holds great potential for enhancing crop resilience in the face of changing climate and environmental challenges. Its successful implementation could be significant in achieving sustainable and resilient agricultural practices.

2 Basics and Action Mechanisms of Nanostimulants

Nanostimulants represent a groundbreaking advancement in crop science, offering an innovative approach to enhancing plant growth and resilience. At their core, nanostimulants are minute particles, typically ranging in size from 1 to 100 nm. Their miniature scale is crucial as it imparts unique physical, chemical, and biological properties not present in their bulk counterparts. This nanoscale size increases surface area and reactivity, making nanostimulants highly efficient in interacting with plants.

Researchers design the composition of nanostimulants based on their application and the specific needs of the plants they intend to support. Commonly, these materials are formed from nanoparticles of essential nutrients like zinc, iron, selenium, and copper or carbon-based materials such as fullerenes or carbon nanotubes. Some nanostimulants also include organic compounds engineered at the nanoscale to enhance their effectiveness. The potential of these materials, generally applied in small quantities through foliar sprays or nutrient solutions, has been recognized for enhancing food crops' composition and nutraceutical quality [18]. Additionally, they have been noted to improve the plant's tolerance to various stress conditions, contributing to more resilient agricultural practices [19]. One of the key differences between nanostimulants and traditional stimulants is their mechanism of action and overall effectiveness. Traditional stimulants like fertilizers and growth hormones often have solubility, stability, and targeted delivery limitations [20]. In this sense, Over 50% of applied chemical fertilizers and pesticides remain unused, accumulating in soil and water through leaching and mineralization, posing environmental risks [21].

In contrast, nanostimulants, owing to their nanoscale size, can easily penetrate plant cells and tissues, facilitating more efficient and targeted delivery of nutrients and growth-promoting substances [19]. Furthermore, nanostimulants can be engineered to have controlled-release properties. This means that they can release their active ingredients slowly over time, providing a sustained benefit to the plants. This contrasts with traditional stimulants, which often release their active components rapidly, leading to a short-lived effect and the potential for environmental runoff. For example, an assay conducted by Valderrama et al. [22] using chitosan nanoparticles loaded with a mass ratio 1:0.25 demonstrated that 100% of indole-3-acetic acid (IAA) was released within 48 h. Moreover, these authors reported that CNPs-IAA applied to hydroponic lettuce crops (*Lactuca sativa* L.) had a beneficial effect on plant growth, resulting in a 30.9% increase in the number of lettuce leaves. This targeted approach reduces the risk of over-application and minimizes environmental impact, as lesser quantities of the stimulant are required to achieve the desired effect.

On the other hand, the efficacy of nanostimulants also extends to their ability to improve plant stress tolerance. By interacting with plant physiological processes, nanostimulants can enhance plant resistance to various abiotic stresses, such as drought, salinity, heavy metal presence, and extreme temperatures. This aspect is valued in the context of climate change and the increasing prevalence of extreme weather conditions. Li et al. [23] demonstrated that silica nanoparticles (SiNPs) benefit plant growth and development, particularly under biotic and abiotic stresses. Their study evaluated the impact of SiNPs on wheat growth, sugar metabolism, and hormone metabolism in the field. The results showed that SiNPs increased tiller numbers and elongated internodes by 66.7 and 27.4%, resulting in a larger biomass. These authors also reported that SiNPs enhanced the net photosynthetic rate by increasing total chlorophyll content. Moreover, they indicate that SiNPs regulated leaf and stem growth by modulating plant hormones and their related pathway genes such as TIR1 (for indol acetic acid) and PYR/PYL, PP2C, SnRK2, and ABF (for abscisic acid) and soluble sugar metabolisms increased gene expression of SPS, SUS, and α-glucosidase. These findings by Li et al. [23] demonstrate that SiNPs can enhance wheat growth and provide a theoretical foundation for their application in field conditions.

In another work, Mahawar et al. [24] evaluated the impact of copper oxide (CuO) and zinc oxide nanoparticles (ZnONPs) on the growth and physiology of radish (*Raphanus sativus* L.) under salinity stress. The foliar application of CuO and ZnONPs at 100 mg/L mitigated adverse effects caused by salinity, with ZnONPs being particularly effective in improving leaf area, photosynthetic efficiency, and mineral content and reducing oxidative stress markers. Nanoestimulants affect plants through physicochemical and molecular interactions, influencing metabolism, growth, and stress tolerance. These interactions depend not on the chemical composition of the nanoparticles or nanomaterials but on their surface properties [25]. This allows nanoestimulants to act regardless of whether the materials are essential for plant nutrition or are engineered structures like carbon nanotubes. The initial interaction between nanoestimulants and plant cells is primarily physicochemical [26]. Their small size enables them to penetrate plant cell walls and membranes more efficiently than larger particles. However, other properties, such as morphology or chemical composition, can affect the interaction and movement inside the nanomaterials plant [27]. Empirical models of the nanoparticle-leaf interaction based on hydrodynamic size and Z potential demonstrated that hydrophilic nanoparticles, including carbon dots, cerium oxide, and SiNPs with a size <20 and 11 nm for cotton and maize, exhibited the highest foliar delivery efficiencies [28]. Once inside the cells, they induce changes in gene expression, metabolic activities, and growth through surface charge interactions [24]. Although the biocompatibility of Ag nanoparticles is frequently considered low, applications on rice plants (*Oryza sativa*) between 10 and 40 ppm caused a significant up-regulation in the expression of CAT and APX genes [29]. However, as with any biostimulant, the positive effects diminish or become negative beyond a specific concentration. In this sense, Dutta Gupta et al. [29] demonstrated that *CuZnSOD* gene decreased with a high Ag nanoparticles concentration [29]. The same authors reported that in Ag nanoparticles treated seedlings, a reduced level of ROS and a lower degree of lipid peroxidation and H_2O_2 content were observed. Additionally, exposure to Ag nanoparticles enhanced catalase, ascorbate peroxidase, and glutathione reductase (GR) activity levels.

Nanoestimulants influence critical physiological processes at the cellular level, such as nutrient uptake and utilization. These materials improve nutrient absorption by carrying essential nutrients, particularly under stress conditions like nutrient-poor soils. Babajani et al. [30] investigated the effects of ZnO (0, 100, and 300 mg/L) and selenium nanoparticles (0, 10, and 50 mg/L) on *Melissa officinalis* plants. The results demonstrated that mixed treatments of selenium (low dosage) and ZnO significantly boosted biomass, lateral bud growth, and lateral root development. However, when selenium nanoparticles were applied at 50 mg/L, they inhibited growth and induced plant toxicity [30]. Additionally, these authors reported that both nanoparticles improved leaf and root concentrations of K, Fe, and Zn, with alterations in leaf thiols, ascorbate, and phenol levels. The exposure also upregulated genes involved in rosmarinic acid production and influenced enzyme activities [30]. Nanoestimulants can also regulate gene expression associated with stress tolerance, enhancing the plant's ability to cope with drought, salinity, or extreme temperatures. Ali et al. [31] investigated the effects of chitosan nanoparticles on *Catharanthus roseus* under drought stress. These authors reported that chitosan nanoparticles (at 1%) mitigated drought's negative impact on plant growth, water content, and chlorophyll levels by enhancing proline accumulation and boosting antioxidant enzymes like catalase and ascorbate peroxidase. These responses in *C. roseus* plants also reduced oxidative damage, limiting H_2O_2 and malondialdehyde levels and thus maintaining membrane integrity [31]. The authors also reported that CSNPs increased alkaloid production under drought conditions, correlating with heightened gene expression related to alkaloid biosynthesis. In another interesting study by Saad-Allah and Ragab [32], it was demonstrated that wheat seeds treated with sulfur nanoparticles at 100 μM showed improved growth rate, pigmentation, and nitrogen metabolism when exposed to salt stress. The authors also reported that plants treated with nanoparticles displayed adequate levels of catalase, ascorbate peroxidase, polyphenol oxidase, and superoxide dismutase, along with reduced levels of proline compared to controls without nanoparticles, thereby enhancing the growth conditions of wheat plants under salinity. Nanoparticles also boost photosynthetic efficiency, ensuring better growth under adverse conditions and aiding in osmotic adjustment by helping plants maintain cellular water balance under drought or salinity stress. For more detailed information on the effects of nanoparticles in mitigating damage caused by abiotic stress, readers can refer to El-Saadony et al. [10], Jalil and Ansari [33], and Kumari et al. [34]. On a molecular level, nanoestimulants modulate enzyme activities or phytohormones in stress response pathways, either enhancing or inhibiting specific enzymes to improve the plants adaptive capacity. Modulation of abscisic acid, gibberellin, cytokinin, ethylene, jasmonic acid, or salicylic acid signaling pathway has been reported in the presence of nanoparticles, as is detailed in Tripathi et al. [15]. These mechanisms collectively enhance plant resilience to various abiotic stresses through intricate interactions at the cellular and molecular levels. As research in this field advances, more profound insights into these processes will emerge, offering promising opportunities for developing more effective and sustainable agricultural practices.

In summary, nanostimulants represent a significant leap forward in agricultural technology. Their unique properties, such as enhanced solubility, targeted delivery, controlled release, and improved stress tolerance, set them apart from traditional stimulants. Nanostimulants can penetrate plant tissues more effectively, ensuring that nutrients and active compounds are delivered directly to the sites where they are most needed. This targeted approach not only maximizes the efficacy of the stimulants but also minimizes waste and environmental impact. Furthermore, the controlled release mechanisms of nanostimulants ensure a steady supply of nutrients over time, which is crucial for sustaining plant health and growth under varying environmental conditions. As research in this area continues to evolve, nanostimulants promise to transform agricultural practices, leading to more sustainable and efficient crop production. Nanostimulants can be pivotal in addressing global food security challenges and promoting sustainable agriculture by enhancing crop resilience to biotic and abiotic stresses.

Nanostimulants, through their distinctive properties, offer innovative ways to enhance crop tolerance to abiotic stresses such as drought or salinity [35, 36]. The mechanisms of action of the nanostimulants at the cellular and molecular levels are multifaceted and complex, involving various pathways and processes. Understanding these mechanisms is crucial for developing efficient and targeted applications in agriculture.

3 Types of Nanostimulants

The field of nano-stimulants encompasses a variety of nanomaterials, each tailored for specific roles in agriculture. These nano-stimulants, including nano-fertilizers, nano-pesticides, and others, are designed to improve crop resilience and productivity. Herein, we bring an overview of the different types of nano-stimulants and their roles in stress tolerance:

3.1 Nano-Fertilizers

The development of nano-fertilizers consists of a strategy to provide nutrients to plants more efficiently than traditional fertilizers. The class nano-fertilizer may refer to nanometric particles derived from large-sized materials commonly used as macronutrients like nitrogen, phosphorus, and potassium, or micronutrients, such as zinc and iron [37], promoting higher stability and controlled release [38]. Due to their nanoscale size, both polymeric and inorganic nanoparticles penetrate plant cells more effectively, ensuring a slow and steady release of nutrients. This precision reduces waste and environmental impact while enhancing nutrient uptake, even under stressful conditions like drought or poor soil quality [39].

One of the most studied nano-fertilizers is based on ZnONPs, as zinc is a primarily used micronutrient, and various simple and cost-effective synthesis routes can obtain ZnO NPs. Zinc is a vital plant micronutrient for enzyme function, protein synthesis, and growth hormone regulation. Zinc deficiency can result in stunted growth, chlorosis, and reduced crop yield [40]. Mehmood et al. [41] have developed an *Azadirachta indica* (Ai)-synthesized ZnO NPs, an eco-friendly nanoparticle that carries characteristic phytochemicals with different biological activities [41]. Maize plants exposed to chromium were treated with 50-nm Ai-ZnONPs, which promoted shoot and fresh weight increase, besides improving chlorophyll and biochemical parameters, alleviating chromium toxicity. A similar pattern was observed for proline-coated ZnO NPs (20 nm, hexagonal NPs) applied in drought-stressed coriander plants [40]. Like Ai-ZnO NPs, proline-coated ZnONPs could increase root and shoot length and fresh and dry weight, besides improving biochemical parameters such as reduced free radical scavenging and antioxidant activity in shoots. Overall, ZnONPs activity has been vastly studied, demonstrating promising potential. Therefore, recent strategies have been developed in the sense of combining ZnONPs with another macro or micronutrient in the same nanomaterial, such as $ZnMgO_2$ NPs and Ca-doped ZnONPs, achieving a more comprehensive range of activity [41, 42].

In addition to ZnONPs-based nano-fertilizers, other nanomaterials have been studied, such as copper, iron oxide, and magnesium oxide nanoparticles [43–46]. All research papers reported an efficient improvement of plants' physiological and biochemical parameters when working in an ideal concentration range. It is essential to highlight that the copper nanoparticles were able to promote an improvement of 80% compared to their bulk form, which indicates that using nano-fertilizers is a promising strategy for reducing cost and environmental pollution [45].

3.2 Nano-Encapsulated Nutrients

Besides the nanoparticles discussed in section a, nutrients can be delivered by nanoparticles that can adsorb or encapsulate these ions/molecules. This adsorption or encapsulation allows a controlled release, beneficial under abiotic stress conditions like extreme temperatures or water scarcity, ensuring that plants receive a consistent nutrient supply [47]. In this scenario, polymeric nanoparticles are preferred for synthesizing nanocapsules, highlighting chitosan and alginate-based nanocapsules. In contrast, hydroxyapatite and silicon-based nanoparticles are commonly selected as adsorbent materials due to their high porosity that enables a high loading of nutrients [48–50].

Regarding polymeric nanocapsules for delivering nutrients, chitosan is one of the main studied materials, especially as a non-toxic, eco-friendly polymer widely used in many applications. It's suitable for active ingredients due to its film-forming properties, safety, biodegradability, and cost-effectiveness [51]. CS-NPs obtained by ionic gelation method with tripolyphosphate (TPP) were loaded with sulfate-supplemented nitrogen, phosphorus, and potassium (NPKS), a commonly employed fertilizer [52]. Maize plants treated with the developed nanomaterial showed a higher number of leaves, superior chlorophyll content, and taller plants, mainly due to a slow-release property of the CS-NPs, promoting nutrient release for periods longer than 1 week. A similar pattern was observed for nitric oxide-loaded CS-NPs employed in the treatment of maize plants under salt stress [35], CuO-loaded CS/alginate NPs applied in Fortunella margarita Swingle seeds [48], and CuO-loaded CS NPs employed in Maize seeds priming [53], confirming the promising potential of chitosan-based nano-fertilizers in delivering different nutrients from the seedling, until the plant growth stage.

Interesting results have also been demonstrated for inorganic adsorbent nanostructures, such as hydroxyapatite and silicon oxide (SiO_2) NPs [50, 54]. Phosphorous-loaded nanohydroxyapatite (HA-NPs) was prepared by dipping the NPs in a humic acid solution, promoting the interaction of phenolic groups from the humic acid with the large surface area of the HA-NPs [50]. A notable improvement in early plant growth, rhizosphere bacteria, and resistance to NaCl-induced abiotic stress was achieved using hydroxyapatite nanoparticles combined with humic substances in Maize plants. Typical NPK fertilizer can also be loaded in adsorbent/porous nanostructures. Mushtaq and coworkers have developed an NPK-controlled release of SiO_2 NPs, double coated with chitosan and sodium alginate/kaolin, for enhanced water absorption and slow NPK release even in drought stress conditions [55]. NPK-loaded SiO_2 NPs could release nutrients gradually and retain large amounts of water, sustaining plants in drought conditions.

Overall, nanocapsules and adsorbent/porous NPs consist of a strategy as interesting as the use of the nutrients in the nanometric size, discussed in topic a. An advantage in the usage of capsules or delivery systems is the ability to use products already commercialized, such as common NPK fertilizers, but improving their efficiency under different conditions, modulating their concentration and release in crops. There is still plenty of room to be explored in the field of nano-fertilizers or nano-encapsulated nutrients, especially considering (a) the scaling-up process, (b) the versatility of the designed product in different conditions and crops, (c) the environmental impact, (d) the cost and stability of the developed nano-fertilizers, to enable their commercialization shortly.

3.3 Nano-Pesticides

Nano-pesticides are designed to protect crops from pests and diseases with higher efficiency and lower environmental impact. These nanoparticles can be engineered to release pesticides in a controlled manner, reducing the frequency of application and the amount of chemicals used [56]. Their targeted delivery system also minimizes the risk of harming non-target organisms. Similar to nano-fertilizers, a specific nanoparticle may also be directly employed as a nano-pesticide, such as zinc and copper-based nanoparticles [57, 58].

Zn-based nanoparticles, such as ZnO NPs, stand out in nanopesticides. Besides being an essential micronutrient, Zn-based nanomaterials present antimicrobial properties and are excellent candidates for nano-pesticides [58]. One of the most recent studies in the field evaluated the nano-pesticide efficacy of ZnO NPs against mealybug (*Puto barberi*) [59]. ZnO NPs were sprayed on a controlled population of this phytophage at 300, 1000, and 3000 ppm. The authors indicated that ZnO NPs could promote approximately 55% of mortality, suggesting a promising potential. Better results were obtained by Abd El-Latef and coworkers, employing Cu-dopped ZnO NPs [60]. The nanoparticles developed contained a maximum proportion of 5% Cu and were tested against a significant pest in Egyptian fields, Spodoptera littoralis. An impressive reduction of 95% in their population was observed with the developed nano-pesticide, suggesting promising application in the field. Suppose the reader wants to deepen in Zn-based nanomaterials and their application as nanopesticides. In that case, we recommend a specific and detailed review focusing only on this class of nano-pesticides by Zhao et al. [58].

Other inorganic nanoparticles have also been developed and evaluated as nano-pesticides. $Cu(OH)_2$ nano-pesticides also stand out due to their antimicrobial broad-spectrum, although recent works have focused on the impact of long-term use of this nano-pesticide and its ecotoxicity [57, 61]. Despite presenting great potential against their target, $Cu(OH)_2$ nano-pesticides have shown a negative impact in beneficial fungal and bacterial communities, besides toxicity against aquatic organisms. Thus, evaluating new pesticides' environmental impact during development is fundamental, as these innovative nanostructures might negatively affect the ecosystem.

For this reason, using nanocapsules is also a promising strategy to minimize the impact of already employed pesticides, enabling the application of lower concentrations with improved effects. Polylactide (PLA) nanocapsules containing validamycin and thifluzamide promoted a 60-fold improvement compared to the commercial non-encapsulated pesticide [62]. Similarly, chlorofenapyr was loaded in O-carboxymethyl chitosan (O-CMCS) nanoparticles and promoted pH-dependent release, which resulted in an improvement of 73% when compared to the non-encapsulated pesticide [63]. In addition, lower levels of pesticide were detected in the acidic gut of worker bees, suggesting that the nanocapsules contributed to reducing the impact of the pesticide on the environment.

3.4 *Nano-Herbicides*

Like nano-pesticides, nano-herbicides are designed for targeted weed control. They ensure the precise delivery of herbicides, reducing the amount of chemical dispersal in the environment. This targeted approach is especially beneficial in stress conditions like high weed density, as it enables effective weed management with minimal herbicide use [64]. In this scenario, similar strategies as those previously described are employed to deliver herbicides, with emphasis on polymer-based nanoparticles able to encapsulate these molecules, such as chitosan and polycaprolactone nanoparticles [64, 65].

Grillo and coworkers encapsulated Paraquat, a widely used herbicide, in chitosan nanoparticles [65]. Results indicated that the effectiveness of the herbicide was maintained, although its release and soil absorption were altered, reducing environmental impacts. The potential of chitosan nanoparticles in delivering herbicides was also evaluated with atrazine-carried nanoparticles [66]. Chitosan and chitosan-alginate nanoparticles could deliver atrazine without harmful effects on non-targeted plants, suggesting promising application of the developed nano-herbicides. Poly(epsilon-caprolactone) nanoparticles also demonstrated potential in delivering atrazine, effective only in targeted plants [65]. In addition, nanoencapsulating the herbicide atrazine promoted reduced soil mobility and genotoxicity, reducing environmental and human health risks.

Recent research has shown that the recently developed nano-herbicides are very effective against various weed species, with great potential to overcome disadvantages commonly found in common herbicides, such as the excessive use and the environmental pollution promoted.

3.5 *Nano-Biostimulant*

These include nanoparticles derived from biological sources or those that stimulate biological processes. They enhance plant growth and stress tolerance by improving nutrient uptake, stimulating growth-related hormones, or enhancing photosynthesis. In stress conditions, nano-biostimulants can help plants better cope with adverse environmental factors [67]. The field of nano-biostimulants is less explored and, most of the time, integrated with the development of nano-fertilizers, as these nutrients directly assist in the development of the plants, assisting in biotic and abiotic stress.

Among different nano-biostimulants, we can mention silicon nanoparticles [67], selenium nanoparticles [68], or more innovative approaches such as quercetin-loaded silicon nanoparticles [69], and salicylic acid loaded chitosan nanoparticles [70]. Generally, studies report an improvement in physical parameters, an increase in antioxidant enzymes, and a different genetic expression concerning biotic and abiotic stress, promoting improved resistance [69, 70]. Although promising, little is known about the mechanism of these innovative nano-biostimulants and their effect on different species. Our chapter motivates research groups to navigate this area further, as promising potential is observed in improving plants' response to adverse conditions with non-toxic and innovative nano-biostimulant formulations.

3.6 *Carbon-Based Nanomaterials*

These include fullerenes, carbon nanotubes, and graphene. They are known for enhancing seed germination, plant growth, and stress tolerance. Their high surface area and reactivity can positively influence plant metabolic processes, especially under stress conditions [71].

Graphene-based nanomaterials have already been shown as an essential alternative in increasing plants' growth and yield, besides improving antioxidative defense and alleviating abiotic stress. Guo and coworkers have demonstrated the potential of graphene oxide on tomato growth in different stages [72]. Results confirmed that graphene oxide improved the development of tomato seedlings and mature plants, upregulating genes related to root development. Likewise, a lysine/methionine-modified graphene oxide improved the growth in pearl millet plants [73]. The nanostructure promoted an outstanding increase in fresh weight threefold, improving plant tolerance and adaptability to salinity stress.

Similar data can be found for plants treated with fullerenes or carbon nanotubes. For example, multi-walled carbon nanotubes (MWCNTs) enhanced the remediation efficiency of *Solanum nigrum* L. under cadmium and arsenic stress [74]. Applying MWCNTs at 100, 500, and 1000 mg kg^{-1} promoted plant growth, enhanced antioxidant enzymatic activity, and decreased cadmium and arsenic accumulation in soil. MWCNTs also promoted beneficial effects on maize plants at a dosage of 100 mg L^{-1}, increasing shoot and root fresh and dry weight and seedling length [75]. The authors have demonstrated that MWCNTs could improve these parameters by regulating critical carbon and nitrogen metabolism enzymes. Ozfidan-Konakci and coauthors also reported that fullerenes ameliorated cobalt toxicity in maize plants through its direct participation in nitrogen metabolism, increasing the levels of defense enzymes and improving nitrogen assimilation [76].

Interestingly, these carbon-based nanomaterials can also be easily dopped with another exciting structure, such as macro or micronutrients, or even pesticides and herbicides, conferring multiple functions for these nanomaterials, with diminished harmful environmental effects. Incorporating these types of nanostimulants into agricultural practices holds the potential to significantly enhance crop resilience to various stresses while also improving yield and reducing environmental impact. Continued research and development in this field are essential to fully harnessing the benefits of nanotechnology in agriculture.

4 The Application Methods of Nanostimulants in Agriculture

In plant science, nanostimulants have emerged as a cutting-edge technology over the past decade, primarily for using various nanoparticles (NPs) as carriers of agrochemicals or fertilizers [77]. These NPs have proven effective in several aspects of plant development, including seedling growth, germination, root development, enhancement of carbohydrate metabolism, and nutrient transport under stress and normal conditions [78]. Additionally, the generation of reactive oxygen species (ROS) has been observed [79]. However, applying nanostimulants in agriculture is a critical aspect of their use, requiring careful consideration of methods, dosage, timing, and frequency to ensure maximum efficacy and safety [80]. The application methods are designed to make the best use of the unique properties of nanostimulants, ensuring that they reach the target sites in plants effectively and efficiently. Various techniques, such as seed priming, foliar spray, soil mixing, and water supply (hydroponic culture and irrigation systems), can be employed, all aimed at sustainable agriculture approaches (Fig. 1).

4.1 Application of NPs by Water Supply

Incorporating nanostimulants into irrigation systems, such as drip irrigation, efficiently applies them over large areas, ensuring even distribution, especially for water-soluble types. Dosage and frequency vary based on crop type, growth stage, and specific stress conditions. Nanostimulants can be directly added to nutrient solutions in hydroponic systems, ensuring continuous plant access while demanding precise dosing to avoid toxicity.

ZnONPs added to Nicotiana tabacum seedlings via nutrient solutions positively influenced growth physiology and metabolite levels, enhancing enzymatic activities and anatomical features under nutrient-deficit conditions [81]. Similarly, ZnONPs added to *Lycopersicon leucocephala* seedling nutrient solutions stimulated the antioxidant system, promoting growth, mitigating lipid peroxidation, and enhancing

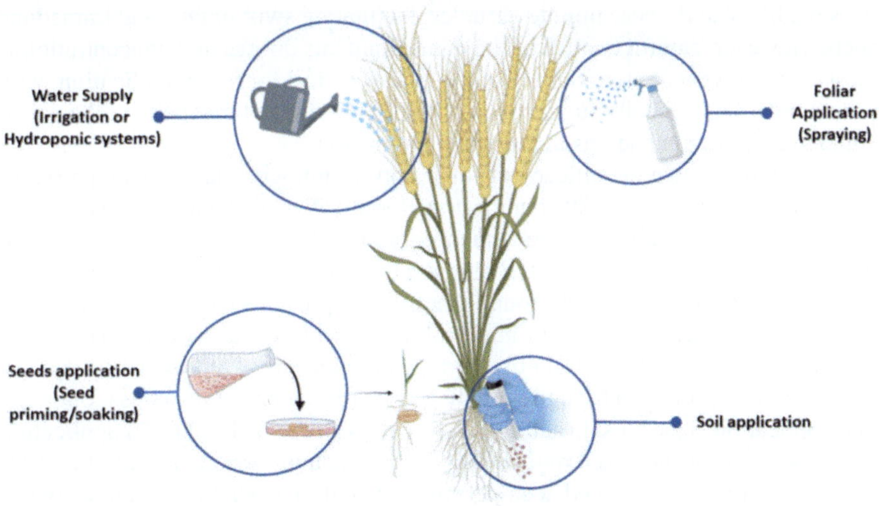

Fig. 1 Principal methods for the applications of NPs to plants

protein and pigment contents under nutrient deficit conditions [82]. Additionally, various other types of nanoparticles have been introduced through liquid solutions. For example, Silicon NPs added to Musa acuminata shoot nutrient solutions-maintained Na + /K + balance, promoted photosynthesis, and increased chlorophyll levels and leaf growth under salinity and drought stress [83]. Also, SiNPs added to irrigation water for Prunus mahaleb increased seed germination and development, photosynthetic parameters, and antioxidant enzyme activity while reducing MDA and H_2O_2 levels, chlorophyll destruction, and oxidative damage under salinity stress [84]. Furthermore, silver nanoparticvle (AgNPs) added to irrigation water for wheat increased leaf number and promoted growth under heat stress [85]. Moreover, AgNPs added via nutrient solution to Licorice decreased lipid peroxidation and H_2O_2 levels while increasing phenolics, total protein, and osmolyte contents under cold stress [86]. Similarly, Titanium dioxide NPs added three times to Hoagland solution two weeks after salinity stress application in *Dracocephalum moldavica*, positively impacted physiochemical properties by inducing antioxidant activities under salinity stress [87]. Iron NPs were added via nutrient media to Grap, which increased total protein content, decreased proline content, antioxidant enzymatic activity, and H_2O_2 levels under salinity stress while decreasing Na^+ and increasing K^+ [88].

4.2 Foliar Application of NPs

Foliar application is a widely adopted method for administering nanostimulants, directly spraying nanoparticles onto plant leaves. This approach proves practical as leaves readily absorb these minute particles, facilitating swift uptake and immediate effects. However, careful control of the nanostimulants dosage and concentration is necessary to prevent phytotoxicity. Timing is also critical for foliar application, with recommendations typically suggesting application during optimal parts of the day to minimize evaporation and ensure maximal absorption.

Studies have shown the efficacy of foliar application with various nanoparticles. For instance, Faizan et al. [89] demonstrated that foliar application of Zn NPs to tomato leaves under salinity stress enhanced antioxidant enzyme levels, promoted root and shoot growth, and increased biomass and photosynthetic pigment contents. Similarly, Noohpisheh et al. [90] found that foliar application to *Trigonella foenum-graecum* leaves increased protein and proline contents while reducing H_2O_2 and MDA levels. Elsheery et al. [91] observed increased levels of antioxidant enzymes, total sugars, and proline in Mangifera indica L. leaves under salinity stress following foliar application of Zn NPs. In addition, Ali et al. [92] reported that foliar application of Si NPs in wheat increased antioxidant enzyme activity, while Wahid et al. [93] showed that Ag NPs enhanced seed germination and induced antioxidant enzymes under salinity stress. Faraji and Sepehri [94] demonstrated that TiO_2 NPs application in wheat improved development, productivity, and stress tolerance. Furthermore, Ye et al. [95] found that Mn NPs boosted seed germination and root growth in bell peppers. Lastly, El-Saadony et al. [96] highlighted the benefits of foliar application in wheat under drought, heat, and fungal infection conditions, maintaining leaf water status and chlorophyll and carotenoid contents, thereby enhancing plant growth and biomass.

4.3 Soil Inoculation of NPs

Direct soil application of nanostimulants offers a practical approach for delivering nutrients or stimulants to plants, focusing on root absorption. This method allows nanoparticles to gradually infiltrate the soil and be assimilated by the root system, ensuring a sustained release of active ingredients. However, it's crucial to carefully manage dosage and application frequency to prevent nanoparticle accumulation in the soil, thus mitigating potential long-term environmental impacts. For instance, Sun et al. [97] conducted a study where they introduced Nano-ZnO suspension into the soil of *Zea mays* L. cv. Jidan 27 plants. This resulted in improved melatonin synthesis and enhanced functioning of the antioxidant system under drought-stress conditions.

Similarly, Suriyaprabha et al. [98] demonstrated the effectiveness of direct soil application by applying Si NPs to maize plants exposed to UV-B radiation stress. This led to enhanced growth and physiological responses. Furthermore, Amini et al. [99] introduced TiO_2 NPs into the soil of chickpea plants to mitigate cold stress, resulting in altered metabolic pathways as observed through transcription profiling. In addition, Rossi et al. [100] utilized CeO_2 NPs in the soil of canola crops experiencing salinity stress. This intervention shortened root apoplastic barriers, facilitated increased Na + transport to shoots, and reduced Na + accumulation, resulting in increased plant biomass and improved photosynthetic apparatus efficiency. Moreover, Behboudi et al. [101] applied chitosan NPs to the soil of wheat crops under drought stress, leading to increased leaf area, relative water content (RWC), chlorophyll content, photosynthetic rate, and activities of catalase (CAT) and superoxide dismutase (SOD), ultimately enhancing crop yield and biomass. Lastly, Manzoor et al. [102] introduced Fe NPs into the soil to improve wheat plants subjected to salinity and cadmium stress. This intervention increased photosynthetic pigments, nitrogen-phosphorus-potassium (NPK) levels, antioxidant enzyme activity, and enhanced growth, plant weight, and biomass.

4.4 Seed Inoculation of NPs

Soaking or priming seeds with nanostimulants before planting is another effective method for enhancing plant growth and providing early-stage stress protection. This approach allows for improved germination and seedling growth, ensuring that emerging seedlings absorb a minimal but effective concentration of the nanostimulants.

For example, Panda [103] conducted a study where rice seeds were soaked in different Zn NPs suspensions, enhancing plant mineral nutrient acquisition, germination, and antioxidant activity. Similarly, Haghighi and Pessarakli [104] and Almutairi [105] demonstrated the efficacy of soaking seeds in silicon nanoparticle suspension of *Solanum lycopersicum* L., resulting in increased seed germination rate, enhanced photosynthetic pigments, improved root growth, and regulated expression of salt stress-associated genes, thereby inhibiting salinity-induced adverse effects on seed germination and development. Moreover, Nejatzadeh [106] showed that soaking seeds in Ag NPs of *Satureja hortensis* L. increased germination percentage, enhanced shoot length, and improved salt stress tolerance. Additionally, Shah et al. [107] utilized TiO_2 NPs through seed priming methods in Zea mays L., leading to enhanced seed germination efficiency, decreased Na^+, proline, and malondialdehyde levels, increased potassium, phenolic and antioxidant contents, relative water content (RWC), fresh weight, and dry weight. Furthermore, Mustafa and Komatsu [108] explored the application of Al_2O_3 NPs added to seed soaking of soybean flooding, which regulated the ascorbate–glutathione pathway, membrane permeability, and tricarboxylic acid cycle activity. Lastly, Maswada et al. [109] utilized Fe NPs through seed soaking of sorghum under salinity stress, resulting in improved photosynthetic rate, chlorophyll index, photosystem II efficiency, RWC, and reduced lipid peroxidation.

In all these methodologies, a comprehensive understanding of the specific requirements of crops, the targeted abiotic stressors, and the characteristics of the employed nano-stimulants are imperative to optimize application efficacy. The timing and frequency of application should be tailored according to the developmental stage of the crop, prevailing environmental conditions, and the nature of the abiotic stress under consideration. Vigilant monitoring and precise management are essential to calibrate application protocols and mitigate potential adverse effects on plant physiology and ecological integrity. As agricultural nanotechnology progresses, these application strategies are refined to enhance efficacy and ensure sustainable farming practices.

5 Future Directions and Research Needs in Nanostimulants Application for Abiotic Stress Tolerance

As the agricultural sector continues to grapple with the challenges posed by climate change, the role of nanostimulants in enhancing crop resilience is increasingly vital. However, further research is crucial for maximizing the potential of these innovative materials in several key areas. First, developing advanced targeting and delivery systems is essential, as future research should focus on creating mechanisms that ensure nanostimulants reach specific plant parts where they are most needed. This approach would increase efficiency and reduce waste. Additionally, long-term impact studies are required to assess the effects of nanostimulants on crops across multiple growing seasons, providing insights into their sustainability and long-term impacts on plant health and soil ecosystems.

Moreover, tailoring nanostimulants to specific crops and stressors—such as drought, salinity, or extreme temperatures—requires a deep understanding of various crops' unique physiological needs and stress responses. Integrating nanostimulants with other stress mitigation strategies, like breeding for stress tolerance, will contribute to a more holistic approach to crop protection. Finally, research must address the synthesis and scalability of nanostimulants by developing cost-effective and environmentally friendly methods for their production.

Furthermore, studies should focus on the scalability of these methods for large-scale agricultural applications. Given the ongoing challenges posed by climate change, research should also investigate how nanostimulants can aid crops in adapting to shifting environmental conditions, thereby supporting food security. In conclusion, nanostimulants offer a promising approach for boosting crop tolerance to abiotic stress. However, ongoing research in these critical areas is essential for their successful and sustainable integration into modern agricultural practices.

In conclusion, exploring nanostimulants in agriculture highlights a significant stride towards enhancing crop resilience and addressing the challenges posed by abiotic stresses. Nanostimulants offer improved efficiency and targeted delivery of nutrients and growth-promoting substances through their innovative application at the nanoscale. Their unique properties, such as increased solubility, precision in targeting, and controlled release, set them apart from traditional agricultural stimulants. These features enable nanostimulants to penetrate plant cells effectively, facilitating improved nutrient uptake and utilization and bolstering plant growth and productivity. The ability of nanostimulants to mitigate the impacts of various abiotic stresses—such as drought, salinity, and extreme temperatures—is particularly noteworthy. Nanostimulants are crucial in sustaining agricultural productivity amid environmental challenges exacerbated by climate change by enhancing plant tolerance to these stresses. This is critical for ensuring food security and the sustainability of farming practices. However, the application of nanostimulants is not without its challenges. Concerns regarding their environmental impact, particularly biodegradability and soil health effects, necessitate careful consideration and responsible management. The potential risks and limitations associated with their use, including safety concerns and the need for comprehensive regulatory frameworks, underline the importance of ongoing research and development in this field. Integrating nanostimulants in agricultural practices presents a promising frontier with the potential to revolutionize crop management and resilience. This technology's continued exploration and responsible implementation can significantly contribute to achieving more sustainable and efficient agricultural systems. As we move forward, it is imperative to balance the benefits of nanostimulants with environmental stewardship, ensuring that their use enhances crop resilience and aligns with the broader goals of sustainable agriculture and ecosystem preservation.

Acknowledgements ANID/FONDECYT 1230529, ANID/FONDECYT 11220070, ANID/FOVI220003, ANID/ATE220038.

References

1. Minhas PS, Rane J, Pasala RK (2017) Abiotic stresses in agriculture: an overview. In: Minhas P, Rane J, Pasala R (eds) Abiotic stress management for resilient agriculture. Springer, Singapore. https://doi.org/10.1007/978-981-10-5744-1_1
2. Ranjan S, Prakash A, Singh RB, Tiwari P, Bhattacharya S, Nongdam P, Al-Tawaha AR, Lal ML, Tiwari RK, Mandal S, Dey A (2023) Effects of drought stress on agricultural plants, and molecular strategies for drought tolerant crop development. In: Aftab T (ed) New frontiers in plant-environment interactions. Environmental science and engineering. Springer, Cham. https://doi.org/10.1007/978-3-031-43729-8_10
3. Zhou H, Shi H, Yang Y, Feng X, Chen X, Xiao F, Lin H, Guo Y (2024) Insights into plant salt stress signaling and tolerance. J Genet Genomics 51(1):16–34
4. Atta K, Mondal S, Gorai S, Singh AP, Kumari A, Ghosh T, Roy A, Hembram S, Gaikwad DJ, Mondal S, Bhattacharya S, Jha UC, Jespersen D (2023) Impacts of salinity stress on crop plants: improving salt tolerance through genetic and molecular dissection. Front Plant Sci 14:1241736. https://doi.org/10.3389/fpls.2023.1241736
5. Trușcă M, Gâdea Ș, Vidican R, Stoian V, Vâtcă A, Balint C, Stoian VA, Horvat M, Vâtcă S (2023) Exploring the research challenges and perspectives in ecophysiology of plants affected by salinity stress. Agriculture 13(734)
6. Zandalinas SI, Mittler R, Balfagon D, Arbona V, Gomez-Cadenas A (2018) Plant adaptations to the combination of drought and high temperatures. Physiol Plant 162:2–12. https://doi.org/10.1111/ppl.12540
7. Banerjee A, Roychoudhury A (2019) Cold stress and photosynthesis. photosynthesis, productivity and environmental stress, 27–37. In: Ahmad P, Ahanger MA, Alyemeni MN, Alam P (eds) Photosynthesis, productivity and environmental stress. John Wiley & Sons Ltd. https://doi.org/10.1002/9781119501800.ch2

8. Hassan MA, Xiang C, Farooq M, Muhammad N, Yan Z, Hui X, Yuanyuan K, Bruno AK, Lele Z, Jincai L (2021) Cold stress in wheat: plant acclimation responses and management strategies. Front Plant Sci 12:676884
9. Mittal D, Kaur G, Singh P, Yadav K, Ali SA (2020) Nanoparticle-based sustainable agriculture and food science: recent advances and future outlook. Front Nanotechnol 2:579954
10. El-Saadony MT, Saad AM, Soliman SM, Salem HM, Desoky E-SM, Babalghith AO, El-Tahan AM, Ibrahim OM, Ebrahim AAM, Abd El-Mageed TA, Elrys AS, Elbadawi AA, El-Tarabily KA, AbuQamar SF (2022) Role of nanoparticles in enhancing crop tolerance to abiotic stress: a comprehensive review. Front Plant Sci 13:946717. https://doi.org/10.3389/fpls.2022.946717
11. Azameti M, Imoro AW (2023) Nanotechnology: a promising field in enhancing abiotic stress tolerance in plants. Crop Design 2(10037)
12. Munir N, Gulzar W, Abideen Z, Hasanuzzaman M, El-Keblawy A, Zhao F (2023) Plant-nanoparticle interactions: transcriptomic and proteomic insights. Agronomy 13:2112
13. Rasheed A, Li H, Tahir MM, Mahmood A, Nawaz M, Shah AN, Aslam MT, Negm S, Moustafa M, Hassan MU, Wu Z (2022) The role of nanoparticles in plant biochemical, physiological, and molecular responses under drought stress: a review. Front Plant Sci 13:976179
14. Wang X, Xie H, Wang P, Yin H (2023) Nanoparticles in plants: uptake, transport and physiological activity in leaf and root. Materials 16(8):3097
15. Tripathi D, Singh M, Pandey-Rai S (2022) Crosstalk of nanoparticles and phytohormones regulate plant growth and metabolism under abiotic and biotic stress. Plant Stress 6(2022):100107. https://doi.org/10.1016/j.stress.2022.100107
16. Martínez G, Merinero M, Pérez-Aranda M, Pérez-Soriano EM, Ortiz T, Villamor E, Begines B, Alcudia A (2021) Environmental impact of nanoparticles' application as an emerging technology: a review. Materials 14(166)
17. Boykov IN, Zhang B (2021) Impact of nanoparticles on plant growth, development, and biomass. Methods Mol Biol 2326:217–224
18. Pelegrino MT, Kohatsu MY, Seabra AB, Monteiro LR, Gomes DG, Oliveira HC, Rolim WR, de Jesus TA, Batista BL, Lange CN (2020) Effects of copper oxide nanoparticles on growth of lettuce (Lactuca sativa L.) seedlings and possible implications of nitric oxide in their antioxidative defense. Environ Monit Assess 192(4):232
19. Juárez-Maldonado A, González-Morales S, Cabrera-De la Fuente M, Medrano-Macías J, Benavides-Mendoza A (2018) Nanometals as promoters of nutraceutical quality in crop plants. In: Grumezescu AM, Holban AM (eds) Impact of nanoscience in food industry. Academic Press, London, UK, pp 277–310
20. Li Z, Zhang R, Xia S, Wang L, Liu C, Zhang R, Fan Z, Chen F, Liu Y (2019) Interactions between N, P and K fertilizers affect the environment and the yield and quality of satsumas. Glob Ecol Conserv 19:e00663
21. Nongbet A, Mishra AK, Mohanta YK, Mahanta S, Ray MK, Khan M, Baek KH, Chakrabartty I (2022) Nanofertilizers: a smart and sustainable attribute to modern agriculture. Plants 11(19):2587
22. Valderrama NA, Jacinto HC, Lay J, Flores EY, Zavaleta CD, Rodríguez Delfín A (2020) Factorial design for preparing chitosan nanoparticles and its use for loading and controlled release of indole-3-acetic acid with effect on hydroponic lettuce crops. Biocatal Agric Biotechnol 26:101640
23. Li Y, Xi K, Liu X, Han S, Han X, Li G, Yang L, Ma D, Fang Z, Gong S, Yin J, Zhu Y (2023) Silica nanoparticles promote wheat growth by mediating hormones and sugar metabolism. J Nanobiotechnol 21(2)
24. Mahawar L, Živčák M, Barboricova M, Kovár M, Filaček A, Ferencova J, Vysoká DM, Brestič M (2024) Effect of copper oxide and zinc oxide nanoparticles on photosynthesis and physiology of Raphanus sativus L. under salinity stress. Plant Physiol Biochem 206:108281
25. Juárez-Maldonado A, Tortella G, Rubilar O, Fincheira P, Benavides-Mendoza A (2021) Biostimulation and toxicity: the magnitude of the impact of nanomaterials in microorganisms and plants. J Adv Res 31:113–126

26. Juárez-Maldonado A, Ortega-Ortíz H, Morales-Díaz AB, González-Morales S, Morelos-Moreno Á, Cabrera-De la Fuente M, Sandoval-Rangel A, Cadenas-Pliego G, Benavides-Mendoza A (2019) Nanoparticles and nanomaterials as plant biostimulants. Int J Mol Sci 20(1):162. https://doi.org/10.3390/ijms20010162
27. Syu Y-Y, Hung J-H, Chen J-C, Chuang H-W (2014) Impacts of size and shape of silver nanoparticles on Arabidopsis plant growth and gene expression. Plant Physiol Biochem 83:57–64
28. Hu P, An J, Faulkner MM, Wu H, Li Z, Tian X, Giraldo JP (2020) Nanoparticle charge and size control foliar delivery efficiency to plant cells and organelles. ACS Nano 14(7)
29. Dutta Gupta S, Agarwal A, Pradhan S (2018) Phytostimulatory effect of silver nanoparticles (AgNPs) on rice seedling growth: an insight from antioxidative enzyme activities and gene expression patterns. Ecotoxicol Environ Saf 161:624–633. https://doi.org/10.1016/j.ecoenv.2018.06.023
30. Babajani A, Iranbakhsh A, Oraghi Ardebili Z, Eslami B (2019) Differential growth, nutrition, physiology, and gene expression in Melissa officinalis mediated by zinc oxide and elemental selenium nanoparticles. Environ Sci Pollut Res Int 26(24):24430–24444. https://doi.org/10.1007/s11356-019-05676-z
31. Ali EF, El-Shehawi AM, Ibrahim OHM, Abdul-Hafeez EY, Moussa MM, Hassan FAS (2021) A vital role of chitosan nanoparticles in improvisation the drought stress tolerance in Catharanthus roseus (L.) through biochemical and gene expression modulation. Plant Physiol Biochem 161:166–175
32. Saad-Allah KM, Ragab GA (2020) Sulfur nanoparticles mediated improvement of salt tolerance in wheat relates to decreasing oxidative stress and regulating metabolic activity. Physiol Mol Biol Plants 26:2209–2223
33. Jalil SU, Ansari MI (2019) Nanoparticles and abiotic stress tolerance in plants: synthesis, action, and signaling mechanisms. In plant signaling molecules: role and regulation under stressful environments. Elsevier, pp 549–561. https://doi.org/10.1016/B978-0-12-816451-8.00034-4
34. Kumari S, Khanna RR, Nazir F, Albaqami M, Chhillar H, Wahid I, Khan MIR (2022) Bio-synthesized nanoparticles in developing plant abiotic stress resilience: a new boon for sustainable approach. Int J Mol Sci 23(8):4452
35. Oliveira HC, Gomes BC, Pelegrino MT, Seabra AB (2016) Nitric oxide-releasing chitosan nanoparticles alleviate the effects of salt stress in maize plants. Nitric Oxide 61:10–19
36. Silveira NM, Prataviera PJC, Pieretti JC, Seabra AB, Almeida RL, Machado EC, Ribeiro RV (2021) Chitosan-encapsulated nitric oxide donors enhance physiological recovery of sugarcane plants after water deficit. Environ Exp Bot 190:104593
37. Dimkpa CO, Bindraban PS (2017) Nanofertilizers: New products for the industry? J Agric Food Chem 66(26). https://doi.org/10.1021/acs.jafc.7b02150
38. Vinzant K, Rashid M, Khodakovskaya MV (2023) Advanced applications of sustainable and biological nano-polymers in agricultural production. Front Plant Sci 13:1081165
39. Rehmanullah, Muhammad Z, Inayat N, Majeed A (2020) Application of nanoparticles in agriculture as fertilizers and pesticides: challenges and opportunities. In: New frontiers in stress management for durable agriculture, pp 281–293
40. Hanif S, Sajjad A, Zia M (2023) Proline coated ZnO NPs as nanofertilizer against drought stress: an in vitro study to Coriandrum sativum. Plant Cell Tissue Organ Cult (PCTOC) 155(2):493–504
41. Mehmood S, Ou W, Ahmed W, Bundschuh J, Rizwan M, Mahmood M, Sultan H, Alatalo JM, Elnahal ASM, Liu W, Li W (2023) ZnO nanoparticles mediated by Azadirachta indica as nano fertilizer: improvement in physiological and biochemical indices of Zea mays grown in Cr-contaminated soil. Environ Pollut 339:122755
42. Patil BM, Patil VL, Bhosale SR, Bhosale RR, Ingavale DR, Patil SS, Kamble PD, Bhosale AG, Mane SM, Lee J, Vanalakar SA (2024) Field application of Ca-doped ZnO nanoparticles to maize and wheat plants. Plant Physiol Biochem 210:108552

43. Elbasuney S, El-Sayyad GS, Abdelaziz AM, Rizk SH, Tolba MM, Attia MS (2024) Stable colloidal iron oxide nanoparticles: a new green nanofertilizer and therapeutic nutrient for eggplant immune response against fusarium wilt disease. J Cluster Sci 35(4):983–997
44. Faiz S, Yasin NA, Khan WU, Shah AA, Akram W, Ahmad A, Ali A, Naveed NH, Riaz L (2022) Role of magnesium oxide nanoparticles in the mitigation of lead-induced stress in Daucus carota: modulation in polyamines and antioxidant enzymes. Int J Phytoremediation 24(4):364–372. https://doi.org/10.1080/15226514.2021.1949263
45. Ibrahim AS, Ali GA, Hassanein A, Attia AM, Marzouk ER (2022) Toxicity and uptake of CuO nanoparticles: evaluation of an emerging nanofertilizer on wheat (Triticum aestivum L.) plant. Sustainability 14(9):4914
46. Mustafa M, Azam M, Bhatti HN, Khan A, Zafar L, Abbasi AMR (2024) Green fabrication of copper nano-fertilizer for enhanced crop yield in cowpea cultivar: a sustainable approach. Biocatal Agric Biotechnol 56:102994
47. Avila-Quezada GD, Ingle AP, Golińska P, Rai M (2022) Strategic applications of nanofertilizers for sustainable agriculture: benefits and bottlenecks. Nanotechnol Rev 11(1):2123–2140
48. Leonardi M, Çaruso GM, Carroccio SC, Boninelli S, Curcuruto G, Zimbone M, Allegra M, Torrisi B, Ferlito F, Miritello M (2021) Smart nanocomposites of chitosan/alginate nanoparticles loaded with copper oxide as alternative nanofertilizers. Environ Sci Nano 8:174–187
49. Sarkar MM, Mathur P, Mitsui T, Roy S (2022) A review on functionalized silica nanoparticle amendment on plant growth and development under stress. Plant Growth Regul 98(3):421–437
50. Yoon HY, Lee JG, Degli Esposti L, Iafisco M, Kim PJ, Shin SG, Jeon J-R, Adamiano A (2020) Synergistic release of crop nutrients and stimulants from hydroxyapatite nanoparticles functionalized with humic substances: Toward a multifunctional nanofertilizer. ACS Omega 5(12):6598–6610
51. DeRosa MC, Monreal C, Schnitzer M, Walsh R, Sultan Y (2010) Nanotechnology in fertilizers. Nat Nanotechnol 5(2):91–91
52. Dhlamini B, Paumo HK, Katata-Seru L, Kutu FR (2020) Sulphate-supplemented NPK nanofertilizer and its effect on maize growth. Mater Res Express 7(9):095011
53. Gomes DG, Pelegrino MT, Ferreira AS, Bazzo JH, Zucareli C, Seabra AB, Oliveira HC (2021) Seed priming with copper-loaded chitosan nanoparticles promotes early growth and enzymatic antioxidant defense of maize (Zea mays L.) seedlings. J Chem Technol Biotechnol 96(8):2176–2184
54. Khaliq M, Hanif MA, Bhatti IA, Mushtaq Z (2023) A novel study for producing complexed and encapsulated nutrients at nanometric scale to enhance plant growth. Sci Rep 13(1):11100
55. Mushtaq A, Jamil N, Rizwan S, Mandokhel F, Riaz M, Hornyak GL, Malghani MN, Shahwani MN (2018) Engineered silica nanoparticles and silica nanoparticles containing controlled release fertilizer for drought and saline areas. IOP Conf Ser Mater Sci Eng 414:012029
56. Yin J, Su X, Yan S, Shen J (2023) Multifunctional nanoparticles and nanopesticides in agricultural application. Nanomaterials 13(7):1255
57. Chen S, Qin Y, Ye X, Liu J, Yan X, Zhou L, Wang X, Martyniuk CJ, Yan B (2023) Neurotoxicity of the $Cu(OH)_2$ nanopesticide through perturbing multiple neurotransmitter pathways in developing Zebrafish. Environ Sci Technol 57(48):19407–19418. https://doi.org/10.1021/acs.est.3c06284
58. Zhao W, Liu Y, Zhang P, Zhou P, Wu Z, Lou B, Jiang Y, Shakoor N, Li M, Li Y, Lynch I, Rui Y, Tan Z (2022) Engineered Zn-based nano-pesticides as an opportunity for treatment of phytopathogens in agriculture. NanoImpact 28:100420
59. Agredo-Gomez AD, Molano-Molano JA, Portela-Patiño MC, Rodríguez-Páez JE (2024) Use of ZnO nanoparticles as a pesticide: in vitro evaluation of their effect on the phytophagous Puto barberi (mealybug). Nanostruct Nano-Objects 37:101095
60. Abd El-Latef EA, Wahba MN, Mousa S, El-Bassyouni GT, El-Shamy AM (2023) Cu-doped ZnO-nanoparticles as a novel eco-friendly insecticide for controlling Spodoptera littoralis. Biocatal Agric Biotechnol 52:102823

61. Peixoto S, Henriques I, Loureiro S (2021) Long-term effects of $Cu(OH)_2$ nanopesticide exposure on soil microbial communities. Environ Pollut 269:116113
62. Cui J, Sun C, Wang A, Wang Y, Zhu H, Shen Y, Li N, Zhao X, Cui B, Wang C et al (2020) Dual-functionalized pesticide nanocapsule delivery system with improved spreading behavior and enhanced bioactivity. Nanomaterials 10(2):220. https://doi.org/10.3390/nano10020220
63. Hou R, Li C, Tan Y, Wang Y, Huang S, Zhao C, Zhang Z (2023) Eco-friendly O-carboxymethyl chitosan base chlorfenapyr nanopesticide for effective pest control and reduced toxicity to honey bees. Int J Biol Macromol 224:972–983
64. Zargar M, Bayat M, Saquee FS, Diakite S, Ramzanovich NM, Akhmadovich KA (2023) New advances in nano-enabled weed management using poly (Epsilon-Caprolactone)-based nanoherbicides: a review. Agriculture 13(10):2031
65. Grillo R, Pereira AE, Nishisaka CS, De Lima R, Oehlke K, Greiner R, Fraceto LF (2014) Chitosan/tripolyphosphate nanoparticles loaded with paraquat herbicide: an environmentally safer alternative for weed control. J Hazard Mater 278:163–171
66. Cordon G, Valino IL, Prieto A, Costa C, Marchi MC, Diz V (2022) Effects of the nanoherbicide made up of atrazine-chitosan on the primary events of photosynthesis. J Photochem Photobiol 12:100144
67. Guerriero G, Sutera FM, Torabi-Pour N, Renaut J, Hausman J-F, Berni R, Pennington HC, Welsh M, Dehsorkhi A, Zancan LR, Saffie-Siebert S (2021) Phyto-courier, a silicon particle-based nano-biostimulant: evidence from Cannabis sativa exposed to salinity. ACS Nano 15(2).
68. Medrano-Macías J, Narvaéz-Ortiz WA (2022) Selenium and nano-selenium as a new frontier of plant biostimulant. Selenium and nano-selenium in environmental stress management and crop quality improvement. Springer International Publishing, Cham, pp 41–54
69. Guerriero G, Sutera FM, Hoffmann J, Leclercq CC, Planchon S, Berni R, Hausman JF, Renaut J, Torabi-Pour N, Pennington HC, Ahmed M, Welsh M, Dehsorkhi A, Saffie-Siebert S (2023) Nanoporous quercetin-loaded silicon-stabilized hybrid lipid nanoparticles alleviate salt stress in tomato plants. ACS Appl Nano Mater 6(5):3647–3660
70. Wang J, Tao M, Xu L, Fan N, Zhao C, Xiao Z, Wang Z (2023) Chitosan nanocarriers loaded with salicylic acid for controlling fall armyworm (Spodoptera frugiperda) and alleviating oxidative stress in maize plants. Environ Sci Nano 10(12):3295–3306
71. Patel DK, Kim HB, Dutta SD, Ganguly K, Lim KT (2020) Carbon nanotubes-based nanomaterials and their agricultural and biotechnological applications. Materials 13(7):1679
72. Guo X, Zhao J, Wang R, Zhang H, Xing B, Naeem M, Yao T, Li R, Xu R, Zhang Z, Wu J (2021) Effects of graphene oxide on tomato growth in different stages. Plant Physiol Biochem 162:447–455
73. Mahmoud NE, Abdelhameed RM (2021) Superiority of modified graphene oxide for enhancing the growth, yield, and antioxidant potential of pearl millet (Pennisetum glaucum L.) under salt stress. Plant Stress 2:100025
74. Chen X, Wang J, Hayat K, Zhang D, Zhou P (2021) Small structures with big impact: Multi-walled carbon nanotubes enhanced remediation efficiency in hyperaccumulator Solanum nigrum L. under cadmium and arsenic stress. Chemosphere 276:130130
75. Hu Y, Zhang P, Zhang X, Liu Y, Feng S, Guo D, Tcyganova N, Song Z, Dang X (2021) Multi-wall carbon nanotubes promote the growth of maize (Zea mays) by regulating carbon and nitrogen metabolism in leaves. J Agric Food Chem 69(17)
76. Ozfidan-Konakci C, Alp FN, Arikan B, Balci M, Parmaksizoglu Z, Yildiztugay E, Cavusoglu H (2022) The effects of fullerene on photosynthetic apparatus, chloroplast-encoded gene expression, and nitrogen assimilation in Zea mays under cobalt stress. Physiol Plant 174(3):e13720
77. Shang Y, Hasan MK, Ahammed GJ, Li M, Yin H, Zhou J (2019) Applications of nanotechnology in plant growth and crop protection: a review. Molecules 24(14):2558
78. Balusamy SR, Joshi AS, Perumalsamy H, Mijakovic I, Singh P (2023) Advancing sustainable agriculture: a critical review of smart and eco-friendly nanomaterial applications. J Nanobiotechnol 21(1):372

79. Marslin G, Sheeba CJ, Franklin G (2017) Nanoparticles alter secondary metabolism in plants via ROS burst. Front Plant Sci 8:832
80. Bulgari R, Franzoni G, Ferrante A (2019) Biostimulants application in horticultural crops under abiotic stress conditions. Agronomy 9:306
81. Tirani MM, Haghjou MM, Ismaili A (2019) Hydroponic grown tobacco plants respond to zinc oxide nanoparticles and bulk exposures by morphological, physiological and anatomical adjustments. Funct Plant Biol 46:360–375. https://doi.org/10.1071/fp18076
82. Venkatachalam P, Jayaraj M, Manikandan R, Geetha N, Rene ER, Sharma NC, Sahi SV (2017) Zinc oxide nanoparticles (ZnONPs) alleviate heavy metal-induced toxicity in Leucaena leucocephala seedlings: a physiochemical analysis. Plant Physiol Biochem 110:59–69
83. Mahmoud LM, Dutt M, Shalan AM, El-Kady ME, El-Boray MS, Shabana YM et al (2020) Silicon nanoparticles mitigate oxidative stress of in vitroderived banana (Musa acuminata' Grand Nain') under simulated water deficit or salinity stress. S Afr J Bot 132:155–163. https://doi.org/10.1016/j.sajb.2020.04.027
84. Ashkavand P, Zarafshar M, Tabari M, Mirzaie J, Nikpour A, Bordbar SK et al (2018) Application of SiO2 nanoparticles as pretreatment alleviates the impact of drought on the physiological performance of Prunus mahaleb (Rosaceae). Bol Soc Argent Bot 53:207–219
85. Iqbal M, Raja NI, Mashwani Z, Hussain M, Muhammad E, Yasmeen F (2019) Effect of silver nanoparticles on growth of wheat under heat stress. Iran J Sci Technol Trans A Sci 43:387–395. https://doi.org/10.1007/s40995-017-0417-4
86. Kardavan Ghabel V, Karamian R (2020) Effects of TiO_2 nanoparticles and spermine on antioxidant responses of Glycyrrhiza glabra L. to cold stress. Acta Bot Croat 79:137–147. https://doi.org/10.37427/botcro-2020-025
87. Gohari G, Mohammadi A, Akbari A, Panahirad S, Dadpour MR, Fotopoulos V, Kimura S (2020) Titanium dioxide nanoparticles (TiO_2 NPs) promote growth and ameliorate salinity stress effects on essential oil profile and biochemical attributes of Dracocephalum moldavica. Sci Rep 10:912
88. Mozafari A, Ghadakchi Asl A, Ghaderi N (2018) Grape response to salinity stress and role of iron nanoparticle and potassium silicate to mitigate salt induced damage under in vitro conditions. Physiol Mol Biol Plants 24:25–35. https://doi.org/10.1007/s12298-017-0488-x
89. Faizan M, Bhat JA, Chen C, Alyemeni MN, Wijaya L, Ahmad P, Yu F (2021) Zinc oxide nanoparticles (ZnO-NPs) induce salt tolerance by improving the antioxidant system and photosynthetic machinery in tomato. Plant Physiol Biochem 161:122–130
90. Noohpisheh Z, Amiri H, Mohammadi A, Farhadi S (2021) Effect of the foliar application of zinc oxide nanoparticles on some biochemical and physiological parameters of Trigonella foenum-graecum under salinity stress. Plant Biosyst. 155:267–280. https://doi.org/10.1080/11263504.2020.1739160
91. Elsheery NI, Helaly MN, El-Hoseiny HM, Alam-Eldein SM (2020) Zinc oxide and silicone nanoparticles to improve the resistance mechanism and annual productivity of salt-stressed mango trees. Agronomy 10:558. https://doi.org/10.3390/agronomy10040558
92. Ali S, Rizwan M, Hussain A, Ur Rehman MZ, Ali B, Yousaf B et al (2019) Silicon nanoparticles enhanced the growth and reduced the cadmium accumulation in grains of wheat (Triticum aestivum L.). Plant Physiol Biochem 140:1–8
93. Wahid I, Kumari S, Ahmad R, Hussain SJ, Alamri S, Siddiqui MH, Khan MIR (2020) Silver nanoparticle regulates salt tolerance in wheat through changes in ABA concentration, ion homeostasis, and defense systems. Biomolecules 10(11):1506
94. Faraji J, Sepehri A (2020) Exogenous nitric oxide improves the protective effects of TiO2 nanoparticles on growth, antioxidant system, and photosynthetic performance of wheat seedlings under droughtstress. J Soil Sci Plant Nutr 20:703–714. https://doi.org/10.1007/s42729-019-00158-0
95. Ye Y, Cota-Ruiz K, Hernández-Viezcas JA, Valdés C, Medina-Velo IA, Turley RS, Peralta-Videa JR, Gardea-Torresdey JL (2020). Manganese nanoparticles control salinity-modulated molecular responses in Capsicum annuum L. through priming: a sustainable approach for agriculture. ACS Sustain Chem Eng 8(3)

96. El-Saadony MT, Saad AM, Najjar AA, Alzahrani SO, Alkhatib FM, Shafi ME et al (2021) The use of biological selenium nanoparticles to suppress Triticum aestivum L. crown and root rot diseases induced by Fusarium species and improve yield under drought and heat stress. Saudi J Biol Sci 28:4461–4471. https://doi.org/10.1016/j.sjbs.2021.04.043
97. Sun L, Song F, Guo J, Zhu X, Liu S, Liu F et al (2020) Nano-ZnO-induced drought tolerance is associated with melatonin synthesis and metabolism in maize. Int J Mol Sci 21:782. https://doi.org/10.3390/ijms21030782
98. Suriyaprabha R, Karunakaran G, Yuvakkumar R, Prabu P, Rajendran V, Kanna N (2012) Growth and physiological responses of maize (Zea mays L.) to porous silica nanoparticles in soil. Nanopart Res 14:1294. https://doi.org/10.1007/s11051-012-1294-6
99. Amini Y, Amel Jamehdar S, Sadri K, Zare S, Musavi D, Tafaghodi M (2017) Different methods to determine the encapsulation efficiency of protein in PLGA nanoparticles. Biomed Mater Eng 28(6):613–620. https://doi.org/10.3233/BME-171705
100. Rossi L, Zhang W, Lombardini L, Ma X (2016) The impact of cerium oxide nanoparticles on the salt stress responses of Brassica napus L. Environ Pollut 219:28–36
101. Behboudi F, Tahmasebi-Sarvestani Z, Kassaee MZ, Modarres-Sanavy SAM, Sorooshzadeh A, Mokhtassi-Bidgoli A (2019) Evaluation of chitosan nanoparticles effects with two application methods on wheat under drought stress. J Plant Nutr 42:1439–1451
102. Manzoor N, Ahmed T, Noman M, Shahid M, Nazir MM, Ali L, Alnusaire TS, Li B, Schulin R, Wang G (2021) Iron oxide nanoparticles ameliorated the cadmium and salinity stresses in wheat plants, facilitating photosynthetic pigments and restricting cadmium uptake. Sci Total Environ 769:145221
103. Panda S (2017) Physiological impact of Zinc nanoparticle on germination of rice (Oryza sativa L) seed. J Plant Sci Phytopathol 1:62–70
104. Haghighi M, Pessarakli M (2013) Influence of silicon and nano-silicon on salinity tolerance of cherry tomatoes (Solanum lycopersicum L.) at early growth stage. Sci Hortic 161:111–117. https://doi.org/10.1016/j.scienta.2013.06.034
105. Almutairi ZM (2016) Effect of nano-silicon application on the expression of salt tolerance genes in germinating tomato (Solanum lycopersicum L.) seedlings under salt stress. Plant Omics 9:106–114
106. Nejatzadeh F (2021) Effect of silver nanoparticles on salt tolerance of Satureja hortensis L. during in vitro and in vivo germination tests. Heliyon 7:e05981
107. Shah T, Latif S, Saeed F, Ali I, Ullah S, Alsahli AA, Jan S, Ahmad P (2021) Seed priming with titanium dioxide nanoparticles enhances seed vigor, leaf water status, and antioxidant enzyme activities in maize (Zea mays L.) under salinity stress. J King Saud Univ Sci 33(1):101207
108. Mustafa G, Komatsu S (2016) Insights into the response of soybean mitochondrial proteins to various sizes of aluminum oxide nanoparticles under flooding stress. J Proteome Res 15:4464–4475. https://doi.org/10.1021/acs.jproteome.6b00572
109. Maswada HF, Djanaguiraman M, Prasad PVV (2018) Seed treatment with nano-iron (III) oxide enhances germination, seeding growth and salinity tolerance of sorghum. J Agro Crop Sci 204:577–587

Improving Nutritional Quality of Food Crops Using Nanomaterials and Nanostimulants

Yolanda González-García and Antonio Juárez-Maldonado

Abstract Nutritional problems in the world are increasing, either because of a lack of food, or because food is not of sufficient quality, i.e., it does not provide the essential elements for the human diet and does not have the appropriate content of bioactive compounds. In either case, it is necessary to develop strategies to improve the quality of food for human consumption. In agriculture, nanotechnology can be an effective tool for solving food quality problems. Through the application of nanomaterials, it is possible to increase the content of essential elements in the consumption organs of plants, which are conventionally difficult to handle. But also, it is possible to increase the content of macromolecules and bioactive compounds, improving the quality of food, and therefore with beneficial effects for human health. It is for this reason that more knowledge of nanotechnology applied to agriculture is required in order to improve the nutritional quality of food. In this sense, the subject of this chapter is addressed, with emphasis on the biofortification of agricultural crops with the application of nanomaterials, and their impact on the content of essential elements and bioactive compounds.

Keywords Agriculture · Antioxidants · Bioactive compounds · Biofortification · Biostimulation · Food production · Nanotechnology

Y. González-García (✉)
Instituto Nacional de Investigaciones Forestales, Agrícolas y Pecuarias, CETODS, La Paz, México
e-mail: gonzalez.yolanda@inifap.gob.mx

A. Juárez-Maldonado
Departamento de Botánica, Universidad Autónoma Agraria Antonio Narro, Saltillo, México

Laboratorio Nacional Conahcyt de Ecofisiología Vegetal y Seguridad Alimentaria (LANCEVSA), Universidad Autónoma Agraria Antonio Narro, Saltillo, México

© The Author(s), under exclusive license to Springer Nature Singapore Pte Ltd. 2025
A. Juárez-Maldonado et al. (eds.), *Plant Biostimulation with Nanomaterials*, Smart Nanomaterials Technology, https://doi.org/10.1007/978-981-96-4648-7_9

1 Introduction

Malnutrition affects almost 10% of the world's population, causing malnutrition and hidden hunger [1]. Therefore providing healthy, nutritionally rich food in sufficient quantities is an important UN Sustainable Development Goal (SDG) to end hunger by 2030 and an imminent challenge for agricultural production [2].

The human body requires a variety of nutrients consisting of approximately 20 minerals, 13 vitamins, nine amino acids that make up proteins and two fatty acids. Most of these nutrients are obtained through the diet [1]. In this sense, plants play an important role since they provide these nutrients directly through different organs of consumption and also serve as food for livestock from which animal products are obtained [1].

Modern agriculture faces critical challenges in solving health and environmental problems associated with macro- and micronutrient deficiencies, food insecurity, low fertilizer use efficiency, overfertilization, climate change, water scarcity, agricultural land reduction and soil degradation. The long-term effects of intensive agronomic practices include significant losses in crop productivity and the nutritional value of agricultural products [3]. This can aggravate the situation of malnutrition of the world's population, so it is necessary to implement strategies to improve the nutritional quality of food, and through these improve nutrition and consequently human health.

In this context, nanotechnology can be a very useful tool to provide solutions to a variety of problems. This technology allows the manipulation of materials, from their creation and modification to their application at the nanometric scale (1–100) [4, 5]. Through the application of nanomaterials (NM), innovative solutions can be proposed to the different problems that affect the agriculture and food system. This is because NM exhibit a number of properties that are different from bulk materials, properties that emerge from the size and others that are only significantly enhanced for the same reason [4, 6]. Thanks to advances in these areas, a huge world of fundamental science and applications has been developed at the nanoscale, making it an essential part of various fields and with a wide range of possibilities [6, 7].

The advantages of nanomaterials to improve the quality of food from agriculture stem from two main characteristics. The first is that they can be a source of minerals when applied to plants, and then potentially be available for human consumption [8]. Therefore, nanomaterials can be used for biofortification of agricultural crops in order to increase the content of some essential elements [9].

The second reason is that nanomaterials have the ability to modify plant metabolism through the process of biostimulation, so they can also induce the increase of bioactive compounds and food quality [10–12]. Hence, they can be used to improve the content of bioactive compounds such as vitamins, antioxidants, etc. and even increase primary macromolecules such as lipids, proteins and sugars in plants, thus improving nutritional quality [13, 14].

This chapter reviews the impact of the use of nanomaterials to improve the nutritional characteristics of agricultural crops that serve as food, with two basic approaches, (1) to increase the content of essential elements for the human diet, and (2) to increase the content of macromolecules, secondary metabolites, and antioxidants in crops.

2 Impact of Nutrition on Human Health

Food and nutrition play a crucial role in the maintenance of human health and the prevention of disease. Nutrients are chemical substances that the body needs to maintain its basic functions and are obtained through a balanced diet [15].

There are six main classes of nutrients essential for human health: fatty acids, lipids, proteins, vitamins, minerals and water [16]. The deficiency of any of these is associated with various physiological, biochemical and molecular disorders that limit the proper functioning of one or multiple organs of the body, affecting health and causing death if they are not adequately treated [9].

Fatty acids are organic molecules made up of long hydrocarbon chains that have a carboxylic acid group at one end and are fundamental constituents of lipids [17]. They can be classified into non-essential fatty acids, which are those that the body can synthesize, and essential fatty acids, which cannot be synthesized and must be acquired through the diet [18]. In humans two are essential in nature, linoleic acid (LA) and α-linoleic acid (ALA), both are involved in various biological activities and are beneficial to health as they serve as energy stores, in the preservation of cell membranes and the prevention of noncommunicable diseases [19]. Both LA and ALA are obtained predominantly through the consumption of vegetable oils (sunflower, canola, safflower, corn, soybean, peanut and palm), seeds (walnuts, flaxseed and chia), some green leafy vegetables such as kale and spinach and to a lesser extent from animal sources such as chicken, eggs and red meat [18].

Lipids are extremely diverse macromolecules, with thousands of subspecies identified to date, categorized into three main classes: glycerophospholipids (GPL), sphingolipids and sterols. They are an essential part of the cell as each compartment has its own lipid composition, which is indispensable for the functional fate of various proteins [20]. They are also involved in other biological functions such as energy storage of vital organs, signal transduction and function as substrates for hormone secretion and metabolic excretion [21]. Deficiency causes various metabolic, hormonal and neuronal disorders [22]. In contrast, excessive accumulation induces diseases such as obesity and metabolic syndrome that represent a significant health risk [23]. The main sources of lipids are staple crops, especially oilseeds such as peanuts, sunflower, flax, rapeseed, canola, camelina, cotton, safflower and castor beans [24], in addition to animal sources such as fish, milk, eggs and red meats [25].

Proteins are one of the six types of macromolecules that are indispensable for humans. They are the main dietary source of nitrogen and indispensable amino acids that the body needs for the formation of new proteins involved in the growth, maintenance of tissues and physiological functions [15]. The class of dietary proteins is extremely diverse and their quality varies substantially depending on their bioavailability, purity, effect of processing prior to ingestion, amino acid profile and digestibility [26].

Both the correct amino acid composition and high digestibility are strictly necessary for proteins to meet the requirements of the human body. Because amino acids are important in the synthesis and function of muscles and organs, as well as in enzymes, hormones and the immune system, amino acids are important in the synthesis and function of muscles and organs, as well as in enzymes, hormones and the immune system [27]. These amino acids are classified as expendable amino acids (EAA), which can be synthesized by the human body, and indispensable amino acids (IAA), which cannot be synthesized by the human body and the only source of these amino acids is through protein intake [15].

The main source of protein for consumption comes from animal origin, however, alternative protein sources are required to meet growing consumer demand and to satisfy the world's protein needs such as those of vegetable origin (cereals, legumes and seeds) [28]. Although these are often recognized as incomplete or nutritionally inferior because they provide a lower content of the required amounts of essential amino acids for human needs, they do not provide the required amounts of essential amino acids for human needs [29].

Vitamins are important nutrients in human nutrition and health of which there is a considerable lack of intake that is not only limited to developing and underdeveloped countries, but also to developed countries [30]. Vitamins can be divided into two categories according to their solubility properties: fat-soluble vitamins (FSV), which include vitamin A, vitamin D, vitamin E and vitamin K, and water-soluble vitamins (HSSV), which include the B vitamins and vitamin C [31]. The biochemical functions derived from vitamins and their forms are enormously varied and include antioxidants, regulators of gene transcription, hormone regulation, among others, so that deficiencies of these vitamins can cause severe, serious and fatal disorders [32].

In general, most vitamins must be obtained through the diet, since the body cannot synthesize them or does not synthesize them in sufficient quantities [31]. The main dietary sources of vitamins are vegetables, mainly leafy vegetables and unprocessed fruits. In addition to vegetable oils such as soybean, sunflower, olive and canola. [33], and to a lesser extent animal sources such as dairy, red meat, eggs and fish [34].

The mineral elements required by humans are classified according to the amount of daily need and their physiological function in macroelements that include calcium (Ca), phosphorus (as PO_4^{3-}), magnesium (Mg), sodium (Na), potassium (K) and chloride (Cl), and trace elements or microelements such as iron (Fe^{2+} and Fe^{3+}), iodine (I), zinc (Zn), selenium (Se), cobalt (Co^{2+} and Co^{3+}), chromium (Cr), copper (Cu^{+} and Cu^{2+}), fluorine (F), molybdenum (Mo), vanadium (V), nickel (Ni), tin (Sn) and arsenic (As) [35].

Macro- and micronutrient deficiencies are a serious health problem, especially in developing countries [36]. According to Ali [35], macroelement deficiency is associated with more than 30 health problems that limit the vital functions of the human body. While micronutrient deficiency, which causes what is known as hidden hunger, is currently suffered by more than half of the world's population, especially deficiencies of zinc, iron, iodine and selenium, it is associated with more than 30 health problems that limit vital functions of the human body [36, 37].

The main source of mineral elements for humans is through the intake of foods of vegetable origin, mainly cereals and fresh vegetables [8], and foods of animal origin, especially milk and its derivatives [38].

As mentioned above, the six main classes of nutrients essential to human health are obtained mainly through the diet, which can be omnivorous, vegan and vegetarian [39]. In this sense, plants play an extremely important role since they provide these nutrients directly through different plant organs of consumption and also serve as food for livestock from which animal products are obtained, so improving the quantity and quality of agricultural crops is an imminent challenge in agriculture [1].

3 Use of Nanomaterials as Source of Minerals

Nanotechnology deals with the application of nanoscience (scientific investigation of particles and structures found on a scale of 1–100 nm) and nanomaterials (materials with a scale of 1–100 nm), focused on the engineering and manufacturing of these nanomaterials for practical applications, as it includes the creation, analysis, production and use of structures, devices, and systems at the nanoscale [4, 5]. Due to the advances in these areas, a huge world of fundamental science and applications has developed at the nanoscale, making it an essential part of several fields and with a wide range of possibilities ranging from quantum physics, chemical and biological mechanisms, device functionality, and materials engineering, nanoscience has become an essential part of several fields [6, 7].

Transformation of bulk materials to nanoscale results in a variety of novel properties that can be classified into two types, those that already exist in the bulk materials but are increased due to the change in size (e.g. surface area); while the second group considers the properties that are new due to the nanoscale size [4, 6]. In any case, these properties that nanomaterials possess have advantages over bulk materials. In the case of plants, nanomaterials can more easily enter and move to the different organs of the plant, which makes them more efficient, improve plant development, promote the activation of the antioxidant defense system and mitigate biotic and abiotic stress, and improve other physiological aspects [10–12].

Due to the nature of nanomaterials, these can be essential elements for the human diet, especially those elements that are difficult to incorporate into agricultural crops and of which the population may present deficiencies such as Zn, Fe, Se, I, etc. [36, 37]. It is for this reason that nanomaterials can be a source of minerals when applied to plants, and then potentially be available for human consumption, since they are mainly obtained from plant foods [8]. In addition to this, nanomaterials have the ability to modify plant metabolism through the process of biostimulation, so they can also induce the increase of bioactive compounds and food quality [10–12].

3.1 *Biofortification of Food Crops*

Biofortification is the process of enriching the nutritional status of food crops by increasing the bioavailable concentrations of nutrients and bioactive compounds in the edible parts of crops through various alternatives such as traditional breeding, biotechnological tools and agronomic methods [9].

Since the beginning of the twenty-first century, biofortification has been used as an additional and complementary strategy to produce healthier foods and is currently ranked as the fifth best and most cost-effective solution to reduce the global problem of malnutrition or hidden hunger [40].

Biofortification differs from conventional fortifications in that it increases nutrient levels in crops through crop management, rather than by artificial means during food processing [41]. In addition to the fact that commercial products obtained from conventional fortification and nutritional supplements are costly and difficult to access for the entire population [9]. Its main objective is to improve the nutritional value of crops by increasing the concentration and bioavailability of nutrients in edible portions without sacrificing agronomic characteristics such as growth, yield or tolerance to biotic and abiotic stressors [42].

3.1.1 Biofortification with Conventional Breeding

Conventional plant breeding is a process of developing new crop varieties using traditional techniques and natural processes without the use of molecular biology tools in plants [43]. It involves the identification of parent plants with desirable characteristics to create favorable combinations in the next generation. This process of plant selection dates back more than 10,000 years and has been substantially refined in the last century [44].

Biofortification of crops through conventional breeding can be achieved through a variety of techniques including the identification and crossing of plants with wild relatives and/or specific varieties of various species with high levels of mineral elements [45]. Or, through cross-pollination, self-pollination or combinations of genetic alleles to produce new progeny varieties or hybrids with superior characteristics with high nutrient content in the different organs of consumption [46, 44].

3.1.2 Genetic Biofortification

Genetic biofortification involves studies of biotechnology, molecular biology and omics sciences to obtain plants with balanced and bioavailable nutritional profiles [9]. This type of biofortification can be achieved through techniques known as new breeding technologies (NBTs) such as transgenic breeding, RNA interference (RNAi), genome editing, clustered regularly interspaced short palindromic repeats (CRISPER), gene overexpression, foreign gene transfer, among others [16, 41].

These techniques allow the development of plants with improved and balanced nutritional profiles in edible organs through four main processes: gene transfer, gene overexpression, gene editing and gene silencing in different species [47].

3.1.3 Agronomic Biofortification

Agronomic biofortification consists of increasing the concentration of mineral elements in foodstuffs through different management practices of plants of agricultural interest [48] (Table 1).

Agronomic biofortification had its beginnings in basic crops, especially grain crops; however, it is currently used in most horticultural plants, including vegetables, fruit, ornamental and aromatic plants [47]. The overall objective of agronomic biofortification is to increase nutrient density in crops through the development of stable processes for the enrichment of nutrients in crops while maintaining and/or improving crop yields so as not to generate losses for farmers [118]. This is achieved through the maintenance and integrated management of the plant, water, soil, fertilization and microorganisms with which the target species interact [114].

Among the different agronomic strategies to obtain biofortified crops are foliar application of nutrients, which is used when minerals cannot be easily translocated to edible tissues [9]. In this technique, nutrients are applied in liquid form or in suspension on the aerial parts of the plants so that they can be absorbed through the stomata and epidermis [43]. Mineral flood fertilization, where elements are supplied to crops along with irrigation and are readily available for absorption and, as a result, increase their accumulation in the edible parts of plants [43].

Seed treatment prior to sowing by means of techniques such as seed priming, which is a process of controlled hydration of seeds in solutions containing mineral elements, biological agents or stimulants [119]. As well as seed coating where different adhesive materials are used to adhere nutrients or desired molecules to the seed surface [120]. Both methods generate changes in seed physiology that improve germination efficiency, increase biomass production, yield and nutrient composition of crops. In addition to inducing tolerance in plants to abiotic factors [121, 120].

Use of bioinoculants and/or growth promoting organisms, mainly beneficial bacteria and fungi with diverse mechanisms of action on plants such as the modulation of phytohormones, release of organic acids, increased nutrient and water absorption and solubilization of some elements such as phosphorus, calcium and iron. In addition to the induction of plant resistance to abiotic stresses [122].

Also, the use of modern agricultural practices such as nanotechnology that through seed priming, foliar, soil or nutrient solution application can increase the content of mineral elements, macromolecules such as sugars, proteins, amino acids and lipids, as well as numerous bioactive compounds derived from secondary plant metabolism in crops [123]. Both non-metallic metal nanoparticles, carbon nanomaterials, organic and combined nanoparticles induce biofortification of crops such as cucumber, bell pepper, chicory, strawberry, tomato, pomegranate, wheat, soybean, sorghum, common bean and many others [123, 124].

Table 1 Application of nanomaterials for crop biofortification

Crop	Nanomaterial	Dose—application route	Increased element	References
Alfalfa (*Medicago sativa*)	Cu NP	80 mg Kg^{-1}—seed priming	Fe and Cu	[49]
Bean (*Phaseolus vulgaris*)	Fe NPs	2 g L^{-1}—foliar spraying	Fe, Zn	[50]
	Se NPs	0.5 μM—foliar spraying	N, S and Se	[51]
	MnO_2	20 mg L^{-1}—foliar spraying	N, P, K, Zn, Mn, Fe and Cu	[52]
	Carbon nanoparticles (C NPs)	10 mg L^{-1}—foliar spraying	N, Fe, Zn and Mn	[53]
Bok choy (*Brassica rapa*)	CuO NPs	75 mg kg^{-1}—soil	Cu	[54]
	ZnO NPs	50 mg kg^{-1}—soil	Ca, P, Fe, Cu and Zn	[55]
Cabbage (*Brassica pekinensis*)	CuO NPs	400 mg kg^{-1}—soil	Cu and Mg	[56]
Carrot (*Daucus carota*)	MgO NPs	5 mmol L^{-1}—soil	Zn, Cu, K, Ca, Mg, S and Mn	[57]
	TiO_2	100 mg kg^{-1}—soil	Fe, Mg, Ca, Zn and K	[58]
Celery (*Apium graveolens*)	Se NPs	5 and10 mg L^{-1}—foliar spraying	Se	[59]
	NH_4NO_3	30 mg L^{-1}—foliar spraying	C, P, Ca and K	[60]
Chinese cabbage (*Brassica rapa*)	Biochar NP	0,5 g L^{-1}—Drench	P	[61]
Coriander (*Coriandrum sativum*)	TiO_2 NPs	50 mg L^{-1}—hydroponic solution	K, Ca, Mg and Fe	[62]
	Zeolite NPs	4 L h^{-1}—Drench	N, P, K, Mg, Ca, Fe, Zn and Mn	[63]
Corn (*Zea mays*)	Zn NPs	0.1%—seed priming	Zn	[64]

(continued)

Table 1 (continued)

Crop	Nanomaterial	Dose—application route	Increased element	References
Corn mint (*Mentha arvensis*)	TiO_2 NPs	100 μg m L^{-1}—hydroponic solution	Fe, Cu and Zn	[65]
Cucumber (*Cucumis sativus*)	Si NPs	200 mg L^{-1}—Foliar spraying	Si	[66]
	SiO NPs	100 mg L^{-1}—Foliar spraying	P, K and Si	[67]
Chickpea (*Cicer arietinum*)	Zn NPs, Fe_2O_3 NP	0.5%—Foliar spraying	Zn and Fe	[68]
Green onion (*Allium fistulosum*)	CuO NPs	150 mg kg $soil^{-1}$—soil	Ca, Mg, Ni and Cu	[69]
Lettuce (*Lactuca sativa*)	ZnO NP	8 mg Zn kg^{-1}—soil	Zn	[70]
	ZnO NP	8.5 mg Zn $plant^{-1}$—foliar spraying	Mg and S	[71]
	TiO_2 NPs	20 mg L^{-1}—foliar spraying	N, P and K	[72]
	Cu NPs	100 mg L^{-1}—soil	Cu and Fe	[73]
	CuO NPs	20 mg L^{-1}—soil irrigation	Ca, K, Mg, S, Fe, Ni and Zn	[74]
	Al_2O_3 NPs	1 mg mL^{-1}—hydroponic solution	P, Ca, Al, Cu and Zn	[75]
Mango (*Mangifera indica*)	Se NP, Ti NP, Si NP	10, 20 mg L^{-1} 60, 80 mg L^{-1} 50, 100, 150 mg L^{-1}—foliar spraying	N, P and K	[76]
Melissa (*Melissa officinalis*)	TiO_2 NPs	50, and 100 mg L^{-1}—supplemented MS medium	K, P, Mg, Mn, Cu, and Mo	[77]
Muskmelon (*Cucumis melo*)	Fe_2O_3 NPs	200 mg L^{-1}—Drench	Fe	[78]
Okra (*Abelmoschus esculentus*)	CeO_2 NPs	200 mg L^{-1}—foliar spraying	K, Mg, Mn, Cu and Zn	[79]

(continued)

Table 1 (continued)

Crop	Nanomaterial	Dose—application route	Increased element	References
Peanuts (*Arachis hypogaea*)	CaO NPs	200 mg L^{-1}—foliar spraying	N, P and K	[80]
Pennyroyal (*Mentha pulegium*)	Ca NPs	250 mg L^{-1}—foliar spraying	Ca and Fe	[81]
Pomegranate (*Punica granatum*)	ZnO NP	250, 500 and1000 mg L^{-1}—foliar spraying	Mg, Ca and Fe	[82]
	Se NPs	2 μM—foliar spraying	N, P, K and Ca	[83]
	Fe NPs	60 mg L^{-1}—foliar spraying	Fe	[84]
	Commercial product Fetrilon contains a mixture of micronutrients, Fe (4%), Zn (4%), Mn (3%), B (1.5%), Mg (2.2%), S (3. 3%), Cu (0.6%), and Mo (0.05%)	Nano micro 1000 µg mL^{-1}—foliar spraying	N, Fe, Zn and Cu	[85]
Potato (*Solanum tuberosum*)	NPK NPs	33% N, 15%, 48% K—foliar spraying	N, P and K	[86]
Purslane (*Portulaca oleracea*)	Ag NPs	2.5 mg L^{-1}—nutritive solution	Ca, Mg, Cu	[87]
Rice (*Oriza sativa*)	Se NP	20, 40, 80 g ha^{-1}—foliar spraying	Mg, P, Mn, S, K, Se, Fe, Zn and Cu	[88]
	FeO_3 NPs	100 mg L^{-1}—seed priming	Fe	[89]
	ZnO NPs	10 mg kg^{-1}—soil	Zn	[90]
	Zn NPs	5 mg kg^{-1}—foliar spraying	Zn and Fe	[91]
	ZnO NPs	25 and 100 mg kg^{-1}—soil	Zn, N, P, and K	[92]
	Nano chelated iron	2.5 g L^{-1}—foliar spraying	N, P, K, Zn, Cu and Mn	[93]
	CuO NPs	75, 150 and 300 mg kg^{-1}—soil	Cu	[94]

(continued)

Table 1 (continued)

Crop	Nanomaterial	Dose—application route	Increased element	References
Soybean (*Glycine max*)	Cu NPs	50, 100, 200, and 500 mg kg^{-1}—soil	Cu	[95]
	Mo NPs	0.15 mg kg^{-1}—soil	N, P, K, Mg Mo and Fe	[96]
	MoS_2 NPs	10 mg Kg^{-1}—soil	P, K, Ca, Mg, Fe, Cu, Zn and Mo	[97]
Spinach (*Spinacia oleracea*)	Cu NPs	400 mg kg^{-1}—soil	Cu	[98]
Squash (*Cucurbita pepo*)	$Mn_{0.5}Zn_{0.5}Fe_2O_4$ NPs	10, 20, 30 mg L^{-1}—foliar spraying	Ca, Zn, Fe and Mn	[99]
Strawberry (*Fragaria* × *ananassa*)	ZnO NPs	200 mg L^{-1}—foliar spraying	N, K and Zn	[100]
Sweet potato (*Ipomoea batatas*)	ZnO NPs	1500 mg L^{-1}—foliar spraying	P, Zn and Cu	[101]
Tomato (*Solanum lycopersicum*)	Se NP	3 mg L^{-1}—foliar spraying	Mg, Fe, Zn and Se	[102]
	Zn NP	20 mg kg^{-1} – Soil	Mg, Cu and Zn	[103]
	Cu NPs Nanocluster	1 mg kg^{-1}—soil	Cu, Fe, Mg, Mn, Ca and Zn	[104]
	CeO_2 NPs	10 mg kg $^{-1}$—Drench	P, S, K, Ca, Mg, Mn, Cu, Zn and Mo	[105]
	Se NPs	25 μM—foliar spraying	Se	[106]
	Se NPs	3 mg L^{-1}—foliar spraying	Mg, Fe, Zn and Se	[102]
	Zn O NPs	3 mg L^{-1}—foliar spraying	K, Fe and Zn	[107]

(continued)

Table 1 (continued)

Crop	Nanomaterial	Dose—application route	Increased element	References
	Si NPs	2 mmol L^{-1}—hydroponic solution	Fe, Ca and N	[108]
	Se NPs	40 mg L^{-1}—foliar spraying	Se	[109]
Wheat (*Triticum aestivum*)	Oxozinc nanofiber (ZnONF)	0.2%—foliar spraying	Zn	[110]
	Zn NPs	1 g L^{-1}—foliar spraying	Zn, N and P	[111]
	Zn and Fe NPs	0.01 and 0.05 mg L^{-1}—foliar spraying	Fe and Zn	[112]
	ZnO NPs	25 mg L^{-1}—foliar spraying	Zn and Fe	[113]
	$ZnSO_4$ NPs	10 mM—foliar spraying	Zn	[114]
	ZnO NPs	75 mg L^{-1}—foliar spraying	Zn	[115]
	ZnO NP	1.7 mg kg^{-1}—soil	N, P and Ca	[116]
	ZnO NPs	1.2 kg ha^{-1}—foliar spraying	Fe, Mn, Cu and Zn	[117]

Therefore, biofortified crops under the different methods represent a sustainable reserve of vital nutrients that are often deficient in diets and generate problems in the health of populations [16].

Table 1 shows several examples where different nanomaterials were used with the objective of biofortifying crops. The observed results are clear, since the application of nanomaterials was efficient in increasing the content of several elements at the same time, not only the element applied at the nanoscale. This is a great advantage, since a more efficient biofortification of crops is achieved.

4 Nanostimulants to Improve Nutritional Quality

A biostimulant is any substance or microorganism that, when applied to plants, improves their ability to absorb and assimilate nutrients, in addition to being able to improve tolerance to abiotic stress or some of their agronomic characteristics [125]. According to this definition, nanomaterials can be considered as biostimulants, due to their ability to induce positive responses in crops that allow them to tolerate stress conditions, but also improve the quality of crops through the production of bioactive compounds potentially beneficial to human health [11, 126].

4.1 Nanostimulation of Crops

The application of nanomaterials induces a series of responses in plants ranging from the morphological, physiological, biochemical, to the genetic level, which will allow them to adapt to various stress conditions, and potentially improve their growth and development [127]. At the biochemical level, the application of nanomaterials to plants can induce changes in the antioxidant defense system that will allow it to more adequately regulate the levels of reactive oxygen species through the production of enzymatic and non-enzymatic antioxidant compounds. This in turn improves the internal conditions of plants resulting in better adaptation of plants to stress [128, 129].

However, it is important to highlight that although nanomaterials can be used as biostimulants on plants to induce tolerance to stress or to improve crop quality, they can induce toxicity when applied in excessive quantities, but also depending on the physical and chemical characteristics, as well as the species, dose, route of application of the nanomaterials and the properties of the medium where it was applied [11, 126]. For this reason, it is essential to adequately define the specific concentrations or quantities of nanomaterials that can be applied to plants in order to induce positive responses without generating negative responses in the plants.

4.2 Impact of Nanostimulants on Primary Metabolism

The plant cell produces thousands of low molecular weight organic compounds classified into three general groups: primary metabolites, which are directly necessary for plant growth; secondary (or specialized) metabolites, which mediate interactions between plants and the environment, and hormones, which regulate the cell's metabolic processes [130].

Specifically, primary metabolites are derived from primary plant metabolism that includes reactions essential for normal growth, development and reproduction, including anabolic and catabolic reactions of carbohydrates, proteins, lipids, carbohydrates and nucleic acids, which are essential elements in the physiological processes mentioned above [131]. In addition, they can serve as building blocks to synthesize other macromolecules and provide energy for plant growth [132].

Also, mineral elements, macro and microelements, are an essential part of primary plant metabolism because they play a crucial role in growth and participate in the metabolism of sugars, proteins and lipids that have numerous functions, including osmotic regulation, immune modulation, antioxidation and anti-aging [133].

Several studies have shown that nanoparticles and/or nanomaterials applied in different doses, generally low and by different routes, exert positive effects on the biosynthesis and accumulation of macromolecules related to the primary metabolism of different plant species [14].

For example, protein content was increased by foliar application of 2 g L^{-1} of Fe NPs [50],1000 g L^{-1} of CeO_2 NPs [134],0.5 μM of Se NPs [51], and 10 mg L^{-1} de ZnO, MnO_2 and MoO_3 in grains of *P. vulgaris* [135]. With 200 mg L^{-1} of CaO NPs, protein in grains of *A. hypogea* was increased [80]. While applying 2.5 g L^{-1} nano chelated iron and 20 mg L^{-1} nano-scale zero valent iron (nZVI) protein was increased in grains of *O. sativa* [136].

It has also been reported the increase of proteins in fruits such as *V. vinifera* when applying 100 mg L^{-1} by foliar spraying of ZnO NPs [137],on *P. dactylifera* by foliar spraying of 1 g L^{-1} of NPK NPs [138], and in *C. melo* with the application via drench of 200 mg L^{-1} de Fe_2O_3 NPs [78].

Other macromolecules such as lipids can also be increased with the application of different nanomaterials. Salama et al. [52, 135] reported the increase of lipids in grains of *P. vulgaris,* with the foliar application of 10 mg L^{-1} de MoO_3 NPS. Osman et al. [139] reported the increase of the content of these macromolecules in the same plant specie with 10 mg L^{-1} de MoO_3 NPs applied via foliar. Guha et al. [136] reported the increase of lipids in grains of *O. sativa,* when applying 20 mg L^{-1} of nano-scale zero valent iron (nZVI) by seed priming. In fruits of *C. pepo* were increased with the foliar application of 10, 20 y 30 mg L^{-1} of $Mn_{0.5}Zn_{0.5}Fe_2O_4$ NPs [99].

Carbohydrates are another macromolecule that can be increased through the application of various nanomaterials, either totally or individually. For example in grains of *L. usitatissimum* Sadak and Bakry [140], reported the increase of total carbohydrate with the foliar application of 20, 40 and 60 mg L^{-1} of ZnO NPs. Hu et al. [106] reported the increase of total carbohydrates in S. lycopersicum fruits with foliar application of 25 μM Se, NPs. Mahmoud et al. [141] reported the increase of total Carbohydrates in *S. tuberosum* tubercles applying to soil 10 mg L^{-1} of nano zeolite.

Javan et al. [142] reported the increase of total carbohydrates with the foliar application of 10 mg L^{-1} of 100 mg L^{-1} of TiO_2 NPs, in fruits of *F. × ananassa.* While in this same plant species, Zahedi et al. [143] reported an increase in soluble carbohydrates with the foliar application of 25 mg L^{-1} of SiO_2 NPs.

In several plant species it has been reported the increase of sugar content in different organs of consumption, especially in fruits, for example in *V. vinifera*, the foliar application of 100 mg L^{-1} of ZnO NPs increased total sugar and reducing sugar [144]. In *P. persica*, foliar application of 4% Ca–Si NPs increased sucrose, turanose and arabinose content [145]. In *C. annuum* the application of Se NPs by foliar application increased the soluble sugar content [146]. In *P. granatum*, 60 mg L^{-1} applied via foliar spraying increased total sugar [84].

In fruits of *S. lycopersicum*, various nanomaterials have been applied and have succeeded in increasing the content of different sugars. 10 mg kg^{-1} of CeO_2 NPs applied via drench increased fructose [105], 50 mg L^{-1} of $MnFe_2O_4$ applied by foliar spraying increased sucrose and total sugar content [147], foliar application of 0.04%

of (Cs) NPs total sugar increment, reducing sugar [148], and 5 mg L^{-1} of Se NPs applied via foliar increased the content of fructose, glucose and sucrose [149].

4.3 Impact of Nanostimulants on Secondary Metabolism

Plant secondary metabolites are a broad class of structurally diverse molecules, consisting mainly of terpenoids, phenolic compounds and nitrogen-containing compounds, which play active roles in plant development and stress responses [150]. They are widely used in various industries, including agriculture, cosmetology and pharmaceuticals because they are extremely beneficial to human health and well-being, exhibiting a wide range of biological activities, including antibacterial, antioxidant and anticarcinogenic capabilities [151].

It has been demonstrated that nanomaterials can increase several secondary metabolites, for example, vitamin C or ascorbic acid, which is a potent antioxidant with high redox capacity that helps neutralize free radicals and reduce oxidative stress in cells [13]. It is also essential for the synthesis of collagen, which is crucial for maintaining tissue integrity and preventing uncontrolled cell growth that causes multiple diseases and strengthens the human immune system [152].

The use of nanomaterials has been shown to be effective in increasing vitamin C in different crops and organs of consumption. The application of Nano-Scale zero valent iron (nZVI, 20 mg L^{-1}) by seed priming increased vitamin C in grains of *O. sativa* [136]. In fruits of *M. domestica*, this antioxidant was increased with foliar application of 250 mg L^{-1} de Zn NPs [153]. In *M. charantia* it was increased by the foliar application of 10 mg L^{-1} de Cs–Se NPs [154]. In *P. dactylifera* with the foliar application of 1 g L^{-1} de NPK NPs [138]. In *P. alkekengi* with the foliar application of Si NPs (200 mg L^{-1}) [155]. In *M. indica* by foliar application of Se NPs (10 and 20 mg L^{-1}), Ti NPs (60 and 80 mg L^{-1}) and Si NPs (50, 100 and 120 mg L^{-1}) [76]. And in *F.* × *ananassa* when applied by foliar application 25 mg L^{-1} of SiO_2 NPs [143], and 200 mg L^{-1} of ZnO NPs [100].

The same effect has been observed in leaf crops. In *C. sativum* with the application via drench of 4 L h^{-1} of Zeolite NPs [63], in *L. sativa* by foliar application of 8.5 mg Zn $plant^{-1}$ of ZnO NPs [71], and 30 mg L^{-1} of GO NPs [156].

It has also been demonstrated that the application of nanoparticles in plants can increase other vitamins such as α-tocopherol or vitamin E. For example, Rady et al. [109] observed that foliar application of 40 mg L^{-1} of Se NPs increases α-tocopherol content in *S. lycopersicum* fruits.

The increase of this vitamin is of utmost importance for health as it is a lipophilic antioxidant capable of inhibiting the formation of lipid peroxidation by two different mechanisms: as an antioxidant that traps radicals, thus interrupting the cascade of chain reactions during lipid peroxidation of membranes and by inhibiting ferroptosis by blocking lipoxygenases, which catalyze the formation of lipid peroxides [157]. Therefore, its intake improves the body's immune responses and reduces the incidence of cardiovascular and chronic degenerative diseases [158].

The increase of carotenoids in various plant species with the application of NPs has been widely demonstrated. Aazami et al. [137] reported that *V. vinifera* fruits increased total carotenoid content with foliar application of 0.5 mM chitosan based salicylic acid nanocomposite. Pejam et al. [107] demonstrated that the foliar application of 3 mg L^{-1} of ZnO NPs on S. lycopersicum plants increased the total carotenoid content in fruits. In this same species, Faizan et al. [159] reported the increase of β-carotene in fruit when 10 mg kg^{-1} de ZnO NPs were applied to the soil. In stems of *A. graveolens*, β-carotene content was increased by foliar application of Se NPs [59]. In *R. sativus* plants, the application of 100 μmol L^{-1} via Soil of TiO_2 NPs increased the content of total carotenoids in bulb [160]. While in *D. carota* the application of 5 mmol L^{-1} via soil of MgO NPs increased β-carotene and α-carotene in bulbs [57].

Carotenoids are a group of lipophilic isoprenoid pigments produced by all plants and other photosynthetic organisms. They are essential for photosynthesis and photoprotection and are precursors of abscisic acid and strigolactones as well as various signaling molecules that regulate plant development, growth and stress response [161]. Humans cannot synthesize carotenoids, however, these pigments are indispensable for health because they are precursors of vitamin A (retinol), which is essential for maintaining health. Carotenoids are also important antioxidants and scavengers of reactive oxygen species, so their intake is correlated with a reduced risk of developing non-communicable diseases [162].

Table 2 presents a review of the application of nanostimulants to improve the nutritional quality of crops in different organs of consumption. Each report describes the macromolecules or antioxidants that were increased by the application of a nanostimulant. In addition, the doses and application routes used in each case are included.

Table 2 Improving nutritional quality of different plant organs using nanostimulants

Crop	Nanostimulant	Dose—application route	Improved nutritional quality	References
Grains				
Bean (*Phaseolus vulgaris*)	Fe NPs	2 g L^{-1}—foliar spraying	Protein	[50]
	CeO_2 NPs	1000 g L^{-1}—foliar spraying	Protein	[134]
	Se NPs	0.5 μM—foliar spraying	Total carbohydrate, crude protein	[51]
	ZnO, MnO_2, MoO_3	10, 10, 10 mg L^{-1}—foliar spraying	Carbohydrate, protein, lipid and total energy	[135]

(continued)

Table 2 (continued)

Crop	Nanostimulant	Dose—application route	Improved nutritional quality	References
	MnO_2	20 mg L^{-1}—foliar spraying	Total energy, protein	[52]
	MoO_3	10 mg L^{-1}—foliar spraying	Carbohydrate, protein, lipid and total energy	[139]
Colza (*Brassica napus*)	Ca NPs	150 mg L^{-1}—foliar spraying	Proline and total soluble proteins	[163]
Cowpea (*Vigna unguiculata*)	Si NPs	50 mg L^{-1}—foliar spraying	Oleic acid, saturated fatty acids and unsaturated fatty acids	[164]
Flax (*Linum usitatissimum*)	ZnO NPs	20, 40 and 60 mg L^{-1}—foliar spraying	Free amino acids and total carbohydrate	[140]
Lentil (*Lens culinaris*) Soybean (*Glycine max*)	Si NPs	10 g L^{-1}—Drench	Protein, free amino acid, total sugar, reducing sugar and free fatty acid	[165]
Peanuts (*Arachis hypogaea*)	CaO NPs	200 mg L^{-1}—foliar spraying	Protein	[80]
Rice (*Oryza sativa*)	Nano chelated iron	2.5 g L^{-1}—foliar spraying	Protein	[93]
	ZnO NPs	20 mg L^{-1}—seed priming	Total soluble sugar and Zn	[68]
	Nano-Scale Zero Valent Iron (nZVI)	20 mg L^{-1}—seed priming	Ascorbic acid, starch, soluble sugar, protein and lipid	[136]
	CuO NPs	75, 150 and 300 mg kg^{-1}—soil	Total sugar, and starch	[94]
Soybean (*Glycine max*)	Mo NPs	0.15 mg kg $^{-1}$—Soil	Fatty acids	[96]
	MoS_2 NPs	10 mg Kg^{-1}—soil	Soluble protein and starch	[97]
Wheat (*Triticum aestivum*)	ZnO NPs	80 mg L^{-1}—foliar spraying	Protein and carotenoids	[166]
	ZnO NPs	1.2 kg ha^{-1}—foliar spraying	Starch	[117]

(continued)

Table 2 (continued)

Crop	Nanostimulant	Dose—application route	Improved nutritional quality	References
Fruits				
Apple (*Malus domestica*)	Zn NP	250 mg L^{-1}—foliar spraying	Ascorbic acid	[153]
Bell pepper (*Capsicum annuum*)	Si NP	1000 mg L^{-1}—foliar spraying	β-carotene	[167]
Bitter melon (*Momordica charantia*)	Cs–Se NPs	10 mg L^{-1}—foliar spraying	Ascorbic acid	[154]
Date palm (*Phoenix dactylifera*)	NPK NPs	1 g L^{-1}—foliar spraying	Proteins and ascorbic acid	[138]
Gooseberry (*Physalis alkekengi*)	Si NPs	200 mg L^{-1}—foliar spraying	Ascorbic acid, palmitic acid, oleic acid and linolenic acid	[155]
Grape (*Vitis vinifera*)	ZnO NPs	100 mg L^{-1}—foliar spraying	Total sugar and reducing sugar	[144]
	Chitosan based salicylic acid nanocomposite	0.5 mM—foliar spraying	Carotenoid and soluble protein	[137]
Mango (*Mangifera indica*)	Se NP Ti NP Si NP	10, 20 mg L^{-1} 60, 80 mg L^{-1} 50, 100, 150 mg L^{-1}—foliar spraying	Vitamin C, total sugars	[76]
	ZnO NPs	100 mg L^{-1}—foliar spraying	Total sugars	[168]
Muskmelon (*Cucumis melo*)	Fe_2O_3 NPs	200 mg L^{-1}—Drench	Soluble protein, ascorbic acid	[78]
Okra (*Abelmoschus esculentus*)	CeO_2 NPs	200 mg L^{-1}—foliar spraying	Ascorbic acid	[79]
Peaches (*Prunus persica*)	Calcium silicate (Ca–Si) NP	4%—foliar spraying	Sucrose, turanose and arabinose	[145]
Pepper (*Capsicum annuum*)	Se NPs	20 mg L^{-1}—foliar spraying	Soluble sugar and ascorbic acid	[146]
Pineapple (*Ananas comosus*)	Ca NPs	0, 12.5, 25, 50, 100, and 200 g ha^{-1}—soil	Ascorbic acid	[169]

(continued)

Table 2 (continued)

Crop	Nanostimulant	Dose—application route	Improved nutritional quality	References
Pomegranate (*Punica granatum*)	Fe NPs	60 mg L^{-1}—foliar spraying	Total sugar	[84]
	Se NPs	20 mg L^{-1}—foliar spraying	Ascorbic acid and proline	[170]
	Commercial product Fetrilon contains a mixture of micronutrients, Fe (4%), Zn (4%), Mn (3%), B (1.5%), Mg (2.2%), S (3. 3%), Cu (0.6%), and Mo (0.05%)	Nano micro 1000 μg mL^{-1}—foliar spraying	Ascorbic acid	[85]
Squash (*Cucurbita pepo*)	$Mn_{0.5}Zn_{0.5}Fe_2O_4$ NPs	10, 20, 30 mg L^{-1}—foliar spraying	Lipids	[99]
Strawberry (*Fragaria* × *ananassa*)	SiO_2 NP Se NP	25 mg L^{-1} 125 mg L^{-1}—foliar spraying	Soluble carbohydrates and ascorbic acid	[143]
	ZnO NPs	200 mg L^{-1}—foliar spraying	Ascorbic acid, total sugar, non-reducing sugar, reducing sugar	[100]
	TiO_2 NPs	10 mg L^{-1}—foliar spraying	Total carbohydrates	[142]
Tomato (*Solanum lycopersicum*)	CeO_2 NPs	10 mg kg $^{-1}$—Drench	Ascorbic acid, fructose, malic acid	[105]
	ZnO NPs	10 mg kg^{-1}—soil	Protein and β-carotene	[159]
	$MnFe_2O_4$	50 mg L^{-1}—foliar spraying	Sucrose and total sugar	[147]
	Se NPs	75 μg Kg^{-1}—foliar spraying	Fatty acids, serine, alanine, valine, aspartic acid, lysine, isoleucine, glutamic acid, glutamine, proline and total aminoacids	[171]

(continued)

Table 2 (continued)

Crop	Nanostimulant	Dose—application route	Improved nutritional quality	References
	Se NPs	25 μM—foliar spraying	Starch, total carbohydrates, proteins, lycopene, total carotenoids	[106]
	Triiron Tetrairon Phosphate ($Fe_7(PO_4)_6$) N NPs	50 mg kg^{-1}-soil	Sucrose content	[172]
	Mn NPs	250 mg L^{-1}—foliar spraying	Total sugar	[173]
	CuO NPs	50 mg kg^{-1}—soil	Lycopene	[174]
	Cu-chitosan (Cs) NPs	0.04%—foliar spraying	Total sugar, reducing sugar and lycopene	[148]
	ZnO NP	100 mg L^{-1}—foliar spraying	Ascorbic acid and lycopene	[175]
	Zn O NPs	3 mg L^{-1}—foliar spraying	Carotenoids, proline	[107]
	Se NPs	5 mg L^{-1}—foliar spraying	Fructose, glucose, sucrose ando ascorbic acid	[149]
	Si NPs	2 mmol L^{-1}—hydroponic solution	Ascorbic acid, lycopene	[108]
	Se NPs	40 mg L^{-1}—foliar spraying	Lycopene, α – tocopherol	[109]
Watermelon (*Citrullus lanatus*)	Ag NPs	5 g mL^{-1}—seed priming	Ascorbic acid	[176]
Leaves				
Basil (*Ocimum basilicum*)	Multiwall carbon nanotubes (MWCNTs)	25, 50 and 100 mg L^{-1}—hydroponic solution	Carotenoids	[177]
Bok choy (*Brassica rapa*)	CuO NPs	75 mg kg^{-1}—soil	Sugar content	[54]
Broccoli (*Brassica oleracea*)	ZnO NPs	200 mg L^{-1}—seed priming	Sugar and proline	[178]
Coriander (*Coriandrum sativum*)	Zeolite NPs	4 L h^{-1}—Drench	Ascorbic acid, proline, thiamine	[63]
	Se NPs	20 mg L^{-1}—foliar spraying	Proline	[179]
	Si NPs	1.5 mg L^{-1}—foliar spraying	Total soluble sugar	[180]

(continued)

Table 2 (continued)

Crop	Nanostimulant	Dose—application route	Improved nutritional quality	References
	Si NPs	50 mg L^{-1}—foliar spraying	Saturated fatty acids	[181]
Corn mint (*Mentha arvensis*)	TiO_2 NPs	100 μg m L^{-1}—hydroponic solution	Cysteine, proline	[65]
Lemon verbena (*Lippia citriodora*)	Se NPs	10 μM—foliar spraying	Soluble sugar and proline	[182]
Lettuce (*Lactuca sativa*)	ZnO NP	8.5 mg Zn $plant^{-1}$—foliar spraying	Ascorbic acid	[71]
	GO NP	30 mg L^{-1}—foliar spraying	Soluble sugar, soluble protein and ascorbic acid	[156]
	Ag NPs	25, 50 mg L^{-1}—Drench	Protein and soluble sugar	[61]
	TiO_2	200 mg L^{-1}—hydroponic solution	Soluble protein	[183]
Melissa (*Melissa officinalis*)	Se NPs	50 mg L^{-1}—foliar spraying	Carotenoids	[184]
	TiO_2 NPs	50, and 100 mg L^{-1}—supplemented MS medium	Carotenoids, protein	[77]
Mint (*Mentha piperita*)	TiO_2 NPs	100 mg L^{-1}—sprayed soil	Carotenoids and proline	[185]
	TiO_2 NPs	150 mg L^{-1}—foliar spraying	Carotenoids	[186]
Pennyroyal (*Mentha pulegium*)	Ca NPs	250 mg L^{-1}—foliar spraying	Carotenoids	[81]
Sweet basil (*Ocimum basilicum*)	Graphene oxide (GO) NPs	100 mg L^{-1}–foliar spraying	Carotenoids and proline	[187]
Swiss chard (*Beta vulgaris*)	ZnO NP	200 mg L^{-1}—foliar spraying	Total soluble proteins	[188]
Stems				
Celery (*Apium graveolens*)	Se NPs	5 and10 mg L^{-1}—foliar spraying	Soluble sugar, protein, fatty acid, β-carotene	[59]
Celery (*Apium graveolens*)	NH_4NO_3	30 mg L^{-1}—foliar spraying	Carotenoids, vitamins	[60]

(continued)

Table 2 (continued)

Crop	Nanostimulant	Dose—application route	Improved nutritional quality	References
Bulbs				
Carrot (*Daucus carota*)	MgO NPs	5 mmol L^{-1}—soil	β-carotene, α-carotene	[57]
Potato (*Solanum tuberosum*)	NPK NPs	33% N, 15%, 48% K—foliar spraying	Starch	[86]
Potato (*Solanum tuberosum*)	Nano zeolite, Zn zinc, boron and silicon	20, 12 y 15 ppm—soil	Carbohydrates, and protein	[141]
Radish (*Raphanus sativus*)	TiO_2 NPs	100 μmol L^{-1}—soil	Carotenoids and total carbohydrates	[160]
Radish (*Raphanus sativus*)	CuO NPs	200 mg kg^{-1}—soil	Nicotinic acid (B3 Vitamin) and biotin	[189]
Sweet potato (*Ipomoea batatas*)	Cu NPs	25 mg kg^{-1}—soil	Sugar and starch	[190]

5 Conclusions

Agriculture is a fundamental activity for the development of the human population, and it becomes even more important if we consider that a large part of the food consumed is obtained from it. Producing these foods with high contents of essential elements for the human diet, as well as with high contents of bioactive compounds is of utmost importance, since they will help to solve the problems of human nutrition. In this sense, it has been demonstrated that the use of nanotechnology to improve the quality of food from agriculture has proved to be a promising tool, since through the application of nanomaterials two fundamental objectives are achieved, (1) increase the content of essential elements for the human diet and, (2) increase the content of bioactive compounds in food. However, the use of nanotechnology in agriculture must be done responsibly. To date, there is a lack of information on the possible impacts of nanomaterials on the environment, on soil microfauna and microflora, and on the flow of nanomaterials through food chains. For this reason, before making large-scale applications in agriculture, it is advisable to continue doing research in these areas to be certain that nanotechnology is safe and that it can be applied without generating problems.

References

1. Kathi S, Laza H, Singh S, Thompson L, Li W, Simpson C (2024) A decade of improving nutritional quality of horticultural crops agronomically (2012–2022): a systematic literature review. Sci Total Environ 911:168665. https://doi.org/10.1016/j.scitotenv.2023.168665
2. Kamble U, Mishra CN, Govindan V, Sharma AK, Pawar S, Kumar S, Krishnappa G, Gupta OP, Singh GP, Singh G (2022) Ensuring nutritional security in india through wheat biofortification: a review. Genes 13:2298. https://doi.org/10.3390/genes13122298
3. Guardiola-Márquez CE, Santos-Ramírez MT, Segura-Jiménez ME, Figueroa-Montes ML, Jacobo-Velázquez DA (2022) Fighting obesity-related micronutrient deficiencies through biofortification of agri-food crops with sustainable fertilization practices. Plants 11:3477. https://doi.org/10.3390/plants11243477
4. Dutta S, Sinelshchikova A, Andreo J, Wuttke S (2024) Nanoscience and nanotechnology for water remediation: an earnest hope toward sustainability. Nanoscale Horiz 9:885–899. https://doi.org/10.1039/d4nh00056k
5. El-Naggar ME, Mahmoud YAG, Abd-Elgawad SE, Zawawy NAE, Hemdan BA (2024) A comprehensive review on nanoscience and nanotechnology with special emphasis on antimicrobial activities. Bionanoscience. https://doi.org/10.1007/s12668-024-01530-w
6. Rajput A, Shevalkar G, Pardeshi K, Pingale P (2023) Computational nanoscience and technology. OpenNano 12:100147. https://doi.org/10.1016/j.onano.2023.100147
7. Yu T, Liu X, Kang H, Ding T, Cheng R, He J, Sun Z, Zeng M, Fu L (2023) Cutting-edge research in nanoscience and nanotechnology: celebrating the 130th Anniversary of Wuhan University. ACS Nano 17:24423–24430. https://doi.org/10.1021/acsnano.3c11892
8. Sultanbawa F, Sultanbawa Y (2023) Mineral nutrient-rich plants—do they occur? Appl Food Res 3:100347. https://doi.org/10.1016/j.afres.2023.100347
9. Sheoran S, Kumar S, Ramtekey V, Kar P, Meena RS, Jangir CK (2022) Current status and potential of biofortification to enhance crop nutritional quality: an overview. Sustainability 14:3301. https://doi.org/10.3390/su14063301
10. Jayasekhar Babu P, Saranya S, Longchar B, Rajasekhar A (2022) Nanobiotechnology-mediated sustainable agriculture and post-harvest management. Curr Res Biotechnol 4:326–336. https://doi.org/10.1016/j.crbiot.2022.07.004
11. Juárez-Maldonado A, Ortega-Ortiz H, González-Morales S, Morelos-Moreno Á, Cabrera-de la Fuente M, Sandoval-Rangel A, Cadenas-Pliego G, Benavides-Mendoza A (2019) Nanoparticles and nanomaterials as plant biostimulants. Int J Mol Sci 20:1–19. https://doi.org/10.3390/ijms20010162
12. Rai-Kalal P, Jajoo A (2021) Priming with zinc oxide nanoparticles improve germination and photosynthetic performance in wheat. Plant Physiol Biochem 160:341–351. https://doi.org/10.1016/j.plaphy.2021.01.032
13. Hee S, Hae J, Eun J, Shin E, Kwak Y (2024) Effect of high-dose vitamin C on renal ischemia-reperfusion injury. Biomed Pharmacother 173:116407. https://doi.org/10.1016/j.biopha.2024.116407
14. Tripathi D, Singh M, Pandey-Rai S (2022) Crosstalk of nanoparticles and phytohormones regulate plant growth and metabolism under abiotic and biotic stress. Plant Stress 6:100107. https://doi.org/10.1016/j.stress.2022.100107
15. Adhikari S, Schop M, de Boer IJM, Huppertz T (2022) Protein quality in perspective: a review of protein quality metrics and their applications. Nutrients 14:947. https://doi.org/10.3390/nu14050947
16. Sandhu R, Chaudhary N, Bindia, Shams R, Singh K, Pandey VK (2023) A critical review on integrating bio fortification in crops for sustainable agricultural development and nutritional security. J Agric Food Res 14:100830. https://doi.org/10.1016/j.jafr.2023.100830
17. Lee-Okada H-C, Xue C, Yokomizo T (2025) Recent advances on the physiological and pathophysiological roles of polyunsaturated fatty acids and their biosynthetic pathway. BBA Mol Cell Biol Lipids 1870:159564. https://doi.org/10.1016/j.bbalip.2024.159564

18. Nayyar D, Said JM, Mccarthy H, Hryciw DH, Keefe LO, Mcainch AJ (2024) Effect of a high linoleic acid diet on pregnant women and their offspring. Nutrients 16:3019. https://doi.org/10.3390/nu16173019
19. Wei G, Wu G, Sun J, Qi Y, Zhao Q, Xu F, Zhang Z, Peng L (2024) Conjugated linoleic acid production in pine uut oil: a *Lactiplantibacillus plantarum* Lp-01 fermentation approach. Foods 13:2472. https://doi.org/10.3390/foods13162472
20. Hamaï A, Drin G (2024) Biochimie specificity of lipid transfer proteins: an in vitro story. Biochimie 6. https://doi.org/10.1016/j.biochi.2024.09.007
21. Wang X, Wu Z, Liu Y, Wu C, Jiang J, Hashimoto K (2024) Neurobiology of disease the role of thyroid-stimulating hormone in regulating lipid metabolism: Implications for body—brain communication. Neurobiol Dis 201:106658. https://doi.org/10.1016/j.nbd.2024.106658
22. Luo J, Wen F, Qiu D, Wang S (2021) Nesfatin-1 in lipid metabolism and lipid-related diseases. Clin Chim Acta 522:23–30. https://doi.org/10.1016/j.cca.2021.08.005
23. Petrovic S, Mouskeftara T, Paunovic M, Deda O, Vucic V, Milosevic M, Gika H (2024) Unveiling lipidomic alterations in metabolic syndrome: a study of plasma, liver, and adipose tissues in a dietary-induced rat model. Nutrients 16:3466. https://doi.org/10.3390/nu16203466
24. Strzemski M, Adamec L, Dresler S, Mazurek B, Dubaj K, Stolarczyk P, Feldo M, Płachno BJ (2024) Shoots and turions of aquatic plants as a source of fatty acids. Molecules 29:2062. https://doi.org/10.3390/molecules29092062
25. Dal Bosco A, Cavallo M, Menchetti L, Angelucci E, Cartoni Mancinelli A, Vaudo G, Marconi S, Camilli E, Galli F, Castellini C, Mattioli S (2024) The healthy fatty index allows for deeper insights into the lipid composition of foods of animal origin when compared with the atherogenic and thrombogenicity indexes. Foods 13:1568. https://doi.org/10.3390/foods13101568
26. Rolands MR, Hackl LS, Bochud M, Lê KA (2024) Protein adequacy, plant protein proportion, and main plant protein sources consumed across vegan, vegetarian, pescovegetarian, and semivegetarian diets: a systematic review. J Nutr. https://doi.org/10.1016/j.tjnut.2024.07.033
27. Lv X, Zhou C, Yan Q, Tan Z, Kang J, Tang S (2022) Elucidating the underlying mechanism of amino acids to regulate muscle protein synthesis: Effect on human health. Nutrition 103–104:111797. https://doi.org/10.1016/j.nut.2022.111797
28. Pastrana-Pastrana ÁJ, Rodríguez-Herrera R, Solanilla-Duque JF, Flores-Gallegos AC (2025) Plant proteins, insects, edible mushrooms and algae: more sustainable alternatives to conventional animal protein. J Fut Foods 5:248–256. https://doi.org/10.1016/j.jfutfo.2024.07.004
29. Sá AGA, Moreno YMF, Carciofi BAM (2020) Plant proteins as high-quality nutritional source for human diet. Trends Food Sci Technol 97:170–184. https://doi.org/10.1016/j.tifs.2020.01.011
30. Afnan F, Kashem MNH, Joshi R, Simpson C, Li W (2024) Growth of romaine lettuce in eggshell powder mixed alginate hydrogel in an aeroponic system for water conservation and vitamin C biofortification. Gels 10:322. https://doi.org/10.3390/gels10050322
31. Duan H, Guo Y, Ren C, Wei R, Yan W (2025) Mechanisms of vitamin and mineral absorption and their potential for eye fatigue relief. J Fut Foods 5:219–228. https://doi.org/10.1016/j.jfutfo.2024.07.001
32. Fitzpatrick TB, Chapman LM (2020) The importance of thiamine (vitamin B1) in plant health: from crop yield to biofortification. J Biol Chem 295:12002–12013. https://doi.org/10.1074/jbc.REV120.010918
33. Rajagopal S, Gupta A, Parveen R, Shukla N, Bhattacharya S, Naravula J, Kumar SA, Mathur P, Simlot A, Mehta S, Bihari C, Mishra AK, Nair BG, Medicherla KM, Reddy GB, Sreenivasulu N, Kishor PBK, Suravajhala P (2022) Vitamin K in human health and metabolism: a nutrigenomics review. Trends Food Sci Technol 119:412–427. https://doi.org/10.1016/j.tifs.2021.12.012
34. Bhat ZF, Bhat HF, Manzoor M, Abdi G, Aadil RM, Hassoun A, Aït-Kaddour A (2024) Enhancing the lipid stability of foods of animal origin using edible packaging systems. Food Chemistry: X 21:101185. https://doi.org/10.1016/j.fochx.2024.101185

35. Ali AAH (2023) Overview of the vital roles of macro minerals in the human body. J Trace ElemTs Miner 4:100076. https://doi.org/10.1016/j.jtemin.2023.100076
36. Singh S, Kaur J, Ram H, Singh J, Kaur S (2023) Agronomic bio-fortification of wheat (*Triticum aestivum* L.) to alleviate zinc deficiency in human being. Rev Environ Sci Bio/Technol 22:505–526. https://doi.org/10.1007/s11157-023-09653-4
37. Younas N, Naveed M, Yaseen M, Younas M, Mumtaz MZ, Babar MH, Nadeem M, Soysal S, Al-Ashkar I, Mustafa A (2024) Processed manures with added zinc improve zinc biofortification in lentils under saline conditions. Agronomy 14:1289. https://doi.org/10.3390/agronomy14061289
38. Amalfitano N, Patel N, Haddi M-L, Benabid H, Pazzola M, Vacca GM, Tagliapietra F, Schiavon S, Bittante G (2024) Detailed mineral profile of milk, whey, and cheese from cows, buffaloes, goats, ewes and dromedary camels, and efficiency of recovery of minerals in their cheese. J Dairy Sci. https://doi.org/10.3168/jds.2023-24624
39. Lee M, Ball L, Hill S, Crowe TC, Walsh H, Cosgrove T, Best T (2025) Omnivore, vegan and vegetarian diet quality associations with depressive symptoms: a comparative cross-sectional analysis of the Australian Longitudinal Study on Women's Health. J Affect Disord 370:18–25. https://doi.org/10.1016/j.jad.2024.10.119
40. Sunic K, Spanic V (2024) Genetic biofortification of winter wheat with selenium (Se). Plants 13:1816. https://doi.org/10.3390/plants13131816
41. Zhu Q, Wang B, Tan J, Liu T, Li L, Liu Y-G (2020) Plant synthetic metabolic engineering for enhancing crop nutritional quality. Plant Commun 1:100017. https://doi.org/10.1016/j.xplc.2019.100017
42. Viltres-Portales M, Sánchez-Martín M-J, Llugany M, Boada R, Valiente M (2024) Selenium biofortification of microgreens: influence on phytochemicals, pigments and nutrients. Plant Physiol Biochem 206:108283. https://doi.org/10.1016/j.plaphy.2023.108283
43. Shahzad R, Jamil S, Ahmad S, Nisar A, Khan S, Amina Z, Kanwal S, Aslam HMU, Gill RA, Zhou W (2021) Biofortification of cereals and pulses using new breeding techniques: current and future perspectives. Front Nutr 8:1–23. https://doi.org/10.3389/fnut.2021.721728
44. Kaiser N, Douches D, Dhingra A, Glenn KC, Herzig PR, Stowe EC, Swarup S (2020) The role of conventional plant breeding in ensuring safe levels of naturally occurring toxins in food crops. Trends Food Sci Technol 100:51–66. https://doi.org/10.1016/j.tifs.2020.03.042
45. Dhanyalakshmi KH, Mohan R, Behera S, Chand Jha U, Moharana D, Behera A, Thomas S, Soumya PR, Sah RP, Beena R (2024) Next generation nutrition: genomic and molecular breeding innovations for iron and zinc biofortification in rice. Rice Sci 31:526–544. https://doi.org/10.1016/j.rsci.2024.04.008
46. Amegbor IK, van Biljon A, Shargie N, Tarekegne A, Labuschagne MT (2022) Heritability and associations among grain yield and quality traits in quality protein maize (QPM) and non-QPM hybrids. Plants 11:713. https://doi.org/10.3390/plants11060713
47. Naik B, Kumar V, Rizwanuddin S, Mishra S, Kumar V, Saris PEJ, Khanduri N, Kumar A, Pandey P, Gupta AK, Khan JM, Rustagi S (2024) Biofortification as a solution for addressing nutrient deficiencies and malnutrition. Heliyon 10:e30595. https://doi.org/10.1016/j.heliyon.2024.e30595
48. Kachinski WD, Ávila FW, dos Reis AR, Muller MML, Mendes MC, Petranski PH (2022) Agronomic biofortification increases concentrations of zinc and storage proteins in common bean (*Phaseolus vulgaris* L.) grains. Food Res Int 155:111105. https://doi.org/10.1016/j.foodres.2022.111105
49. Cota-Ruiz K, Ye Y, Valdes C, Deng C, Wang Y, Hernández-Viezcas JA, Duarte-Gardea M, Gardea-Torresdey JL (2020) Copper nanowires as nanofertilizers for alfalfa plants: understanding nano-bio systems interactions from microbial genomics, plant molecular responses and spectroscopic studies. Sci Total Environ 742:140572. https://doi.org/10.1016/j.scitotenv.2020.140572
50. Fatollahpour Grangah M, Rashidi V, Mirshekari B, Khalilvand Behrouzyar E, Farahvash F (2020) Effects of nano-fertilizers on physiological and yield characteristics of pinto bean cultivars under water deficit stress. J Plant Nutr 43:2898–2910. https://doi.org/10.1080/01904167.2020.1799000

51. Abouelhamd N, Gharib FAEL, Amin AA, Ahmed EZ (2023) Impact of foliar spray with Se, nano-Se and sodium sulfate on growth, yield and metabolic activities of red kidney bean. Sci Rep 13:17102. https://doi.org/10.1038/s41598-023-43677-8
52. Salama DM, Abd El-Aziz ME, Osman SA, Abd Elwahed MSA, Shaaban EA (2022a) Foliar spraying of MnO 2 -NPs and its effect on vegetative growth, production, genomic stability, and chemical quality of the common dry bean. Arab J Basic Appl Sci 29:26–39. https://doi.org/10.1080/25765299.2022.2032921
53. Salama DM, El-Aziz ME, El-Naggar ME, Shaaban EA, Abd El-Wahed MS (2021) Synthesis of an eco-friendly nanocomposite fertilizer for common bean based on carbon nanoparticles from agricultural waste biochar. Pedosphere 31:923–933. https://doi.org/10.1016/S1002-0160(21)60024-3
54. Deng C, Wang Y, Cota-Ruiz K, Reyes A, Sun Y, Peralta-Videa J, Hernandez-Viezcas JA, Turley RS, Niu G, Li C, Gardea-Torresdey J (2020) Bok choy (Brassica rapa) grown in copper oxide nanoparticles-amended soils exhibits toxicity in a phenotype-dependent manner: translocation, biodistribution and nutritional disturbance. J Hazard Mater 398:122978. https://doi.org/10.1016/j.jhazmat.2020.122978
55. Shen M, Liu W, Zeb A, Lian J, Wu J, Lin M (2022) Bioaccumulation and phytotoxicity of ZnO nanoparticles in soil-grown *Brassica chinensis* L. and potential risks. J Environ Manag 306:114454. https://doi.org/10.1016/j.jenvman.2022.114454
56. Zhang J, Wang Q, Wang Y, Xu Y, Du W, Guo H (2023) Joint effects of CuO nanoparticles and perfluorooctanoic acid on cabbage (*Brassica pekinensis* L.). Environ Sci Pollut Res 30:66745–66752. https://doi.org/10.1007/s11356-023-26862-0
57. Faiz S, Yasin NA, Khan WU, Shah AA, Akram W, Ahmad A, Ali A, Naveed NH, Riaz L (2022) Role of magnesium oxide nanoparticles in the mitigation of lead-induced stress in *Daucus carota*: modulation in polyamines and antioxidant enzymes. Int J Phytorem 24:364–372. https://doi.org/10.1080/15226514.2021.1949263
58. Wang Y, Deng C, Cota-Ruiz K, Peralta-Videa JR, Hernandez-Viezcas JA, Gardea-Torresdey JL (2021b) Soil-aged nano titanium dioxide effects on full-grown carrot: dose and surface-coating dependent improvements on growth and nutrient quality. Sci Total Environ 774:145699. https://doi.org/10.1016/j.scitotenv.2021.145699
59. Li D, An Q, Wu Y, Li JQ, Pan C (2020a) Foliar application of selenium nanoparticles on celery stimulates several nutrient component levels by regulating the α-Linolenic acid pathway. ACS Sustain Chem Eng 8:10502–10510. https://doi.org/10.1021/acssuschemeng.0c02819
60. Abdelkader M, Zargar M, Bayat M, Pakina E, Shehata ASA, Suliman AA (2024) Biogenic nano-fertilizers as a sustainable approach to alleviate nitrate accumulation and enrich quality traits of vegetable crops. Horticulturae 10:789. https://doi.org/10.3390/horticulturae10080789
61. Hasan M, Mehmood K, Mustafa G, Zafar A, Tariq T, Hassan SG, Loomba S, Zia M, Mazher A, Mahmood N, Shu X (2021) Phytotoxic evaluation of phytosynthesized silver nanoparticles on lettuce. Coatings 11:225. https://doi.org/10.3390/coatings11020225
62. Hu J, Wu X, Wu F, Chen W, White JC, Yang Y, Wang B, Xing B, Tao S, Wang X (2020a) Potential application of titanium dioxide nanoparticles to improve the nutritional quality of coriander (*Coriandrum sativum* L.). J Hazard Mat 389:121837. https://doi.org/10.1016/j.jhazmat.2019.121837
63. Mahmoud AWM, Rashad HM, Esmail SEA, Alsamadany H, Abdeldaym EA (2023) Application of silicon, zinc, and zeolite nanoparticles—a tool to enhance drought stress tolerance in coriander plants for better growth performance and productivity. Plants 12:2838. https://doi.org/10.3390/plants12152838
64. Choukri M, Abouabdillah A, Bouabid R, Abd-Elkader OH, Pacioglu O, Boufahja F, Bourioug M (2022) Zn application through seed priming improves productivity and grain nutritional quality of silage corn. Saudi J Biol Sci 29:103456. https://doi.org/10.1016/j.sjbs.2022.103456
65. Kumar S, Dwivedi A, Pandey AK, Vajpayee P (2023) TiO2 nanoparticles alter nutrients acquisition, growth, biomacromolecules, oil composition and modulate antioxidant defense system in *Mentha arvensis* L. Plant Nano Biol 3:100029. https://doi.org/10.1016/j.plana.2023.100029

66. Shalaby TA, Abd-Alkarim E, El-Aidy F, Hamed E-S, Sharaf-Eldin M, Taha N, El-Ramady H, Bayoumi Y, dos Reis AR (2021) Nano-selenium, silicon and H_2O_2 boost growth and productivity of cucumber under combined salinity and heat stress. Ecotoxicol Environ Saf 212:111962. https://doi.org/10.1016/j.ecoenv.2021.111962
67. Tajik Khademi H, Khodadadi M, Hassanpanah D, Hajainfar R (2024) Changes in fruit yield, biochemical attributes, and leaf minerals of different cucumber (*Cucumis sativus* L.) cultivars under foliar application of silicon nanoparticles. Environ Sci Pollut Res 31:42012–42022. https://doi.org/10.1007/s11356-024-33890-x
68. Dhaliwal SS, Sharma V, Shukla AK, Verma V, Behera SK, Singh P, Alotaibi SS, Gaber A, Hossain A (2021) Comparative efficiency of mineral, chelated and nano forms of zinc and iron for improvement of zinc and iron in Chickpea (*Cicer arietinum* L.) through biofortification. Agronomy 11:2436. https://doi.org/10.3390/agronomy11122436
69. Wang Y, Deng C, Cota-Ruiz K, Peralta-Videa JR, Sun Y, Rawat S, Tan W, Reyes A, Hernandez-Viezcas JA, Niu G, Li C, Gardea-Torresdey JL (2020) Improvement of nutrient elements and allicin content in green onion (*Allium fistulosum*) plants exposed to CuO nanoparticles. Sci Total Environ 725:138387. https://doi.org/10.1016/j.scitotenv.2020.138387
70. Francisco MD, Fernandes-Silva P, Durães L, Romeiro A, Álvarez-Torrellas S, Almendros P (2024) Influence of the application of different zinc oxide nanoparticles on a lettuce crop grown in an acidic mediterranean soil. Horticulturae 10:681. https://doi.org/10.3390/horticulturae10070681
71. Garza-Alonso CA, Juárez-Maldonado A, González-Morales S, Cabrera-De la Fuente M, Cadenas-Pliego G, Morales-Díaz AB, Trejo-Téllez LI, Tortella G, Benavides-Mendoza A (2023) ZnO nanoparticles as potential fertilizer and biostimulant for lettuce. Heliyon 9:e12787. https://doi.org/10.1016/j.heliyon.2022.e12787
72. Ahmed GO, Halshoy HS, Mahmood CH, Hama JR (2024) Titanium nanoparticle and humic acid applications improve seed germination, growth development, and phytochemical contents of lettuce (*Lactuca sativa*) plants. BioNanoScience. https://doi.org/10.1007/s12668-024-01545-3
73. Xiong T, Zhang T, Xian Y, Kang Z, Zhang S, Dumat C, Shahid M, Li S (2021) Foliar uptake, biotransformation, and impact of CuO nanoparticles in *Lactuca sativa* L. var. ramosa Hort. Environ Geochem Health 43:423–439. https://doi.org/10.1007/s10653-020-00734-9
74. Kohatsu MY, Pelegrino MT, Monteiro LR, Freire BM, Pereira RM, Fincheira P, Rubilar O, Tortella G, Batista BL, de Jesus TA, Seabra AB, Lange CN (2021) Comparison of foliar spray and soil irrigation of biogenic CuO nanoparticles (NPs) on elemental uptake and accumulation in lettuce. Environ Sci Pollut Res 28:16350–16367. https://doi.org/10.1007/s11356-020-12169-x
75. Hayes KL, Mui J, Song B, Sani ES, Eisenman SW, Sheffield JB, Kim B (2020) Effects, uptake, and translocation of aluminum oxide nanoparticles in lettuce: a comparison study to phytotoxic aluminum ions. Sci Total Environ 719:137393. https://doi.org/10.1016/j.scitotenv.2020.137393
76. Almutairi KF, Górnik K, Awad RM, Ayoub A, Abada HS, Mosa WFA (2023) Influence of selenium, titanium, and silicon nanoparticles on the growth, yield, and fruit quality of mango under drought conditions. Horticulturae 9:1231. https://doi.org/10.3390/horticulturae9111231
77. Razavizadeh R, Adabavazeh F, Mosayebi Z (2023) Titanium dioxide nanoparticles improve element uptake, antioxidant properties, and essential oil productivity of *Melissa officinalis* L. seedlings under in vitro drought stress. Environ Sci Pollut Res 30:98020–98033. https://doi.org/10.1007/s11356-023-29384-x
78. Wang Y, Wang S, Xu M, Xiao L, Dai Z, Li J (2019) The impacts of γ-Fe_2O_3 and Fe_3O_4 nanoparticles on the physiology and fruit quality of muskmelon (*Cucumis melo*) plants. Environ Pollut 249:1011–1018. https://doi.org/10.1016/j.envpol.2019.03.119
79. Ogunkunle CO, Balogun GY, Olatunji OA, Han Z, Adeleye AS, Awe AA, Fatoba PO (2023) Foliar application of nanoceria attenuated cadmium stress in okra (*Abelmoschus esculentus* L.). J Hazard Mater 445:130567. https://doi.org/10.1016/j.jhazmat.2022.130567

80. Abdelghany AM, El-Banna AAA, Salama EAA, Ali MM, Al-Huqail AA, Ali HM, Paszt LS, El-Sorady GA, Lamlom SF (2022) The individual and combined effect of nanoparticles and biofertilizers on growth, yield, and biochemical attributes of peanuts (*Arachis hypogea* L.). Agronomy 12:398. https://doi.org/10.3390/agronomy12020398
81. Roosta HR, Samadi A, Bikdeloo M (2023) Different cultivation systems and foliar application of calcium nanoparticles affect the growth and physiological characteristics of pennyroyal (*Mentha pulegium* L.). Sci Rep 13:20334. https://doi.org/10.1038/s41598-023-47855-6
82. Abd El-wahed AEN, Khalifa SM, Alqahtani MD, Abd –Alrazik AM, Abdel-Aziz H, Mancy A, Elnaggar IA, Alharbi BM, Hamdy A, Elkelish A (2024) Nano-enhanced growth and resilience strategies for Pomegranate cv. Wonderful: Unveiling the impact of zinc and boron nanoparticles on fruit quality and abiotic stress management. J Agric Food Res 15:100908. https://doi.org/10.1016/j.jafr.2023.100908
83. Zahedi SM, Hosseini MS, Daneshvar Hakimi Meybodi N, Teixeira da Silva JA (2019) Foliar application of selenium and nano-selenium affects pomegranate (*Punica granatum* cv. Malase Saveh) fruit yield and quality. South Afric J Bot 124:350–358. https://doi.org/10.1016/j.sajb.2019.05.019
84. Davarpanah S, Tehranifar A, Zarei M, Aran M, Davarynejad G, Abadía J (2020) Early season foliar iron fertilization increases fruit yield and quality in pomegranate. Agronomy 10:832. https://doi.org/10.3390/agronomy10060832
85. Hussein AS, Abeed AHA, Usman ARA, Abou-Zaid EAA (2024) Conventional vs. nano-micronutrients as foliar fertilization for enhancing the quality and nutritional status of pomegranate fruits. J Saudi Soc Agric Sci 23:112–122. https://doi.org/10.1016/j.jssas.2023.09.008
86. Abd El-Azeim MM, Sherif MA, Hussien MS, Tantawy IAA, Bashandy SO (2020) Impacts of nano- and non-nanofertilizers on potato quality and productivity. Acta Ecol Sin 40:388–397. https://doi.org/10.1016/j.chnaes.2019.12.007
87. Zare Z, Pishkar L, Iranbakhsh A, Talei D (2020) Physiological and molecular effects of silver nanoparticles exposure on purslane (*Portulaca oleracea* L.). Russ J Plant Physiol 67:521–528. https://doi.org/10.1134/S1021443720030231
88. de Lima Lessa JH, Raymundo JF, Branco Corguinha AP, Dias Martins FA, Araujo AM, Melo Santiago FE, Pereira de Carvalho HW, Guimarães Guilherme LR, Lopes G (2020) Strategies for applying selenium for biofortification of rice in tropical soils and their effect on element accumulation and distribution in grains. J Cereal Sci 96:103125. https://doi.org/10.1016/j.jcs.2020.103125
89. Ahmad S, Ahmad N, Islam MS, Ahmad MA, Ercisli S, Ullah R, Bari A, Munir I (2024) Rice seeds biofortification using biogenic ıron oxide nanoparticles synthesized by using *Glycyrrhiza glabra*: a study on growth and yield ımprovement. Sci Rep 14:12368. https://doi.org/10.1038/s41598-024-62907-1
90. Mazhar Z, Akhtar J, Alhodaib A, Naz T, Zafar MI, Iqbal MM, Fatima H, Naz I (2023) Efficacy of ZnO nanoparticles in Zn fortification and partitioning of wheat and rice grains under salt stress. Sci Rep 13:2022. https://doi.org/10.1038/s41598-022-26039-8
91. Adhikary S, Mandal N, Rakshit R, Das A, Kumar V, Kumari N, Choudhary SK, Homa F (2020) Field evaluation of Zincated nanoclay polymer composite (ZNCPC): impact on DTPA-extractable Zn, sequential Zn fractions and apparent Zn recovery under rice rhizosphere. Soil Tillage Res 201:104607. https://doi.org/10.1016/j.still.2020.104607
92. Yang G, Yuan H, Ji H, Liu H, Zhang Y, Wang G, Chen L, Guo Z (2021) Effect of ZnO nanoparticles on the productivity, Zn biofortification, and nutritional quality of rice in a life cycle study. Plant Physiol Biochem 163:87–94. https://doi.org/10.1016/j.plaphy.2021.03.053
93. Fakharzadeh S, Hafizi M, Baghaei MA, Etesami M, Khayamzadeh M, Kalanaky S, Akbari ME, Nazaran MH (2020) Using nanochelating technology for biofortification and yield increase in rice. Sci Rep 10:4351. https://doi.org/10.1038/s41598-020-60189-x
94. Deng C, Wang Y, Navarro G, Sun Y, Cota-Ruiz K, Hernandez-Viezcas JA, Niu G, Li C, White JC, Gardea-Torresdey J (2022) Copper oxide (CuO) nanoparticles affect yield, nutritional quality, and auxin associated gene expression in weedy and cultivated rice (*Oryza sativa* L.) grains. Sci Total Environ 810:152260. https://doi.org/10.1016/j.scitotenv.2021.152260

95. Yusefi-Tanha E, Fallah S, Rostamnejadi A, Pokhrel LR (2020) Root system architecture, copper uptake and tissue distribution in soybean (*Glycine max* (L.) Merr.) grown in copper oxide nanoparticle (CuONP)-amended soil and implications for human nutrition. Plants 9:1326. https://doi.org/10.3390/plants9101326
96. Zheng X, Shou K, Hu C, Wu S, Sun J, Tan Q, Sun X (2024) Molybdenum oxide nanoparticles improve soybean yield and enhance nutritional quality. J Food Compos Anal 132:106307. https://doi.org/10.1016/j.jfca.2024.106307
97. Li M, Zhang P, Guo Z, Cao W, Gao L, Li Y, Tian CF, Chen Q, Shen Y, Ren F, Rui Y, White JC, Lynch I (2023) Molybdenum nanofertilizer boosts biological nitrogen fixation and yield of soybean through delaying nodule senescence and nutrition enhancement. ACS Nano 17:14761–14774. https://doi.org/10.1021/acsnano.3c02783
98. Rawat S, Cota-Ruiz K, Dou H, Pullagurala VLR, Zuverza-Mena N, White JC, Niu G, Sharma N, Hernandez-Viezcas JA, Peralta-Videa JR, Gardea-Torresdey JL (2021) Soil-weathered CuO nanoparticles compromise foliar health and pigment production in spinach (*Spinacia oleracea*). Environ Sci Technol 55:13504–13512. https://doi.org/10.1021/acs.est.0c06548
99. Shebl A, Hassan AA, Salama DM, Abd El-Aziz ME, Abd Elwahed MSA (2020) Template-free microwave-assisted hydrothermal synthesis of manganese zinc ferrite as a nanofertilizer for squash plant (Cucurbita pepo L). Heliyon 6:e03596. https://doi.org/10.1016/j.heliyon.2020.e03596
100. Saini S, Kumar P, Sharma NC, Sharma N, Balachandar D (2021) Nano-enabled Zn fertilization against conventional Zn analogues in strawberry (*Fragaria* × *ananassa* Duch.). Sci Hortic 282:110016. https://doi.org/10.1016/j.scienta.2021.110016
101. Awad AAM, Sweed AAA, Rady MM, Majrashi A, Ali EF (2021) Rebalance the nutritional status and the productivity of high $CaCO_3$-stressed sweet potato plants by foliar nourishment with zinc oxide nanoparticles and ascorbic acid. Agronomy 11:1443. https://doi.org/10.3390/agronomy11071443
102. Neysanian M, Iranbakhsh A, Ahmadvand R, Oraghi Ardebili Z, Ebadi M (2020) Comparative efficacy of selenate and selenium nanoparticles for improving growth, productivity, fruit quality, and postharvest longevity through modifying nutrition, metabolism, and gene expression in tomato; potential benefits and risk assessment. PLoS ONE 15:e0244207. https://doi.org/10.1371/journal.pone.0244207
103. Almendros P, González D, Fernández MD, García-Gomez C, Obrador A (2022) Both Zn biofortification and nutrient distribution pattern in cherry tomato plants are influenced by the application of ZnO nanofertilizer. Heliyon 8:e09130. https://doi.org/10.1016/j.heliyon.2022.e09130
104. Wang C, Liu X, Li J, Yue L, Yang H, Zou H, Wang Z, Xing B (2021a) Copper nanoclusters promote tomato (*Solanum lycopersicum* L.) yield and quality through improving photosynthesis and roots growth. Environ Pollut 289:117912. https://doi.org/10.1016/j.envpol.2021.117912
105. Feng Y, Wang C, Chen F, Cao X, Wang J, Yue L, Wang Z (2022) Molecular mechanisms of CeO_2 nanomaterials improving tomato yield, fruit quality, and postharvest storage performance. Environ Sci Nano 9:4382–4392. https://doi.org/10.1039/D2EN00783E
106. Hu W, Su Y, Yang R, Xie Z, Gong H (2023) Effect of foliar application of silicon and selenium on the growth, yield and fruit quality of tomato in the field. Horticulturae 9:1126. https://doi.org/10.3390/horticulturae9101126
107. Pejam F, Ardebili ZO, Ladan-Moghadam A, Danaee E (2021) Zinc oxide nanoparticles mediated substantial physiological and molecular changes in tomato. PLoS ONE 16:e0248778. https://doi.org/10.1371/journal.pone.0248778
108. Costan A, Stamatakis A, Chrysargyris A, Petropoulos SA, Tzortzakis N (2020) Interactive effects of salinity and silicon application on *Solanum lycopersicum* growth, physiology and shelf-life of fruit produced hydroponically. J Sci Food Agric 100:732–743. https://doi.org/10.1002/jsfa.10076
109. Rady MM, Belal HEE, Gadallah FM, Semida WM (2020) Selenium application in two methods promotes drought tolerance in *Solanum lycopersicum* plant by inducing the

antioxidant defense system. Sci Hortic 266:109290. https://doi.org/10.1016/j.scienta.2020.109290

110. Ahmad W, Zou Z, Awais M, Munsif F, Khan A, Nepal J, Ahmad M, Akbar S, Ahmad I, Khan MS, Qamar Z, Khan H (2023) Seed-primed and foliar oxozinc nanofiber application increased wheat production and Zn biofortification in calcareous-alkaline soil. Agronomy 13:400. https://doi.org/10.3390/agronomy13020400
111. Jalal A, Oliveira CES, Fernandes GC, da Silva EC, da Costa KN, de Souza JS, da Leite GS, Biagini ALC, Galindo FS, Teixeira Filho MCM (2023) Integrated use of plant growth-promoting bacteria and nano-zinc foliar spray is a sustainable approach for wheat biofortification, yield, and zinc use efficiency. Front Plant Sci 14:1–14. https://doi.org/10.3389/fpls.2023.1146808
112. Kiran A, Wakeel A, Sultana R, Khalid A, Qurrat-ul-Ain Mubaraka R, Shahzad AN, Khan SJ, Noor M (2021) Concentration and localization of Fe and Zn in wheat grain as affected by its application to soil and foliage. Bull Environ Contam Toxicol 106:852–858. https://doi.org/10.1007/s00128-021-03183-x
113. Hussain A, Rizwan M, Ali S, ur Rehman MZ, Qayyum MF, Nawaz R, Ahmad A, Asrar M, Ahmad SR, Alsahli AA, Alyemeni MN (2021) Combined use of different nanoparticles effectively decreased cadmium (Cd) concentration in grains of wheat grown in a field contaminated with Cd. Ecotoxicol Environ Saf 215:112139. https://doi.org/10.1016/j.ecoenv.2021.112139
114. Sattar A, Wang X, Ul-Allah S, Sher A, Ijaz M, Irfan M, Abbas T, Hussain S, Nawaz F, Al-Hashimi A, Al Munqedhi BM, Skalicky M (2022) Foliar application of zinc improves morpho-physiological and antioxidant defense mechanisms, and agronomic grain biofortification of wheat (*Triticum aestivum* L.) under water stress. Saudi J Biol Sci 29:1699–1706. https://doi.org/10.1016/j.sjbs.2021.10.061
115. Doolette CL, Read TL, Howell NR, Cresswell T, Lombi E (2020) Zinc from foliar-applied nanoparticle fertiliser is translocated to wheat grain: a 65Zn radiolabelled translocation study comparing conventional and novel foliar fertilisers. Sci Total Environ 749:142369. https://doi.org/10.1016/j.scitotenv.2020.142369
116. Dimkpa CO, Andrews J, Sanabria J, Bindraban PS, Singh U, Elmer WH, Gardea-Torresdey JL, White JC (2020) Interactive effects of drought, organic fertilizer, and zinc oxide nanoscale and bulk particles on wheat performance and grain nutrient accumulation. Sci Total Environ 722:137808. https://doi.org/10.1016/j.scitotenv.2020.137808
117. Sun H, Du W, Peng Q, Lv Z, Mao H, Kopittke PM (2020) Development of ZnO nanoparticles as an efficient Zn fertilizer: using synchrotron-based techniques and laser ablation to examine elemental distribution in wheat grain. J Agric Food Chem 68:5068–5075. https://doi.org/10.1021/acs.jafc.0c00084
118. Fassinou Hotegni NV, Sohindji FS, Salaou MAB, Agbandou PC, Azonhoumon LWS, Tchokponhoué D, Houdegbe C, Adjé CAO, Achigan-Dako EG (2024) Agronomic biofortification of cereals and legumes with iron, zinc, calcium and magnesium for food and nutrition security: available options for farmers in Sub-Saharan Africa. J Agric Food Res 18:101391. https://doi.org/10.1016/j.jafr.2024.101391
119. Tolay I (2024) Seed priming and soil application of Zinc improved yield and shoot Zn concentration of corn (*Zea mays* L.). J King Saud Univ Sci 36:103433. https://doi.org/10.1016/j.jksus.2024.103433
120. Hayat HS, Rehman A, Farooq S, Naveed M, Ali HM, Hussain M (2023) Boron seed coating combined with seed inoculation with boron tolerant bacteria (*Bacillus* sp. MN-54) and maize stalk biochar improved growth and productivity of maize (*Zea mays* L.) on saline soil. Heliyon 9:e22075. https://doi.org/10.1016/j.heliyon.2023.e22075
121. Adeel MA, Hussain S, Basit A, Hussain MB, Aon M (2024) Biofortification of wheat in salt-affected soil through seed priming and soil application of zinc. J Trace ElemTs Miner 8:100159. https://doi.org/10.1016/j.jtemin.2024.100159
122. Guidinelle RB, Burak DL, Rangel OJP, Peçanha AL, Passos RR, da Rocha LO, Olivares FL, de Mendonça ES (2024) Impact of historical soil management on the interaction of plant-growth-promoting bacteria with maize (*Zea mays* L.). Heliyon 10:e28754. https://doi.org/10.1016/j.heliyon.2024.e28754

123. El-ramady H, Abdalla N, Elbasiouny H, Elbehiry F, Elsakhawy T, Omara AE, Amer M, Bayoumi Y, Shalaby TA, Eid Y, Rehman MZ (2021) Ecotoxicology and environmental safety nano-biofortification of different crops to immune against COVID-19: a review. Ecotoxicol Environ Saf 222:112500. https://doi.org/10.1016/j.ecoenv.2021.112500
124. Rathee S, Nayak V, Rb K, Ojha A (2022) Food Chemistry: molecular sciences nanofortification of vitamin B-complex in food matrix: need, regulations, and prospects. Food Chem Mol Sci 4:100100. https://doi.org/10.1016/j.fochms.2022.100100
125. du Jardin P (2015) Plant biostimulants: definition, concept, main categories and regulation. Sci Hortic 196:3–14. https://doi.org/10.1016/j.scienta.2015.09.021
126. Juárez-Maldonado A, Tortella G, Rubilar O, Fincheira P, Benavides-Mendoza A (2021) Biostimulation and toxicity: the magnitude of the impact of nanomaterials in microorganisms and plants. J Adv Res. https://doi.org/10.1016/j.jare.2020.12.011
127. Anjum S, Anjum I, Hano C, Kousar S (2019) Advances in nanomaterials as novel elicitors of pharmacologically active plant specialized metabolites: current status and future outlooks. RSC Adv 9:40404–40423. https://doi.org/10.1039/C9RA08457F
128. Gao M, Chang J, Wang Z, Zhang H, Wang T (2023) Advances in transport and toxicity of nanoparticles in plants. J Nanobiotechnology. https://doi.org/10.1186/s12951-023-01830-5
129. Verma KK, Song XP, Singh M, Huang HR, Bhatt R, Xu L, Kumar V, Li YR (2022) Influence of nanosilicon on drought tolerance in plants: an overview. Front Plant Sci. https://doi.org/10.3389/fpls.2022.1014816
130. Erb M, Kliebenstein DJ (2020) Plant secondary metabolites as defenses, regulators, and primary metabolites: the blurred functional trichotomy. Plant Physiol 184:39–52. https://doi.org/10.1104/pp.20.00433
131. Djanaguiraman M, Anbazhagan V, Dhankher OP, Prasad PVV (2024) Uptake, translocation, toxicity, and impact of nanoparticles on plant physiological processes. Plants 13:3137. https://doi.org/10.3390/plants13223137
132. Shiade SRG, Zand-Silakhoor A, Fathi A, Rahimi R, Minkina T, Rajput VD, Zulfiqar U, Chaudhary T (2024) Plant metabolites and signaling pathways in response to biotic and abiotic stresses: exploring bio stimulant applications. Plant Stress 12:100454. https://doi.org/10.1016/j.stress.2024.100454
133. Zhu X, Abudouaini M, Geng Z, Liu N, Peng T, He Q, Li Y (2024) Metabolite and mineral contents in root, seed, testa, stem and leaf of *Peganum harmala* L. Heliyon 10:e40009. https://doi.org/10.1016/j.heliyon.2024.e40009
134. Salehi H, Chehregani Rad A, Raza A, Chen J-T (2021) Foliar application of CeO2 nanoparticles alters generative components fitness and seed productivity in bean crop (*Phaseolus vulgaris* L.). Nanomaterials 11:862. https://doi.org/10.3390/nano11040862
135. Salama DM, Abd El-Aziz ME, Shaaban EA, Osman SA, Abd El-Wahed MS (2022b) The impact of nanofertilizer on agro-morphological criteria, yield, and genomic stability of common bean (*Phaseolus vulgaris* L.). Sci Rep 12:18552. https://doi.org/10.1038/s41598-022-21834-9
136. Guha T, Mukherjee A, Kundu R (2022) Nano-scale zero valent iron (nZVI) priming enhances yield, alters mineral distribution and grain nutrient content of *Oryza sativa* L. cv. Gobindobhog: a field study. J Plant Growth Regul 41:710–733. https://doi.org/10.1007/s00344-021-10335-0
137. Aazami MA, Maleki M, Rasouli F, Gohari G (2023) Protective effects of chitosan based salicylic acid nanocomposite (CS-SA NCs) in grape (*Vitis vinifera* cv. 'Sultana') under salinity stress. Sci Rep 13:883. https://doi.org/10.1038/s41598-023-27618-z
138. Shareef HJ, Al-Yahyai RA, Omar AE-DK, Barus WA (2021) Foliar nano-fertilization enhances fruit growth, maturity, and biochemical responses of date palm. Can J Plant Sci 101:299–306. https://doi.org/10.1139/cjps-2020-0081
139. Osman SA, Salama DM, Abd El-Aziz ME, Shaaban EA, Abd Elwahed MS (2020) The influence of MoO3-NPs on agro-morphological criteria, genomic stability of DNA, biochemical assay, and production of common dry bean (*Phaseolus vulgaris* L.). Plant Physiol Biochem 151:77–87. https://doi.org/10.1016/j.plaphy.2020.03.009

140. Sadak MS, Bakry BA (2020) Zinc-oxide and nano ZnO oxide effects on growth, some biochemical aspects, yield quantity, and quality of flax (*Linum uitatissimum* L.) in absence and presence of compost under sandy soil. Bull Natl Res Centre 44:98. https://doi.org/10.1186/s42269-020-00348-2
141. Mahmoud AWM, Abdeldaym EA, Abdelaziz SM, El-Sawy MBI, Mottaleb SA (2019) Synergetic effects of zinc, boron, silicon, and zeolite nanoparticles on confer tolerance in potato plants subjected to salinity. Agronomy 10:19. https://doi.org/10.3390/agronomy10010019
142. Javan M, Selahvarzi Y, Sayyad-Amin P, Rastegar S (2024) Potential application of TiO_2 nanoparticles to improve the nutritional quality of strawberry cv Camarosa under drought stress. Sci Hortic 330:113055. https://doi.org/10.1016/j.scienta.2024.113055
143. Zahedi SM, Moharrami F, Sarikhani S, Padervand M (2020) Selenium and silica nanostructure-based recovery of strawberry plants subjected to drought stress. Sci Rep 10:17672. https://doi.org/10.1038/s41598-020-74273-9
144. Abou El-Nasr MK, El-Hennawy HM, Samaan MSF, Salaheldin TA, Abou El-Yazied A, El-Kereamy A (2021) Using zinc oxide nanoparticles to improve the color and berry quality of table grapes cv. crimson seedless. Plants 10:1285. https://doi.org/10.3390/plants10071285
145. Maletsika P, Liava V, Sarrou E, Titeli VS, Nasiopoulou E, Martens S, Karagiannis E, Grigoriadou K, Molassiotis A, Nanos GD (2023) Foliar calcium effects on quality and primary and secondary metabolites of white-fleshed 'Lemonato' peaches. Horticulturae 9:299. https://doi.org/10.3390/horticulturae9030299
146. Li D, Zhou C, Zhang J, An Q, Wu Y, Li J-Q, Pan C (2020b) Nanoselenium foliar applications enhance the nutrient quality of pepper by activating the capsaicinoid synthetic pathway. J Agric Food Chem 68:9888–9895. https://doi.org/10.1021/acs.jafc.0c03044
147. Yue L, Feng Y, Ma C, Wang C, Chen F, Cao X, Wang J, White JC, Wang Z, Xing B (2022) Molecular mechanisms of early flowering in tomatoes induced by manganese ferrite ($MnFe_2O_4$) nanomaterials. ACS Nano 16:5636–5646. https://doi.org/10.1021/acsnano.1c10602
148. Meena M, Pilania S, Pal A, Mandhania S, Bhushan B, Kumar S, Gohari G, Saharan V (2020) Cu-chitosan nano-net improves keeping quality of tomato by modulating physio-biochemical responses. Sci Rep 10:21914. https://doi.org/10.1038/s41598-020-78924-9
149. Liu Y, Liu R, Cheng L, Yu S, Nie Y, Zhang H, Li J-Q, Pan C, Zhu W, Diao J, Zhou Z (2023) Improvement by application of three nanomaterials on flavor quality and physiological and antioxidant properties of tomato and their comparison. Plant Physiol Biochem 201:107834. https://doi.org/10.1016/j.plaphy.2023.107834
150. Deng K, Li Z, Huang T, Huang J (2024) Noncoding RNAs in regulation of plant secondary metabolism. Plant Physiol Biochem 211:108718. https://doi.org/10.1016/j.plaphy.2024.108718
151. Al Aboud NM (2024) Unlocking the genetic potential: strategies for enhancing secondary metabolite biosynthesis in plants. J Saudi Soc Agric Sci. https://doi.org/10.1016/j.jssas.2024.06.004
152. Sakhawat A, Awan SJ, Khan MU (2024) In silico and in vitro analyses to investigate the effects of vitamin C on VEGF protein. J Taibah Univ Med Sci 19:775–789. https://doi.org/10.1016/j.jtumed.2024.06.008
153. Montaño-Herrera A, Santiago-Saenz YO, López-Palestina CU, Cadenas-Pliego G, Pinedo-Guerrero ZH, Hernández-Fuentes AD, Pinedo-Espinoza JM (2022) Effects of edaphic fertilization and foliar application of Se and Zn nanoparticles on yield and bioactive compounds in *Malus domestica* L. Horticulturae 8:542. https://doi.org/10.3390/horticulturae8060542
154. Sheikhalipour M, Esmaielpour B, Behnamian M, Gohari G, Giglou MT, Vachova P, Rastogi A, Brestic M, Skalicky M (2021) Chitosan-Selenium nanoparticle (Cs–Se NP) foliar spray alleviates salt stress in bitter melon. Nanomaterials 11:684. https://doi.org/10.3390/nano11030684
155. Abdi MJ, Ghanbari Jahromi M, Mortazavi SN, Kalateh Jari S, Nazarideljou MJ (2023) Foliar-applied silicon and selenium nanoparticles modulated salinity stress through modifying yield, biochemical attribute, and fatty acid profile of *Physalis alkekengi* L. Environ Sci Pollut Res 30:100513–100525. https://doi.org/10.1007/s11356-023-29450-4

156. Gao M, Xu Y, Chang X, Dong Y, Song Z (2020) Effects of foliar application of graphene oxide on cadmium uptake by lettuce. J Hazard Mater 398:122859. https://doi.org/10.1016/j.jhazmat.2020.122859
157. Coornaert I, Breynaert A, Hermans N, De Meyer GRY (2024) α -Tocopherol inhibits atherogenesis and improves cardiac function in mice independently of its antioxidant properties. Vasc Biol 6. https://doi.org/10.1530/VB-24-0002
158. Ratib FL, Arif M, Zafendi I, Aizuddin M, Lazaldin M (2024) The use of vitamin E in ocular health: bridging omics approaches with tocopherol and tocotrienol in the management of glaucoma. Food Chem Mol Sci 9:100224. https://doi.org/10.1016/j.fochms.2024.100224
159. Faizan M, Faraz A, Hayat S (2020) Effective use of zinc oxide nanoparticles through root dipping on the performance of growth, quality, photosynthesis and antioxidant system in tomato. J Plant Biochem Biotechnol 29:553–567. https://doi.org/10.1007/s13562-019-00525-z
160. Vatankhah A, Aliniaeifard S, Moosavi-Nezhad M, Abdi S, Mokhtarpour Z, Reezi S, Tsaniklidis G, Fanourakis D (2023) Plants exposed to titanium dioxide nanoparticles acquired contrasting photosynthetic and morphological strategies depending on the growing light intensity: a case study in radish. Sci Rep 13:5873. https://doi.org/10.1038/s41598-023-32466-y
161. Torres-montilla S, Rodriguez-concepcion M (2021) Progress in lipid research making extra room for carotenoids in plant cells: new opportunities for biofortification. Prog Lipid Res 84:101128. https://doi.org/10.1016/j.plipres.2021.101128
162. Plant M, Zheng X, Zhang Y, Balakrishna A, Liew KX, Kuijer HNJ, Xiao TT, Blilou I, Al-babili S (2023) Installing the neurospora carotenoid pathway in plants enables cytosolic formation of provitamin A and its sequestration in lipid droplets. Mol Plant 16:1066–1081. https://doi.org/10.1016/j.molp.2023.05.003
163. Ayyaz A, Fang R, Ma J, Hannan F, Huang Q, Athar H-R, Sun Y, Javed M, Ali S, Zhou W, Farooq MA (2022) Calcium nanoparticles (Ca-NPs) improve drought stress tolerance in *Brassica napus* by modulating the photosystem II, nutrient acquisition and antioxidant performance. NanoImpact 28:100423. https://doi.org/10.1016/j.impact.2022.100423
164. Zadegan K, Monem R, Pazoki A (2023) Silicon dioxide nanoparticles improved yield, biochemical attributes, and fatty zcid profile of cowpea (*Vigna unguiculata* [L.] Walp) under different irrigation regimes. J Soil Sci Plant Nutr 23:3197–3208. https://doi.org/10.1007/s42729-023-01297-1
165. Sarkar MM, Sarkar A, Roy S (2024) Fertigation of NaCl-stressed lentil and soybean plants with silica nanoparticles improves seed yield and nutritional attributes. Plant Nano Biol 8:100068. https://doi.org/10.1016/j.plana.2024.100068
166. Sheoran P, Grewal S, Kumari S, Goel S (2021) Enhancement of growth and yield, leaching reduction in *Triticum aestivum* using biogenic synthesized zinc oxide nanofertilizer. Biocatal Agric Biotechnol 32:101938. https://doi.org/10.1016/j.bcab.2021.101938
167. González-García Y, Cárdenas-Álvarez C, Cadenas-Pliego G, Benavides-Mendoza A, Cabrera-de-la-Fuente M, Sandoval-Rangel A, Valdés-Reyna J, Juárez-Maldonado A (2021) Effect of three nanoparticles (Se, Si and Cu) on the bioactive compounds of bell pepper fruits under saline stress. Plants 10:217. https://doi.org/10.3390/plants10020217
168. Elsheery NI, Helaly MN, El-Hoseiny HM, Alam-Eldein SM (2020) Zinc oxide and silicone nanoparticles to improve the resistance mechanism and annual productivity of salt-stressed mango trees. Agronomy 10:558. https://doi.org/10.3390/agronomy10040558
169. Teixeira GCM, Junior JSP, Mattiuz B-H, de Mello Prado R, Corrêa AJ, Rocha AMS, do Vale DW (2022) Spraying of calcium carbonate nanoparticles on pineapple fruit reduces sunburn damage. South Afric J Bot 148:643–651. https://doi.org/10.1016/j.sajb.2022.04.004
170. Zahedi SM, Hosseini MS, Daneshvar Hakimi Meybodi N, Peijnenburg W (2021) Mitigation of the effect of drought on growth and yield of pomegranates by foliar spraying of different sizes of selenium nanoparticles. J Sci Food Agric 101:5202–5213. https://doi.org/10.1002/jsfa.11167

171. Cheng B, Wang C, Cao X, Yue L, Chen F, Liu X, Wang Z, Xing B (2022) Selenium nanomaterials induce flower enlargement and improve the nutritional quality of cherry tomatoes: pot and field experiments. Environ Sci Nano 9:4190–4200. https://doi.org/10.1039/D2EN00599A
172. Wang Z, Le X, Cao X, Wang C, Chen F, Wang J, Feng Y, Yue L, Xing B (2022) Triiron tetrairon phosphate ($Fe_7(PO_4)_6$) nanomaterials enhanced flavonoid accumulation in tomato fruits. Nanomaterials 12:1341. https://doi.org/10.3390/nano12081341
173. Cantu JM, Ye Y, Hernandez-Viezcas JA, Zuverza-Mena N, White JC, Gardea-Torresdey JL (2022a) Tomato fruit nutritional quality is altered by the foliar application of various metal oxide nanomaterials. Nanomaterials 12:2349. https://doi.org/10.3390/nano12142349
174. Cantu JM, Ye Y, Valdes C, Cota-Ruiz K, Hernandez-Viezcas JA, Gardea-Torresdey JL (2022b) Citric acid-functionalized CuO nanoparticles alter biochemical responses in candyland red tomato (*Solanum lycopersicum*). ACS Agric Sci Technol 2:359–370. https://doi.org/10.1021/acsagscitech.1c00259
175. Ahmed R, Uddin MK, Quddus MA, Samad MYA, Hossain MAM, Haque ANA (2023) Impact of foliar application of zinc and zinc oxide nanoparticles on growth, yield, nutrient uptake and quality of tomato. Horticulturae 9:162. https://doi.org/10.3390/horticulturae9020162
176. Acharya P, Jayaprakasha GK, Crosby KM, Jifon JL, Patil BS (2020) Nanoparticle-mediated seed priming improves germination, growth, yield, and quality of watermelons (*Citrullus lanatus*) at multi-locations in Texas. Sci Rep 10:5037. https://doi.org/10.1038/s41598-020-61696-7
177. Gohari G, Safai F, Panahirad S, Akbari A, Rasouli F, Dadpour MR, Fotopoulos V (2020) Modified multiwall carbon nanotubes display either phytotoxic or growth promoting and stress protecting activity in *Ocimum basilicum* L. in a concentration-dependent manner. Chemosphere 249:126171. https://doi.org/10.1016/j.chemosphere.2020.126171
178. Awan S, Shahzadi K, Javad S, Tariq A, Ahmad A, Ilyas S (2021) A preliminary study of influence of zinc oxide nanoparticles on growth parameters of *Brassica oleracea* var italic. J Saudi Soc Agric Sci 20:18–24. https://doi.org/10.1016/j.jssas.2020.10.003
179. Babashpour-Asl M, Farajzadeh-Memari-Tabrizi E, Yousefpour-Dokhanieh A (2022) Foliar-applied selenium nanoparticles alleviate cadmium stress through changes in physio-biochemical status and essential oil profile of coriander (*Coriandrum sativum* L.) leaves. Environ Sci Pollut Res 29:80021–80031. https://doi.org/10.1007/s11356-022-19941-1
180. Afshari M, Pazoki A, Sadeghipour O (2021) Foliar-applied silicon and its nanoparticles stimulate physio-chemical changes to improve growth, yield and active constituents of coriander (*Coriandrum sativum* L.) essential oil under different irrigation regimes. Silicon 13:4177–4188. https://doi.org/10.1007/s12633-021-01101-8
181. Amiripour A, Jahromi MG, Soori MK, Torkashvand AM (2021) Changes in essential oil composition and fatty acid profile of coriander (*Coriandrum sativum* L.) leaves under salinity and foliar-applied silicon. Indus Crops Prod 168:113599. https://doi.org/10.1016/j.indcrop.2021.113599
182. Ghanbari F, Bag-Nazari M, Azizi A (2023) Exogenous application of selenium and nano-selenium alleviates salt stress and improves secondary metabolites in lemon verbena under salinity stress. Sci Rep 13:5352. https://doi.org/10.1038/s41598-023-32436-4
183. Hu J, Wu X, Wu F, Chen W, Zhang X, White JC, Li J, Wan Y, Liu J, Wang X (2020b) TiO_2 nanoparticle exposure on lettuce (*Lactuca sativa* L.): dose-dependent deterioration of nutritional quality. Environ Sci Nano 7:501–513. https://doi.org/10.1039/C9EN01215J
184. Ghasemian S, Masoudian N, Saeid Nematpour F, Safipour Afshar A (2021) Selenium nanoparticles stimulate growth, physiology, and gene expression to alleviate salt stress in *Melissa officinalis*. Biologia 76:2879–2888. https://doi.org/10.1007/s11756-021-00854-2
185. Mohammadi H, Parviz L, Beyrami A, Anosheh-Bonab F, Ghorbanpour M (2024) Exposure to TiO_2 nanoparticles (NPs) and zeolite stimulates growth, physiology, and phytochemical characteristics and elevates *Mentha piperita* L. tolerance to salinity stress. Indus Crops Prod 211:118228. https://doi.org/10.1016/j.indcrop.2024.118228

186. Mohammadi H, Kazemi Z, Aghaee A, Hazrati S, Golzari Dehno R, Ghorbanpour M (2023) Unraveling the influence of TiO_2 nanoparticles on growth, physiological and phytochemical characteristics of *Mentha piperita* L. in cadmium-contaminated soil. Sci Rep 13:22280. https://doi.org/10.1038/s41598-023-49666-1
187. Ganjavi AS, Oraei M, Gohari G, Akbari A, Faramarzi A (2021) Glycine betaine functionalized graphene oxide as a new engineering nanoparticle lessens salt stress impacts in sweet basil (*Ocimum basilicum* L.). Plant Physiol Biochem 162:14–26. https://doi.org/10.1016/j.plaphy.2021.02.028
188. Miliauskienė J, Brazaitytė A, Sutulienė R, Urbutis M, Tučkutė S (2022) ZnO nanoparticle size-dependent effects on Swiss Chard growth and nutritional quality. Agriculture 12:1905. https://doi.org/10.3390/agriculture12111905
189. Xu Y, Du W, Yin Y, Sun Y, Ji R, He H, Yang S, Li S, Wu J, Guo H (2022) CuO nanoparticles modify bioaccumulation of perfluorooctanoic acid in radish (*Raphanus sativus* L.). Environ Pollut Bioavailab 34:34–41. https://doi.org/10.1080/26395940.2021.2025150
190. Bonilla-Bird NJ, Ye Y, Akter T, Valdes-Bracamontes C, Darrouzet-Nardi AJ, Saupe GB, Flores-Marges JP, Ma L, Hernandez-Viezcas JA, Peralta-Videa JR, Gardea-Torresdey JL (2020) Effect of copper oxide nanoparticles on two varieties of sweetpotato plants. Plant Physiol Biochem 154:277–286. https://doi.org/10.1016/j.plaphy.2020.06.009

186. Mohammadi H, Kazemi Z, Aghaee A, Hazrati S, Golzari Dehno R, Ghorbanpour M (2023) Unravelling the influence of TiO_2 nanoparticles on growth, physiological and phytochemical characteristics of *Mentha piperita* L. in cadmium-contaminated soil. Sci Rep 13:22280. https://doi.org/10.1038/s41598-023-49666-1
187. [illegible] (2021) Glycine betaine functionalized graphene oxide as a new engineered nanoparticle lessens salt stress impacts in sweet basil (*Ocimum basilicum* L.). Plant Physiol Biochem 162:14–26. https://doi.org/10.1016/j.plaphy.2021.02.028
188. [illegible], Sutulienė R, Urbutis M, Tučkutė S (2022) ZnO nanoparticle size-dependent effects on Swiss Chard growth and nutritional quality. Agriculture 12:1905. https://doi.org/10.3390/agriculture12111905
189. Xu Y, Du W, Yin Y, Sun Y, Ji R, He H, Yang S, [illegible] (2022) CeO_2 nanoparticles [illegible] in radish (*Raphanus* [illegible]) [illegible]. https://doi.org/10.1080/[illegible]
190. Bonilla-Bird NJ, Ye Y, Akter T, Valdes-Bracamontes C, Darrouzet-Nardi AJ, Saupe GB, Flores-Marges JP, Ma L, Hernandez-Viezcas JA, Peralta-Videa JR, Gardea-Torresdey JL (2020) Effect of copper oxide nanoparticles on two varieties of sweetpotato plants. Plant Physiol Biochem 154:277–286. https://doi.org/10.1016/j.plaphy.2020.06.009

Use of Nanomaterials and Nanofertilizers to Increase Fertilizer Use Efficiency

André Paganotti, Keith Dayane Leite Lira, Lidiane Maria dos Santos Lima, Maiara Gonçalves Rodrigues, Roberta Albino dos Reis, Vitor Gonçalves Vital, and Suzan Pantaroto de Vasconcellos

Abstract The agricultural practice is one of the main reasons to keep humans in our planet. It is marked by the collection and care of seeds and plants for human consumption, playing a crucial role in sustaining life. The use of additives to improve crop yields began alongside this process and came to be known as fertilization. In the current days, to meet the demand put on the agricultural system by the growing population, fertilizers have been used to enhance food production, and overall efficiency for farming practices by enriching the natural quality of the soil. However, this increase in production is not linear, as the uptake of this additives has limits, with the excessive use already in place, consequences as the deterioration of soil structure, decline of soil microorganisms, bioaccumulation and biomagnification of substances or metals in plants and soil and water pollution are taking place. To ensure the global food production and decrease the impact caused by the overuse of fertilizers, nanofertilizers emerged as a promising way to increase the efficiency of fertilization. Nanotechnologies were defined as a promising tool to have more efficient and selective delivery systems, that can enhance the plant growth and its nutrient uptake, promoting a better and more sustained growth.

Keywords Agricultural system · Efficiency of fertilization · Food production · Nanotechnology · Soil pollution · Sustainable agriculture · Water pollution

---

A. Paganotti (✉) · K. D. L. Lira · L. M. dos Santos Lima · V. G. Vital · S. P. de Vasconcellos
Instituto de Ciências Ambientais, Químicas e Farmacêuticas, Universidade Federal de São Paulo, Diadema, São Paulo, Brazil
e-mail: andre.paganotti@unifesp.br

M. G. Rodrigues · R. A. dos Reis
Centro de Ciências Naturais e Humanas, Universidade Federal do ABC, Santo André, São Paulo, Brazil

© The Author(s), under exclusive license to Springer Nature Singapore Pte Ltd. 2025
A. Juárez-Maldonado et al. (eds.), *Plant Biostimulation with Nanomaterials*, Smart Nanomaterials Technology, https://doi.org/10.1007/978-981-96-4648-7_10

1 Fertilizers History and Overview

Ever since man went from nomad to hunter-gatherer, the use of substances began as a form to facilitate or increase agricultural production. Initially, primitive societies used sulfur from natural deposits as a pesticide and additive in their agricultural properties [1]. One of the first substances found to be used is elemental sulfur, occurring naturally and being used both as fertilizer and pesticide. With the advancement of technology, the sulfur gave way to the superphosphate, a substance made in the early years of the industrial revolution by treating bones with sulfuric acid. Lather on, the newly discovered minerals took places as source of nutrients for fertilizers. With World War I and II and the demand for explosives, the fertilizer industry developed new process and routes, giving birth to the modern fertilizer industry [2].

Nowadays the fertilizer industry supplies a market of more than US$ 212 billion, a number which will grow each year at around four per cent (4%) [3]. In this scenario most of the market is represented by solid inorganic fertilized based on nitrogen, phosphate and potassium and potash (soluble potassium salts), being that much of this strategy was set atop of *Freiherr* Justus von Liebig work published in 1840 entitled "Organic chemistry in its application to agriculture and physiology" [4]. In this manuscript, the authors were concerned with the elemental necessities of plants needed and farm practices that helped restore these. This work was published almost thirty years before the Mendeleev periodic table and would perdure for over a century, when the synthetic fertilizers were produced in appreciable quantities after the Saskatchewan deposits discovery [2, 5].

As conceptualized by Liebig, the nutritional necessities of plant organisms go far beyond the water from rain and light from the sun. Today with modern techniques and knowledge of atoms undiscovered at the time we can fully comprehend Liebig idea. It is possible to categorize as fertilizer any material that is applied to the soil to supply nutrients, it is important here that the idea of fertilization is to promote the necessary nutrient for the adequate growth of the plant, plants without nutrients grow less, but adding fertilizer beyond the plant necessities will not increase its yield [6]. The essential nutrients for plant organisms can be seen in the schematic representation in Fig. 10.1, constructed based on the data found on [7–9]. These elements can be subdivided into those who constitutes most organic compounds, the inorganic elements and the salt bases and metals, as shown in the mentioned Figure. Traditionally, both Liebig and the industry in the twentieth century focused in replenishing the inorganics and salt compounds, that present very slow cycles of renovation, with low solubility or replenishing by natural process. The known NPK triad composed by nitrogen, phosphorus and potassium as mentioned before is still the breadwinner of the industry, however, the importance of micronutrients such as metallic elements as growth regulators and stress protection.

The importance of making plants grow more in less space is brought to the limelight more and more with the current population growth, which, if maintained, the world will reach 10 billion people by 2050. According to the Organization for Economic Co-operation and Development and Food and Agriculture Organization

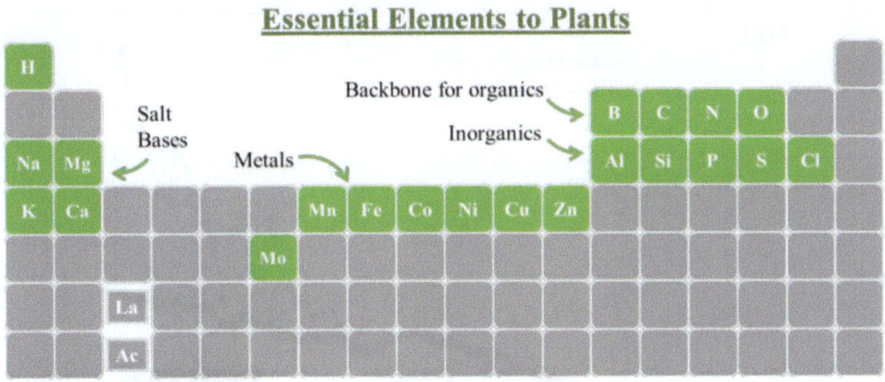

Fig. 10.1 Schematic periodic table showing the essential chemical elements necessary for plant organisms

of the United Nations, global cereal production is expected to increase by 13% by 2027 (OECD/FAO [10]. However, citing Guo et al.: "increased grain production may lead to chemical fertilizer overuse, which threatens the sustainable development of agriculture." [11]. To evidence that, Brazil is the largest country in terms of arable land, ranking in the five biggest producers of at least thirty agricultural commodities, and the largest net exporter [12, 13]. To supply this growing demand, Brazil use three times more fertilizers than the global average, as shown in Fig. 10.2.

By Fig. 10.2 it is possible to notice that the use of nitrogen fertilizers meets the global means, however the fertilizers based on both phosphorous and potassium are overused, dwarfing the world average. The overuse of fertilizers can leach into ground water and promote the eutrophication of the waterbodies, decreasing the water quality and even promoting the overgrowth of algae. Ongoing examples of this type of issue are both Lake Erie and Lough Neagh [14–17]. Even so fertilizers promoted the great improvement in plant growth and yield, as well as other agrochemicals not in the scope of this chapter, the actual use of this substances is reaching both its enviromental and efficiency limits for use.

To increase the use efficiency of fertilizers it is possible to use a group of materials that changes the chemistry area, by promoting huge gains in terms of process and activity of common materials.

2 Nanomaterials: Small but Powerful

Nanotechnology is a revolutionizing trend in science that in the last four decades or so [18–20]. These materials revolutionized the computer science industry, chemistry, and material science in general. In the last decade this revolution impacted the agricultural science by means of a new class of fertilizers, one which could do a lot more with a lot less [21]. Nanofertilizers compose a fast growing and promising field of

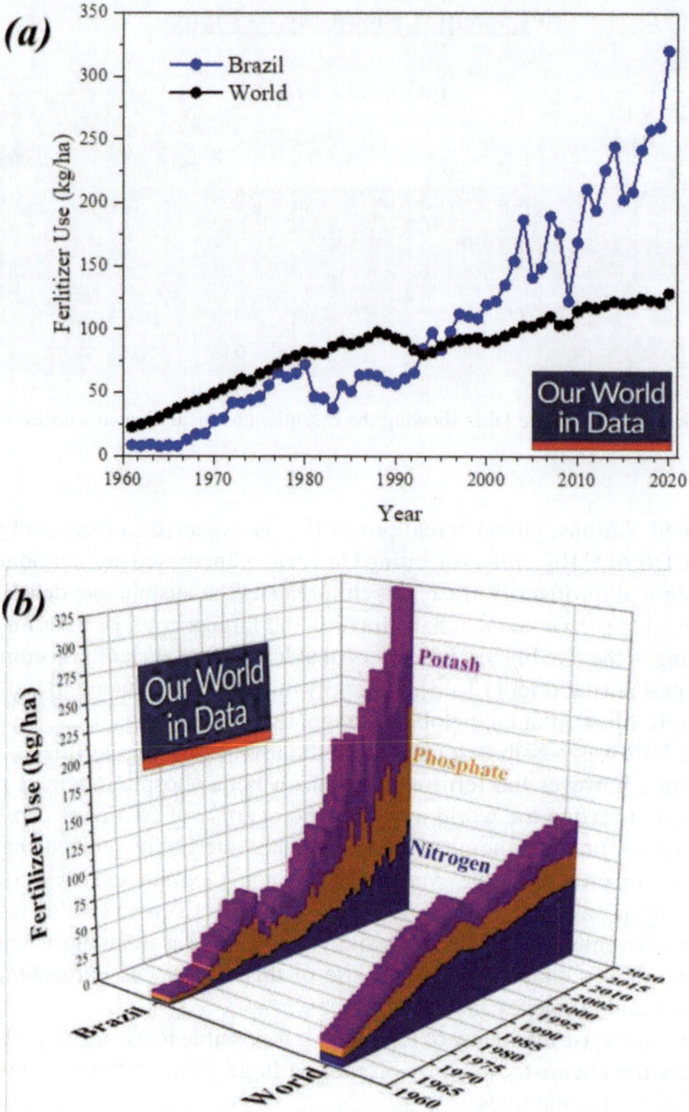

Fig. 10.2 Fertilizer use by unit of land in the world and in Brazil in **a** total amounts and **b** by mains constituting

research, emerging from the application of nanomaterials in agriculture, this class of product showcase the huge increase in performance and new properties that occurs during the scale transition between micro and nano meters.

This area is relatively new to the science, the first use of the word nanofertilizer according to web of science was in the article "Nanofertilizers, Pee For Your Plants"

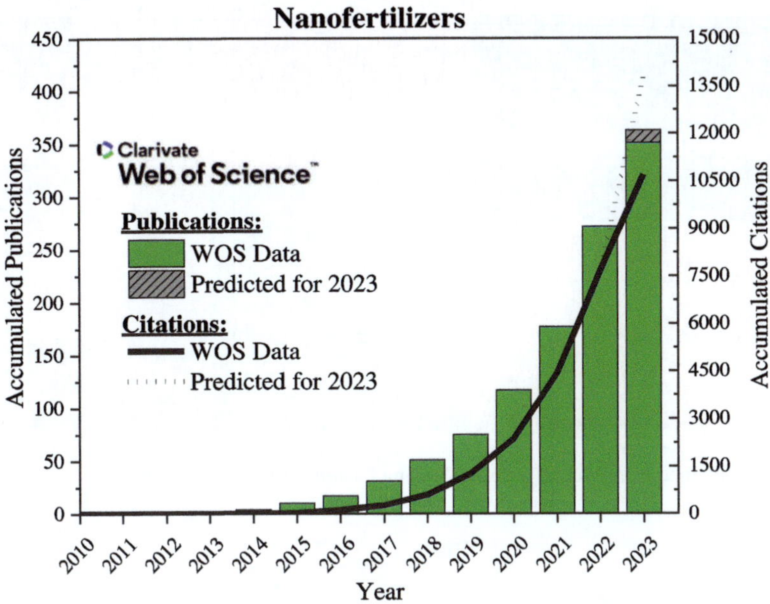

Fig. 10.3 Published articles and citations information obtained from Web of Science™ using "nanofertilizer" as key entry

by Kenneth Moore published in 2010 [22]. And the most cited paper to this date is "Potentials of engineered nanoparticles as fertilizers for increasing agronomic productions" by Liu&Lal, published in 2015 [23]. After a couple of decades, the manuscripts using the term quickly grew, starting an exponential growth after 2014 as can be seen in Fig. 10.3.

This area has many technical challenges, matching materials science to develop the nanomaterials, biological sciences to help understand and evaluate the impact of the produced materials, and environmental sciences to understand the use and impact in crop productions. Overall, most of the papers are related to plant and environmental sciences, allied with chemistry and materials science, as shown in Fig. 10.4.

This multidisciplinarity can be well represented by the word cloud made by using the titles and abstracts of the published articles, as shown in Fig. 10.5. This figure was made using Scopus ® database, showing the must recurrent words and their connection. It is possible to notice both the material characterization related words, as well as words associated with evaluating the impact of the material. By these pictures, it is possible to notice that the words phosphate and potash are not present, and the nitrogen bubble is not significative as expected by their use in today's fertilizer industry. These words give their place to words like chitosan, zinc oxide, copper, and unclassified drug.

Although a moderate quantity of papers is already published, in the patent side of the intellectual property, few documents have been deposited. Overall, 56 documents,

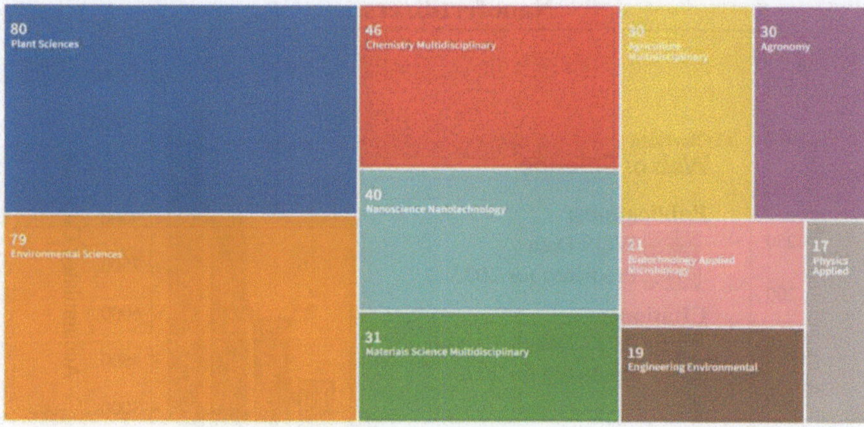

Fig. 10.4 Obtained from Web of Science™ using "nanofertilizer" as key entry

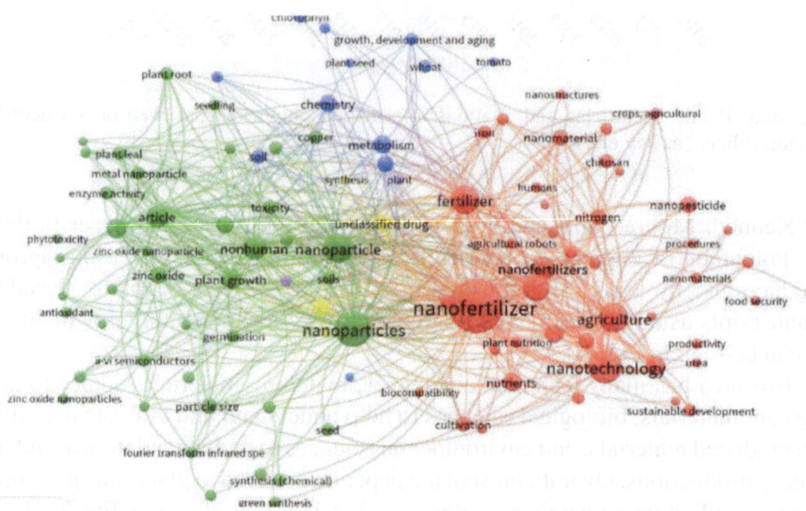

Fig. 10.5 Bibliometric network constructed by VOSviwer data obtained from Scopus® [24]

accounting for granted patents and applications were found to this chapter date. Being that only ten patents are active, as shown in Fig. 10.6.

As the change to nanometric scale increase the efficiency of use and yield, the nanofertilizers products usually present low concentration of the main fertilizing element, this is reflected by the most used Cooperative Patent Classification number: C05D9/02, this code is inside inorganic fertilizers and is described by "Other inorganic fertilizers containing trace elements". Also, this resonates with the word cloud presented in Fig. 10.5. with the use of new, non-traditional elements to promote an increase in the plant growth.

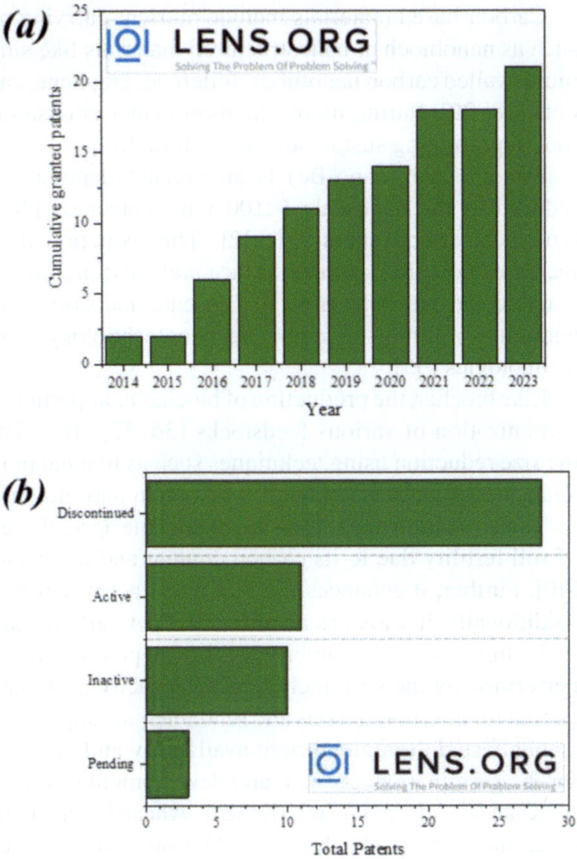

Fig. 10.6 Total of **a** granted patents and **b** overall status of the patents. Data from [25]

Due to this complexity of both acting as direct fertilizers and efficiency boosters to traditional fertilizers, nanofertilizers are a broad and promising class of materials. To summarize the findings in the literatures this chapter will be divided into four topics regarding the origin of the nanofertilizer, being those: Carbon based, Inorganics, Metal, Composite. After these topics, a topic regarding the Environmental impacts of nanofertilizers will be presented.

3 Carbon Based Products

Carbon is a ubiquitous element in all forms of life in the terrestrial biosphere; it is found in organic compounds and a variety of different molecular structures due to its allotropic characteristic [26, 27]. In this chapter section it will be considered as carbon base both elemental materials based on carbon allotropes, as well as polymers and other biomolecules mainly constituted by carbon.

Carbon-based materials include nutrient-carrying nanocarriers based on biochar, such as nanobiochar, nanostructured materials like single-walled carbon nanotubes, multi-walled carbon nanotubes, fullerene, graphene, carbon nanofibers, and quantum dots [28, 29]. Furthermore, the biopolymer chitosan has emerged as a noteworthy tool in promoting sustainable agriculture [30].

Nanobiochar (nano-Bc) is an organic material rich in carbon, with particles reduced to the nanoscale (<100 nm), enhanced physicochemical properties, and biological effectiveness [31, 32]. The reduction in the size of biochar particles increases porosity, specific surface area, and mobility in the soil [33, 34]. Recently, nanobiochar has been used to fabricate nanocomposites, primarily for sustainable agriculture, aiming to reconcile nanotechnology with biomass waste conversion technologies [35].

Like biochar, the production of biochar nanoparticles involves the thermochemical carbonization of various feedstocks [36, 37]. The distinctive feature of nano-Bc is the size reduction using techniques such as manual or ball milling, as well as sieving and aqueous suspension, aiming to obtain particles at the nanoscale [38, 39].

The utilization of nano-Bc in sustainable agriculture contributes to the modulation of soil fertility due to its carbon content and the presence of macro/micronutrients [40]. Further, it enhances the biological activity of the microbial community [34]. Additionally, it serves as an alternative for carbon capture and sequestration [37].

In light of this, nanobiochar has a positive impact on the several functions performed by the soil, including its capacity as a water reservoir, nutrient storage, habitat for microorganisms and fundamental support for plant life [41]. This influence is manifested through nutrient availability and the promotion of optimal conditions for the germination, growth, and development of plants [34].

Khan et al. [31] investigated wheat-based nanobiochar incorporated with macronutrients (NO_3, K_2O e P_2O_5) and micronutrient (Na_2O) as a slow-release nanofertilizer to enhance soil fertility. Experimental results from the slow-release studies demonstrated rapid nutrient desorption at the beginning of the study and gradually constant desorption of nutrients.

In a recent investigation by Mahmoud et al. [42] focusing on batch sorption, it unveiled the modification of nano-Bc within salt-corrected soil, increasing the adsorption of phosphorus. According to Rajput et al. [43], the limited availability and poor absorption of phosphorus by plants can impede growth, induce osmotic stress, and cause ionic imbalance.

Considering the beneficial impacts of nanobiochar on soil fertility and plant growth, Safdar et al. [44] emphasize the need for additional studies aiming at a deeper understanding of the mobility and environmental dispersion of nanobiochar in the soil, soil–plant system, and on a large scale in agriculture.

Additionally, nanobiochar has been explored in the production of nanocomposites for various purposes. Currently, different materials such as clay, zeolite, natural polymers and synthetic polymers have been used in the preparation of coated or encapsulated nanocomposites for the delivery of agrochemicals [32]. This approach is highly promising for waste management and to solve environmental issues, such fertilizer leaching into the soil [45].

About carbon nanotubes (CNTs), these nanomaterials exhibit remarkable applicability in the agricultural, environmental, industrial, and biotechnological fields. Starting from the carbon atom (C), it is possible to obtain a variety of molecular structures [29, 44]. Single-walled carbon nanotubes (SWCNTs) are cylindrical structures composed of a single carbon layer. In contrast, multi-walled carbon nanotubes (MWCNTs) consist of multiple concentric layers of graphene forming a tube [46]. The synthesis of CNTs involves different techniques (arc discharge, laser ablation, and chemical vapor deposition) and carbon precursors [26].

Fullerene is another molecular form of C atoms arranged on a spherical surface. Graphene is an allotrope of C formed by two-dimensional layers of carbon atoms in a hexagonal crystalline structure [47]. Furthermore, carbon nanofibers (CNFs) are formed by stacking multiple layers of graphene sheets in a cone-shaped structure [48]. In recent years, quantum dots (CQD) have stood out in the scientific community. It's a fluorescent carbon nanoparticle (<10 nm) with unique optical and electronic properties [47].

In this context, CNTs exhibit remarkable electrical conductivity, thermal conductivity, and high mechanical strength [46]. These properties allow the development of agriculturally efficient products in smart systems for delivering fertilizers, soil amendments, phytopathogen control agents (antifungals and antibacterials), and plant growth promoters [29, 26].

Intelligent or guided delivery systems based on the targeted distribution of encapsulated agrochemicals, demonstrated stability and protection against degradation. Sarlak et al. [49] investigated the enhanced toxicity of a fungicide encapsulated in citric acid-functionalized MWCNTs compared to the same free fungicide against the fungus *Alternaria alternata.*

Similar instances are reported in the literature employing graphene oxide films for the encapsulation of macronutrients. Zhang et al. [50] developed a fertilizer using graphene oxide films for encapsulating potassium nitrate. Following desorption analysis, the fertilizer exhibited prolonged action behavior.

The study conducted by Haghighi and Teixeira da Silva [51] evaluated the impact of carbon nanotubes on the germination and growth of tomato, onion, turnip, and radish seedlings. Seeds were germinated at four concentrations of CNTs (0, 10, 20, and 40 mg L^{-1}) under laboratory conditions. The authors reported a positive effect on the germination of all evaluated vegetable seeds at the concentration of 10 mg L^{-1} of CNTs. Higher concentrations of CNTs indicated a decrease in seedling weight and a reduction in plant weight due to diminished growth, suggesting a potential toxic effect of elevated CNT concentrations.

Recently, fluorescent carbon dots have been investigated in agriculture due to their significant potential in enhancing the photosynthesis of food crops. Wang et al. [52] examined the cultivation of mung bean seeds in an aqueous solution of nitrogen-doped carbon dots (N-CDs) at different concentrations (0, 0.2, or 4.0 mg mL^{-1}) as nanofertilizers. The results revealed that a 0.2 mg mL^{-1} aqueous solution of N-CDs could increase the growth of mung bean sprouts by 200% in the average length of both the aboveground and root parts and by 17.46% in wet weight. Following the bean growth, weakened fluorescence of the N-CD cultivation solution was observed due to the absorption of nanofertilizers by the plants during germination. This example illustrates the significant potential of these luminescent nanofertilizers compared to conventional fertilizers, given that carbon dots exhibit excellent luminescent properties, and their wavelength is adjustable (covering almost the entire visible range), thereby enhancing the photosynthetic efficiency of plants [53].

Another carbon-derived nanomaterial is chitosan, a naturally occurring amino polymer resulting from the partial deacetylation of chitin [54, 55]. This biopolymer is sourced from crustaceans, insects, and more recently, fungi [56, 57]. The amine functional groups inherent in chitosan enhance its solubility in neutral to acidic environments [54, 56]. Moreover, chitosan exhibits biodegradability, biocompatibility, non-toxicity, and remarkable antimicrobial properties [30].

In the agricultural context, the use of chitosan nanomaterials (CNMs) has seen substantial growth in the past decade, demonstrating their utility in enhancing agricultural production, phytopathogen control, and the slow, controlled, and targeted release of nutrients [54]. Consequently, chitosan nanomaterials have served as a primary means of nutrient distribution for plants, contributing to their growth through significant quantities of carbon, oxygen, nitrogen, and phosphorus [58].

Moreover, chitosan-based nanoparticles (CNPs) have been formulated for application in agriculture. CNPs are synthesized through a simple and low-cost process, ionic gelation [57]. This method relies on the electrostatic interaction between the polymer units of the amino group in chitosan and the poly-anion group of sodium tripolyphosphate under conditions of mechanical stirring, forming the hydrogel [30]. CNPs can be modified to broaden their application scope, particularly in agro systems, through the incorporation of additives, pesticides, metal ions, and fertilizers or through functionalization [55].

Investigative studies on the use of chitosan nanomaterials in economically significant or non-significant cultivars have been reported in scientific research conducted at laboratory, greenhouse, and field scales [57, 59, 60, 30, 54].

Chitosan-based NPK nanofertilizers show significant effects on the germination, growth, and development of plants [57, 61, 62]. Consequently, the slow and controlled release of fertilizer systems have been designed using CNPs [59] to minimize nutrient losses in the soil and environmental issues. These nanofertilizers can be considered an alternative or substitute for traditional fertilizers and hold great potential in sustainable agriculture [54].

Chitosan-based delivery systems can be utilized to carry and protect active ingredients, releasing them at the target absorption site of plants to achieve controlled action against phytopathogenic fungi [63]. It is important to highlight that both the direct and indirect application of CNPs contribute to the management of plant diseases. The small size and substantial contact area for the transport of active ingredients can facilitate their penetration and passage through the membranes of phytopathogens, thereby triggering an efficient defense response [30]. Additionally, studies investigating the efficacy of salicylic acid-functionalized nano-chitosan and copper-chitosan nanoparticles revealed increased defense responses (higher antioxidant activities) and growth in corn plants [58, 64].

Summarizing carbon-based nanofertilizers in agriculture represent a promising innovation capable of significantly enhancing the efficiency and sustainability of the agricultural sector. Characteristics such as reduced size and specific properties offer advantages such as increased solubility and controlled release of nutrients, resulting in optimized absorption by plants. Their formulation can minimize nutrient loss to the soil and water, thereby reducing pollution caused by conventional fertilizers. The improved effectiveness of these nanofertilizers may also lead to a reduction in the overall quantity of fertilizers required, positively impacting the cost-effectiveness of agricultural production.

Nevertheless, it is crucial to continue investigating and assessing the long-term effects of carbon nanofertilizers, ensuring environmental safety and food chain security through phytotoxicity studies. The successful implementation of this technology requires a cautious approach and constant monitoring to prevent potential adverse impacts. The new era of agriculture demands innovative and revolutionary approaches that provide efficiency and sustainability benefits while contributing to a more efficient and environmentally conscious agricultural future.

4 Inorganics Products

Inorganic nanofertilizers represent an innovative and rapidly evolving domain in agricultural sciences, emerging as a pivotal solution to several challenges faced by conventional fertilization methods. Inorganic nanofertilizers offer numerous advantages over traditional fertilizers, including enhanced nutrient efficiency, reduced environmental impact, and improved plant growth and health [65–68]. Their development and application in agriculture hold great promise for sustainable and precision farming practices.

Beginning by phosphorus approach, a novel phosphorus nanofertilizer (P-NF) was evaluated for its impact on soil health and microbial communities. This study diverged from traditional fertilizers, like triple superphosphate (TSP), by showing a reduced release of soluble phosphorus and no adverse effects on soil microbial life. The findings are promising, suggesting that P-NF could be a viable option for environmentally conscious fertilization. However, it also points to the need for more comprehensive studies to fully understand the long-term effects of these nanomaterials on plant and soil health [65]. Another study by Priyam et al. explored the application of phosphorus-based nanomaterials (NMs) in agriculture, employing green nanosynthesis methods. This research focused on biologically synthesized nanohydroxyapatite (nHAP) and nanophosphorus (nP), along with their chemically synthesized counterparts, to examine their effects on the growth of Japonica rice. The study's innovative approach involved using a fertigation method in soils with varying calcareous conditions. The results showed that both nHAPs and nP significantly improved germination, root and shoot lengths, and the overall biomass of rice plants, marking a significant advancement in the field of nanofertilizer [66]. A groundbreaking study involving hydroxyapatite nanoparticles (HANPs) and dicalcium phosphate anhydride (DCPA) showcased the potential of these materials as controlled-release phosphorus (P) fertilizers. The nanoparticles were synthesized under various conditions, adjusting factors like temperature, pH, and cation substitution, to optimize their efficacy. In trials with lettuce, both HANPs and DCPA notably improved plant growth, demonstrating the direct link between nanoparticle properties and their agricultural benefits. The study revealed that DCPA could double crop yields, although it led to the highest phosphorus loss in leachates. In contrast, HANPs synthesized at neutral pH significantly increased yields with minimal P loss, highlighting the importance of fine-tuning nanoparticle properties for sustainable agriculture [67].

Advancing the concept of precision agriculture, calcium nanoparticles conjugated with phosphorous are a good mode of nutrient releases. Its potential was further exemplified in a study where non-toxic calcium phosphate nanoparticles doped with urea (U-ACP) were developed for use as fertilizers on durum wheat. These nanoparticles maintained the crop yields and quality even with reduced nitrogen dosage, demonstrating their effectiveness in nutrient delivery. The study's findings on the rapid uptake of these nanoparticles by the plant's roots compared to the leaves indicated the potential for more sustainable wheat fertilization practices using nanotechnology (Ramírez-Rodríguez et al. [69]. Yet in the calcium approach, I was developed biomimetic calcium phosphate nanoparticles (CaP) doped with potassium (K) and nitrogen (N) to create nanoU-NPK. These nanoparticles, capable of a slow and gradual release of essential macronutrients, showed promise in enhancing fertilization efficiency. Preliminary tests on durum wheat revealed that nanoU-NPK could reduce the nitrogen requirement by 40% without affecting kernel weight, aligning with precision agriculture principles [70].

Optimizing a multiple inorganic method, it was investigated the effects of combining phosphorus, zinc, and silicon nanoparticles (PNPs, ZnNPs, and SiNPs) with mineral nitrogen, phosphorus, and potassium (NPK) fertilizers on the Egyptian Giza 179 rice variety. This study, spanning two growing seasons, employed a randomized complete block design with various treatment combinations. The findings were substantial, showing significant improvements in grain yield, straw yield, yield attributes, and chemical compositions of the plant. Notably, certain combinations of NPK with ZnNPs and SiNPs showed remarkable enhancement in all studied characteristics, underscoring the potential of nanofertilizers in improving crop yields and quality [71].

The utilization of nanobiotechnology in plant sciences, especially through nanomaterials like nanosilica, has shown great promise in enhancing plant growth and offering protection against various stresses. Nanosilica, with its increased absorptivity, has been found to facilitate the uptake of silica in plants, reduce the accumulation of reactive oxygen species (ROS), and limit the entry of harmful ions such as sodium and heavy metals into the plant system. Furthermore, it enhances plant defenses against pathogens, suggesting a multifaceted role in plant health and productivity [72]. In a study focusing on maize, researchers employed high throughput silica nanoparticles (SiO2-NPs) prepared via the sol–gel technique, which were used as both, nanofertilizers and as pesticides. The study assessed the response of maize hybrids to mineral NPK fertilizers in combination with varying concentrations of SiO2-NPs. The results indicated that using the recommended dose of mineral NPK, along with SiO2-NPs, significantly enhanced plant growth and yield characteristics. This dual role of SiO2-NPs as a growth promoter and an unconventional pesticide underscores their versatility and potential in sustainable agriculture [73]. A greenhouse pot culture experiment with rice cultivars grown in soil artificially polluted with heavy metals like cadmium, lead, copper, and zinc provided insights into the efficacy of organic and inorganic nano-silicon (nano-Si) fertilizers. The study found that nano-Si positively affected the growth and yields of at least four rice cultivars, with both forms of nano-Si increasing average spike weight and reducing heavy metal concentrations in the grains. This research highlights the potential of nano-Si fertilizers in mitigating heavy metal stress in crops [74]. In a hydroponic pot experiment, the role of foliar application with nano-silicon in alleviating cadmium (Cd) stress in rice seedlings was studied. The findings showed that nano-Si improved growth and nutrition of the rice seedlings under Cd stress, highlighting its potential as a mitigative agent against heavy metal toxicity in crops [75].

In agriculture, boron (B) nanofertilizers have shown great potential as an efficient alternative to traditional fertilizers. A study focusing on their impact on green bean plants revealed that foliar application of B nanofertilizers at varying concentrations improved biomass yield, nitrogen assimilation, and photosynthetic activity. The study suggested that B nanofertilizers could be a promising solution for enhancing crop growth and productivity, although further research is needed to compare their effects with conventional B fertilizers and assess potential phytotoxic effects [76].

Finally, the use of mycogenic selenium nanoparticles (SeNPs) synthesized from *Trichoderma atroviride* for their antifungal activities was explored. The study demonstrated strong in vitro antifungal activity of SeNPs against various phytopathogens, offering a sustainable solution for managing plant diseases. The unique aggregation and binding properties of these nanoparticles to the zoospores of pathogens underscored their potential as eco-friendly antifungal agents in agriculture [77]. The environmental behavior and impact of selenium nanoparticles (SeNPs), particularly alginate-coated (Alg-SeNPs) and polyvinyl alcohol-coated (PVA-SeNPs), were investigated in aquatic systems. The study highlighted the complex dynamics of SeNPs in natural waters and the need for further research on their interactions with natural colloids, particularly in the context of their use as nanofertilizers [68].

In conclusion, the advent of inorganic nanofertilizers in agricultural sciences marks a significant leap towards sustainable and efficient agricultural practices. These nanomaterials, through their unique properties and mechanisms, offer a range of benefits from enhanced nutrient use efficiency to controlled release of nutrients and reduced environmental impact. The studies reviewed here provide a comprehensive understanding of the various applications and potential of inorganic nano-fertilizers in addressing the challenges of modern agriculture, paving the way for more targeted and sustainable crop management strategies.

5 Metal Products

Of the one hundred and eighteen known elements on the periodic table, forty comprise the d-block, also known as transition metals, and overall, eighty per cent of the elements are considered metal in some extent–metalloids tend to blur the lines [5]. These elements present unique electronic properties, exchanging electrons with relatively ease, which was used by organisms to transfer electrons and catalyze reactions. This incorporation into biological systems has played a crucial role in shaping the course of evolutionary processes [78]. *Exempli gratia*, at least a third of all known proteins are metalloproteins, proteins which present a metal ion coordinated into cofactor easing the electronic transportation [78–80]. Consequently, many essential processes to life involve these proteins, including photosynthesis, respiration, and nitrogen fixation.

Due to the reactivity and catalytic capabilities of metals inside living organisms, these elements are necessary only in trace amounts [81]. When developing what was necessary to the grow of vegetable organisms, the mais focus was on the main building blocks of the plant, like nitrogen and phosphorus. Furthermore, in dry soils the biodistribution of nutrients is impaired, reducing the efficiency of some fertilizers. It is known that synthetic fertilizers have low effectiveness of nitrogen (N) 30–35%, phosphorus (P) 18–20% and potassium (K) 35–40% indicating that more than half of the nutrients are lost before reaching their target locations, becoming biologically unavailable to plants [82]. However, the supplementation of trace elements can benefit the vegetable growth, allowing the increased uptake and overall increase in biological functions, such as photosynthesis.

However, it is important to notice that exactly due to this reactivity metallic elements present relatively low toxic doses, meaning that if more than trace amounts are made available, this leads to the decrease in biological process [83], [81]. Alongside that, these elements tend to be of low solubility and due its density, of low absorption by organisms. These factors have always factored in the use of metallic elements in fertilizations, copper being except, since copper sulphate presents a lower toxicity and high solubility.

During the transition to the nanoscale the increase in surface area and lowering of the apparent density of the materials present the ultima pathway to the ease in metal absorption. As example of this phenomenon, it is possible to cite the CosmoFer®, in 1974 it was the first Food and Drug Administration approved drug based on nanotechnology [84]. This drug consisted of Fe nanoparticles to treat patients with iron deficiency, by providing an Fe source with higher bioavailability [85].

The idea behind metallic nanofertilizers is the same, to provide source of bioavailable metal in trace concentrations such that all the metabolic process of the plant can be fully functioning. This supplementation leads to a higher growth rate and yield helping in the photosynthesis and accelerating enzymatic process such as those involved in seed germination, generating fast-growing and strong plants.

Due to the different oxidation potentials and stages, each metal tends to act on slightly different pathways inside plants, and when taken into consideration the different types of particles with a plethora of sizes and possible structures, metallic nanofertilizers compose an eclectic and complex field of study [86]. So, a rule of thumb is not available to understand the effect of these formulations, being necessary the investigation of the general role of the metal in the plant and the available data in the literature. A subdivision to better organize the matter can be the natural abundancy of the metals in plant, *id est*, separate the metals by the necessity of the plant, into the most needed to those not necessary to the metabolism but have impacts if available.

The most abundant metal in plants is iron, followed by zinc [87]. Iron plays a pivotal role in essential metabolic processes like DNA synthesis, respiration, and photosynthesis, besides activating numerous metabolic pathways [88]. Meanwhile zinc is the only metal represented in all six enzyme classes, regulating most of the enzymatic function and reactivity and in contrast with Cu and Fe is redox-stable under physiological conditions [87]. Table 1 showcases some of the finding in the literature regarding the use of nanofertilizers based on iron and zinc. These findings are broad, but generally regards the increase in seed germination, yield and increase in photosynthesis.

Table 1 Metal nanofertilizers based on iron and zinc and their applications

Metal element	Action described	Organism	References
Fe, Mn	Photosynthesis promoters	Black-eyed pea	[89]
Fe_2O_3	Increased in growth, photosynthetic pigments, and total protein contents	*Chlorella vulgaris*	[90]
Fe_2O_3	Ameliorate chlorosis	*Citrus reticulata*	[91]
ZnO	Stimulation of seed germination and plant growth	*Arachis hypogea,* tomato, alfalfa, cucumber and soybean	[92]
$ZnSO_4$ and Chitosan	10 folds decrease in zinc usage	*Triticum aestivum*	[93]
ZnO and Zn salts	Higher yield, plant mass and quality	Pinto bean	[94]
Fe_2O_3	Root growth	*Pisum sativum L. Cicer arientinum, Vigna radiate,* and *Phaseolus vulgaris*	[95]
ZnO	Increase in germination and vigor	*Capsicum chinense*	[96]
Fe_2O_3	Improved leaf area, leaves per plant and chlorophyll content	*Zea mays*	[97]
Zn-Fe	Increased proline, soluble sugars, and enzyme activities	*Triticum aestivum*	[98]
ZnO	Increase in drought tolerance	*Dracocephalum kotschyi*	[99]

Table 2 Metal nanofertilizers and their applications

Metal element	Action	Organism	References
Mg	Improve in the nutrition and nutraceutical quality	Green beans	[105]
Mg	Increase in mass and nitrate reductase	Green beans	[106]
Cu	Enhance in root length, shoot length, and biomass	Chickpeas	[107]
$Ca_{10}(PO_4)_6(OH)_2$	Root elongation	*Solanum lycopersicum*	[108]
Cu	Increase in size and pigment concentration	Basil	[109]
Mn	Increase in mass and P and K uptake	*Triticum aestivum*	[110]
Co	Increase in height and growth index	*Mentha Longifolia*	[111]
$CoFe_2O_4$	Increase in nutrient uptake	*Lycopersicon lycopersicum*	[112]

Besides Fe and Zn, other metals are also needed to the plant metabolism. Ni is involved in ureolysis, hydrogen metabolism, methane biogenesis and acetogenesis [83]. Co is a major component of cobalamin (B12 vitamin) and accelerates the nitrogen fixation [100]. Mn is involved in redox reactions, phosphorylation, decarboxylation, and hydrolysis [101]. Cu is very important to enzymatic activity [102] and can play a major role also combating diseases, especially against fungal phytopathogens [103]. And new discoveries reveal that even rare earth elements present biological relevance Lanthanides [104]. Table 2 shows some exemples found in the literatures regarding these metals.

6 Composite Products

A nanocomposite consists of different elemental material combined to form hybridizations based on the nanometric scale and has presented itself as a promising candidate in comparison to conventional strategies since they (i) increase the absorption efficiency, (ii) can have a slower release of the desired active ingredients and (iii) improve the solubility, bioavailability and dispersion of the compounds [113], due to these properties, the nanofertilizers are considered "smart nutrient systems" [82].

As can be observed in Table 3, several strategies and different nanomaterials have been used to develop candidates for application as nanofertilizers. For instance, the titanium dioxide nanoparticles (TiO2-NPs) can act in plant nutrition and growth, even in lower concentrations [113]. Chitosan NPs can cross-link nanometals [113, 114]. Zn are involved in responses to water stress as well as maintaining the integrity of biological membranes, essential in absorption tests. Cu and B participate in the strengthening and functioning of the cell wall, with Cu being necessary for the synthesis of lignin and B being fundamental for the structural integrity of the plant cytoskeleton [114]. Colloidal forms of Fe (III) minerals can be associated with nanoforms of phosphate (P) fertilizers [115].

Several approaches aim to increase the absorption of micronutrients and/or reduce environmental nutrient losses [116]. The nanocomposites are being widely used to solve challenges such as (i) scarcity of available land due to soil degradation as erosion, mineral deficiency, drought and salinity, (ii) water eutrophication by fertilizers with low absorption by plants and high absorption by the soil, (iii) toxic concentrations of micronutrients that affect the normal development of plants [117]. This scenario highlights some of the challenges faced by agriculture nowadays and how nanocomposites can be employed as smart nutrient systems for agriculture.

7 Environmental Impacts

The concept of sustainable agriculture is based on using less agrochemicals to mitigate adverse environmental effects. Nanotechnology presents itself as a viable alternative to precision farming, given its minimal application in crops, thereby serving as a substitute for conventional agrochemicals [123].

In addition to the use of nanotechnology in agriculture for the making of nanofertilizers, other agricultural products are in investigation to improve the production of food in a sustainable way and in sufficient amounts to meet the demands of a growing population. Such products are nanoherbicides, nanofungicides and nanoinsecticides. The nanoherbicides operate by targeting receptors in the roots of specific weeds, resulting in the inhibition of the weeds' nutrient reserves. This inhibition ultimately culminates in the death of the plants [124]. A study conducted by Sousa et al. revealed that nano formulations of poly(ε-caprolactone) that also contained the herbicide atrazine were effective in inhibiting the growth of *Brassica sp,* being not too toxic to untargeted organisms when the herbicide alone is applied [125]. A similar study conducted by Namasivayam et al. encapsulated AgNP in chitosan, which showed greater herbicidal activity against *Eichhornia crassipes* with the advantage of not reducing macro and micronutrients, as well as not affecting the soil microflora [126].

Other examples in literature also illustrate the use of nanoparticles being more effective and sustainable as conventional herbicides: CuNPs for *Cucurbita pepo* [127] and CuO NPs for *Cucumis sativus, Lolium rigidum* and *Lolium perenne* [128–130].

Nano fungicides, representing another agricultural product with the potential for enhanced sustainability, have been under investigation. Given that fungal diseases pose significant threats to crops, contributing to considerable economic impacts, the exploration of nano fungicides holds promise in addressing these agricultural challenges. However, controlling fungal diseases is not an easy task taken that some fungi are more resistant to fungicides, like Fusarium and Verticillium [123]. The importance of fungicide relies on its ability to control fungal diseases at their development,

Table 3 Nanocomposites employed as smart nutrient systems for agriculture

Nanomaterial	Objective	Plant	Delivery mechanism	Action mechanism	Efficacy	References
ZnO, B2O3, and CuO, and their salts:	Mitigation of drought stress	Soybean	Soil and foliar applied	Cell wall strengthening and functioning	Micronutrient nanoparticles boosted crop performance and the uptake under drought stress	[114]
Phytosynthesis of TiO2-NP (extract of pomegranate rind) and chitosan (from *Aspergillus brasiliensis* biomass)	Nanofertilizers to deliver TiO2-NPs	Not defined	Liquid spray for foliar feeding or powder for soil amendment	Cell wall strengthening and functioning; support of beneficial microbial activities	Not tested. Possible candidates for application as nanofertilizers	[113]
Chitosan/Alginate NPs loaded with CuO	Nanofertilizers to deliver CuO-NPs	*Fortunella margarita* Swingle seeds (kumquat tree)	Suspension applied in seed	Plant growth and development; slow nutrient release	Deedling growth with a synergic effect in developing the epigean and hypogean parts. Possible candidates for application as nanofertilizers	[117]
Nitrate-doped amorphous calcium phosphate (ACP) nanoparticles with urea	Sustainable nanofertilizers	Cucumber plants (*Cucumis sativus*)	Soil and foliar application and soilless (hydroponic) cultivation	Increase micronutrient absorption; reduce environmental nutrient losses	Urea fertilization can induce an increase in the allocation of biomass in plants	[118]

(continued)

Table 3 (continued)

Nanomaterial	Objective	Plant	Delivery mechanism	Action mechanism	Efficacy	References
Nanoporous SiO2 (NanoSi) to enhance the adhesion of carbon dots (CDs)	Increase the adhesion of CDs on crop leaf surfaces	Corn	Foliar application	Increase the bioavailability of nanomaterials in sustainable agriculture; reduce environmental nutrient losses	Photosynthetic rate, fresh weight and dry weight of corn increased	[119]
TPP-Chitosan and TPP-Chitosan-ZnO nanofertilizers	Control leaching loss and application in crop fertilization through the reuse and nano-functionalization of tripolyphosphate (TPP)	Wheat	Soil application	Reuse of a non-fertilizing material for agricultural applications; reduce environmental nutrient losses	Improves overall nutritional quality and nutrient mobilization of wheat	[120]

(continued)

Table 3 (continued)

Nanomaterial	Objective	Plant	Delivery mechanism	Action mechanism	Efficacy	References
Nanobiochar and CuO-NPs	Influence soil microbes, nutrient content, and wheat productivity	Wheat	Soil application	The organic nature of nanobiochar decreases toxicity levels and maintains the beneficial effects of CuO-NPs	Improves soil microbial activity, nutrient content, and wheat production; increased wheat Cu uptake being recommended to clayey loam soil quality and increasing Cu uptake and crop productivity	[121]
Chitosan-modified Se nanoparticles (Ch-SeNPs)	Evaluate the biotransformation of chitosan-modified Se nanoparticles (Ch-SeNPs) and their effect on the metabolism of essential metals (Fe, Cu, Zn and Mo) when applied to hydroponic cultivation	*R. sativus* and *B. juncea*	Soil application	Transformation of Ch-SeNPs to seleno-amino acids: Selenomethionine (SeMet), Semethylselenocysteine (SeMetSeCys) and γ-glutamyl-Se-MetSeCys	Ch-SeNPs supplementation affected the absorption of Zn in both species	[122]

allowing the crops to flourish properly which will result in better harvest [131]. Very similar to herbicides, nanofungicides have to be applied carefully in order to prevent interaction with untargeted organisms, which could cause environmental impacts and disturb soil fertility [123]. Examples of the use of nanofungicides are AgNPs against *Sclerotinia sclerotiorum, Botrytis cinerea, R. solani, Curvularia lunata* [132] CuNPs with chitosan has also shown potential in inhibiting the growth of species like *A. alternata, M. phaseolina* and *R. solani* [133] Fe_2O_3 and Fe_3O_4 nanoparticles proved to be effective against *Trichothecium roseum, Cladosporium herbarum, Penicillium chrysogenum, Alternaria alternata,* and *Aspergillus niger* [134].

Nano insecticides play a crucial role in the realm of agricultural products, given the significance of insect control for crop productivity. However, insecticides are usually made of substances that threaten ecosystems. Insecticides exert significant impact on the environment, for example, causing surface water pollution, which poses great threat to aquatic biodiversity [135]. Furthermore, conventional insecticides not only induce severe toxicity in non-targeted organisms but also contribute to the development of resistance in targeted organisms. This impact poses harm to ecosystems and carries economic consequences [136]. Nanotechnology presents an alternative since it can make nano insecticides engineered to release active substances in insects' stomach, for example, offering controlled and more effective use of pesticides [137]. Examples in literature are Al_2O_3 against *Sitophilus oryzae* and *Rhyzopertha dominica* [138], CuNPs against *Anopheles stephensi, Aedes aegypti, Culex quinquefasciatus* and *Tenebrio molitor* [139] and *Spodoptera frugiperda* [140].

Nanofertilizers are designed to release nutrients in the right quantities, providing appropriate conditions for the plants to develop properly with minimized waste of nutrients [141, 142] also they are site-specific, minimizing excess run off [143]. Additionally, the utilization of nanofertilizers not only reduces the quantity of materials required but also results in an increase in efficiency, when compared to conventional fertilizers. This holds significant promise in mitigating environmental impacts, as plants absorb the nutrients from nanofertilizers without leaving residual elements in the ecosystem [144]. This stands in contrast to conventional fertilizers, which can potentially enter water bodies and contribute to eutrophication [143]. Conventional fertilizers are applied by spraying into the fields, which results in limited absorption by the plants and lost of nutrients by leaching, leading to soil and water pollution [145]. It is estimated that 50–70% of agrochemicals applied are lost by leaching, bioconversion or mineralization, highlighting the impact they cause in the environment [146]. Additionally, even though conventional fertilizers have increased food production in the past decades, they have diminished soil fertility and crops quality [146].

The CeO_2 NP, when used as a nanofertilizer, has shown potential in improving the growth of plants [147], with minimal toxicity in crops and biotransformation [143, 104]. A study conducted by Mukherjee et al. [46] has shown, though, that high concentrations of CeO_2 NP can alter the biochemical and molecular aspects of certain crops and can also induce soil contamination. Iron NP have demonstrated no toxicity or stress to maize seedlings,on the contrary, they have been found to enhance growth and increase chlorophyll content. Nevertheless, it is important to note that with other plant species, iron NPs may adhere to the roots and inhibit germination [148].

It is noteworthy that each nanofertilizer exhibits distinct behaviors with different plants, indicating a wide range of variables to take into consideration regarding nanomaterials usage for the establishment of policies and guidelines. Each case presents a unique situation, requiring specific precautions and tailored approaches. Some studies have shown that nano agrochemicals could cause phytotoxicity to edible and medical plants, which could have antagonistic results of what is expected [145]. Essentially, the toxicity of this kind of material is context dependent, meaning that nanomaterials will not be toxic to all plants [149]. Moreover, nanotoxicity is not fully understood. The assumptions suggest that the chemical composition and structure, as well as particle size, enhanced surface area and charge, shape and aggregation might influence the behavior of nanomaterials in the ecosystems, generating unintended impacts [149, 150]. On the other hand, soil physicochemical and biological characteristics also play a role on the nanofertilizers interaction with the environment, for instance, pH has been described by Servin et al. [142] to influence nutrients availability [142]. Other factors like inorganic and organic compounds, as well as plant exudates and microrganisms, are also important to determine how nanofertilizers will interact with that environment [149].

Limited information on nanofertilizer fate, including elemental analysis of plants and soil, release kinetics, and leaching, has been described in literature. Notably, detailed nanomaterial characterization was generally insufficient, highlighting the need for standardized assessments, particularly in the quantification of freely dissolved ions for potentially toxic micronutrients. This type of information, although crucial for studies assessing the environmental risk of nanomaterials, has not yet been thoroughly investigated [151]. Kah et al. [151] argues that merely altering nanoformulations may not suffice to reduce environmental impact. Many studies lack nanospecific quality assurance, and experiments are not representative of large-scale production and real field conditions, impeding a critical assessment of potential impacts on non-target organisms. This knowledge gap raises concerns about adverse effects on human and ecosystem health, necessitating further research to assess the risks and benefits associated with nanoagrochemicals [151, 152].

The understanding of these agricultural nano products gives an overview of what is being applied to ecosystems and what kind of environmental impact they might lead to. Although not the primary focus of this chapter, nano herbicides, nano fungicides, and nano insecticides studies might give light to what kind of environmental impacts nanofertilizers could also have given the recent nature of this research field. It is possible to see how promising nanotechnology applied to agriculture regarding the economic factors, production to accommodate the growing population, and sustainability, envisioning the reduction of harmful agrichemicals to the environment. However, little is known and described in literature as to how these nano agricultural products are impacting the environment. Considering this, the field of ecotoxicology provides valuable insights into the behavior of nanomaterials in the environment and underscores the necessary precautions to be taken during their development [123]. This is very important to establish comprehensive policies and guidelines concerning the development and use of nanomaterials in agriculture [153]. Furthermore, it is crucial for government agencies, the scientific community and industries to work together in order to develop nanomaterials safe to ecosystems and human beings.

Currently, global initiatives are underway to address this issue through the implementation of early legislative measures. For example, in 2009, the European Food Safety Authority (EFSA) released a guideline highlighting the potential toxicity of nanomaterials and in the United States Environmental Protection Agency (EPA) granted approval for the nanoparticle-based antimicrobial pesticide HeiQ AGS-20, however, with inconclusive regulations on nano agrichemicals [142]. It is important to emphasize that climate change and a growing human population lead to an increased demand for innovation in agriculture [143]. Nanomaterials can serve as a viable alternative to enhance production with fewer environmental impacts compared to conventional agrichemicals. Given the circumstances, agriculture demands infrastructure and precision farming, in order to provide quality food in sufficient quantities. Moreover, surface runoff, excessive pesticide use, soil erosion, and changes in soil fertility, moisture, and temperature, collectively impact the ecosystem quality. Therefore, the development of new technology is imperative to create an enhanced system that mitigates environmental hazards and addresses nutrient deficiencies [145].

8 Conclusions

The exploration of nanofertilizers in this chapter sheds light on their promising potential to revolutionize agriculture by significantly enhancing fertilizer efficiency. The use of new technologies to increase food production is a reality fueled by the population growth and environmental protection. The use of nanomaterials as fertilizers has demonstrated notable improvements in nutrient delivery, uptake, and utilization by plants, thereby maximizing crop yields. However, it is imperative to acknowledge the environmental implications associated with nanofertilizer applications. While nanofertilizers have the capacity to reduce nutrient runoff and minimize environmental degradation, their long-term impacts on soil health and ecosystems warrant careful consideration. Striking a balance between reaping the benefits of enhanced efficiency and mitigating potential environmental risks is crucial for the sustainable implementation of nanofertilizers. Moreover, a comprehensive understanding of the physicochemical properties of nanomaterials is essential for optimizing their design and ensuring their safe integration into agricultural practices. As we navigate the evolving landscape of nanotechnology in agriculture, continued research and collaboration are essential to refine nanofertilizer technologies, addressing both productivity and environmental sustainability for the future of global food security.

Acknowledgements The authors would like to thank FAPESP (grants number 2023/12466-6, 2023/01487-2 and 2021/11443-7) for financial support. This study was financed in part by the Coordenação de Aperfeiçoamento de Pessoal de Nível Superior - Brasil (CAPES) - Finance Code 001.

References

1. Tudi M, Daniel Ruan H, Wang L et al (2021) Agriculture development, pesticide application and its impact on the environment. Int J Environ Res Public Health 18:1112. https://doi.org/10.3390/ijerph18031112
2. Russel DA, Williams GG (1977) History of chemical fertilizer development. Soil Sci Soc Am J 41:260–265. https://doi.org/10.2136/sssaj1977.03615995004100020020x
3. Mordor Intelligence (2023a) Fertilizer market size & share analysis–growth trends & forecasts (2023–2028)
4. Liebig J, Playfair LP (1840) Organic chemistry in its applications to agriculture and physiology. Printed for Taylor and Walton, London
5. Schwerdtfeger P, Smits OR, Pyykkö P (2020) The periodic table and the physics that drives it. Nat Rev Chem 4:359–380. https://doi.org/10.1038/s41570-020-0195-y
6. Rahman K, Zhang D (2018) Effects of fertilizer broadcasting on the excessive use of inorganic fertilizers and environmental sustainability. Sustainability 10:759. https://doi.org/10.3390/su10030759
7. Merchant SS (2010) The elements of plant micronutrients. Plant Physiol 154. https://doi.org/10.1104/pp.110.161810
8. Alloway BJ (2013) Heavy metals and metalloids as micronutrients for plants and animals
9. Njinga RL, Moyo MN, Abdulmaliq SY (2013) Analysis of essential elements for plants growth using instrumental neutron activation analysis. Int J Agron 2013. https://doi.org/10.1155/2013/156520
10. OECD/FAO (2023) OECD-FAO agricultural outlook 2023–2032
11. Guo L, Li H, Cao X et al (2021) Effect of agricultural subsidies on the use of chemical fertilizer. J Environ Manage 299:113621. https://doi.org/10.1016/j.jenvman.2021.113621
12. Valdes C (2022) Brazil's momentum as a global agricultural supplier faces headwinds
13. Mordor Intelligence (2023b) Brazil fertilizers market size & share analysis–growth trends & forecasts up to 2028
14. Wood RB, Gibson CE (1973) Eutrophication and Lough Neagh. Water Res 7:173–187. https://doi.org/10.1016/0043-1354(73)90161-9
15. Bunting L, Leavitt PR, Gibson CE et al (2007) Degradation of water quality in Lough Neagh, Northern Ireland, by diffuse nitrogen flux from a phosphorus-rich catchment. Limnol Oceanogr 52:354–369. https://doi.org/10.4319/lo.2007.52.1.0354
16. Elliott JA, McElarney YR, Allen M (2016) The past and future of phytoplankton in the UK's largest lake, Lough Neagh. Ecol Indic 68:142–149. https://doi.org/10.1016/j.ecolind.2015.07.015
17. McElarney Y, Rippey B, Miller C et al (2021) The long-term response of lake nutrient and chlorophyll concentrations to changes in nutrient loading in Ireland's largest lake, Lough Neagh. Biol Environ Proc R Irish Acad 121B:47–60. https://doi.org/10.1353/bae.2021.0002
18. Bayda S, Adeel M, Tuccinardi T et al (2019) The history of nanoscience and nanotechnology: from chemical-physical applications to nanomedicine. Molecules 25:112. https://doi.org/10.3390/molecules25010112
19. Drexler KE (1981) Molecular engineering: an approach to the development of general capabilities for molecular manipulation. Proc Natl Acad Sci 78:5275–5278. https://doi.org/10.1073/pnas.78.9.5275
20. Drexler KE (1986) Engines of creation: the coming era of nanotechnology, 1st edn. Doubleday
21. Usman M, Farooq M, Wakeel A, et al (2020) Nanotechnology in agriculture: current status, challenges and future opportunities. Sci Total Environ 721:137778. https://doi.org/10.1016/j.scitotenv.2020.137778
22. Moore K (2010) Nanofertilizers, pee for your plants. Chem Eng News Arch 88:1–1. https://doi.org/10.1021/cen-v088n009.p088
23. Liu R, Lal R (2015) Potentials of engineered nanoparticles as fertilizers for increasing agronomic productions. Sci Total Environ 514:131–139. https://doi.org/10.1016/j.scitotenv.2015.01.104

24. Scopus (2023) Scopus search–nanofertilizer. https://www.scopus.com/results/results.uri?sort=plf-f&src=s&st1=nanofertilizer&sid=adce9c563a9f5bbda2870062baf32872&sot=b&sdt=b&sl=29&s=TITLE-ABS-KEY%28nanofertilizer%29&origin=searchbasic&editSaveSearch=&yearFrom=Before+1960&yearTo=Present&sessionSearchId=adce9c563a9f5bbda2870062baf32872&limit=10. Accessed 26 Oct 2023
25. Lens.org (2023) Lens search for "nanofertilizers"
26. Patel DK, Kim HB, Dutta SD, et al (2020) Carbon nanotubes-based nanomaterials and their agricultural and biotechnological applications. Materials 13
27. Yadav A, Yadav K, Abd-Elsalam KA (2023) Nanofertilizers: types, delivery and advantages in agricultural sustainability. Agrochemicals 2. https://doi.org/10.3390/agrochemicals2020019
28. Song G, Pandorf M, Westerhoff P, Ma Y (2019) Carbon nanomaterial-based fertilizers can improve plant growth. In: Nanotechnology applications in the food industry
29. Zaytseva O, Neumann G (2016) Carbon nanomaterials: production, impact on plant development, agricultural and environmental applications. Chem Biol Technol Agric 3
30. Hoang NH, Le Thanh T, Sangpueak R, et al (2022) Chitosan nanoparticles-based ionic gelation method: a promising candidate for plant disease management. Polymers (Basel) 14
31. Khan HA, Naqvi SR, Mehran MT, et al (2021) A performance evaluation study of nano-biochar as a potential slow-release nano-fertilizer from wheat straw residue for sustainable agriculture. Chemosphere 285. https://doi.org/10.1016/j.chemosphere.2021.131382
32. Ramanayaka S, Tsang DCW, Hou D, et al (2020) Green synthesis of graphitic nanobiochar for the removal of emerging contaminants in aqueous media. Sci Total Environ 706. https://doi.org/10.1016/j.scitotenv.2019.135725
33. Oleszczuk P, Ćwikła-Bundyra W, Bogusz A, et al (2016) Characterization of nanoparticles of biochars from different biomass. J Anal Appl Pyrolysis 121. https://doi.org/10.1016/j.jaap.2016.07.017
34. Sani MNH, Amin M, Siddique AB, et al (2023) Waste-derived nanobiochar: a new avenue towards sustainable agriculture, environment, and circular bioeconomy. Sci Total Environ 905
35. Song B, Cao X, Gao W, et al (2022) Preparation of nano-biochar from conventional biorefineries for high-value applications. Renew Sustain Energy Rev 157
36. Oni BA, Oziegbe O, Olawole OO (2019) Significance of biochar application to the environment and economy. Ann Agric Sci 64
37. Patel MR, Panwar NL (2023) Biochar from agricultural crop residues: environmental, production, and life cycle assessment overview. Resourc, Conserv Recycl Adv 19
38. Dong X, He L, Hu H, et al (2018) Removal of 17B-estradiol by using highly adsorptive magnetic biochar nanoparticles from aqueous solution. Chem Eng J 352. https://doi.org/10.1016/j.cej.2018.07.025
39. Kumar M, Xiong X, Wan Z, et al (2020) Ball milling as a mechanochemical technology for fabrication of novel biochar nanomaterials. Bioresour Technol 312
40. Cui S, Kong F, Li Y, et al (2021) Effect of mineral loaded biochar on the leaching performances of nitrate and phosphate in two contrasting soils from the coastal estuary area. Sci Total Environ 779. https://doi.org/10.1016/j.scitotenv.2021.146346
41. Chausali N, Saxena J, Prasad R (2021) Nanobiochar and biochar based nanocomposites: advances and applications. J Agric Food Res 5. https://doi.org/10.1016/j.jafr.2021.100191
42. Mahmoud E, El Baroudy A, Ali N, Sleem M (2020) Spectroscopic studies on the phosphorus adsorption in salt-affected soils with or without nano-biochar additions. Environ Res 184. https://doi.org/10.1016/j.envres.2020.109277
43. Rajput VD, Minkina T, Ahmed B, et al (2022) Nano-biochar: a novel solution for sustainable agriculture and environmental remediation. Environ Res 210. https://doi.org/10.1016/j.envres.2022.112891
44. Safdar M, Kim W, Park S, et al (2022) Engineering plants with carbon nanotubes: a sustainable agriculture approach. J Nanobiotechnol 20
45. Lateef A, Nazir R, Jamil N, et al (2019) Synthesis and characterization of environmental friendly corncob biochar based nano-composite—a potential slow release nano-fertilizer for sustainable agriculture. Environ Nanotechnol Monit Manag 11. https://doi.org/10.1016/j.enmm.2019.100212

46. Mukherjee A, Majumdar S, Servin AD, et al (2016) Carbon nanomaterials in agriculture: a critical review. Front Plant Sci 7
47. Chowmasundaram YAP, Tan TL, Nulit R, et al (2023) Recent developments, applications and challenges for carbon quantum dots as a photosynthesis enhancer in agriculture. RSC Adv 13
48. Azwar E, Wan Mahari WA, Chuah JH, et al (2018) Transformation of biomass into carbon nanofiber for supercapacitor application—a review. Int J Hydrogen Energy 43. https://doi.org/10.1016/j.ijhydene.2018.09.111
49. Sarlak N, Taherifar A, Salehi F (2014) Synthesis of nanopesticides by encapsulating pesticide nanoparticles using functionalized carbon nanotubes and application of new nanocomposite for plant disease treatment. J Agric Food Chem 62. https://doi.org/10.1021/jf404720d
50. Zhang M, Gao B, Chen J, et al (2014) Slow-release fertilizer encapsulated by graphene oxide films. Chem Eng J 255. https://doi.org/10.1016/j.cej.2014.06.023
51. Haghighi M, Teixeira Da Silva JA (2014) The effect of carbon nanotubes on the seed germination and seedling growth of four vegetable species. J Crop Sci Biotechnol 17. https://doi.org/10.1007/s12892-014-0057-6
52. Wang A, Kang F, Wang Z, et al (2019) Facile synthesis of nitrogen-rich carbon dots as fertilizers for mung bean sprouts. Adv Sustain Syst 3. https://doi.org/10.1002/adsu.201800132
53. Guo B, Liu G, Wei H, et al (2022) The role of fluorescent carbon dots in crops: mechanism and applications. SmartMat 3
54. Prajapati D, Pal A, Dimkpa C, et al (2022) Chitosan nanomaterials: a prelim of next-generation fertilizers; existing and future prospects. Carbohydr Polym 288
55. Van Bavel N, Issler T, Pang L, et al (2023) A simple method for synthesis of chitosan nanoparticles with ionic gelation and homogenization. Molecules 28. https://doi.org/10.3390/molecules28114328
56. Aranaz I, Alcántara AR, Civera MC, et al (2021) Chitosan: an overview of its properties and applications. Polymers (Basel) 13
57. Ha NMC, Nguyen TH, Wang SL, Nguyen AD (2019) Preparation of NPK nanofertilizer based on chitosan nanoparticles and its effect on biophysical characteristics and growth of coffee in green house. Res Chem Intermediates 45. https://doi.org/10.1007/s11164-018-3630-7
58. Choudhary RC, Kumaraswamy RV, Kumari S, et al (2017) Cu-chitosan nanoparticle boost defense responses and plant growth in maize (Zea mays L.). Sci Rep 7. https://doi.org/10.1038/s41598-017-08571-0
59. Kubavat D, Trivedi K, Vaghela P, et al (2020) Characterization of a chitosan-based sustained release nanofertilizer formulation used as a soil conditioner while simultaneously improving biomass production of Zea mays L. Land Degrad Dev 31. https://doi.org/10.1002/ldr.3629
60. Hartoyo APP, Octaviani EA, Syamani FA, et al (2022) Potential of chitosan/carbon nanoparticles and chitosan/lignocellulose nanofiber composite as growth media for peatland paddy seeds. Environ Res 212. https://doi.org/10.1016/j.envres.2022.113235
61. Abdel-Aziz H, Hasaneen MN, Omar A (2018) Effect of foliar application of nano chitosan NPK fertilizer on the chemical composition of wheat grains. Egypt J Bot https://doi.org/10.21608/ejbo.2018.1907.1137
62. Abdel-Aziz HMM, Hasaneen MNA, Omer AM (2016) Nano chitosan-NPK fertilizer enhances the growth and productivity of wheat plants grown in sandy soil. Spanish J Agric Res 14. https://doi.org/10.5424/sjar/2016141-8205
63. Maluin FN, Hussein MZ (2020) Chitosan-based agronanochemicals as a sustainable alternative in crop protection. Molecules 25
64. Kumaraswamy R V., Kumari S, Choudhary RC, et al (2019) Salicylic acid functionalized chitosan nanoparticle: a sustainable biostimulant for plant. Int J Biol Macromol 123. https://doi.org/10.1016/j.ijbiomac.2018.10.202
65. Ciurli A, Giagnoni L, Pastorelli R, et al (2022) A novel P nanofertilizer has no impacts on soil microbial communities and soil microbial activity. Applied Soil Ecology 178. https://doi.org/10.1016/j.apsoil.2022.104570
66. Priyam A, Yadav N, Reddy PM, et al (2022) Uptake and benefits of biogenic phosphorus nanomaterials applied via fertigation to Japonica Rice (Taipei 309) in low- and high-calcareous soil conditions. ACS Agric Sci Technol 2. https://doi.org/10.1021/acsagscitech.1c00244

67. Sakhno Y, Ma C, Borgatta J, et al (2022) Role of cation substitution and synthesis condition in a calcium phosphate-based novel nanofertilizer on Lettuce (Lactuca sativa) yield. ACS Sustain Chem Eng 10. https://doi.org/10.1021/acssuschemeng.2c04451
68. Yu S, Liu H, Yang R, et al (2023) Aggregation and stability of selenium nanoparticles: complex roles of surface coating, electrolytes and natural organic matter. J Environ Sci (China) 130. https://doi.org/10.1016/j.jes.2022.10.025
69. Ramírez-Rodríguez GB, Miguel-Rojas C, Montanha GS, et al (2020b) Reducing nitrogen dosage in triticum durum plants with urea-doped nanofertilizers. Nanomaterials 10. https://doi.org/10.3390/nano10061043
70. Ramírez-Rodríguez GB, Dal Sasso G, Carmona FJ, et al (2020a) Engineering biomimetic calcium phosphate nanoparticles: a green synthesis of slow-release multinutrient (NPK) nanofertilizers. ACS Appl Bio Mater 3. https://doi.org/10.1021/acsabm.9b00937
71. Elekhtyar NM, AL-Huqail AA (2023) Effect of foliar application of phosphorus, zinc, and silicon nanoparticles along with mineral NPK fertilization on yield and chemical compositions of rice (Oryza sativa L.). Agriculture (Switzerland) 13. https://doi.org/10.3390/agriculture13051061
72. Mathur P, Roy S (2020) Nanosilica facilitates silica uptake, growth and stress tolerance in plants. Plant Physiol Biochem 157
73. El-Naggar ME, Abdelsalam NR, Fouda MMG, et al (2020) Soil application of nano silica on maize yield and its insecticidal activity against some stored insects after the post-harvest. Nanomaterials 10. https://doi.org/10.3390/nano10040739
74. Wang S, Wang F, Gao S, Wang X (2016b) Heavy metal accumulation in different rice cultivars as influenced by foliar application of nano-silicon. Water Air Soil Pollut 227. https://doi.org/10.1007/s11270-016-2928-6
75. Wang S, Wang F, Gao S (2015) Foliar application with nano-silicon alleviates Cd toxicity in rice seedlings. Environ Sci Pollut Res 22. https://doi.org/10.1007/s11356-014-3525-0
76. Franco-Lagos CL, Sánchez E, Palacio-Márquez A, et al (2023) Efficacy of the application of boron nanofertilizer on biomass, yield, nitrogen assimilation and photosynthetic activity in green beans. Not Bot Horti Agrobot Cluj Napoca 51. https://doi.org/10.15835/nbha51112795
77. Joshi SM, De Britto S, Jogaiah S, Ito SI (2019) Mycogenic selenium nanoparticles as potential new generation broad spectrum antifungal molecules. Biomolecules 9. https://doi.org/10.3390/biom9090419
78. Butler A (1998) Acquisition and utilization of transition metal ions by marine organisms. Science 1979:281
79. Thomson AJ, Gray HB (1998) Bio-inorganic chemistry. Curr Opin Chem Biol 2:155–158. https://doi.org/10.1016/S1367-5931(98)80056-2
80. Yannone SM, Hartung S, Menon AL et al (2012) Metals in biology: defining metalloproteomes. Curr Opin Biotechnol 23:89–95. https://doi.org/10.1016/j.copbio.2011.11.005
81. Karges J (2020) Combining inorganic chemistry and biology: the underestimated potential of metal complexes in medicine. ChemBioChem 21:3044–3046. https://doi.org/10.1002/cbic.202000397
82. Jakhar AM, Aziz I, Kaleri AR, et al (2022) Nano-fertilizers: A sustainable technology for improving crop nutrition and food security. NanoImpact 27. https://doi.org/10.1016/j.impact.2022.100411
83. Chen C, Huang D, Liu J (2009) Functions and toxicity of nickel in plants: recent advances and future prospects. Clean (Weinh) 37
84. Huang H, Feng W, Chen Y, Shi J (2020) Inorganic nanoparticles in clinical trials and translations. Nano Today 35:100972. https://doi.org/10.1016/j.nantod.2020.100972
85. Auerbach M, Goodnough LT, Shander A (2013) Iron: the new advances in therapy. Best Pract Res Clin Anaesthesiol 27:131–140. https://doi.org/10.1016/j.bpa.2012.12.001
86. Zhang C, Sun R, Xia T (2020) Adaption/resistance to antimicrobial nanoparticles: will it be a problem? Nano Today 34
87. Broadley MR, White PJ, Hammond JP et al (2007) Zinc in plants. New Phytol 173:677–702. https://doi.org/10.1111/j.1469-8137.2007.01996.x

88. Rout GR, Sahoo S (2015) Role of iron in plant growth and metabolism. Rev Agric Sci 3. https://doi.org/10.7831/ras.3.1
89. Delfani M, Baradarn Firouzabadi M, Farrokhi N, Makarian H (2014) Some physiological responses of black-eyed pea to iron and magnesium nanofertilizers. Commun Soil Sci Plant Anal 45. https://doi.org/10.1080/00103624.2013.863911
90. Askary M, Amirjani MR, Saberi T (2017) Comparison of the effects of nano-iron fertilizer with iron-chelate on growth parameters and some biochemical properties of Catharanthus roseus. J Plant Nutr 40. https://doi.org/10.1080/01904167.2016.1262399
91. Li J, Hu J, Xiao L, et al (2017) Physiological effects and fluorescence labeling of magnetic iron oxide nanoparticles on citrus (Citrus reticulata) Seedlings. Water Air Soil Pollut 228. https://doi.org/10.1007/s11270-016-3237-9
92. Singh A, Singh NB, Afzal S, et al (2018) Zinc oxide nanoparticles: a review of their biological synthesis, antimicrobial activity, uptake, translocation and biotransformation in plants. J Mater Sci 53
93. Dapkekar A, Deshpande P, Oak MD, et al (2018) Zinc use efficiency is enhanced in wheat through nanofertilization. Sci Rep 8. https://doi.org/10.1038/s41598-018-25247-5
94. Mahdieh M, Sangi MR, Bamdad F, Ghanem A (2018) Effect of seed and foliar application of nano-zinc oxide, zinc chelate, and zinc sulphate rates on yield and growth of pinto bean (Phaseolus vulgaris) cultivars. J Plant Nutr 41. https://doi.org/10.1080/01904167.2018.1510517
95. Palchoudhury S, Jungjohann KL, Weerasena L, et al (2018) Enhanced legume root growth with pre-soaking in α-Fe2O3 nanoparticle fertilizer. RSC Adv 8. https://doi.org/10.1039/c8ra04680h
96. Israel García-López J, Lira-Saldivar RH, Zavala-García F, et al (2018) Effects of zinc oxide nanoparticles on growth and antioxidant enzymes of Capsicum chinense. Toxicol Environ Chem 100. https://doi.org/10.1080/02772248.2018.1550781
97. Hasan M, Rafique S, Zafar A, et al (2020) Physiological and anti-oxidative response of biologically and chemically synthesized iron oxide: Zea mays a case study. Heliyon 6. https://doi.org/10.1016/j.heliyon.2020.e04595
98. Seyed Sharifi R, Khalilzadeh R, Pirzad A, Anwar S (2020) Effects of biofertilizers and nano zinc-iron oxide on yield and physicochemical properties of wheat under water deficit conditions. Commun Soil Sci Plant Anal 51. https://doi.org/10.1080/00103624.2020.1845350
99. Karimian Z, Samiei L (2023) ZnO nanoparticles efficiently enhance drought tolerance in Dracocephalum kotschyi through altering physiological, biochemical and elemental contents. Front Plant Sci 14. https://doi.org/10.3389/fpls.2023.1063618
100. Akeel A, Jahan A (2020) Role of cobalt in plants: its stress and alleviation. Contaminants in agriculture. Springer International Publishing, Cham, pp 339–357
101. Schmidt H (2019) The biochemical properties of manganese in plants. Plants 8:381. https://doi.org/10.3390/plants8100381
102. Yruela I (2005) Copper in plants. Braz J Plant Physiol 17
103. Bramhanwade K, Shende S, Bonde S, et al (2016) Fungicidal activity of Cu nanoparticles against Fusarium causing crop diseases. Environ Chem Lett 14. https://doi.org/10.1007/s10311-015-0543-1
104. Cotruvo JA (2019) The chemistry of lanthanides in biology: recent discoveries, emerging principles, and technological applications. ACS Cent Sci 5:1496–1506. https://doi.org/10.1021/acscentsci.9b00642
105. Amaya-Olivas NI, Sánchez E, Hernández-Ochoa L, et al (2023) Biofortification with magnesium nanofertilizer on bioactive compounds and antioxidant capacity in green beans. Not Bot Horti Agrobot Cluj Napoca 51. https://doi.org/10.15835/nbha51112830
106. Salcido-Martínez A, Sánchez E, Lorena Licón-Trillo P, et al (2020) Impact of the foliar application of magnesium nanofertilizer on physiological and biochemical parameters and yield in green beans. Not Bot Horti Agrobot Cluj Napoca 48. https://doi.org/10.15835/48412090

107. Hanif S, Javed R, Khan A, et al (2023) IAA-decorated CuO nanocarriers significantly improve Chickpea growth by increasing antioxidative activities. 3 Biotech 13. https://doi.org/10.1007/s13205-023-03516-z
108. Marchiol L, Filippi A, Adamiano A, et al (2019) Influence of hydroxyapatite nanoparticles on germination and plant metabolism of tomato (Solanum lycopersicum L.): preliminary evidence. Agronomy 9. https://doi.org/10.3390/agronomy9040161
109. Abbasifar A, Shahrabadi F, ValizadehKaji B (2020) Effects of green synthesized zinc and copper nano-fertilizers on the morphological and biochemical attributes of basil plant. J Plant Nutr 43. https://doi.org/10.1080/01904167.2020.1724305
110. Dimkpa CO, Singh U, Adisa IO, et al (2018) Effects of manganese nanoparticle exposure on nutrient acquisition in wheat (Triticum aestivum L.). Agronomy 8. https://doi.org/10.3390/agronomy8090158
111. Talankova-Sereda TE, Liapina K V, Shkopinskij EA, et al (2016) The influence of Cu и Co nanoparticles on growth characteristics and biochemical structure of Mentha Longifolia in vitro. In: Springer proceedings in physics
112. López-Moreno ML, Avilés LL, Pérez NG, et al (2016) Effect of cobalt ferrite (CoFe2O4) nanoparticles on the growth and development of Lycopersicon lycopersicum (tomato plants). Sci Total Environ 550. https://doi.org/10.1016/j.scitotenv.2016.01.063
113. El-Hefnawy ME (2020) Biodegradable films from phytosynthesized TiO^2 nanoparticles and nanofungal chitosan as probable nanofertilizers. Int J Polym Sci 2020. https://doi.org/10.1155/2020/6727132
114. Dimkpa CO, Bindraban PS, Fugice J, et al (2017) Composite micronutrient nanoparticles and salts decrease drought stress in soybean. Agron Sustain Dev 37. https://doi.org/10.1007/s13593-016-0412-8
115. Bollyn J, Faes J, Fritzsche A, Smolders E (2017) Colloidal-bound polyphosphates and organic phosphates are bioavailable: a nutrient solution study. J Agric Food Chem 65. https://doi.org/10.1021/acs.jafc.7b01483
116. Zhu X, Liu W, Chen J, et al (2020) Reductions in water, soil and nutrient losses and pesticide pollution in agroforestry practices: a review of evidence and processes. Plant Soil 453. https://doi.org/10.1007/s11104-019-04377-3
117. Leonardi M, Caruso GM, Carroccio SC, et al (2021) Smart nanocomposites of chitosan/alginate nanoparticles loaded with copper oxide as alternative nanofertilizers. Environ Sci Nano 8. https://doi.org/10.1039/d0en00797h
118. Carmona FJ, Dal Sasso G, Ramírez-Rodríguez GB, et al (2021) Urea-functionalized amorphous calcium phosphate nanofertilizers: optimizing the synthetic strategy towards environmental sustainability and manufacturing costs. Sci Rep 11. https://doi.org/10.1038/s41598-021-83048-9
119. Yao Y, Yue L, Cao X, et al (2023) Carbon dots embedded in nanoporous SiO_2 nanoparticles for enhancing photosynthesis in agricultural crops. ACS Appl Nano Mater 6. https://doi.org/10.1021/acsanm.2c03843
120. Dimkpa CO, Deng C, Wang Y, et al (2023) Chitosan and zinc oxide nanoparticle-enhanced tripolyphosphate modulate phosphorus leaching in soil. ACS Agric Sci Technol 3. https://doi.org/10.1021/acsagscitech.3c00054
121. Rashid MI, Shah GA, Sadiq M, et al (2023) Nanobiochar and copper oxide nanoparticles mixture synergistically increases soil nutrient availability and improves wheat production. Plants 12. https://doi.org/10.3390/plants12061312
122. Moreno-Martín G, Sanz-Landaluze J, León-González ME, Madrid Y (2020) Insights into the accumulation and transformation of Ch-SeNPs by Raphanus sativus and Brassica juncea: effect on essential elements uptake. Sci Total Environ 725. https://doi.org/10.1016/j.scitotenv.2020.138453
123. Jampílek J, Kráľová K (2015) Application of nanotechnology in agriculture and food industry, its prospects and risks. Ecol Chem Eng S 22:321–361. https://doi.org/10.1515/eces-2015-0018
124. Nair R, Varghese SH, Nair BG et al (2010) Nanoparticulate material delivery to plants. Plant Sci 179:154–163. https://doi.org/10.1016/j.plantsci.2010.04.012

125. Sousa GFM, Gomes DG, Campos EVR, et al (2018) Post-emergence herbicidal activity of nanoatrazine against susceptible weeds. Front Environ Sci 6. https://doi.org/10.3389/fenvs.2018.00012
126. Namasivayam SKR, Aruna A (2014) Evaluation of silver nanoparticles-chitosan encapsulated synthetic herbicide paraquat (AgNp-CS-PQ) preparation for the controlled release and improved herbicidal activity against Eichhornia crassipes. Res J Biotechnol 9:19–27
127. Musante C, White JC (2012) Toxicity of silver and copper to Cucurbita pepo: differential effects of nano and bulk-size particles. Environ Toxicol 27:510–517. https://doi.org/10.1002/tox.20667
128. Atha DH, Wang H, Petersen EJ et al (2012) Copper oxide nanoparticle mediated DNA damage in terrestrial plant models. Environ Sci Technol 46:1819–1827. https://doi.org/10.1021/es202660k
129. Kim S, Lee S, Lee I (2012) Alteration of phytotoxicity and oxidant stress potential by metal oxide nanoparticles in Cucumis sativus. Water Air Soil Pollut 223:2799–2806. https://doi.org/10.1007/s11270-011-1067-3
130. Lee S, Chung H, Kim S, Lee I (2013) The genotoxic effect of ZnO and CuO nanoparticles on early growth of buckwheat, Fagopyrum Esculentum. Water Air Soil Pollut 224:1668. https://doi.org/10.1007/s11270-013-1668-0
131. McGrath MT (2004) What are fungicides? Plant Health Instruct. https://doi.org/10.1094/PHI-I-2004-0825-01
132. Krishnaraj C, Ramachandran R, Mohan K, Kalaichelvan PT (2012) Optimization for rapid synthesis of silver nanoparticles and its effect on phytopathogenic fungi. Spectrochim Acta A Mol Biomol Spectrosc 93:95–99. https://doi.org/10.1016/j.saa.2012.03.002
133. Saharan V, Mehrotra A, Khatik R et al (2013) Synthesis of chitosan based nanoparticles and their in vitro evaluation against phytopathogenic fungi. Int J Biol Macromol 62:677–683. https://doi.org/10.1016/j.ijbiomac.2013.10.012
134. Cruz-Luna AR, Vásquez-López A, Rojas-Chávez H et al (2023) Engineered metal oxide nanoparticles as fungicides for plant disease control. Plants 12:2461. https://doi.org/10.3390/plants12132461
135. Stehle S, Schulz R (2015) Agricultural insecticides threaten surface waters at the global scale. Proc Natl Acad Sci 112:5750–5755. https://doi.org/10.1073/pnas.1500232112
136. Rai M, Ingle A (2012) Role of nanotechnology in agriculture with special reference to management of insect pests. Appl Microbiol Biotechnol 94:287–293. https://doi.org/10.1007/s00253-012-3969-4
137. Guan H, Chi D, Yu J, Li X (2008) A novel photodegradable insecticide: preparation, characterization and properties evaluation of nano-Imidacloprid. Pestic Biochem Physiol 92:83–91. https://doi.org/10.1016/j.pestbp.2008.06.008
138. Stadler T, Buteler M, Weaver DK (2010) Novel use of nanostructured alumina as an insecticide. Pest Manag Sci 66:577–579. https://doi.org/10.1002/ps.1915
139. Vivekanandhan P, Swathy K, Thomas A et al (2021) Insecticidal efficacy of microbial-mediated synthesized copper nano-pesticide against insect pests and non-target organisms. Int J Environ Res Public Health 18:10536. https://doi.org/10.3390/ijerph181910536
140. Rahman A, Pittarate S, Perumal V, et al (2022) Larvicidal and antifeedant effects of copper nano-pesticides against Spodoptera frugiperda (J.E. Smith) and its immunological response. Insects 13:1030. https://doi.org/10.3390/insects13111030
141. Iavicoli I, Leso V, Beezhold DH, Shvedova AA (2017) Nanotechnology in agriculture: opportunities, toxicological implications, and occupational risks. Toxicol Appl Pharmacol 329:96–111. https://doi.org/10.1016/j.taap.2017.05.025
142. Servin A, Elmer W, Mukherjee A et al (2015) A review of the use of engineered nanomaterials to suppress plant disease and enhance crop yield. J Nanopart Res 17:92. https://doi.org/10.1007/s11051-015-2907-7
143. Chhipa H (2017) Nanofertilizers and nanopesticides for agriculture. Environ Chem Lett 15:15–22. https://doi.org/10.1007/s10311-016-0600-4

144. Wang P, Lombi E, Zhao F-J, Kopittke PM (2016a) Nanotechnology: a new opportunity in plant sciences. Trends Plant Sci 21:699–712. https://doi.org/10.1016/j.tplants.2016.04.005
145. Pradhan S, Mailapalli DR (2017) Interaction of engineered nanoparticles with the agri-environment. J Agric Food Chem 65:8279–8294. https://doi.org/10.1021/acs.jafc.7b02528
146. Bollag J-M, Myers CJ, Minard RD (1992) Biological and chemical interactions of pesticides with soil organic matter. Sci Total Environ 123–124:205–217. https://doi.org/10.1016/0048-9697(92)90146-J
147. Mohammadi MHZ, Panahirad S, Navai A, et al (2021) Cerium oxide nanoparticles (CeO2-NPs) improve growth parameters and antioxidant defense system in Moldavian Balm (Dracocephalum moldavica L.) under salinity stress. Plant Stress 1:100006. https://doi.org/10.1016/j.stress.2021.100006
148. Zuverza-Mena N, Martínez-Fernández D, Du W et al (2017) Exposure of engineered nanomaterials to plants: insights into the physiological and biochemical responses-a review. Plant Physiol Biochem 110:236–264. https://doi.org/10.1016/j.plaphy.2016.05.037
149. Dimkpa CO, Bindraban PS (2018) Nanofertilizers: new products for the industry? J Agric Food Chem 66:6462–6473. https://doi.org/10.1021/acs.jafc.7b02150
150. Freitas DC, de Andrade AM, da Costa LF et al (2021) There is plenty of room at the plant science: a review of nanoparticles applied to plant cultures. Ann Appl Biol 178:149–168. https://doi.org/10.1111/aab.12640
151. Kah M, Kookana RS, Gogos A, Bucheli TD (2018) A critical evaluation of nanopesticides and nanofertilizers against their conventional analogues. Nat Nanotechnol 13:677–684. https://doi.org/10.1038/s41565-018-0131-1
152. Makarenko NA, Makarenko VV (2019) Nanotechnologies in crop cultivation: ecotoxicological aspects. Biosyst Divers 27:148–155. https://doi.org/10.15421/011920
153. Feregrino-Perez AA, Magaña-López E, Guzmán C, Esquivel K (2018) A general overview of the benefits and possible negative effects of the nanotechnology in horticulture. Sci Hortic 238:126–137. https://doi.org/10.1016/j.scienta.2018.03.060

Nanostimulants to Manage Environmental Stresses in Crops

Hipólito Hernández-Hernández, Mari Carmen López-Pérez, and Fabián Pérez-Labrada

Abstract Environmental stresses negatively affect crop development, and these scenarios induce an oxidative burst in the plant that can lead to cell death. Nanostimulation is a window of opportunity to promote plant tolerance. Nanostimulants (NSs), such as nanochitosan, nanoparticles, algae and plant nanoextracts and nanobiostimulants can be applied to the seed and/or plant (via foliar or edaphic routes) and enter the plant differentially and interact with cellular structures, or they can be mobilized by vascular bundles. These NSs activate signaling processes such as the promotion of reactive oxygen species (ROS), antioxidant defense systems (enzymatic and non-enzymatic), synthesis of secondary metabolites, and transcript modifications. The present chapter highlights the main results of the use of NSs in the promotion of abiotic stress tolerance and discusses the mechanisms that are promoted in plants under different stressful environments. NSs are an emerging strategy to promote abiotic stress tolerance in diverse crops; their response depends on the type, size, and dose of the nanomaterial, as well as stress conditions. Future studies should be conducted to deepen and standardize the mechanisms of action of NSs and to verify their impact on plant conditions under different stresses simultaneously.

Keywords Nanomaterials · Nanoparticles · Nanochitosan · Antioxidant system · Reactive oxygen species

H. Hernández-Hernández
Institute of Agroengineering, University of Papaloapan, Loma Bonita, Oaxaca, Mexico

M. C. López-Pérez
Navojoa Academic Unit, Sonora State University, Navojoa, Sonora, Mexico

F. Pérez-Labrada (✉)
Department of Botany, Autonomous Agrarian University Antonio Narro, Saltillo, Coahuila, Mexico
e-mail: fabian.perezl@uaaan.edu.mx

© The Author(s), under exclusive license to Springer Nature Singapore Pte Ltd. 2025

A. Juárez-Maldonado et al. (eds.), *Plant Biostimulation with Nanomaterials*, Smart Nanomaterials Technology, https://doi.org/10.1007/978-981-96-4648-7_11

1 Introduction

Plants converge daily with environmental conditions that influence their growth; however, when environmental conditions exceed the permissible limits that plants can tolerate, they induce alterations in their physiology and metabolism [1]. Abiotic stresses (water deficit, drought, salinity, alkalinity, temperature, irradiance, heavy metal toxicity, and microplastics, among others) can cause oxidative bursts (high production of ROS such as hydrogen peroxide and H_2O_2), causing alterations in the functioning of cells and oxidative damage 2]. If the enzyme system is not adequate, the tolerance capacity of plants to stress is significantly reduced [3], reducing the growth, development, yield, and quality of the final products of agricultural crops, depending on the magnitude and time of stress exposure [4].

The main damage caused by stress is the increase in lipid peroxidation of the cell membrane and the oxidation of cellular components and proteins, derived from the accumulation of ROS and malondialdehyde (MDA). In the face of these perturbations, plants activate physiological, biochemical, and transcriptional mechanisms [5] with the intention of mitigating damage. The main mitigation strategies are the adjustment of osmotic (Ψ_s) and hydric (Ψ) potentials by the synthesis, accumulation, and/or mobilization of osmolytes (proline, glycine, betaine glycine, and free amino acids), repair of the electron transport system, especially photosystem II (PSII), activation of enzymatic and non-enzymatic antioxidant systems, catalase (CAT), peroxidase (POD), superoxide dismutase (SOD), ascorbate peroxidase (APX), glutathione reductase (GR), guaiacol peroxidase (GPx), ascorbate (AsA), and glutathione (GSH), among others, in addition to lipoxygenases (LOX), which minimize the impact of oxidative stress indicators by reducing lipid peroxidation.

Sometimes, the mechanisms to cope with abiotic stress may be insufficient, and stimulation is a feasible strategy because it promotes one or several characteristics in the spermosphere, plant, rhizosphere, and/or phyllosphere, inducing tolerance to abiotic stress [6, 7]. Some stimulants include humic substances and complex organic materials, beneficial elements, protein hydrolysates, amino acids and N-containing compounds, algal extracts and botanicals, beneficial microorganisms, biopolymers (chitosan and lignin), and inorganic compounds [8, 9]. However, the bulk use of these stimulants is sometimes ineffective [10] and expensive. Therefore, their application in nanoforms (nanostimulants, NSs) maximizes tolerance to abiotic stresses compared to their bulk form owing to their nanometric characteristics [11, 12], which alter the holobiont (Fig. 1) and promote a rapid response.

As a novel and innovative alternative NSs (1–100 nm) can help in promoting food security in the face of strong abiotic stresses, since their application induces a series of ROS signaling, kinases, and transcription factors that positively alter certain morphological, physiological, biochemical, and molecular parameters of the plant owing to their broad reactivity and higher surface-to-volume ratio [13, 14]. Consequently, plants show more efficient mechanisms to cope with stress, thanks to the harmonization of the oxidative burst, antioxidant defense systems (enzymatic and non-enzymatic) are activated and/or exacerbated, promoting the redox balance

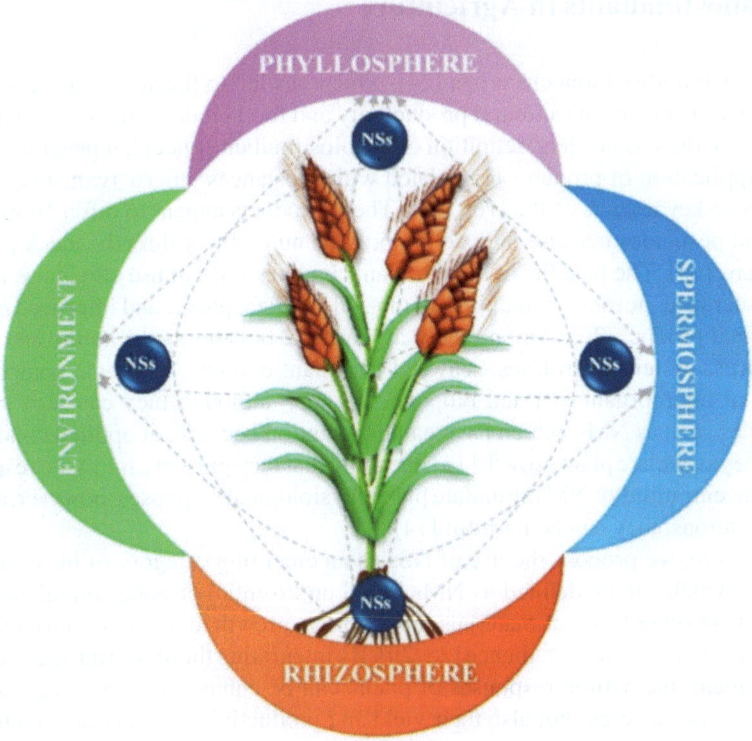

Fig. 1 The comprehensive response of crop nanostimulation under abiotic stress; to cope with environmental conditions, the use of NSs in crops promotes alterations in the plant by modifying it endogenously as well as by promoting association patterns in the rhizosphere, phyllosphere, and/or spermosphere

and improving the photosynthetic process (CO_2 assimilation). They can also increase the synthesis and concentration of plant hormones and secondary metabolites, which contribute to the promotion of tolerance to abiotic stress.

The use of NSs in crops under abiotic stress conditions can stimulate metabolic processes (primary and secondary) by inducing a series of biomolecules that can act as protectors and regulators of environmental stress [15, 16]. Several studies corroborate the positive and encouraging impact of these NSs in plants under different environmental stresses [17–20], in general, an elimination or balance in ROS content avoiding lipid peroxidation and protecting cell membrane integrity [21], promoted by the plant defense system [22, 23].

In the present chapter, we analyze how the use of NSs affects the induction of abiotic stress tolerance. Initially, we start from the concept of stimulation and the main biostimulants, and then approach it from the nanoscale, analyzing the main mechanisms by which NSs interfere in the plant under stressful conditions.

2 Nanostimulants in Agriculture

Weather variability threatens crops [24], which has led to the search for innovative alternatives to ensure agricultural productivity and food security, such as biostimulation. While there is no clear definition of the biostimulant concept, it generally refers to the application of products formulated with substances, microorganisms, or ions to improve key aspects of the crop [25]. These products appear to differ from fertilizers and pesticides because they do not provide nutrients or directly attack pests or pathogens [25]. The benefits of biostimulants include nutrient use efficiency, abiotic stress tolerance, nutrient availability in the soil or rhizosphere, and improved quality of the final product [26]. Among these agricultural biostimulants, humic and fulvic substances, protein hydrolases, seaweed and plant extracts, inorganic compounds, beneficial microorganisms, and biopolymers [27], although they can also include molecules such as NMs, which have gained attention because, at optimal concentrations, they stimulate plant growth [13], meaning that they present a hormetic response. Low concentrations of NMs stimulate plant physiological responses; however, at high concentrations, they can be harmful [14].

Therefore, we propose the use of NSs as an emerging category of biostimulants (Fig. 2), which can be defined as NMs based on essential or nonessential elements for plants or other types of materials that improve growth, efficiency, nutrient availability, and abiotic stress resilience [28, 29]. By integrating these NSs into agricultural management, the natural responses of plants can be enhanced, improving not only their resilience to stress but also their yield and productivity in scenarios of climatic disturbances.

2.1 Nanostimulants Characteristics

NSs have unique characteristics that confer stimulatory properties, which are dependent on factors such as concentration, size, area-to-surface ratio, morphology, and surface charge [28]. Generally, these materials can exert biostimulant effects on crops at low concentrations [13, 30]. Among these factors, particle size is crucial for crop stimulation, with a typical range between 1 and 100 nm. Recent research has shown particular interest in studying NPs of different sizes, highlighting that those between 1 and 40 nm have greater ease in penetrating plant tissues, whereas those between 41 and 100 nm present greater difficulty in entering cells [31–33]. This difference is key because it directly influences the degree of biostimulation that can be achieved. In addition, it should be noted that smaller NSs possess a larger specific surface area, which favors greater adhesion to plant surfaces [34]. However, the morphology of the nanomaterial affects the area-to-surface ratio and contact angle with plant tissues, influencing its absorption. For example, in *Nicotiana benthamiana* leaves, rod-shaped gold NPs are absorbed and internalized more efficiently than spherical NPs [35]. In terms of surface charge, both positively and negatively charged NSs

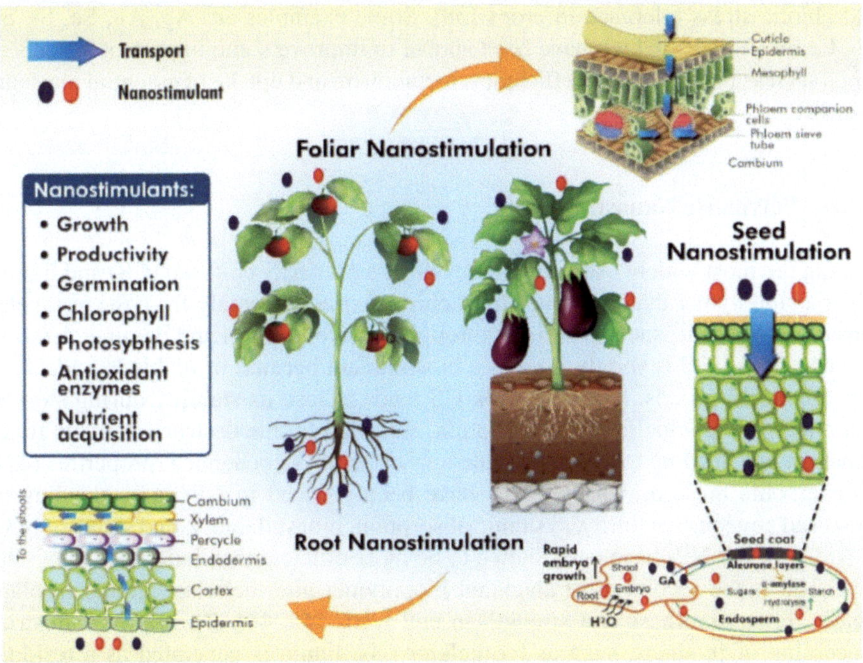

Fig. 2 Nanostimulant-induced responses in plants. The application of NSs by foliar, edaphic (root), or seed application promotes differentiated responses in plants, these NSs can interact directly with cells or be transported to different organelles where they induce a specific response, stimulating mechanisms such as photosynthetic activity and enzymatic activity

can be adsorbed by plants; however, positively charged NSs tend to adsorb more intensively because of the electrostatic attraction generated by the negative charge of the cell walls (13, 36, 37].

2.2 *Main Nanostimulants*

NSs can be categorized into four major groups: inorganic, polymeric, plant extracts (green synthesis), and nanomaterials combined with conventional biostimulants (nano-biostimulants). A general description of each group is provided below.

2.2.1 Inorganic Nanostimulants

Inorganic NSs include metallic, metal-oxide, metalloid, metalloid, and nonessential magnetic metal NPs for plants that have the ability to promote growth. These types of NPs are among the most studied in the field of agriculture for increasing productivity and abiotic stress tolerance in crops [38]. Some examples are Ag, Au, Se, Si, Si, Co, Ce and Ti NPs, which have been shown to improve some important aspects in crops such as photosynthetic efficiency, metabolism and uptake of essential nutrients [38–40].

2.2.2 Polymeric Nanostimulants

Among the most widely used polymers, NSs are primarily chitosan (CS) and lignin. CS is a biopolymer that originates from chitin, extracted mainly from the exoskeletons of crustaceans, such as shrimp, through deproteinization and demineralization processes [41]. CS is widely used as a biostimulant because of its biodegradability and functional groups. However, bulk CS tends to lose its stability during storage and undergoes gradual chain degradation, followed by the destruction of its functional groups, leading to an irreversible loss of its physicochemical properties [42]. To overcome these drawbacks, NPs have been reduced in NPs sizes, leading to increased aqueous solubility, systemic absorption, bioavailability, and stability [43]. In addition, CS-NPs have been shown to be more efficient than bulk CS [44].

Lignin is the second most abundant biopolymer after cellulose, found in plant tissues, giving stiffness and strength to cell walls [45, 46]. During the industrial processing of products, such as biofuels or oils, lignin is generated as a residual by-product. Instead of being discarded, it can be valorized and reused in various fields such as agriculture, where it has shown promising applications, especially as a biostimulant [47, 11]. However, its relatively large particle size, together with its irregular and heterogeneous structure, have limited its efficacy in some agricultural applications. To overcome these limitations, lignin was modified and used in the form of NPs, resulting in the creation of more efficient agricultural NSs [48].

2.2.3 Nanostimulants of Plant and Algae Extracts

To reduce environmental pollution, NMs are being developed using plant or seaweed extracts, a process known as green synthesis, or by using these extracts directly in their nanoform. The use of plant extracts for the green synthesis of NPs tends to be faster than the use of microorganisms, such as bacteria and fungi [49]. The extraction process in plants consists of collecting the part of interest (leaf, stem, root, flower, and fruit), rinsing it with distilled water, and drying it at room temperature or in a drying oven. Subsequently, it underwent boiling, and the resulting solution was filtered to remove insoluble particles [50, 49]. The process of extracting the seaweed extract was similar to that of the plants [51]. The extract was then used as a biological reducing agent, which, in combination with ionic solutions such as silver or gold, resulted in the formation of NPs [52, 51]. During the synthesis process, factors such as the type of plant or seaweed, extract concentration, temperature, metal salt, and pH must be considered [49, 51]. Once NPs are synthesized, they can be used as NSs in crops, among which Ag, Au, Se, Ti, Si NPs are the most important,

2.2.4 Nano-Biostimulants

This group refers to the combination of NMs with conventional biostimulants such as humic and fulvic acids and beneficial microorganisms. Examples include CS-NPs in combination with fulvic acid, which is used to stimulate growth and biomass accumulation in maize and rice plants under drought stress [53, 54]. In addition, fulvic acid is used to improve the stability of NPs-ZnO and promote the growth of mung beans [55]. Humic substances are used to functionalize NP-hydroxyapatite and improve the growth and productivity of maize [56]. In the particular case of microorganisms, the co-application of fungi (arbuscular mycorrhizae and *Trichoderma*) and growth-promoting bacteria with engineered NMs result in a promising strategy to improve nutrient uptake, crop growth and yield, and crop resilience to abiotic stress [57–59]. This is because NMs can enhance the efficiency of beneficial fungi and bacteria. However, these combinations are in their infancy and require critical evaluation.

2.3 Nanostimulant-Plant Interaction

NSs can be applied via foliar, soil, or seed application. In the leaves, NSs enter through the cuticle or stomatal pores and are then transported by apoplastic and symplastic pathways until they reach the phloem and are then translocated to shoots, flowers, fruits, and roots [34]. By contrast, in the root, NSs can enter through pores, carrier proteins, endocytosis, and plasmodesmata. Subsequently, they are transported through apoplastic and symplastic pathways until they reach the root xylem, and are then transported long distances to the leaves 60]. In seeds, NMs have the ability to pass through the testa or coat, with entry being easier in seeds that have coats with large intercellular spaces. This induces faster water uptake, activation of gibberellic acid biosynthesis, and hydrolysis of starch into soluble sugars, which are exploited for rapid embryo growth, thus enhancing the germination process [61]. As mentioned above, smaller NPs can enter plant cells more easily, and larger NPs can remain attached to the plant surface. This is an advantage of NSs over ionic biostimulants, as it has been shown that ions tend to leach more rapidly from plant surfaces [34]. In addition, NPs attached and adsorbed on leaves, roots, or seeds may undergo transformations due to physical or chemical changes, such as oxidative dissolution, chlorination, sulfidation, and complexation with organic ligands [62]. For example, NPs-ZnO is gradually transformed to Zn^{2+} ions owing to oxidative dissolution [32], which could represent a slow and controlled release of zinc.

Regardless of the route of NS application, they cause positive changes that help improve crop growth and yield under both normal and abiotic stress conditions. These NSs act at the molecular and cellular levels in a very precise manner, improving nutrient uptake and utilization and activating defense mechanisms to counteract the effects caused by different types of abiotic stresses [28, 29]. One of the key actions of NSs is the activation of ROS, which function as signaling molecules. This leads to post-translational oxidative modifications of stress-related kinases, hormones, and transcription factors [29]. In addition, NSs stimulate chlorophyll biosynthesis and photosynthetic activity, modulate oxidative stress by activating antioxidant enzymes, and enhance stomatal opening, facilitating gas exchange and increasing the rate of CO_2 assimilation. They also promote the activity of enzymes, such as amylase, nitrate reductase (NR), phosphatase, phytase, and carbonic anhydrases (CA), all of which are involved in metabolism and nutrient acquisition [63, 22].

3 Nanostimulants as Stress Tolerance Inducers

Owing to the nanometer scale, NSs can enhance their interaction with cells and maximize signaling and response to environmental stimuli, promoting differential responses that reduce the damage caused by abiotic stress (Fig. 3). In addition, NSs allow a reduction in application rates compared to the bulk forms.

Several studies have shown positive responses to the application of NSs (mainly derived from complex organic materials, algal and plant nanoextracts,

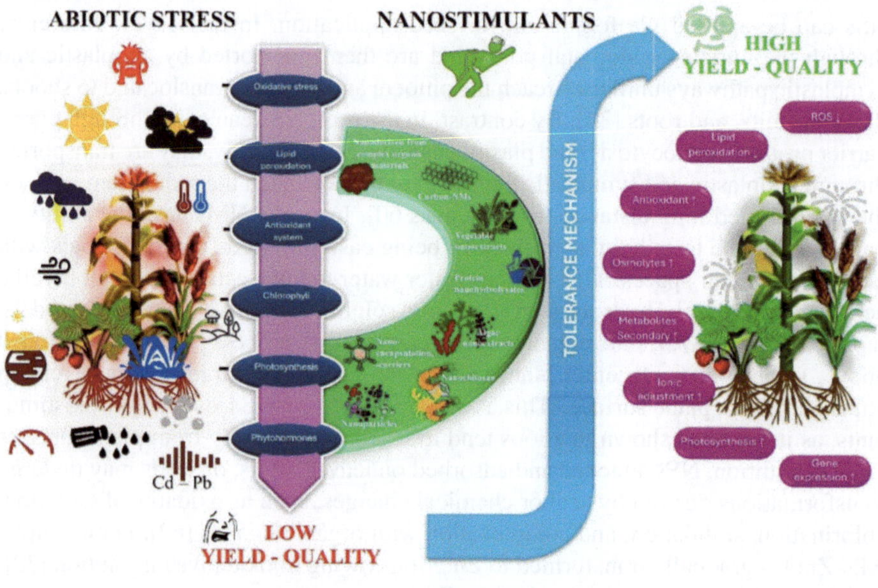

Fig. 3 Plant responses to abiotic stress and tolerance induction by nanostimulants. When abiotic stress conditions exceed the level of plasticity of plants, an oxidative burst occurs, causing lipid peroxidation, which is usually accompanied by a disruption in the photosynthetic process, stomatal closure, ionic and osmotic imbalance, and alterations in hormonal balance, reducing crop yield and quality. The use of NSs promotes the antioxidant system and synthesis of secondary metabolites by suppressing the load of oxidative radicals, thus minimizing damage to cell membranes and maintaining an adequate photosynthetic process

protein nanohydrolysates, nanoparticles, carbon NMs, nanochitosan, and nano-biostimulants) that differentially alter abiotic stress tolerance in crops depending on the dose, application route, type of NSs, timing of application, and stress level of the plant (Table 1).

3.1 Nanostimulants Derived from Complex Organic Materials

Examples of NSs include agro-industrial and urban waste, sewage sludge extracts, compost and manure, and biochar or humic substances in their nanometric form. In this sense, applications of nano-vermicompost (100 mg kg^{-1} of soil) palliate damage in tomatoes under drought stress by improving chlorophyll content, osmolytes (proline, glycine betaine, and sugars), photosynthesis, PSII activity, and NR activity. In conjunction with the reduction of ROS, stabilization of the cell membrane allows the metabolism of nitrogen, P, and K^+, improving dry matter accumulation under abiotic stress [130]. In wheat, the use of nanobiochar (0.5, 1.00, 1.5%) induced higher POD, SOD, CAT, and APX activity, as well as a promotion of Ψ_s mitigating drought stress [131], and NSs increased leaf area due to the stimulation of carbohydrate, proline, and chlorophyll content [132]. Tolerance by this type of NSs has also been reported in *Spinacia oleracea* L. (under salinity and drought) by stimulating the net assimilation rate, leaf area, photosynthetic pigments, and nutrient content, in addition to increasing CAT, POD, and SOD enzyme activity, in line with lower lipid peroxidation and electrolyte leakage [133]. Nanobiochar, which has greater penetration into the cell, probably enhances the responses by stimulating the proton pump (H^+-ATPase) in the cell membrane, promoting a signaling response that allows morphophysiological improvements in the cultures.

3.2 Nanostimulants from Seaweed and Plant Extracts

Seaweeds (organisms rich in biocompounds such as polysaccharides, antioxidants, hormones, and macro- and micronutrients) promote abiotic stress tolerance and can enhance the response by improving crop production [134, 135]. The use of nanoextracts of *Marrubium alysson* and *Torilis arvensis* (containing NPs-Ag) to imprint *Triticum aestivum* L. seeds showed increased CAT and GPX activity, soluble proteins, phenolic compounds, flavonoids, soluble sugars, and starch, which promoted tolerance to drought conditions. This response may be due to the fact that these compounds act as osmoprotectors of cellular tissues, allowing the establishment of an osmotic balance and reducing oxidative stress caused by H_2O_2 and MDA [65]. The authors pointed out that phenolic and flavonoid compounds play an important role in radical scavenging as they can be used to synthesize lignin and flavolignans. They also

Table 1 Responses to the promotion of abiotic stress tolerance by the application of nanostimulants to crops

Nanostimulant	Abiotic stress	Response	References
Nano-water hyacinth protein	Common bean seedlings under cadmium pollution	↑ Antioxidant activity ↓ Lipid peroxidation and H_2O_2	[64]
Nano extracts of *Marrubium alysson* and *Torilis arvensis* (containing NPs-Ag)	*Triticum aestivum* L. seeds under drought stress	↑ CAT, GPX, soluble proteins, phenolic compounds, flavonoids, soluble sugars, starch, flavolignan and osmotic balance ↓ H_2O_2 and MDA	[65]
Nano-biomass *Carthamus oxyacantha* (<250 ppm)	Wheat plants under rainfed condition	↑ Growth rate and grain yield	[66]
NPs-ascorbic, benzoic, salicilic acid (0.50 y 0.75 mM)	*Solanum tuberosum* L. under drought stress	↑ Seedling length, # leaves, roots and lateral branches per plant	[67]
NPs-Mg (0.3–0.5 g L^{-1})	*Achillea millefolium* L. plants under drought stress	↑ Terpenoids ↓ Electrolyte leakage	[68]
NPs-Zn (0.1%)	*Brassica rapa* L. plants under drought stress	↑ Clorophyll, SOD, POD, CAT and osmotic balance	[69]
NPs-Fe_2O_3 (75 mg L^{-1})	*Zea mays* L. seeds under drought stress	↑ Ψ_w, water use efficiency, chlorophyll, total flavonoids, AsA, SOD and POD ↓ MDA and H_2O_2	[70]
NPs-SiO_2 (100 mg L^{-1})	*Rosa damascena* Mill plants under drought stress	↑ Chlorophyll, biomass, carbohydrates, flavonoids, total phenols, anthocyanins, P, K, Mn, Fe, Zn and Cu uptake	[71]
Oleylamine-coated calcium hydroxide-NPs (15 and 30 mg L^{-1})	*Lycopersicon esculentum Mill. cv Galli* plants under high photosynthetic photon flux density	↑ Φ_{PSII}, q*p*, Fv'/Fm', electron transport rate ↓ ROS	[72]
Humic acid (2 kg $tree^{-1}$) + NPs ZnO_2 (300 mg) + TiO_2 (70 mg)	Ten-year-old Picual olive trees under salinity stress	↑ Clorophyll, fruit set percentage and fruit quality, N, P and K in the leaf	[73]

(continued)

Table 1 (continued)

Nanostimulant	Abiotic stress	Response	References
NPs-Si (rice husk; 0,15%)	*Zea mays* L. plants under drought stress	↑ Clorophyll, relative water content, SOD ↓ Electrolyte leakage and MDA	[74]
Nano-SiO_2 (80 mg L^{-1}) and mycorrhiza	*Triticum* under late-season drought	↑ Yield grain and regulation of stomatal conductance	[75]
NPs-SiO_2 (150 y 300 mg L^{-1})	Maize seedlings under low light stress	↑ SOD, NR, AsA	[76]
NPs-SiO_2 (0.4 g L^{-1})	Rice plants under water deficit conditions	↑ Clorophyll, carotenoids, CAT, POD	[77]
NPs-SiO_2	*Ehretia macrophylla* plants under drought stress	↑ Auxin signal transduction, MKP signaling gene expression, fatty acid and α-linolenic acid metabolism ↓ MDA and H_2O_2	[29]
Zinc nanochelate	Wheat genotypes under water deficit stress	↑ Clorophyll, sugars, CAT, SOD, cytokinin y auxins ↓ ABA and MDA	[78]
Nano-nitrogen fertilizers	Guinea grass plants grown under water deficiency	↑ Macro and micronutrients, total carbohydrates, crude protein, GA_3, ABA, SOD and POD, stomatic conductance	[79]
Arbuscular mycorrhiza fungus plus NPs-ZnO and NPs-Se	Chili pepper under cold-stress conditions	↑ AsA, capsaicin, POD, thickness of grana thylakoids and chloroplasts ↓ H_2O_2 and MDA	[80]
Nano-hydroxyapatite	*Coriandrum sativum* L. growth under Cd, Pb stress	↑ Dry biomass of shoots ↓ Cd and Pb uptake	[81]
NPs-SiO_2	*Spinacia oleracea* L. plants under heavy metal–contaminated soil	↑ Photosynthetic pigments, total soluble sugars, free proline, membrane stability, AsA, relative water content, α-tocopherol, glycine betaine, POD, GR, CAT, SOD and APX ↓ Pb y Cd uptake, MDA, $O_2^{\bullet-}$, H_2O_2, electrolyte leakage, lipid peroxidation	[82]
NPs-Si	Rice under water deficit	↑ CAT, POD, proline	[83]
NPs-TiO_2 (50 µg mL^{-1})	Wheat under salinity stress	↑ Chlorophyll, free amino acids, soluble sugar, proline, SOD and POD	[84]

(continued)

Table 1 (continued)

Nanostimulant	Abiotic stress	Response	References
NPs-Si (60 ppm)	Wheat under drought stress	↑ CAT, SOD, chlorophyll, proline, relative water content, *DREB2*, *MYB33m* *MYB3R*, *WRKY 19* and *SnRK 2.4* gene expression ↓ H_2O_2 and MDA	[85]
Nano-fertilizers (MgO-4.0%, CaO-2.0%, Ca-36.0%, Fe-0.02%, Mn-0.01%, Zn-0.002%)	Maize plants under cold stress (5–6 °C)	↑ Effective quantum yield of photosystem II, maximum photosynthetic efficiency of PSII and electron transport rate, maximum net photosynthetic rate	[86]
Nano-CeO_2 (0.1 mg L^{-1})	*Nicotiana tabacum* L. plants under salt stress	↑ *IPUT1*, *SOS3*, *SOS2*, *SOS1* and *NHX1* gene expression	[87]
Poly(acrylic) acid-coated nanoceria (8.36 nm)	Rapeseed plants under salinity stress	↑ Membrane integrity, *Cu–Zn SOD-I* and *Cu–Zn SOD-II* gene expression ↓ *LOX-IV* gene expression	[88]
Carbon dots (<20 nm; 5 mg L^{-1})	*Lactuca sativa* L. plants under cold stress (4 °C by 12 h)	↑ CAM, CRLK and MAPK ↓*CBF 8*, *CBF 12* and *COR 413* gene expression	[89]
NPs-Zn (20 ppm)	Wheat plants under salinity stress	↑ Photosynthetic rate, free amino acids, GA_3, IAA, Ca^{2+}, K^+, Mg^{2+}, N and P uptake ↓ ABA, MDA, H_2O_2, electrolyte leakage, Na^+ and Cl^- uptake	[90]
NPs-TiO_2 -ZnO (10 mg L^{-1})	Wheat plants under water stress	↑ Photosynthesis and osmoprotectants	[91]
Nano-selenium	*Capsicum annuum* L. plants under Cd stress	↑ Synapyl alcohol, p-coumaryl alcohol, caffeoyl alcohol, coniferaldehyde, brassinolide, ABA. JA, SeMet and MeSeCys, *PAL, CAD, 4CL, COMT*, *BZR1*, *LOX3* and *NCDE1* genes expression	[92]
NPs-SiO_2 (50 ppm)	*Pisum sativum* L. var. 'Respect' under drought stress	↑ Relative water content, CAT, APX, SOD, GR, antioxidant capacity and total phenolic ↓ H_2O_2 and lipid peroxidation	[23]

(continued)

Table 1 (continued)

Nanostimulant	Abiotic stress	Response	References
NPs-ZnO y -Se (25 μM)	*Glycine max* L. plants under AsV stress	↑ GSH, AsA, CAT, SOD, POD, GR, APX, *GmSOD*, *GmPOD*, *GmCAT*, *GmAPX*, *GmGR*, *GmGST* and *GmP5CS* gene expression ↓ As, oxidized glutathione, ROS and MDA	[93]
NPs-SiO_2 (25 y 50 mg L^{-1})	Gerbera plants under salt stress	↑ APX, GPx, Ca^{2+} and K^+ uptake ↓ Na^+ absorption, electrolyte leakage, MDA and H_2O_2	[94]
Nano-silicon (1 mM)	Sugar beet plants under water stress conditions	↑ Chlorophyll, net photosynthesis, glycine betaine, quercetin, rutin, SOD, CAT and GPx ↓ H_2O_2 and MDA	[95]
Nano-TiO_2	*Vicia faba* L. plants under water stress	↑ SOD, POD, CAT, APX, GR, NR, AC, proline and glycine betaine ↓ ROS, electrolyte leakage and lipid peroxidation	[96]
NPs-CeO_2 (100 y 500 mg L^{-1})	Rice plants under low nitrogen (LN) and high nitrogen (HN) stress	↑ N uptake, GS, GOGAT, GDH, proline, AG_3, ABA, zeatin riboside and indole-3-acetic acid ↓ N uptake, GS and GDH, oxidative damage	[97]
NPs-Zero-valent iron	Tomato plants under Cd stress	↑ *SlEFR1* gene expression ↓ Cd uptake and accumulation	[98]
NPs-SiO_2, -CeO_2, -ZnO y -S (100 y 500 mg kg^{-1})	*Brassicca rapa chinensis* plants under As stress	↑ SOD y POD ↓ As on soil pore water and leaf tissue	[99]
Fe nanocomplexes (from almond and pomegranate)	Almond tree plants under alkaline stress	↑ Clorophyll and Fe recovery, CAT ↓ Oxidative stress	[100]
NPs–CeO_2 (50 mg L^{-1})	*Vitis vinifera* L., cv. Superior Seedless under drought	↑ Photosynthesis, *VvCLH1*, *VvCu/ZnSOD*, *VvRD29A* and *VvRBOHA* gene expression ↓ MDA and H_2O_2	[101]
NPs–ZnO and nanohidroxiapatita	Wheat plants under Cd stress	↑ SOD, CAT and POD ↓ Cd and MDA	[102]

(continued)

Table 1 (continued)

Nanostimulant	Abiotic stress	Response	References
NMs-SiO_2	Bitter gourd under Cd stress	↑ Antioxidant activity ↓ MDA and Cd	[103]
NPs-Fe_3O_4 (15 y 30 mg L^{-1})	*Brassica oleracea* var. *italica* plants under Cd stress	↑ Chlorophylls, carotenoids, soluble sugars, proteins y NR ↓ Electrolyte leakage, lipid peroxidation, MDA and H_2O_2	[104]
Nano-SiO_2 (5–40 mg L^{-1})	Barley under Cd stress	↑ Chlorophylls, proteins, net photosynthetic rate ↓ Cd and MDA	[105]
NPs-iron oxide (25, 50, 100 mg kg^{-1} soil)	Cherry tomatoes under saline-alkaline stress	↑ Chlorophylls, carotenoids, photosynthetic efficiency, osmoregulators, *SlF3H*, *SlF´3H*, *SlF3´5´H*, *SlFLS*, *SlDFR* and *SlANS* gene expression ↓ NPQ, H_2O_2, O_2^- and MDA	[106]
NPs-silica (60 ppm)	Wheat plants under drought stress	↑ *ABC1*, *Wdhn13*, *CHP*, y *EXP2* gene expression	[107]
NPs-TiO_2 (20 y 40 mg L^{-1})	Plantas de garbanzo bajo estrés por sequía	↑ APX, POD, SOD, total phenols, DPPH, proline, chlorophyll and carotenoids ↓ MDA	[108]
Nanosilica (2 mM L^{-1})	Scarlet eggplant under P toxicity	↑ C:N homeostasis, phenols and AsA, efficiency in the use of Fe, Cu, Zn ↓ Electrolyte leakage, oxidative stress	[109]
NPs-SiO_2 (25–100 mg L^{-1})	*Solanum lycopersicum* plants under Cd stress	↑ AsA, GSH, SOD, proline, GABA, *GR* and *GST* genes expression ↓ H_2O_2 and MDA	[110]
NPs-Ca^{2+} (100 mg L^{-1})	*Brassica napus* L. plants under drought stress	↑ PSI and PSII efficiency, SOD, POD, CAT, APX, protease, LOX, proline, proteins ↓ H_2O_2 and MDA	[111]
NPs-ZnO (75 y 100 mg L^{-1})	*Cucumis melo* L. plants drought stress	↑ *SOD*, *POD*, *CAT*, *APX*, *DREB2D* and *DREB3* genes expression, sugars and proteins ↓ Oxidative stress	[112]

(continued)

Table 1 (continued)

Nanostimulant	Abiotic stress	Response	References
NPs-iron (1.5 y 3 mg L^{-1})	Canola plants under drought stress	↑ Ψs regulation, proline, soluble sugar, phenol, flavonoids, CAT, PPO, succinate dehydrogenase and aconitase ↓ Lipid peroxidation	[113]
NPs-CuO (5 μM)	Maize plants under salinity stress (5 dS m^{-1})	↑ Photosynthetic parameters, N, P, K, Ca and Mg	[114]
CS-NPs (20 mg L^{-1})	*Vicia faba* plants under drought stress	↑ IAA, osmolytes, phenolic compounds, GSH, α-tocopherol, CAT, POX, SOD, ASP and GR ↓ H_2O_2, MDA and LOX	[115]
CS-NPs-NO (50 μM)	*Glycine max* L. under drought stress	↑ Proline, CAT, AsA. POD, *GmDREB*$_{1a}$, *GmP5CS*, *GmDEFENSIN*, *GmGSNOR*$_1$ and *GmNOX*$_1$ genes expression ↓ Electrolyte leakage, MDA and H_2O_2	[5]
Fulvic acid releasing CS-NPs (0.1 mM)	Rice under drought stress	↑ Chlorophyll, soluble protein, APX, *OsDREB1A*, *OsCIPK3* y *OsNAC77* genes expression ↓ H_2O_2 and MDA	]53]
CS-NPs-fulvic acid (100 μM)	Maize plants under drought stress	↑ Chlorophyll, APX, *ZmDREB1A*, *ZmbZIP1*, *ZmNAC28* and *ZmCIPK3* genes expression ↓ Electrolyte leakage, MDA and H_2O_2	[54]
CS-loaded NPs encapsulated in glycine betaine (20 μg)	Maize plants under drought stress (20%)	↑ Photosynthetic pigments, CAT, SOD, APX, proline, *ZmAPX*, *ZmCAT* y *ZmSOD* genes expression ↓ MDA	[116]
Nano-Si encapsulated in nano-CS	*Triticum aestivum* L. plants under salinity stress	↑ Antioxidant activity., redox balance, osmolytes ↓ Lipid peroxidation and H_2O_2	[117]
CS-NPs	*Solanum lycopersicum* L. plants under Cd stress	↑ CAT, POD, SOD, GSH, AsA and proline ↓ Cd uptake and accumulation. H_2O_2 and MDA	[118]
CS-NPs (60 y 90 ppm)	Barley plants under late season drought stress	↑ CAT, SOD and proline	[119]

(continued)

Table 1 (continued)

Nanostimulant	Abiotic stress	Response	References
CS-NPs incorporated with salicylic acid	*Gossypium hirsutum* plants under heat stress	↑ RuBisCO, invertase and sucrose phosphate synthase, sucrose and sucrose translocation	[120]
CS-NPs	*Phaseolus vulgaris* L. under salinity stress	↑ Photosynthetic pigments, carbohydrates, proline, CAT and APX ↓ Lipid peroxidation, electrolyte leakage and H_2O_2	[121]
CS-NPs (90 ppm)	Wheat plants under drought stress	↑ Photosynthetic rate. CAT and SOD	[122]
CS-NPs (300 a 500 mg L^{-1})	*Zea mays* L. under water stress	↑ Ψs, phenols and flavonoids, CAT, POD and SOD	[123]
CS-NPs (100 ppm)	*Phaseolus vulgaris* L. plants under drought stress	↑ Modulation of stearic and palmitic fatty acids ↓ MDA	[124]
CS-NPs	*Phaseolus vulgaris* L. under salinity stress	↑ Carbohydrates, proline and antioxidants ↓ Lipid peroxidation, electrolyte leakage and H_2O_2	[121]
NPs-MgO and CS (100 mg L^{-1})	Rice plants under high salinity stress (200 mM)	↑ Free amino acids, tetrapyrrole biosynthesis, CAT, POD, SOD and K^+/Na^+ ratio ↓ MDA, H_2O_2 and Na^+	[125]
NMs of CS with selenocystamine hydrochloride (300 mg L^{-1})	Rice plants under salinity stress	↑ CAT, POD, SOD, gene transcription and MAPK signaling ↓ MDA	[126]
CS-NPs (50 mg L^{-1})	*Lupine termis* L. plants under drought stress	↑ Photosynthetic pigments, proline and free amino acids	[127]
NPs O-Carboxymethylchitosan	Maize plants under water stress (waterlogging–drought)	↑ SOD, CAT, POD, PPO. Positive expression of *ZmSOD*, *ZmCAT*, *ZmPOD*, *ZmTIFY*, *ZmACO*, *ZmPYL2*, *ZmNF-YC12* and *ZmEREB180* genes ↓ MDA	[128]

(continued)

Table 1 (continued)

Nanostimulant	Abiotic stress	Response	References
NPs-Zn-CS-salicylic acid (100 mg L^{-1})	Wheat under drought stress	↑ Relative water content, proline, AsA, flavonoids, SOD, APX, GR and GPx	[129]
Nano-NPK (2 mL L^{-1}) + nano-CS (2 mL L^{-1})	*Kigelia africana* L. under salinity stress	↑ Clorophyll, carotenoids, N, P, K, CAT, POD and SOD	[21]

pointed out that fluctuations in sugar content may be due to the reallocation or consumption of starch synthesis, which allows carbon fixation.

Another study reported that exogenous application of *Carthamus oxyacantha* nano-biomass (<250 ppm) improves drought tolerance in wheat by promoting growth rate and grain yield because the nano-biomass has a high content of alcohols, phenols, alkanes, 1-amines, aromatic amines, alkyl halides, phenolic compounds, lignin, flavonoids, carotenoids, salicylic acid, tannins, phytoalexins, aromatics, aromatic amines, alkyl halides, phenolic compounds, lignin, flavonoids, carotenoids, salicylic acid, tannins, and phytoalexins, which could act as antioxidants, structural polymers, attractants, UV reflectors, signaling, or defense systems [66].

3.3 *Nanostimulants of Protein Hydrolysates*

Tolerance to Cd stress has been reported with applications of nano-water hyacinth protein (derived from water impermecin protein invomilized in nanochitosan), this NSs promotes plant antioxidant activity and soil enzymatic activity and reduces oxidative stress in common bean seedlings because Cd is adsorbed by NSs, reducing its availability [64]. Recently, it was reported that okra seed protein nanoemulsions are rich in glutamic acid and show stability to changes in pH, temperature, and ionic strength [136], and could be an alternative for nanostimulation of crops under stress conditions.

3.4 *Inorganic Nanostimulants*

Within NMs, carbon-based NMs and NPs can be considered biostimulants as they promote plant growth by inducing signaling [96, 137]. However, this only occurs under a specific range of concentrations (generally low) and is dependent on the type and properties of NPs [138].

3.4.1 Induction of Crop Tolerance Under Water Deficit Stress

Water deficit stress reduces chlorophyll and sugar contents, alters hormone balance, decreases antioxidant activity, and reduces PSII quantum efficiency and relative water content by increasing stress markers (MDA and H_2O_2) [80]. Some NPs have been reported to induce tolerance to this type of stress; for example, NPs-Si applied to rice under water deficit reduces oxidative damage by improving proline, chlorophyll, and carotenoid content, as well as CAT and POD activity [83, 77]. Si is absorbed rapidly and more efficiently in its nanoform, suggesting the promotion of photosynthetic activity and plant growth under stressful conditions. Similarly, nano-Si in sugar beet plants under deficit irrigation promoted chlorophyll content, net photosynthesis, glycine betaine, and flavonols (quercetin and rutin), and increased SOD, CAT, and guaiacol peroxidase (GPx) activity, which reduced H_2O_2 and MDA content [95].

On the other hand, NPs-TiO_2 applications under water stress conditions promote plant tolerance by stimulating SOD, POD, CAT, APX, GR, NR, and AC activities. Likewise, an increase in osmolytes accompanied by a decrease in H_2O_2 and $O_2^{\bullet-}$ managed to avoid electrolyte leakage and lipid peroxidation, leading to higher relative water content and chlorophyll content [96]. The authors pointed out that increased NR activity catalyzes nitric oxide (NO) biosynthesis, promoting further stress signaling. These NPs-TiO_2 and NPs-ZnO in wheat promote osmoprotectants to improve cytoplasmic Ψs to cope with water deficit, in addition to improving protein content that allows preservation of cellular structures [91].

A zinc nano-chelate in wheat promotes tolerance to water deficit by increasing chlorophylls, sugars, auxin, cytokinin, CAT and SOD accompanied by a reduction in the concentration of abscisic acid (ABA) and MDA [78], while iron nano-oxide (0.9 g L^{-1}) promotes proline content in *Hordeum vulgare* L. [80]. In guinea grass, NPs-nitrogen increased macro- and micronutrient content, total carbohydrates, plant pigments, and crude protein, in addition to higher hormone content and SOD and POD activity, allowing higher water use efficiency, stomatal conductance, and intercellular CO_2 concentration [79].

Carbon NMs [C60 fullerene, multi-walled carbon nanotubes, and graphene nanoplatelets) differentially promote water stress tolerance in sweet bell pepper plants by stimulating the accumulation of fruit biomass, gas exchange, photosynthetic rate, flavonoids, and phenols [139].

3.4.2 Induction of Crop Tolerance Under Drought-Stress Conditions

In *Rosa damascena* Mill, the use of NPs-SiO_2 reduces the damage caused by this type of stress by promoting the leaf area index, chlorophyll content, and biomass accumulation, as well as the concentration of compatible solutes and nutrient uptake [71]. In maize, such NPs increase plant height and biomass, and reduce membrane damage. The same authors cited an increase in relative water content, higher SOD activity, lower MDA content, and electrolyte leakage [74]. When deposited and accumulated in plants, these NPs generate a polymerized gel that strengthens the cells, avoiding the loss of electrolytes and promoting an increase in total carbohydrates, which act as compatible osmolytes. In wheat, the same NPs promoted grain yield, probably by improving the stomatal conductance [75]. In the case of pea, the NPs-SiO_2 upon entering the cells interact in the cytosol owing to their high contact surface, and their zeta potential promotes the activity of antioxidant enzymes, total phenolics, and relative water content, causing a reduction in H_2O_2 and lipid peroxidation [23].

In *Ehretia macrophylla*, NPs-SiO_2 differentially modified genes associated with auxin signal transduction, mitogen-activated protein kinase (MKP) signaling, redox system, and antioxidant enzyme activity. Interestingly, they increase fatty acid metabolism (α-linolenic acids), allowing the maintenance of chloroplast structure, thus improving tolerance to this stress [29]. Similarly, NPs-Si in wheat showed an increase in CAT and SOD activity, chlorophyll, proline, and relative water content, as well as a decrease in H_2O_2 and MDA. Similarly, increased expression of genes (*DREB2*, *MYB33m MYB3R*, *WRKY 19*, and *SnRK 2.4*) associated with stress was found [107, 85]. In another study carried out in the same culture, this type of NPs improved stress tolerance by increasing morphological parameters and *ABC1, Wdhn13, CHP,* and *EXP2* genes expression [107].

In contrast, root application of NPs-SiO_2, in addition to protecting rice plants against *Magnaporthe oryzae*, promotes root development under drought conditions owing to an increased expression of salicylic acid-associated genes (*PR1A*, *PR1B*, *PR5*, *PR8*, *PR10*, and *PAD4*), which act as signaling mechanisms under drought stress conditions [140].

NPs-TiO_2 promoted drought stress tolerance in chickpea by increasing antioxidant activity (APX, POD, and SOD), total phenols, and DPPH, as well as proline, chlorophyll, and carotenoids, while reducing MDA content [108]. In contrast, NPs-CeO_2 in grapes reduced drought stress by promoting agronomic traits, photosynthetic parameters, SOD, CAT, and APX activity, reducing H_2O_2 and MDA levels, and minimizing electrolyte leakage. These responses may be due to the promotion of *VvCLH1*, *VvCu/ZnSOD*, *VvRD29A*, and *VvRBOHA* genes expression [101].

In *Cucumis melo* L., NPs-ZnO tolerance to this type of stress may be due to differential stimulation of *SOD*, *POD*, *CAT*, *APX*, *DREB2D, and DREB3* genes as well as improved pigment, soluble sugar, and protein content. Increased tolerance to oxidative drought stress might result from improved nutrient acquisition and alterations in chlorophyll fluorescence [112]. In the case of golden turnip plants, drought reduces their growth and chemical attributes; however, the foliar use of NPs-Zn improves height, biomass, root length, and diameter, and promotes the antioxidant defense system (flavonoids and ascorbic acid), photosynthetic pigments, and osmotic adjustment in the root and foliar system of the plant, because Zn in its nanoform regulates in greater proportion the metabolism and biosynthesis of photosynthetic pigments [69]. In addition, increasing the activities of PSI and PSII improves carbon yield and increases sucrose generation, which stimulates plant metabolism by activating the antioxidant system and reducing free radicals.

The use of NPs-Fe has been reported in canola plants under drought stress, where their application allows the regulation of Ψ_s (thanks to the increase of proteins, proline, and soluble sugars), as well as activates CAT and PPO enzymes as well as non-enzymatic antioxidants (phenol, flavonol, and flavonoid), which reduce lipid peroxidation and improve membrane stability. Interestingly, these NPs induced protoporphyrin, magnesium protoporphyrin, and protochlorophyllide, in addition to stimulating succinate dehydrogenase and aconitase enzymes [113].

In contrast, priming maize seeds with NPs-Fe_2O_3 improves maize leaf water relations, relative water content, leaf water potential (Ψ_w), and efficient water use. It also improves chlorophyll content, total flavonoids, and ascorbic acid, and promotes SOD and POD activity by reducing MDA and H_2O_2 levels [70]. The authors pointed out that these NPs modify the lipid composition of membranes, altering membrane fluidity and promoting energetic forces that allow the entry of water into the cell, thereby reducing the impact of drought stress. In addition, Fe is associated with the synthesis of RuBisCO and chlorophyll, which enhances electron transport between photosystems and CO_2 assimilation. These NPs-Fe_2O_3 promote Ca^{2+}-mediated signaling by enhancing the production of flavonoids, glucosinolates, and phenols, which, together with the accumulation of soluble sugars, increases tolerance to this type of abiotic stress.

In *Achillea millefolium* L. under drought, NPs-Mg alleviates stress by minimizing electrolyte loss and enhancing the content of terpenoid compounds such as α-pinene, 1,8-cineole, camphor, α-bisabolol, chamazulene, and caryophyllene oxide [68], whereas in lettuce, these NPs (-MgO and -$MgCO_3$) induce the accumulation of chlorophylls (a and b) and carotenoids [141]. In the case of foliar applications of NPs-Ca^{2+} in canola, photoprotection in PSII is documented by increasing non-photochemical quenching (NPQ) and the yield of regulated energy dissipation, Y(NPQ), and improvements in mesophyll ultrastructure. These responses lead to reduced ROS, increased antioxidant defense, and increased protease, LOX, proline, and soluble protein levels [111].

Under the same stress applications, multi-walled carbon nanotubes (50 or 100 mg L^{-1}) seem to palliate the damage in camellia plants, probably by altering the fatty acid composition [142]. In peanuts, the use of carbon nanotubes (40 mg L^{-1}) stimulates the production of indole acetic acid (IAA), total soluble carbohydrates, proline, free amino acids, and phenolics, as well as β-carotene, lycopene, and flavonoids in seeds [143] improving its tolerance.

3.4.3 Induction of Crop Tolerance Under Irradiance Stress

Aspersions of oleylamine-coated $Ca(OH)_2$ NPs on tomato plants grown under 1000 μmol photons m^{-2} s^{-1} conditions reduced the photochemical efficiency of PSI, in addition to promoting higher excess excitation energy in PSII (at 90 min); however, at 72 h, the effective quantum yield of PSII electron transport (ΦPSII) was improved by an increase in the fraction of PSII reaction centers open (*qp*) and by improving the excitation capture efficiency (Fv'/Fm'). Interestingly, reduction in non-photochemical quenching (NPQ) promoted ROS generation by $^1O_2$ and H_2O_2. By regulating NPQ, these oleylamine-coated $Ca(OH)_2$ NPs promote the electron transport rate and decrease the excess excitation energy. Importantly, increased ROS leads to the upregulation of signaling processes that enhance PSII activity, so these NPs appear to promote tolerance by biostimulating photosynthetic activity [72]. In contrast, low radiation stress leads to a reduction in the antioxidant system of plants; therefore, NPs-silica increases SOD and NR activity and improves ascorbic acid content, which increases maize biomass [76].

3.4.4 Induction of Crop Tolerance Under Temperature Stress

Temperature stress leads to alterations in membrane fluidity, which causes ROS production. The use of NPs is an attractive strategy to mitigate this type of stress; in corn under cold stress, some nanofertilizers (MgO, CaO, Ca, Fe, Mn, and Zn) increase the effective quantum yield of PSII, the maximum photosynthetic efficiency of PSII, and the rate of electron transport, thereby improving the maximum net photosynthetic rate [86], which prevents the inhibition of photosynthetic parameters in the light phase. Under the same stress (4 °C), the use of carbon dots (<20 nm) promotes tolerance in lettuce by eliminating ROS, reducing MDA, and increasing sucrose, jasmonic acid (JA), salicylic acid, and proline contents [89]. Apparently, this nanomaterial activates the signaling mechanisms of calmodulin (CAM), calmodulin-regulated receptor kinase signaling chain (CRLK), and mitogen activated protein kinase (MAPK) by 12.8, 11.9, and 21.9, respectively, in addition to activating the expression of the cold tolerance genes C-repeat-binding factors *CBF 8* and *CBF 12*, and cold-regulated gene (*COR 413*) by 29.7, 48.2, and 93.8%, respectively. Similarly, NPs-Ag and NPs-Zn promote heat stress tolerance in wheat plants by improving growth and yield [144].

Sesamum indicum L. plants show tolerance to heat stress in response to spraying of multi-walled carbon nanotubes + NPs-TiO_2, due to the reduction of MDA and H_2O_2, the increase in POD, and the content and proportion of unsaturated fatty acids [145].

3.4.5 Induction of Tolerance of Crops Under Salt Stress

Another very strong stress worldwide is salinity, in this respect the use of NPs-SiO_2 in gerbera reduces the loss of electrolytes and minimizes the content of stress signaling molecules (MDA and H_2O_2) in addition to promoting greater APX and GPx activity [94]. The same authors found increased Ca^{2+} and K^+ uptake, together with a reduction in Na^+ uptake, which could improve the osmotic state of plants and promote tolerance to this stress. An increase in SOD and proline levels with NPs-SiO_2 has also been reported [146]. Their use in tomatoes leads to an increase in phenylalanine ammonia lyase (PAL) and vitamin C activity [147]. Natural siliceous NMs (natural zeolites and diatomaceous earth) function as NSs in plants under abiotic stress conditions because of their composition and porous structure, which stimulates the photosynthetic process, balances ROS, and reduces lipid peroxidation [148, 149].

NPs-TiO_2, they ameliorates the damage caused by salinity in wheat by promoting chlorophyll content and morphological parameters derived from an increase in osmolytes and greater SOD and POD activity [84]. In *Nicotiana tabacum* L. NaCl stress causes significant damage at the cellular level, to ameliorate this deterioration, the application of nano-CeO_2 and nitrogen-doped carbon dots has been studied. These NSs increased the expression of *IPUT1*, *SOS3*, *SOS2*, *SOS1*, and *NHX1*, genes associated with Na^+ perception and extrusion (into the vacuole). They also modify Ca^{2+} influx and Na^+ efflux, which promote metabolic alterations by regulating the content of fatty acids, amino acids, and carbohydrates [87]. On the other hand, oilseed rape treated with poly(acrylic) acid-coated nanoceria showed an increase in chlorophyll and a higher rate of carbon assimilation, accompanied by a lower presence of O_2^-, H_2O_2, and MDA. In addition to these responses, positive differential expression of *Cu–Zn SOD-I* (81%) and *Cu–Zn SOD-II* (40%) genes was documented, along with negative repression of the *LOX-IV* gene associated with the LOX enzyme. These results suggest that the modulation of SOD and LOX enzyme activity minimizes oxidative stress while maintaining cell membrane integrity [88].

In the case of NPs-Zn sprayed on wheat, it seems to promote tolerance to this stress by altering the concentration of ABA, MDA, H_2O_2, and electrolyte leakage [90], in addition to altering the differential uptake of ions ($<Na^+$ and Cl^-, $>Ca^{2+}$, K^+, Mg^{2+}, N, and P), which, together with the accumulation of free amino acids, led to osmotic stabilization and thus higher relative water content and photosynthetic rate. Likewise, NPs-ZnO differentially reduced MDA and H_2O_2 content in *Pisum sativum* L. as a function of the dose applied [150].

In a study carried out on cherry tomatoes under saline-alkaline stress, it was documented that NPs-FeO promote the concentration of ions (Fe, Cu, Zn, and Mn), chlorophylls, carotenes, and osmoregulators. Likewise, a reduction in NPQ was found to improve photosynthetic efficiency, in addition to decreasing the contents of H_2O_2, O_2^-, and MDA [106]. The authors also found an increased differential expression of genes associated with flavonoid synthesis (*SlF3H, SlF'3H, SlF3'5'H, SlFLS, SlDFR,* and *SlANS*). Interestingly, Fe nanocomplexes seem to alleviate bicarbonate stress in almond trees (soilless culture) by inducing chlorophyll recovery, H_2O_2 reduction, CAT increase, and Fe recovery at a similar level to FeEDDHA applications [100].

Other NPs studied are NPs-CuO, which modulate the tolerance of corn plants to salt stress by promoting photosynthetic parameters and N, P, K, Ca, and Mg content in roots and shoots, thereby improving plant growth [114]. In contrast, nano-K_2SO_4 promoted changes in morphological parameters and relative water content in alfalfa under salinity stress. Fe–Mn nanocomposites doped with graphene quantum dots on *Triticum aestivum* alleviate salt stress by enhancing CAT, POD, GR, and NADPH oxidase activity and reducing lipid peroxidation, suggesting increased plasma membrane integrity by osmolyte accumulation and redox homeostasis [151].

Sugar-terminated carbon nanodots (trehalose and glucose) have been reported to reduce the negative effects of salinity stress by enhancing sugar uptake and minimizing oxidative damage in plants [152]. Applications of multi-walled carbon nanotubes (200 $\mu g\ mL^{-1}$) in *Mentha piperita* promote higher relative water content, essential oil percentage, and CAT, SOD, and POD activity [153], promoting tolerance. Some carbon NSs (240 and 320 $mg\ L^{-1}$) improved gas exchange in *Vigna unguiculata* L. plants [154] or when used in tomato seed priming, increased the antioxidant defense system (CAT, APX, glutathione peroxidase, and phenylalanine ammonia lyase) [155], promoting tolerance to salinity. In grapes, nanosilica with multi-walled carbon nanotubes (90 $\mu g\ mL^{-1}$) reduced MDA and improved antioxidant activity (SOD, POD, CAT, dehydroascorbate reductase, monodehydroascorbate reductase, APX, glutathione transferase, and GR) [156].

3.4.6 Induction of Tolerance of Crops Under Heavy Metal Stress

Heavy metals are a strong abiotic stressor because many soils can present a high presence of these ions, which causes an oxidative burst that alters the stability of the plasma membrane. Foliar applications of NPs-SiO_2 in spinach promoted morphological parameters and photosynthetic pigments, allowing the accumulation of osmolytes and α-tocopherol, GSH, and AsA in conjunction with higher POD, GR, CAT, SOD, and APX activity. These responses seem to promote greater cell membrane stability, improve the relative water content, and allow greater tolerance to Pb and Cd stress, as well as reduce the content of these metalloids in the plant, thereby reducing MDA content, electrolyte leakage, and ROS [82]. Under Cd stress, NPs-Si promotes tolerance in rice by improving the polysaccharide content and the antioxidant defense system; interestingly, an increase in the concentration of Cd in the cell wall was observed, a phenomenon that leads to a low translocation of Cd in different cellular tissues [157]. The application of NPs-SiO_2 in bitter gourd promotes tolerance to Cd stress because it improves morphological parameters and stimulates the content of Zn, Na, and Mg in leaves, together with an increase in antioxidant enzyme activity and a reduction in MDA and Cd in the plant [103]. In tomato, the same NPs increased the content of AsA, GSH, and SOD, as well as the intermediary metabolites of the ascorbate–glutathione cycle, proline, and γ-aminobutyric acid, and increased differential expression of *GR* (NM_001321393.1) and *GST* (XM_019212803.2) genes. As a consequence of this response, a lower content of H_2O_2 and MDA was induced, which allowed stabilization of the cell membrane [110]. This may be due to the rapid incorporation of nano-Si into the cell wall and the induction of lignin synthesis. In wheat grown in Cd-contaminated soils, NPs-Se and -SiO_2 induced lower concentrations of this metalloid in cereal grains, probably because of alterations in the differential expression of genes, such as lower expression of Cd transporters (*TaNramp5* and *TaLCT1*) and higher expression of Cd-binding vacuolar proteins (*TaHMA3* and *TaTM20*). In addition, there is a higher content of antioxidants, lower abundance of rhizospheric organic acids, lower bioavailability of Cd in the soil, and an improvement in carbon- and nitrogen-associated microorganisms [158].

In the case of Al stress, the use of NP-silica improves tolerance in *Cicer arietinum* plants because these NSs generate redox homeostasis by increasing the expression of antioxidant genes (*SOD, CAT, and APX*) and reducing the cytotoxic products of lipid peroxidation [159]. Similarly, NPs-Si promotes the expression of genes associated with photosynthetic proteins (PSI and PSII), scavenges ROS, improves the antioxidant profile of plants, and stimulates endogenous hormone balance [ABA, IAA, cytokinin (CK), ethylene (ET), GA_3, and JA] in addition to interacting with signaling molecules (Ca^{2+}, K^+, Na^+, NO, soluble sugars, and TF transcription factors) [160].

Some NPs-SiO_2, -CeO_2, -ZnO, and -S reduced the damage in pakchoi under As stress in a differential way, since the dissolved NPs-S releases SO_4^{2-}, which reduces the pH of the soil and the As content in the pore water, minimizing the As content in the shoot. On the other hand, NPs-ZnO inhibits As uptake because it increases dissolved organic carbon and because the hydrated particle size is smaller and promotes higher SOD and POD activity [99]. Thus, NPs-ZnO and nano-hydroxyapatite reduced Cd damage in wheat plants by reducing the Cd content and increasing SOD, CAT, and POD activity, consequently reducing MDA, which promoted tolerance and improved grain yield. In addition, these NPs promote the transformation of exchangeable Cd to organically bound Cd, which is more stable in the soil, thus reducing its availability [102]. In barley, NPs-SiO_2 have been documented to reduce Cd damage by promoting a higher net photosynthetic rate and reducing the concentration of Cd and MDA in cell tissue [105].

In the case of Se-NPs applied to bell pepper under Cd stress, an increase in the synthesis of selenomethionine (SeMet) and methylselenocysteine (MeSeCys) has been documented, promoting genes associated with lignin synthesis (*PAL, CAD, 4CL* and *COMT*), with metabolite content (synapyl alcohol, phenylalanine, p-coumaryl alcohol, caffeoyl alcohol and coniferaldehyde) as well as increased signaling of signaling genes (*BZR1, LOX3* and *NCDE1*) and metabolites (brassinolide, ABA and JA) in roots and leaves [92]. These responses indicated an adequate level of protection against the bell pepper root system under these stresses by stimulating the integrity of the cell wall (lignin). The same NPs, as well as NPs-ZnO, in soybeans with AsV ameliorate stress [93] due to an increase in reduced glutathione content, ascorbic acid, photosynthetic parameters, and higher antioxidant activity. This was associated with increased expression of the *GmSOD, GmPOD, GmCAT, GmAPX, GmGR, GmGST*, and *GmP5CS* defense genes. Furthermore, the reduction of As in the leaf tissue minimizes the disruption of carbon assimilation, improving the efficiency of PSII by minimizing the content of oxidized glutathione, ROS, and MDA. These NPs-Se, in addition to improving antioxidant activity, promote amino acid content (L-histidine, L-serine, L-threonine, L-valine, L-isoleucine, L-leucine, L-tyrosine, L-phenylalanine, L-proline, and L-arginine) in leaves and fruits, as well as IAA, total flavonoids, total phenols, rutin, quercetin, and caffeic acid, which are compounds that regulate phenylpropanoid metabolism [161].

Similarly, the application of NPs-Fe_3O_4 to broccoli (under Cd stress) promoted sprouting, root length, plant biomass, chlorophyll content, carotenoids, total soluble sugars, protein content, and NR activity. Similarly, less electrolyte leakage, lipid peroxidation, and Cd accumulation were observed, which may be due to improved nutrient uptake [104]. Similarly, the use of zero valent iron-NPs on tomato improves antioxidant activity and reduces Cd accumulation in the plant by promoting differential expression of the *SlEFR1* gene, which interacts with proteins associated with hormones and stress signaling [98]. Similarly, nano-hydroxyapatite improves shoot dry biomass and reduces Cd and Pb uptake in *Coriandrum sativum* plants under stress caused by this metalloid [81].

Some studies have found that seed priming with NPs-TiO_2 and -Si tends to minimize the damage caused by Cu stress, which may be due to a decrease in Cu content in plant tissues, a reduction in oxidative stress, and the stimulation of organic acid exudation [162] that could supplement the excess Cu by storing it in the plant. Likewise, priming with NPs-ZnO induced tolerance to Cd stress in maize seeds because it improved the total chlorophyll content, ascorbic acid, and DPPH activity, as well as increased Mg^{2+}, K^{+}, and Zn^{2+} uptake [163].

Under Cd toxicity, priming maize seeds with multi-walled carbon nanotubes ameliorated damage and promoted tolerance by stimulating POD, SOD, and CAT and reducing MDA [164]. The same NSs applied to *Calendula officinalis* L. plants (250 mg L^{-1}) under Pb and Cd stress alleviated toxicity by increasing the enzymatic and non-enzymatic antioxidant defense system, in addition to promoting the accumulation of these metals in the soil [165], which may be due to the ability of this NSs to reduce the bioavailable forms (Pb^{2+} and Cd^{2+}) and allow their uptake into the plant. Under Cu^{2+} stress, the use of carbon NPs in maize promotes tolerance by increasing the activity of the antioxidant system, improving photochemical efficiency, and allowing a greater accumulation of N and K [166]. Finally, in perfumed rice grown under Cd and Pb stress, the use of multi-walled carbon nanotubes (100 and 300 mg kg^{-1}) promotes biomass accumulation and grain yield by reducing H_2O_2 content, regulating stressor translocation, and exacerbating the soil microbiological load [167].

3.4.7 Induction of Crop Tolerance Under Mineral Nutrient Stresses

Nutrient deficit leads to strong environmental stress on plants, reducing their productivity; therefore, studies show that application of NPs-CeO_2 promotes tolerance in rice plants under low nitrogen (LN) and excess nitrogen (HN) stress. These NPs improved assimilation and N content in roots and shoots under LN conditions due to increased glutamine synthetase (GS), glutamine oxoglutarate aminotransferase (GOGAT), and glutamate dehydrogenase (GDH) activity. In the case of HN stress NPs-CeO_2 reduce the level of N derived from the negative regulation of GS and GDH enzymes. Similarly, NPs reduce oxidative damage to the membrane and DNA, and promote tolerance by regulating the antioxidant enzyme system, proline levels, and phytohormones (AG_3, ABA, zeatin riboside, and IAA). It should be noted that the authors emphasize that the supply of these NPs under appropriate N conditions modifies the transformation of NPs at the root interface, and eventually the uptake and translocation of Ce [97].

In forage crops, phosphorus toxicity causes stress; therefore, the use of NPs-Si in *Megathyrsus maximus* via fertigation promotes greater Si absorption, improving C:N:P homeostasis, as well as nutritional efficiency (both in deficiency stress and P toxicity) [168]. This may be associated with an increase in the energy required for plant metabolism by improving the antioxidant mechanism. However, P toxicity (8 to 13 mmol L^{-1} of P) leads to oxidative stress and nutritional alterations in scarlet eggplant plants, and the use of nanosilica promotes tolerance by reducing P uptake, improving C:N homeostasis, and increasing the efficiency of the use of Fe, Cu, and Zn. These responses lead to a decrease in oxidative stress and electrolyte loss, in addition to an increase in phenols and ascorbic acid [109].

3.4.8 Induction of Tolerance of Crops Under Herbicide Stress

NPs-Se appears to stimulate the phenylpropanoid biosynthesis pathway, leading to increased levels of flavonoids (quercetin) and phenolic acids (vanillic acid) in leaves and roots of wheat plants, in addition to soil urease and fluorescein diacetate enzyme activity. These results suggest the potential use of nano-Se to mitigate herbicide stress in soils [169] by enhancing plant development.

3.4.9 Induction of Crop Tolerance to Microplastic Stresses

Recently, stress by micro- and nanoplastics (mainly polyethylene, 50 nm–10 μm) in agricultural soil, irrigation water, and even in the phyllosphere has been reported and should be considered a strong abiotic stressor. We consider that the negative impact of this type of stress is due to "hidden damage" to crops; that is, indirect impacts that can be mistaken for other types of stress. Among the damage caused by micro- and nanoplastics, we can cite the effects of the soil-leaf microbiome, alteration in water and nutrient uptake, induction of micro-drought, ROS burst, alterations in ionome, hormonal effects, anomalies in enzymatic activity, and reduction of chlorophyll and photosynthesis [170–172].

In this context, the use of NPs-ZnO in maize seedlings reduces the accumulation of polystyrene nanoplastics in the plant, thereby protecting chlorophyll and photosynthesis. From a deeper perspective, the promotion of tolerance to this type of stress can be explained by a differential expression of genes associated with GSH, signal transduction pathways at the hormonal level and photosynthesis; highlighting genes associated with photosystem II (*PetH*), chlorophyll synthesis (*CHLH*, *CHLD*, *CHLM*, *DVR*, *GTR* and *POR*) and carbon fixation enzymes (*pepc*, *rbcL* and *rbcS*) [173]. Another study revealed how NPs-Fe_3O_4 minimize the damage caused by polyethylene chloride and polyvinyl chloride microplastics in ryegrass seeds by increasing the content of carotenoids and chlorophyll, while also disturbing POD activity differentially depending on the type of microplastic. Similarly, the authors demonstrated an improvement in the microbiological load and its connectivity with the plant phyllosphere [174]. The use of NSs to promote tolerance to this type of stress is emerging but seems promising, however since micro- and nanoplastics form a "bio-crown" similar to NPs future studies should discern what implications this may entail.

3.5 Nanopolymers: Nanochitosan

Owing to its high biocompatibility, biodegradability, and versatility, chitosan (CS) at the nanometer scale (CS-NPs) promotes biostimulation [175]. Owing to their larger contact surface area, surface functional groups, high dispersity, and high zeta potential, they can penetrate barriers present in cuticles, trichomes, stomata, stigma, and hydathodes, allowing greater interaction with plant cells, which can induce abiotic and biotic stress tolerance responses [176–178]. In addition, the amino group (-NH_2) of CS provides the net charge of CS-NPs, which allows them to interact with anionic molecules such as cell wall phospholipids and functional and structural anionic proteins [177]. Some studies have shown that the use of lecithin/chitosan-based NPs formulations as carriers of a protein hydrolysate of *Arthrospira platensis* [179], CS-NMs [180], CS-NPs-S-nitrosoglutathione [181], CS-NMs-nanourea [182] are able to stimulate plants by improving the antioxidant system, increasing nutrient uptake and osmolyte content, modifying the genes expression associated with phytohormones, and reducing ROS, making their use under abiotic stress conditions.

3.5.1 Induction of Crop Tolerance Under Water Stress

Water stress significantly reduces crop productivity, in barley plants under this stress it has been reported that CS-NPs promote tolerance by substantially improving parameters such as leaf area, proline content and CAT and SOD activity causing higher relative water content and grain yield [119]. In maize, foliar use of CS-loaded NPs ameliorated water stress by improving Ψ_s and stabilizing the membrane, in addition to increasing CAT, POD, and SOD activity and phenolic and flavonoid content [123]. This may be due to the fact that the functional groups present on the surface of CS-NPs acted as reducing agents, leading to stress tolerance and eventually improving ear weight and kernel number and reducing MDA. Under water stress (waterlogging and drought), CS-NPs-O-Carboxymethyl promoted tolerance in maize seedlings by the accumulation of NPs in the seed embryo and root tip that stimulated enzymes (SOD, CAT, POD, and polyphenols), CAT, POD, and polyphenol oxidase (PPO), allowing MDA reduction, reflected in an alteration in the differential expression of *ZmSOD*, *ZmCAT*, *ZmPOD*, *ZmTIFY*, *ZmACO*, *ZmPYL2*, *ZmNF-YC12*, and *ZmEREB180* genes [128].

3.5.2 Induction of Crop Tolerance Under Drought Stress

Drought stress is one of the greatest concerns in agriculture, and several studies have shown that CS-NPs promote tolerance. In this regard, it was found that in wheat, NPs applied during tillering, stem elongation, and bolting promoted tolerance to this stress by improving relative water content and leaf area, thus improving photosynthetic rate and CAT and SOD activity, which led to an increase in biomass and yield [122]. In *Catharanthus roseus* plants, CS-NPs stimulate growth, proline accumulation, and CAT and APX activity, leading to lower H_2O_2 and MDA contents, together these data suggesting lower oxidative stress and preservation of cell membrane integrity. Likewise, an increase in alkaloid content was associated with a higher differential expression of strictosidine synthase (*STR*), deacetylvindoline-4-O-acetyltransferase (*DAT*), peroxidase 1 (*PRX1*), and geissoschizine synthase (*GS*) genes [183]. The same response has been reported in tomato when applying these CS-NPs, since an increase in the expression of drought-responsive genes *HsfA1a*, *SlAREB1*, *LeNCED1*, and *LePIP1* has been reported [184]. These NPs increase the transcripts encoding detoxifying proteins that limit radical production and promote tolerance to stress conditions [129].

In broad bean, a promotion of tolerance to this stress has been reported when applying CS-NPs because they improve the content of osmolytes, phenolic compounds, GSH, and α-tocopherol, as well as the enzymatic activity that reduces H_2O_2, MDA, and LOX [115]. The same authors found a higher IAA content, which explains the high rate of cell division and root and aerial growth. The hydrophilic properties of CS-NPs reduce drought stress levels by increasing the water and nutrient uptake capacity, allowing osmotic adjustment and photosynthetic activity. In *L. termis*, tolerance to this type of stress by CS-NPs is due to an increase in photosynthetic pigments, IAA, total soluble sugars, proline, and free amino acid content, which leads to improved growth and yield [127]. This type of stress seems to alter the fatty acid ratio (increase in polyunsaturated and reduction in monounsaturated); however, CS-NPs protect proline, stearic acid, and palmitic acid, thereby ameliorating this environmental stress. In bean cultivars under drought stress, foliar sprays of CS-NPs modulated the relative water content and reduced MDA [124].

Other studies have shown that the use of CS-coated iron oxide NPs in peppermint plants modifies the chlorophyll index, stomatal conductance, maximum fluorescence (Fm), and variable fluorescence (Fv), as well as increases menthol content [1]. In the case of wheat, NPs-Zn-CS-salicylic acid applied during the vegetative stage ameliorates the physiological and biochemical damage caused by drought, increasing the relative water content and enzyme activity (SOD, APX, GR, and GPx), as well as increasing proline, AsA, and flavonoid content. These parameters improve the osmotic state and ROS scavenging, allowing a yield improvement of up to 63% [129]. Interestingly, the use of CS-NPs-NO by constantly releasing NO to the plant increased drought tolerance in soybean because of a reduction in electrolyte loss accompanied by an increase in proline, CAT, and APX activity and lower MDA and H_2O_2. Likewise, a higher differential expression of *GmDREB1a*, *GmP5CS*, *GmDEFENSIN*, *GmGSNOR1*, and *GmNOX1* genes was found [5].

In rice, the use of fulvic acid-releasing CS-NPs improves the accumulation of fresh and dry biomass in the roots and shoots of plants, as well as increases leaf area and relative water content [53]. The same authors found that NSs improved stomatal control, chlorophyll, soluble protein, and APX levels, and reduced H_2O_2 and MDA levels. Drought marker genes (*OsDREB1A*, *OsCIPK3*, and *OsNAC77*) showed higher differential expression, reducing the negative effects of stress. Similarly, the use of CS-NPs-fulvic acid improved the growth of maize plants under drought conditions due to a reduction in H_2O_2 content, MDA, and electrolyte leakage concomitant with an increase in chlorophyll content and APX activity, which reduced lipid peroxidation levels and maintained redox homeostasis [54]. Similarly, an increase in the expression of transcription factors associated with drought tolerance (*ZmDREB1A*, *ZmbZIP1*, *ZmNAC28*, and *ZmCIPK3*) was observed. Likewise, the use of CS-loaded

NPs encapsulated in glycine betaine attenuated this type of stress by increasing photosynthetic pigments, enzymatic activity, and proline content, in addition to positively modulating the relative expression of *the ZmAPX*, *ZmCAT*, and *ZmSOD* genes [116].

3.5.3 Induction of Tolerance of Temperature-Stressed Crops

Exogenous application of CS-NPs incorporated with salicylic acid to cotton plants under heat stress promoted RuBisCO, invertase, and sucrose phosphate synthase enzyme activity. Modification of these enzymes improved the sucrose content and pollen tube length. These results indicate that under high-temperature stress, CS-NPs with salicylic acid enhance sucrose translocation from source to sink, that is, this NSs seems to restore sucrose metabolism in the reproductive structure of cotton [120].

3.5.4 Induction of Crop Tolerance Under Salt Stress

In maize plants under saline conditions, the application of CS-NPs improves CAT and POD activity and promotes stress tolerance [185]. Likewise, the use of CS-NPs allows buffering and mitigating salinity stress in beans by improving photosynthetic pigments, carbohydrate content, proline, and antioxidant systems (CAT, APX), which reduces H_2O_2, lipid peroxidation, and electrolyte leakage, which may be due to the structural composition of chitosan and the wide range of functional groups (C = O, OH, and COOH) [121]. Foliar sprays of nano-CS on *Kigelia africana* L. promoted morphological parameters by improving chlorophyll and carotenoid content and increasing CAT, POD, and SOD activity, allowing greater membrane stability [21]. The same authors found an increase in N, P, and K contents, which together promoted greater biomass accumulation in these plants.

The efforts in the promotion of tolerance to salinity stress with the use of CS-NPs derived in a study where nano-Si was encapsulated in nano-CS and NSs when applied in wheat under salinity induced higher antioxidant activity in conjunction with a reduction of H_2O_2. In addition, stress tolerance is also promoted by an increase in the content of osmotic compounds (proline, free amino acids, betaine glycine, and sugars), which protect the lipid membrane, proteins, and photosynthetic apparatus [117]. The application of NPs-MgO-CS in rice plants under strong salt stress (200 mM NaCl) improved morphological parameters, free amino acid content, tetrapyrrole biosynthesis, and CAT, POD, and SOD enzyme activity, reducing MDA and H_2O_2 content. In addition, in leaf tissues, these NPs maintained adequate K^+ levels, reduced Na^+, and improved the K^+/Na^+ ratio [125]. Another study highlighted the use of CS-NMs oxidized with selenocystamine hydrochloride as a salinity tolerance promoter in rice by promoting CAT, POD, and SOD activity, along with lower MDA content. These CS-NMs also increased the transcription of genes associated with oxidative stress and MAPK signaling, in addition to inducing a greater abundance and variety of rhizobacteria in the soil [126].

3.5.5 Induction of Crop Tolerance Under Heavy Metal Stresses

In the case of heavy metal stress, CS-NPs induced tolerance by minimizing the absorption and accumulation of Cd and reducing the content of H_2O_2 and MDA. In addition, greater proline accumulation and enzymatic (CAT, POD, SOD) and non-enzymatic (GSH and AsA) antioxidant activities were observed [118].

3.6 *Nano-Biostimulants*

Studies are being conducted in this area considering stimulation tools and nanometric science. For example, in *Nigella sativa* under drought conditions, joint applications of humic acid (500 mg L^{-1}) and iron nano-chelate (2 g L^{-1}) minimize damage and promote tolerance owing to a greater assimilation of nutrients and by increasing the content of essential oils [186]. Similarly, soil application of humic acid (2 kg $tree^{-1}$) in conjunction with NPs-ZnO_2 + –TiO_2 sprays promotes tolerance to salt stress in olive trees by improving chlorophyll content, fruit set percentage, fruit quality (yield, weight, size, fruit firmness, total soluble solids, and oil content), and leaf N, P, and K content [73]. Under this same type of stress, a synergistic effect was documented between the use of biochar, nanozeolite and nanosilicon in *Thymus vulgaris* as morphological, photosynthetic parameters, nutrient content (N, P, K, Na, Ca, Mg, Zn, Fe, Mn), proline, phenols and total flavonoids as well as antioxidant enzymes and vital hydrocarbons and oxygenated compounds are improved [187], as in olive trees, this integrated management seems to induce a greater ease in osmotic adjustment allowing to cope with salinity.

In *Rosa* × *damascena*, the application of *Ascophyllum nodosum* extract (2 and 3 g L^{-1}) and NPs-SiO_2 (100 mg L^{-1}) induced salinity tolerance by improving photosynthetic parameters and antioxidant activity (SOD, POD, CAT, and GPx) as well as a reduction in MDA and H_2O_2 [188]. Similarly, seaweed extracts (0.5 g L^{-1}) in conjunction with nano-NPK (1.0 g L^{-1}) on onion [189], demonstrate an increase in agronomic parameters allowing greater tolerance to abiotic stress. In coriander, under salinity (in vitro), NPs-ZnO coated with glycine betaine induced greater plant length and biomass [190]. This type of seaweed extract has also been studied in conjunction with nonohumica in wheat grown in soil at a pH of 7.2, stimulating grain yield [191].

In bean plants the spraying of NSs nanosilicon (SiO_2; 2 mM), nano polyamine “Spermidine” (1 mM), nano seaweed compound (0. 5 mL L^{-1} containing 7% salicylic acid 5% + proline 4%), nano-biofertilizer (containing a mixture of nano-symbiotic N_2-fixing bacteria of the genera *Azospirillum*, *Azotobacter*, and *Kelebsilla*) reduced salt stress damage (6 dS m^{-1}), and improved tolerance by stimulating a wide range of morphological and physiological parameters (net assimilation rate, relative growth, relative water content, membrane integrity, chlorophylls, and carotenoids). Likewise, these NSs induced higher enzymatic (POD and PPO) and non-enzymatic (carotenoids, total soluble sugars, total amino acids, and proline) antioxidant activities. In particular, N, P, and K contents are improved, and Na^+ is reduced [192], which allows the regulation of water status and preserves osmotic homeostasis.

The combined use of nano-ferrihydrite and fulvic acid on spinach plants in calcareous soils promotes tolerance by increasing nutrient uptake, chlorophyll content, growth, and yield (193). This response may be because the nanoform of Fe penetrates more easily into the leaf, allowing its rapid assimilation, whereas fulvic acid could induce greater solubility and ease of absorption when applied foliarly [193].

One aspect that cannot be ignored is the use of microorganisms in conjunction with NMs to induce tolerance to abiotic stress. In this regard, the use of arbuscular fungi and NPs-ZnO and NPs-Se stimulates ascorbic acid and capsaicin content in chili bell pepper fruits subjected to cold stress. In addition, a higher content of POD and nitrogen glutamate dehydrogenase associated with lower levels of H_2O_2 and MDA has been documented. This may be due to the fact that spraying of NMs increased the thickness of grana thylakoids and the number of chloroplasts [194]. In another study, indigenous arbuscular mycorrhizal fungi (*Glomus* spp.) in conjunction with nano-TiO_2 (200 mg kg^{-1}) promoted chlorophyll content and reduced Cd uptake and translocation by minimizing lipid peroxidation in cowpea plants under Cd stress [195]. Along the same lines, applications of NPs-Fe together with *Staphylococcus aureus* inocula on rice plants under Cr stress allow the microorganisms to convert Cr^{6+} to Cr^{3+}, reducing its adsorption and accumulation in plants while minimizing oxidative damage, promoting photosynthesis, and stimulating the root system [196].

It has also been reported that the combined use of phosphorus and phosphate-solubilizing bacteria applied in calcareous soils (22% $CaCO_3$) promotes stress tolerance in *Phaseolus vulgaris* L. by increasing P levels, improving PSII efficiency, Fv/Fm, gas exchange, relative water content, antioxidant enzyme activity [AsA, GSH, SOD, CAT, APX, GR, glutathione S-transferase, monodehydroascorbate reductase, and dehydroascorbate reductase], and metabolites (soluble sugars, glycine betaine). Coupled with the above, a reduction in oxidative stress (electrolyte leakage, lipid peroxidation, O_2^-, and H_2O_2) was found to promote greater membrane stability [197]. Another study reported that nano-ZnO and nano-SiO_2 (50 mg L^{-1}) with biochar inoculated with *Bacillus subtilis* (2%) stimulated the tolerance of wheat plants to salinity stress by increasing leaf area and improving the synthesis of total soluble sugars and proline [198]. Although several studies have demonstrated the collation and stimulating power of nanocapsules cross-linked with CS and alginate with humic acid (loaded with nano-NPK and/or *Pseudomonas fluorescence*) [199], studies under abiotic stress conditions are needed.

3.7 Other Nanostimulants

There is evidence that clarifies the stimulating effect of the joint use of NPs with other biostimulants or delves into other alternatives at the nanoscale. In this sense, it has been documented that NPs (17.34 nm) coated with glycine betaine applied to coriander promote tolerance to drought stress by increasing plant length and fresh biomass accumulation together with an increase in defense systems [200]. Similarly, a nanofertilizer rich in betaine (50 mM) in *Zea mays* var. *saccharata* and *Arabidopsis thaliana* stimulate morphological variables under salinity, allowing an adequate level of stress tolerance [201]. In contrast, NPs-ascorbic acid, -benzoic acid, and -salicylic acid (0.50–0.75 mM) promote drought tolerance in potato plants by increasing parameters such as length, number of leaves, roots, and lateral branches per plant and root length [67].

Similarly, under heat stress (35 °C), nanoencapsulated 5-aminolevulinic acid (δ-aminolevulinic acid) is an oxygen- and nitrogen-containing hydrocarbon, a common precursor of all tetrapyrrole compounds, including chlorophyll, which promotes tolerance in bell pepper, probably by increasing the concentration of K^+, proline, glucose, fructose, and photosynthetic activity, in addition to reducing electrolyte leakage [202]. This may be due to the fact that these molecules are part of plant metabolism under normal and stress conditions, so their exogenous and nano application may maximize the signaling mechanism.

Finally, the application of nano-phytocarriers effectively stimulates plants; for example, Si-stabilized nanoporous hybrid lipid-NPs show a suitable delivery system for the flavonoid quercetin in tomato plants under salinity (200 mM NaCl). These NSs induce alterations in the content of proteins associated with cytoprotection against oxidative stress and proteins involved in chromatin remodeling [203], thereby inducing stress tolerance.

4 Conclusion

Currently, crop development and quality food production face great challenges because of the increasingly strong and prevalent abiotic stresses caused by environmental fluctuations, as well as poor agricultural practices. This scenario causes damage to crops, causing pressures on the global agri-food system, and to overcome this challenge, the use of nanostimulants is a great ally. Because of the particular characteristics of the nanoscale, the use of these stimulants promotes differentiated stress signaling (depending on the type of nanostimulant, size, dose, application route, and stress level) that leads to greater antioxidant activity, particularly improving enzymatic and non-enzymatic antioxidant activity, osmoprotectants, endogenous hormone balance, synthesis of secondary metabolites, and photosynthetic efficiency.

Nanostimulation (nanochitosan, nanoparticles, algae and plant nanoextracts, nano-biostimulants) can reduce oxidative stress and lipid peroxidation of cells, allowing the stabilization and integrity of the cell membrane, which leads to an adequate ionic-osmotic state, redox homeostasis, nutrient absorption-translocation, and good photosynthetic activity. Therefore, the use of nanostimulants is an appropriate strategy to induce abiotic stress tolerance in crops, promoting global food security; however, the increasing and unpredictable climatic conditions and unsustainable production systems require more studies to standardize the use of these nanostimulants worldwide. In addition, as nanostimulants continue to evolve, it is imperative to understand the opportunities to exploit their full potential, not to mention their impacts on human safety, so it is important to generate useful information to expand future research.

References

1. Giglou MT, Giglou RH, Esmaeilpour B, Azarmi R, Padash A et al (2022) A new method in mitigation of drought stress by chitosan-coated iron oxide nanoparticles and growth stimulant in peppermint. Ind Crops Prod 187:115286. https://doi.org/10.1016/j.indcrop.2022.115286
2. Wang P, Liu W, Han C, Wang S, Bai M et al (2024b) Reactive oxygen species: multidimensional regulators of plant adaptation to abiotic stress and development. J Integr Plant Biol 66(3):330–367. https://doi.org/10.1111/jipb.13601
3. Saini N, Modolo LV, Deswal R, Sehrawat A, Yadav N et al (2024) Expanding roles of cross-talk between hydrogen sulfide and nitric oxide under abiotic stress in plants. Plant Physiol Biochem 214:108852. https://doi.org/10.1016/j.plaphy.2024.108852
4. Pramanik B, Sar P, Bharti R, Gupta RK, Purkayastha S et al (2023) Multifactorial role of nanoparticles in alleviating environmental stresses for sustainable crop production and protection. Plant Physiol Biochem 201:107831. https://doi.org/10.1016/j.plaphy.2023.107831
5. Methela NJ, Pande A, Islam MS, Rahim W, Hussain A et al (2023) Chitosan-GSNO nanoparticles: a positive modulator of drought stress tolerance in soybean. BMC Plant Biol 23(1):1–15. https://doi.org/10.1186/s12870-023-04640-x
6. Pooja, Munjal R (2020) Use of biostimulants in conferring tolerance to environmental stress. In: Plant ecophysiology and adaptation under climate change: mechanisms and perspectives II: mechanisms of adaptation and stress amelioration. Springer, Singapore, pp 213–244
7. Etesami H (2024) Enhancing soil microbiome resilience: the mitigating role of silicon against environmental stresses. Front Agron 6:1465165. https://doi.org/10.3389/fagro.2024.1465165
8. García-Sánchez F, Simón-Grao S, Navarro-Pérez V, Alfosea-Simón M (2022) Scientific advances in biostimulation reported in the 5th biostimulant world congress. Horticulturae 8(7):665. https://doi.org/10.3390/horticulturae8070665
9. Singh NB, Kumar B, Usman UL, Susan MABH (2024) Nano revolution: exploring the frontiers of nanomaterials in science, technology, and society. Nano-Struct Nano-Objects 39:101299. https://doi.org/10.1016/j.nanoso.2024.101299
10. Su C, Chen A, Liang W, Xie W, Xu X et al (2024) Copper-based nanomaterials: opportunities for sustainable agriculture. Sci Total Environ 926:171948. https://doi.org/10.1016/j.scitotenv.2024.171948
11. Magnabosco P, Masi A, Shukla R, Bansal V, Carletti P (2023) Advancing the impact of plant biostimulants to sustainable agriculture through nanotechnologies. Chem Biol Technol Agric 10(1):117. https://doi.org/10.1186/s40538-023-00491-8
12. Bisht A, Chhabra R (2024) Biostimulants: paving way towards sustainable agriculture and food security. Theor Exp Plant Physiol 36(2):139–163. https://doi.org/10.1007/s40626-024-00329-0
13. Juárez-Maldonado A, Ortega-Ortíz H, Morales-Díaz AB, González-Morales S, Morelos-Moreno Á et al (2019) Nanoparticles and nanomaterials as plant biostimulants. Int J Mol Sci 20(1):162. https://doi.org/10.3390/ijms20010162
14. Jabín BBJ, Luis SCJ, Eucario MÁ (2024) Hormesis in plant tissue culture. Plant Cell Tissue Organ Cult 159(1):16
15. Shoukat A, Pitann B, Zafar MM, Farooq MA, Haroon M et al (2024) Nanotechnology for climate change mitigation: enhancing plant resilience under stress environments. J Plant Nutr Soil Sci 187(5):604–620. https://doi.org/10.1002/jpln.202300295

16. Shiran Z, Esmaeilzadeh Bahabadi S, Razmara Z, Beluri K, Easmin N et al (2024) Elucidation of biochemical and physiological modulations in Triticum aestivum induced by green synthesized nitrogen-enriched zinc nano-complexes. Plant Nano Biol 10:100094. https://doi.org/10.1016/j.plana.2024.100094
17. Khedr R, Aboukhadrah S, El-Hag D, Elmohamady E, Abdelaal K (2023) Ameliorative effects of nano silica and some growth stimulants on water relations, biochemical and productivity of wheat under saline soil conditions. Fresenius Environ Bull 32(1):375–384
18. Joody AT, Jebur ZI (2024) Role of NPK, root stimulator and nano micronutrients on growth of citrus rootstock Rangpur. Kufa J Agric Sci 16(3):11–23
19. Khan AR, Salam A, Li G, Iqbal B, Ulhassan Z et al (2024) Nanoparticles and their crosstalk with stress mitigators: a novel approach towards abiotic stress tolerance in agricultural systems. Crop J. https://doi.org/10.1016/j.cj.2024.09.010
20. Swedan EA, Shible K, Yassin YM, Glowacka A, Ahmed MAA (2024) Synergistic effect of nano-potassium and chitosan as stimulants inducing growth and yield of bird of paradise (Sterlitiza reginae L.) in newly lands. Commun Integr Biol 17(1):2406754. https://doi.org/10.1080/19420889.2024.2406754
21. Nofal E, Menesy F, Elbably S, Abd-El Rahman M, El-Ramady H, et al (2024) Response of Kigelia africana (Lam.) benth transplants to nano-NPK and nano-chitosan under salinity stress. Egypt J Soil Sci 64(2):661–672. https://doi.org/10.21608/ejss.2024.267464.1718
22. Khalid MF, Iqbal Khan R, Jawaid MZ, Shafqat W, Hussain S et al (2022) Nanoparticles: the plant saviour under abiotic stresses. Nanomaterials 12(21):3915. https://doi.org/10.3390/nano12213915
23. Sutulienė R, Ragelienė L, Samuolienė G, Brazaitytė A, Urbutis M, et al (2022) The response of antioxidant system of drought-stressed green pea (Pisum sativum L.) affected by watering and foliar spray with silica nanoparticles. Horticulturae 8(1):35. https://doi.org/10.3390/horticulturae8010035
24. Kopeć P (2024) Climate change—the rise of climate-resilient crops. Plants 13(4):490. https://doi.org/10.3390/plants13040490
25. du Jardin P, Xu L, Geelen D (2020) Agricultural functions and action mechanisms of plant biostimulants (PBs). Chem Biol Plant Biostimulants. Wiley. 1–30
26. Martínez-Lorente SE, Martí-Guillén JM, Pedreño MÁ, Almagro L, Sabater-Jara AB (2024) Higher plant-derived biostimulants: mechanisms of action and their role in mitigating plant abiotic stress. Antioxidants 13(3):318. https://doi.org/10.3390/antiox13030318
27. Grammenou A, Petropoulos SA, Thalassinos G, Rinklebe J, Shaheen SM et al (2023) Biostimulants in the soil-plant interface: agro-environmental implications—a review. Earth Syst Environ 7(3):583–600. https://doi.org/10.1007/s41748-023-00349-x
28. Silva S, Dias MC, Silva AMS (2022) Titanium and zinc based nanomaterials in agriculture: a promising approach to deal with (A)biotic stresses? Toxics 10(4):172. https://doi.org/10.3390/toxics10040172
29. Chen M, Jiao SQ, Xie L, Geng X, Qi S, et al (2024a) Integrated physiological, transcriptomic, and metabolomic analyses of drought stress alleviation in Ehretia macrophylla Wall. seedlings by SiO2 NPs (silica nanoparticles). Front Plant Sci 15:1260140. https://doi.org/10.3389/fpls.2024.1260140
30. Méndez-López A, González-García Y, Juárez-Maldonado A (2022) Stimulatory role of nanomaterials on agricultural crops. In: Nano-enabled Agrochemicals in Agriculture. Elsevier, pp 219–246
31. Hu T, Li H, Li J, Zhao G, Wu W, et al (2018) Absorption and bio-transformation of selenium nanoparticles by wheat seedlings (Triticum aestivum L.). Front Plant Sci 9:9. https://doi.org/10.3389/fpls.2018.00597
32. Lv Z, Sun H, Du W, Li R, Mao H et al (2021) Interaction of different-sized ZnO nanoparticles with maize (Zea mays): accumulation, biotransformation and phytotoxicity. Sci Total Environ 796:148927. https://doi.org/10.1016/j.scitotenv.2021.148927
33. Sembada AA, Lenggoro IW (2024) Transport of nanoparticles into plants and their detection methods. Nanomaterials 14(2):131. https://doi.org/10.3390/nano14020131

34. Avellan A, Yun J, Morais BP, Clement ET, Rodrigues SM et al (2021) Critical review: role of inorganic nanoparticle properties on their foliar uptake and in planta translocation. Environ Sci Technol 55(20):13417–13431. https://doi.org/10.1021/acs.est.1c00178
35. Zhang H, Goh NS, Wang JW, Pinals RL, González-Grandío E et al (2022) Nanoparticle cellular internalization is not required for RNA delivery to mature plant leaves. Nat Nanotechnol 17(2):197–205. https://doi.org/10.1038/s41565-021-01018-8
36. Hu P, An J, Faulkner MM, Wu H, Li Z et al (2020) Nanoparticle charge and size control foliar delivery efficiency to plant cells and organelles. ACS Nano 14(7):7970–7986. https://doi.org/10.1021/acsnano.9b09178
37. Zhu J, Wang J, Zhan X, Li A, White JC et al (2021) Role of charge and size in the translocation and distribution of zinc oxide particles in wheat cells. ACS Sustain Chem Eng 9(34):11556–11564. https://doi.org/10.1021/acssuschemeng.1c04080
38. Sarraf M, Vishwakarma K, Kumar V, Arif N, Das S et al (2022) Metal/metalloid-based nanomaterials for plant abiotic stress tolerance: an overview of the mechanisms. Plants 11(3):316. https://doi.org/10.3390/plants11030316
39. Badawy SA, Zayed BA, Bassiouni SMA, Mahdi AHA, Majrashi A, et al (2021) Influence of nano silicon and nano selenium on root characters, growth, ion selectivity, yield, and yield components of rice (Oryza sativa L.) under salinity conditions. Plants 10(8):1657. https://doi.org/10.3390/plants10081657
40. Dam P, Paret ML, Mondal R, Mandal AK (2023) Advancement of noble metallic nanoparticles in agriculture: a promising future. Pedosphere 33(1):116–128. https://doi.org/10.1016/j.pedsph.2022.06.026
41. Younes I, Rinaudo M (2015) Chitin and chitosan preparation from marine sources. structure, properties and applications. Mar Drugs 13(3):1133–1174. https://doi.org/10.3390/md13031133
42. Szymańska E, Winnicka K (2015) Stability of chitosan—a challenge for pharmaceutical and biomedical applications. Mar Drugs 13(4):1819–1846. https://doi.org/10.3390/md13041819
43. Ahmed T, Aljaeid B (2016) Preparation, characterization, and potential application of chitosan, chitosan derivatives, and chitosan metal nanoparticles in pharmaceutical drug delivery. Drug Des Dev Ther 10:483. https://doi.org/10.2147/DDDT.S99651
44. Balusamy SR, Rahimi S, Sukweenadhi J, Sunderraj S, Shanmugam R et al (2022) Chitosan, chitosan nanoparticles and modified chitosan biomaterials, a potential tool to combat salinity stress in plants. Carbohydr Polym 284:119189. https://doi.org/10.1016/j.carbpol.2022.119189
45. Bajwa DS, Pourhashem G, Ullah AH, Bajwa SG (2019) A concise review of current lignin production, applications, products and their environmental impact. Ind Crops Prod 139:111526. https://doi.org/10.1016/j.indcrop.2019.111526
46. Moretti C, Corona B, Hoefnagels R, Vural-Gürsel I, Gosselink R et al (2021) Review of life cycle assessments of lignin and derived products: lessons learned. Sci Total Environ 770:144656. https://doi.org/10.1016/j.scitotenv.2020.144656
47. Ariyanta HA, Sari FP, Sohail A, Restu WK, Septiyanti M et al (2023) Current roles of lignin for the agroindustry: applications, challenges, and opportunities. Int J Biol Macromol 240:124523. https://doi.org/10.1016/j.ijbiomac.2023.124523
48. Schneider WDH, Dillon AJP, Camassola M (2021) Lignin nanoparticles enter the scene: a promising versatile green tool for multiple applications. Biotechnol Adv 47:107685. https://doi.org/10.1016/j.biotechadv.2020.107685
49. Alabdallah NM, Hasan MM (2021) Plant-based green synthesis of silver nanoparticles and its effective role in abiotic stress tolerance in crop plants. Saudi J Biol Sci 28(10):5631–5639. https://doi.org/10.1016/j.sjbs.2021.05.081
50. Jadoun S, Arif R, Jangid NK, Meena RK (2021) Green synthesis of nanoparticles using plant extracts: a review. Environ Chem Lett 19(1):355–374. https://doi.org/10.1007/s10311-020-01074-x
51. Roseline TA, Sudhakar MP, Arunkumar K (2023) Aqueous extraction of red seaweed bioactive compounds and synthesis of silver nanoparticles for agriculture applications. J Agric Food Res 14:100769. https://doi.org/10.1016/j.jafr.2023.100769

52. Castillo-Henríquez L, Alfaro-Aguilar K, Ugalde-Álvarez J, Vega-Fernández L, Montes de Oca-Vásquez G et al (2020) Green synthesis of gold and silver nanoparticles from plant extracts and their possible applications as antimicrobial agents in the agricultural area. Nanomaterials 10(9):1763. https://doi.org/10.3390/nano10091763
53. Faluku M, Al-Azawi TNI, Methela NJ, Khan M, Huy VN et al (2024) Fulvic acid-releasing chitosan nanoparticles promote the growth and drought stress tolerance of rice plants. J Crop Heal 76(3):739–751. https://doi.org/10.1007/s10343-024-00979-9
54. Brown A, Al-Azawi TNI, Methela NJ, Rolly NK, Khan M et al (2024) Chitosan-fulvic acid nanoparticles enhance drought tolerance in maize via antioxidant defense and transcriptional reprogramming. Physiol Plant 176(4):e14455. https://doi.org/10.1111/ppl.14455
55. Guo J, Fu Q, Tang M, Bai J, Liu R et al (2024) Fulvic acid modified ZnO nanoparticles improve nanoparticle stability, mung bean growth, grain zinc content, and soil biodiversity. Sci Total Environ 913:169840. https://doi.org/10.1016/j.scitotenv.2023.169840
56. Yoon HY, Lee JG, Esposti LD, Iafisco M, Kim PJ et al (2020) Synergistic release of crop nutrients and stimulants from hydroxyapatite nanoparticles functionalized with humic substances: toward a multifunctional nanofertilizer. ACS Omega 5(12):6598–6610. https://doi.org/10.1021/acsomega.9b04354
57. Shah IH, Sabir IA, Rehman A, Hameed MK, Albashar G et al (2023) Co-application of copper oxide nanoparticles and Trichoderma harzianum with physiological, enzymatic and ultrastructural responses for the mitigation of salt stress. Chemosphere 336:139230. https://doi.org/10.1016/j.chemosphere.2023.139230
58. Joel JM, Johnson R, Puthur JT (2023) Co-application of arbuscular mycorrhizal fungi and engineered nanomaterials: a promising strategy for crop resilience against abiotic stresses. South African J Bot 162:314–323. https://doi.org/10.1016/j.sajb.2023.09.022
59. Verma KK, Joshi A, Song X-P, Singh S, Kumari A, et al (2024) Synergistic interactions of nanoparticles and plant growth promoting rhizobacteria enhancing soil-plant systems: a multigenerational perspective. Front Plant Sci 15. https://doi.org/10.3389/fpls.2024.1376214
60. Gao M, Chang J, Wang Z, Zhang H, Wang T (2023) Advances in transport and toxicity of nanoparticles in plants. J Nanobiotechnol 21(1):75. https://doi.org/10.1186/s12951-023-01830-5
61. Khan MN, Fu C, Li J, Tao Y, Li Y et al (2023) Seed nanopriming: how do nanomaterials improve seed tolerance to salinity and drought? Chemosphere 310:136911. https://doi.org/10.1016/j.chemosphere.2022.136911
62. Huang D, Dang F, Huang Y, Chen N, Zhou D (2022) Uptake, translocation, and transformation of silver nanoparticles in plants. Environ Sci Nano 9(1):12–39. https://doi.org/10.1039/D1EN00870F
63. Landa P (2021) Positive effects of metallic nanoparticles on plants: overview of involved mechanisms. Plant Physiol Biochem 161:12–24. https://doi.org/10.1016/j.plaphy.2021.01.039
64. Alzandi AA, Al Hijab LYA, Alzahrani ZI, Naguib DM (2024) Nano-water hyacinth protein adsorbent as soil amendment alleviates cadmium stress in common bean seedlings by improving soil enzymes and mitigating oxidative stress. Adsorption 30(6):1419–1435. https://doi.org/10.1007/s10450-024-00511-5
65. El-Berawey DY, Eldebawy EMM (2024) The effects of Marrubium alysson and Torilis arvensis natural and nano extracts on priming of wheat seeds in response to drought. Cereal Res Commun 1–15. https://doi.org/10.1007/s42976-024-00539-x
66. Ali S, Manzoor A, Nawaz A, Mahmood S, Rana RM, et al (2024) Exogenous utilization of nano-biomass of Carthamus oxyacantha to improve yield and drought tolerance of wheat under rainfed conditions. Pakistan J Agric Sci 61(1):1–11. https://doi.org/10.21162/PAKJAS/24.207
67. Alowaiesh BF, Awad NS, Eldenary ME, El Moneim DA (2024) Enhancement of drought tolerance in potato employing nanoparticles of different biostimulants. Chil J Agric Res 84(2):246–259. https://doi.org/10.4067/s0718-58392024000200246

68. Ojagh SE, Moaveni P (2022) Foliar-applied magnesium nanoparticles modulate drought stress through changes in physio-biochemical attributes and essential oil profile of yarrow (Achillea millefolium L.). Environ Sci Pollut Res 29(39):59374–59384. https://doi.org/10.1007/s11356-022-19559-3
69. Li J, Zafar S, Javaid A, Perveen S, Hasnain Z, et al (2023) Zinc nanoparticles (ZnNPs): high-fidelity Amelioration in Turnip (Brassica rapa L.) production under drought stress. Sustainbility 15(8):6512. https://doi.org/10.3390/su15086512
70. Mazhar MW, Ishtiaq M, Maqbool M, Muzammil K, Mohieldin A et al (2024) Optimizing water relations, gas exchange parameters, biochemical attributes and yield of water-stressed maize plants through seed priming with iron oxide nanoparticles. BMC Plant Biol 24(1):1–17. https://doi.org/10.1186/s12870-024-05324-w
71. Seyed Hajizadeh H, Azizi S, Rasouli F, Kaya O (2023b) Evaluation of nano-silicon efficiency on compatible solutes and nutrient status of Damask rose affected by in vitro simulated drought stress. Chem Biol Technol Agric 10(1):1–19. https://doi.org/10.1186/s40538-023-00397-5
72. Tryfon P, Sperdouli I, Moustaka J, Adamakis IDS, Giannousi K et al (2024) Hormetic response of photosystem II function induced by nontoxic calcium hydroxide nanoparticles. Int J Mol Sci 25(15):8350. https://doi.org/10.3390/ijms25158350
73. Al-Saif AM, Sas-Paszt L, Mosa WFA (2024) Olive performance under the soil application of humic acid and the spraying of titanium and zinc nanoparticles under soil salinity stress. Horticulturae 10(3):295. https://doi.org/10.3390/horticulturae10030295
74. Bekkam R, Thiyagarajan C (2024) Foliar application of rice husk derived nanosilica for mitigating drought stress in hybrid maize by boosting antioxidant defense system. J Hazard Mater Adv 16:100482. https://doi.org/10.1016/j.hazadv.2024.100482
75. Direkvandy S, Eisvand HR, Azizi K, Akbarpour O, Smith DL (2024) Correction to: Mitigation impact of SiO_2 nanoparticles and mycorrhiza on wheat growth and yield under late-season drought stress (Cereal Research Communications, (2024), https://doi.org/10.1007/s42976-024-00526-2). Cereal Res Commun 1–14. https://doi.org/10.1007/s42976-024-00536-0
76. Li G, Ma Z, Zhang N, Li M, Li W et al (2024) Foliar application of silica nanoparticles positively influences the early growth stage and antioxidant defense of maize under low light stress. J Soil Sci Plant Nutr 24(2):2276–2294. https://doi.org/10.1007/s42729-024-01789-8
77. Abd-El-Aty MS, Kamara MM, Elgamal WH, Mesbah MI, Abomarzoka ESA et al (2024) Exogenous application of nano-silicon, potassium sulfate, or proline enhances physiological parameters, antioxidant enzyme activities, and agronomic traits of diverse rice genotypes under water deficit conditions. Heliyon 10(5):e26077. https://doi.org/10.1016/j.heliyon.2024.e26077
78. Habibi R, Sharfi S, Mohammadi S, Mahmoudi TM, Yazdanseta S (2024) Physiological, biochemical and hormonal response of wheat cultivars to foliar application of growth stimulants and zinc nano-chelate under water deficit stress. Plant Sci Today 11(3):245–256. https://doi.org/10.14719/pst.2946
79. Rashad HM, Mahmoud AWM, Alsamadany H, Alzahrani Y, Seleem EA et al (2023) Evaluation of nano-nitrogen fertilizers and other nitrogen sources on the performance of Guinea grass plants grown in newly reclaimed soil under water deficiency. Plant Stress 10:100282. https://doi.org/10.1016/j.stress.2023.100282
80. Seyed Sharifi R, Khalilzadeh R, Dadashzadeh S (2024) Effect of biofertilizer and nano-Fe oxide foliar application on alleviation of water deficit in yield and some physico-chemical properties of barley. Russ J Plant Physiol 71(1):1–12. https://doi.org/10.1134/S1021443723602938
81. Mi N, Hao W, Zhou Z, Li L, Wang F, et al (2022) Effects of amendments and indigenous microorganisms on the growth and Cd and Pb uptake of coriander (Coriandrum sativum L.) in heavy metal-contaminated soils. Toxics 10(8):408. https://doi.org/10.3390/toxics10080408
82. Eltahawy AMAE, Awad ESAM, Ibrahim AH, Merwad ARMA, Desoky ESM (2022) Integrative application of heavy metal–resistant bacteria, moringa extracts, and nano-silicon improves spinach yield and declines its contaminant contents on a heavy metal–contaminated soil. Front Plant Sci 13:1019014. https://doi.org/10.3389/fpls.2022.1019014

83. Abd El-Aty MS, Kamara MM, Elgamal WH, Mesbah MI, Behiry SI, et al (2023) Influence of foliar supplied of some biostimulants on physiological, agronomic characters and water productivity of rice under water deficit and normal conditions. Egypt J Soil Sci 63(4):455–464. https://doi.org/10.21608/ejss.2023.236602.1662
84. Badshah I, Mustafa N, Khan R, Mashwani ZUR, Raja NI et al (2023) Biogenic titanium dioxide nanoparticles ameliorate the effect of salinity stress in wheat crop. Agronomy 13(2):352. https://doi.org/10.3390/agronomy13020352
85. Boora R, Sheoran P, Rani N, Kumari S, Thakur R et al (2023b) Biosynthesized silica nanoparticles (Si NPs) helps in mitigating drought stress in wheat through physiological changes and upregulation of stress genes. Silicon 15(13):5565–5577. https://doi.org/10.1007/s12633-023-02439-x
86. Ratajczak K, Sulewska H, Panasiewicz K, Faligowska A, Szymańska G (2023) Phytostimulator application after cold stress for better maize (Zea mays L.) plant recovery. Agriculture 13(3):569. https://doi.org/10.3390/agriculture13030569
87. Chen F, Zhu L, Tang J, Wang C, Yue L et al (2022) Nanomaterial-modulated cellular sodium extrusion and vacuolar sequestration for salt tolerance. Environ Sci Nano 9(10):4018–4026. https://doi.org/10.1039/d2en00623e
88. Li Y, Liu J, Fu C, Khan MN, Hu J et al (2022) CeO2 nanoparticles modulate Cu-Zn superoxide dismutase and lipoxygenase-IV isozyme activities to alleviate membrane oxidative damage to improve rapeseed salt tolerance†. Environ Sci Nano 9(3):1116–1132. https://doi.org/10.1039/d1en00845e
89. Cheng B, Ding Z, Yue L, Chen F, Cao X, et al (2023) Carbon dots enhanced cold tolerance of lettuce (Lactuca sativa L.): scavenging reactive oxygen species, modulating hormones and up-regulating gene expression. Environ Sci Nano 10(10):2849–2860. https://doi.org/10.1039/d3en00257h
90. Chattha MU, Amjad T, Khan I, Nawaz M, Ali M et al (2022) Mulberry based zinc nanoparticles mitigate salinity induced toxic effects and improve the grain yield and zinc biofortification of wheat by improving antioxidant activities, photosynthetic performance, and accumulation of osmolytes and hormones. Front Plant Sci 13:920570. https://doi.org/10.3389/fpls.2022.920570
91. El-Bassiouny HMS, Mahfouze HA, Abdallah MMS, Bakry BA, El-Enany MAM (2022) Physiological and molecular response of wheat cultivars to titanium dioxide or zinc oxide nanoparticles under water stress conditions. Int J Agron 2022(1):3806574. https://doi.org/10.1155/2022/3806574
92. Li D, Zhou C, Ma J, Wu Y, Kang L et al (2021) Nanoselenium transformation and inhibition of cadmium accumulation by regulating the lignin biosynthetic pathway and plant hormone signal transduction in pepper plants. J Nanobiotechnol 19(1):1–14. https://doi.org/10.1186/s12951-021-01061-6
93. Zeeshan M, Hu YX, Iqbal A, Salam A, Liu YX et al (2021) Amelioration of AsV toxicity by concurrent application of ZnO-NPs and Se-NPs is associated with differential regulation of photosynthetic indexes, antioxidant pool and osmolytes content in soybean seedling. Ecotoxicol Environ Saf 225:112738. https://doi.org/10.1016/j.ecoenv.2021.112738
94. Hajizadeh HS, Asadi M, Zahedi SM, Hamzehpour N, Rasouli F et al (2021) Silicon dioxide-nanoparticle nutrition mitigates salinity in gerbera by modulating ion accumulation and antioxidants. Folia Hortic 33(1):91–105. https://doi.org/10.2478/fhort-2021-0007
95. Namjoyan S, Sorooshzadeh A, Rajabi A, Aghaalikhani M (2020) Nano-silicon protects sugar beet plants against water deficit stress by improving the antioxidant systems and compatible solutes. Acta Physiol Plant 42(10):1–16. https://doi.org/10.1007/s11738-020-03137-6
96. Khan MN, AlSolami MA, Basahi RA, Siddiqui MH, Al-Huqail AA et al (2020) Nitric oxide is involved in nano-titanium dioxide-induced activation of antioxidant defense system and accumulation of osmolytes under water-deficit stress in Vicia faba L. Ecotoxicol Environ Saf 190:110152. https://doi.org/10.1016/j.ecoenv.2019.110152
97. Wang Y, Zhang P, Li M, Guo Z, Ullah S et al (2020) Alleviation of nitrogen stress in rice (Oryza sativa) by ceria nanoparticles. Environ Sci Nano 7(10):2930–2940. https://doi.org/10.1039/d0en00757a

98. Anwar A, Wang Y, Chen M, Zhang S, Wang J et al (2024) Zero-valent iron (nZVI) nanoparticles mediate SlERF1 expression to enhance cadmium stress tolerance in tomato. J Hazard Mater 468:133829. https://doi.org/10.1016/j.jhazmat.2024.133829
99. Wang Y, Chen W, Gu X, Zhou D (2024a) Comparison of the arsenic protective effects of four nanomaterials on pakchoi in an alkaline soil. Sci Total Environ 912:168918. https://doi.org/10.1016/j.scitotenv.2023.168918
100. Mohamadi S, Karimi S, Tavallali V (2024) Differential responses of green-synthesized iron nano-complexes in mitigating bicarbonate stress in almond trees. Heliyon 10(3). https://doi.org/10.1016/j.heliyon.2024.e25322
101. Daler S (2024) Improving grapevine (Vitis vinifera L., cv. Superior Seedless) drought tolerance with cerium oxide nanoparticles: agronomic and molecular insights. Sci Hortic (Amsterdam). 338:113606. https://doi.org/10.1016/j.scienta.2024.113606
102. Li G, Duan R, Liang X, Liu H, Qin S et al (2024) Zinc oxide nanoparticles and nano-hydroxyapatite enhanced Cd immobilization, activated antioxidant activity, improved wheat growth, and minimized dietary health risks in soil-wheat system. J Environ Chem Eng 12(5):113574. https://doi.org/10.1016/j.jece.2024.113574
103. Sun H, Zhang B, Rong Z, He S, Gao Y et al (2023) Effects of nano-silicon dioxide on minerals, antioxidant enzymes, and growth in bitter gourd seedlings under cadmium stress. Acta Physiol Plant 45(10):1–9. https://doi.org/10.1007/s11738-023-03610-y
104. Singh N, Singh MK, Yadav RK, Azim Z (2023) Role of green synthesized nano iron oxide in alleviating the cadmium toxicity in Brassica oleracea var. italica seedlings. Plant Nano Biol 6:100055. https://doi.org/10.1016/j.plana.2023.100055
105. He S, Lian X, Zhang B, Liu X, Yu J, et al (2023) Nano silicon dioxide reduces cadmium uptake, regulates nutritional homeostasis and antioxidative enzyme system in barley seedlings (Hordeum vulgare L.) under cadmium stress. Environ Sci Pollut Res 30(25):67552–67564. https://doi.org/10.1007/s11356-023-27130-x
106. Shahzad R, Harlina PW, Khan SU, Koerniati S, Hastilestari BR et al (2024) Iron oxide nanoparticles alleviate salt-alkaline stress and improve growth by modulating antioxidant defense system in cherry tomato. J Plant Interact 19(1):2375508. https://doi.org/10.1080/17429145.2024.2375508
107. Boora R, Rani N, Kumari S, Yashveer S, Kumari N et al (2023a) Efficacious role of silica nanoparticles in improving growth and yield of wheat under drought stress through stress-gene upregulation. Plant Nano Biol 6:100051. https://doi.org/10.1016/j.plana.2023.100051
108. Ghorbani R, Ganjeali A, Movafeghi A, Nabati J (2023) Exposure to TiO_2 nanoparticles improves the physiological characteristics of drought-challenged chickpeas (Cicer arietinum L.). Legum Sci 5(4):e208. https://doi.org/10.1002/leg3.208
109. Alves DMR, de Oliveira JN, de Mello Prado R, Ferreira PM (2023) Silicon in the form of nanosilica mitigates P toxicity in scarlet eggplant. Sci Rep 13(1):1–13. https://doi.org/10.1038/s41598-023-36412-w
110. Rahmatizadeh R, Jamei R, Arvin MJ (2024) Silicon nanoparticles (SiNPs) mediate GABA, SOD and ASA-GSH cycle to improve cd stress tolerance in Solanum lycopersicum. Sci Rep 14(1):1–12. https://doi.org/10.1038/s41598-024-72908-9
111. Ayyaz A, Fang R, Ma J, Hannan F, Huang Q et al (2022) Calcium nanoparticles (Ca-NPs) improve drought stress tolerance in Brassica napus by modulating the photosystem II, nutrient acquisition and antioxidant performance. NanoImpact 28:100423. https://doi.org/10.1016/j.impact.2022.100423
112. Rehman A, Weng J, Li P, Shah IH, ur Rahman S, et al (2023) Green synthesized zinc oxide nanoparticles confer drought tolerance in melon (Cucumis melo L.). Environ Exp Bot 212:105384. https://doi.org/10.1016/j.envexpbot.2023.105384
113. Rezayian M, Niknam V, Arabloo M (2023) Iron nanoparticle regulate succinate dehydrogenase activity in canola plants under drought stress. Sci Rep 13(1):1–14. https://doi.org/10.1038/s41598-023-36105-4
114. Shafiq H, Shani MY, Ashraf MY, De Mastro F, Cocozza C, et al (2024) Copper oxide nanoparticles induced growth and physio-biochemical changes in maize (Zea mays L.) in saline soil. Plants 13(8):1080. https://doi.org/10.3390/plants13081080

115. Dawood MG, sayed El-Awadi ME, Sadak MS (2024) Chitosan and its nanoform regulates physiological processes and antioxidant mechanisms to improve drought stress tolerance of Vicia faba plant. J Soil Sci Plant Nutr 24(3):5696–5709. https://doi.org/10.1007/s42729-024-01934-3
116. Jabeen Z, Mahmood B, Rehman S, Butt AM, Hussain N (2024) Biodegradable glycine betaine encapsulated chitosan nanoparticles induce the expression of antioxidant enzyme genes to improve drought tolerance in maize. South African J Bot 171:571–582. https://doi.org/10.1016/j.sajb.2024.06.012
117. Hajihashemi S, Kazemi S (2022) The potential of foliar application of nano-chitosan-encapsulated nano-silicon donor in amelioration the adverse effect of salinity in the wheat plant. BMC Plant Biol 22(1):1–15. https://doi.org/10.1186/s12870-022-03531-x
118. Faizan M, Rajput VD, Al-Khuraif AA, Arshad M, Minkina T et al (2021) Effect of foliar fertigation of chitosan nanoparticles on cadmium accumulation and toxicity in solanum lycopersicum. Biology (Basel) 10(7):666. https://doi.org/10.3390/biology10070666
119. Behboudi F, Sarvestani TT, Kassaee ZZ, Sanavi MM, Sorooshzadeh S, et al (2018) Evaluation of chitosan nanoparticles effects on yield and yield components of barley (Hordeum vulgare L.) under late season drought stress. J Water Environ Nanotechnol 3(1):22–39. https://doi.org/10.22090/JWENT.2018.01.003
120. Savani KR, Gajera HP, Hirpara DG, Savaliya DD, Kandoliya UK (2023) Salicylic acid-functionalised chitosan nanoparticles restore impaired sucrose metabolism in the developing anther of cotton (Gossypium hirsutum) under heat stress. Funct Plant Biol 50(9):736–751. https://doi.org/10.1071/FP22309
121. Alenazi MM, El-Ebidy AM, El-shehaby OA, Seleiman MF, Aldhuwaib KJ et al (2024) Chitosan and chitosan nanoparticles differentially alleviate salinity stress in Phaseolus vulgaris L. Plants. Plants 13(3):398. https://doi.org/10.3390/plants13030398
122. Behboudi F, Tahmasebi-Sarvestani Z, Kassaee MZ, Modarres-Sanavy SAM, Sorooshzadeh A et al (2019) Evaluation of chitosan nanoparticles effects with two application methods on wheat under drought stress. J Plant Nutr 42(13):1439–1451. https://doi.org/10.1080/01904167.2019.1617308
123. Inam A, Javad S, Naseer I, Alam P, Almutairi ZM et al (2024) Efficacy of chitosan loaded zinc oxide nanoparticles in alleviating the drastic effects of drought from corn crop. Plant Stress 14:100617. https://doi.org/10.1016/j.stress.2024.100617
124. Dolatkhah Dashtmian A, Hosseini Mazinani SM, Pazoki A (2023) Exogenous chitosan nanoparticles modulated drought stress through changing yield, biochemical attributes, and fatty acid profile of common bean (Phaseolus Vulgaris L.) cultivars. Gesunde Pflanz 75(6):2463–2476. https://doi.org/10.1007/s10343-023-00912-6
125. Song Y, Zheng C, Li S, Chen J, Jiang M (2023) Chitosan-magnesium oxide nanoparticles improve salinity tolerance in rice (Oryza sativa L.). ACS Appl Mater Interfaces 15(17):20649–20660. https://doi.org/10.1021/acsami.3c00043
126. Fang J, Peng Y, Zheng L, He C, Peng S et al (2024) Chitosan-Se engineered nanomaterial mitigates salt stress in plants by scavenging reactive oxygen species. J Agric Food Chem 72(1):176–188. https://doi.org/10.1021/acs.jafc.3c06185
127. Bakhoum GS, Sadak MS, Tawfik MM (2022) Chitosan and chitosan nanoparticle effect on growth, productivity and some biochemical aspects of Lupinus termis L plant under drought conditions. Egypt J Chem 65(5):537–549. https://doi.org/10.21608/ejchem.2021.97832.4563
128. Wu H, Du P, Miao X, Hou R, Li S, et al (2024) O-carboxymethyl chitosan nanoparticles: a novel approach to enhance water stress tolerance in maize seedlings. Int J Biol Macromol 277:134459. https://doi.org/10.1016/j.ijbiomac.2024.134459
129. Das D, Bisht K, Chauhan A, Gautam S, Jaiswal JP et al (2023) Morpho-physiological and biochemical responses in wheat foliar sprayed with zinc-chitosan-salicylic acid nanoparticles during drought stress. Plant Nano Biol 4:100034. https://doi.org/10.1016/j.plana.2023.100034
130. Ahanger MA, Qi M, Huang Z, Xu X, Begum N et al (2021) Improving growth and photosynthetic performance of drought stressed tomato by application of nano-organic fertilizer involves up-regulation of nitrogen, antioxidant and osmolyte metabolism. Ecotoxicol Environ Saf 216:112195. https://doi.org/10.1016/j.ecoenv.2021.112195

131. Sammar Raza MA, Shah AN, Shahid MA, Nawaz M, Ibrahim MA, et al (2023) Nano-biochar enhances wheat crop productivity by vindicating the effects of drought: in relation to physiological and phenological stages. ACS Omega 8(41):37808–37819. https://doi.org/10.1021/acsomega.3c01629
132. Tourajzadeh O, Piri H, Naserin A, Cahri MM (2024) Effect of nano biochar addition and deficit irrigation on growth, physiology and water productivity of quinoa plants under salinity conditions. Environ Exp Bot 217:105564. https://doi.org/10.1016/j.envexpbot.2023.105564
133. Rasheed A, Anwar S, Shafiq F, Zaib-un-Nisa SK et al (2024) Physiological and biochemical effects of biochar nanoparticles on spinach exposed to salinity and drought stresses. Environ Sci Pollut Res 31(9):14103–14122. https://doi.org/10.1007/s11356-024-31953-7
134. Aljana MHN, Noaema AH, Alhasany AR, Leiby HR, Jazea RM (2022) Effect of using technology spraying by nano-seaweed extract and bio-nano fertilizer on growth of barley (Hordeum vulgare L.). Int J Agric Stat Sci 18:1461–1466. Accessed on 7 Oct 2024
135. Jaafar HS, Abd AA, Al-Bayati AS (2022) Vicia faba L. assessment for growth and productivity traits under magnetized irrigation water and nano-seaweed extract. Sabrao J Breed Genet 54(5):1216–1222. https://doi.org/10.54910/sabrao2022.54.5.22
136. Bai L, Geng S, Zhou Y, Ma H, Liu B (2024) Ultrasound-assisted fabrication and stability evaluation of okra seed protein stabilized nanoemulsion. Ultrason Sonochem 104:106807. https://doi.org/10.1016/j.ultsonch.2024.106807
137. Gontrani L, Bauer EM, Casoli L, Ricci C, Lembo A et al (2024) Inulin-coated ZnO nanoparticles: a correlation between preparation and properties for biostimulation purposes. Int J Mol Sci 25(5):2703. https://doi.org/10.3390/ijms25052703
138. Fu L, Wang Z, Dhankher OP, Xing B (2020) Nanotechnology as a new sustainable approach for controlling crop diseases and increasing agricultural production. J Exp Bot 71(2):507–519. https://doi.org/10.1093/jxb/erz314
139. Ahmadi SZ, Zahedi B, Ghorbanpour M, Mumivand H (2024) Comparative morpho-physiological and biochemical responses of Capsicum annuum L. plants to multi-walled carbon nanotubes, fullerene C60 and graphene nanoplatelets exposure under water deficit stress. BMC Plant Biol 24(1):1–19. https://doi.org/10.1186/s12870-024-04798-y
140. Du J, Liu B, Zhao T, Xu X, Lin H et al (2022) Silica nanoparticles protect rice against biotic and abiotic stresses. J Nanobiotechnol 20(1):1–18. https://doi.org/10.1186/s12951-022-01420-x
141. Silva S, Dias MC, Silva AMS (2023) Potential of MgO and $MgCO_3$ nanoparticles in modulating lettuce physiology to drought. Acta Physiol Plant 45(2):1–9. https://doi.org/10.1007/s11738-022-03507-2
142. Wu HZ, Gao Y, Zhang Y, Yu J, Kim DS et al (2023) Exogenous application of multi-walled carbon nanotubes (MWCNTs) and nano-selenium (Nano-Se) alleviated the PEG-induced water deficit stress and improved the crop performance of camelina. Agronomy 13(4):979. https://doi.org/10.3390/agronomy13040979
143. Bakry BA, Sadak MS, Al Ashkar NM, Ibrahim OM, Okla MK et al (2024) The role of carbon nanotubes in improving drought tolerance via upregulation of the physiological processes of peanut plants grown in sandy soils. Agronomy 14(3):611. https://doi.org/10.3390/agronomy14030611
144. Shukla G, Singh A, Chaudhary N, Singh S, Basnal N, et al (2024) Metal nanoparticles to improve the heat resilience in wheat (Triticum aestivum L.). Nanotechnology 35(20):205101. https://doi.org/10.1088/1361-6528/ad27af
145. Mahmoud NE, Abdelhameed RM (2023) Use of titanium dioxide doped multi-wall carbon nanotubes as promoter for the growth, biochemical indices of Sesamum indicum L. under heat stress conditions. Plant Physiol Biochem 201:107844. https://doi.org/10.1016/j.plaphy.2023.107844
146. El-Shal RM, El-Naggar AH, El-Beshbeshy TR, Mahmoud EK, El-Kader NIA et al (2022) Effect of nano-fertilizers on alfalfa plants grown under different salt stresses in hydroponic system. Agriculture 12(8):1113. https://doi.org/10.3390/agriculture12081113

147. Pinedo-Guerrero ZH, Cadenas-Pliego G, Ortega-Ortiz H, González-Morales S, Benavides-Mendoza A et al (2020) Form of silica improves yield, fruit quality and antioxidant defense system of tomato plants under salt stress. Agriculture 10(9):1–21. https://doi.org/10.3390/agriculture10090367
148. Constantinescu-Aruxandei D, Lupu C, Oancea F (2020) Siliceous natural nanomaterials as biorationals—plant protectants and plant health strengtheners. Agronomy 10(11):1791. https://doi.org/10.3390/agronomy10111791
149. Mathur P, Roy S (2020) Nanosilica facilitates silica uptake, growth and stress tolerance in plants. Plant Physiol Biochem 157:114–127. https://doi.org/10.1016/j.plaphy.2020.10.011
150. Mustafa G, Chaudhari SK, Manzoor M, Batool S, Hatami M et al (2024) Zinc oxide nanoparticles mediated salinity stress mitigation in Pisum sativum: a physio-biochemical perspective. BMC Plant Biol 24(1):1–15. https://doi.org/10.1186/s12870-024-05554-y
151 Haydar MS, Ali S, Mandal P, Roy D, Roy MN et al (2023) Fe–Mn nanocomposites doped graphene quantum dots alleviate salt stress of Triticum aestivum through osmolyte accumulation and antioxidant defense. Sci Rep 13(1):1–26. https://doi.org/10.1038/s41598-023-38268-6
152. Sarkar MM, Pradhan N, Subba R, Saha P, Roy S (2022) Sugar-terminated carbon-nanodots stimulate osmolyte accumulation and ROS detoxification for the alleviation of salinity stress in Vigna radiata. Sci Rep 12(1):1–17. https://doi.org/10.1038/s41598-022-22241-w
153. Fallah R, Gerami M, Ramezani M (2023) Beneficial role of multi-walled carbon nanotubes on physiological and phytochemical responses of Mentha piperita L under salinity stress. J Essent Oil-Bearing Plants 26(2):323–342. https://doi.org/10.1080/0972060X.2023.2182708
154. de Oliveira LKB, da Costa RS, de Silva JS, da Silva BA, de Lima KVG, et al (2024) Morphophysiology of cowpea under salt stress and application of carbon-based nanobiostimulant in the vegetative stage. Rev Bras Eng Agric Ambient 28(4):e279070. https://doi.org/10.1590/1807-1929/agriambi.v28n4e279070
155. González-García Y, López-Vargas ER, Pérez-Álvarez M, Cadenas-Pliego G, Benavides-Mendoza A et al (2022) Seed priming with carbon nanomaterials improves the bioactive compounds of tomato plants under saline stress. Plants 11(15):1984. https://doi.org/10.3390/plants11151984
156. Li Y, Sheng Y, Shu J, Hao S, Wang J et al (2024) Effects of nano-silica and multi-walled carbon nanotubes on grape seedlings under salt stress. Agronomy 14(3):622. https://doi.org/10.3390/agronomy14030622
157. Riaz M, Kamran M, Fahad S, Wang X (2023) Nano-silicon mediated alleviation of Cd toxicity by cell wall adsorption and antioxidant defense system in rice seedlings. Plant Soil 486(1–2):103–117. https://doi.org/10.1007/s11104-022-05588-x
158. Wang M, Mu C, Lin X, Ma W, Wu H, et al (2024c) Foliar application of nanoparticles reduced cadmium content in wheat (Triticum aestivum L.) grains via long-distance "Leaf-Root-Microorganism" regulation. Environ Sci Technol 58(16):6900–6912. https://doi.org/10.1021/acs.est.3c10506
159. Chandra J, Chauhan R, Korram J, Satnami ML, Keshavkant S (2020) Silica nanoparticle minimizes aluminium imposed injuries by impeding cytotoxic agents and over expressing protective genes in Cicer arietinum. Sci Hortic (Amsterdam) 260:108885. https://doi.org/10.1016/j.scienta.2019.108885
160. Mukarram M, Petrik P, Mushtaq Z, Khan MMA, Gulfishan M et al (2022) Silicon nanoparticles in higher plants: uptake, action, stress tolerance, and crosstalk with phytohormones, antioxidants, and other signalling molecules. Environ Pollut 310:119855. https://doi.org/10.1016/j.envpol.2022.119855
161. Zhou C, Miao P, Xu Z, Yi X, Yin X et al (2024b) Exploring the mechanism of nano-selenium treatment on the nutritional quality and resistance in plum plants. Ecotoxicol Environ Saf 284:116957. https://doi.org/10.1016/j.ecoenv.2024.116957
162. Alshegaihi RM, Saleem MH, Saleem A, Ali B, Aziz H, et al (2024) Silicon and titanium dioxide mitigate copper stress in wheat (Triticum aestivum L.) through regulating antioxidant defense mechanisms. J Plant Growth Regul 43(5):1519–1535. https://doi.org/10.1007/s00344-023-11204-8

163. Ahmed S, Ashraf S, Yasin NA, Sardar R, Al-Ashkar I, et al (2024) Exogenously applied nano-zinc oxide mitigates cadmium stress in Zea mays L. through modulation of physiochemical activities and nutrients homeostasis. Int J Phytoremediation https://doi.org/10.1080/15226514.2024.2383657
164. Chen J, Zeng X, Yang W, Xie H, Ashraf U et al (2021) Seed priming with multiwall carbon nanotubes (MWCNTs) modulates seed germination and early growth of maize under cadmium (Cd) toxicity. J Soil Sci Plant Nutr 21(3):1793–1805. https://doi.org/10.1007/s42729-021-00480-6
165. Sharifi P, Bidabadi SS, Zaid A, Abdel Latef AAH (2021) Efficacy of multi-walled carbon nanotubes in regulating growth performance, total glutathione and redox state of Calendula officinalis L. cultivated on Pb and Cd polluted soil. Ecotoxicol Environ Saf 213:112051. https://doi.org/10.1016/j.ecoenv.2021.112051
166. Xin X, Farid G, Nepal J, He S, Yang X, et al (2024) Comparative effectiveness of carbon nanoparticles and biochar in alleviating copper stress in corn (Zea mays L.). Chemosphere 355:141745. https://doi.org/10.1016/j.chemosphere.2024.141745
167. Chen Z, Ma Y, Ren Y, Ma L, Tang X et al (2024) Multi-walled carbon nanotubes affect yield, antioxidant response, and rhizosphere microbial community of scented rice under combined cadmium-lead (Cd–Pb) stress. Plant Physiol Biochem 213:108826. https://doi.org/10.1016/j.plaphy.2024.108826
168. de Faria Melo CC, Silva Amaral D, de Mello Prado R, de Moura Zanine A, de Jesus Ferreira D et al (2023) Nanosilica modulates C:N: P stoichiometry attenuating phosphorus toxicity more than deficiency in Megathyrsus maximus cultivated in an Oxisol and Entisol. Sci Rep 13(1):1–15. https://doi.org/10.1038/s41598-023-37504-3
169. Zhou C, Miao P, Dong Q, Li D, Pan C (2024a) Multiomics explore the detoxification mechanism of nanoselenium and melatonin on bensulfuron methyl in wheat plants. J Agric Food Chem 72(8):3958–3972. https://doi.org/10.1021/acs.jafc.3c08429
170. Jia L, Liu L, Zhang Y, Fu W, Liu X et al (2023) Microplastic stress in plants: effects on plant growth and their remediations. Front Plant Sci 14:1226484. https://doi.org/10.3389/fpls.2023.1226484
171. Shi R, Liu W, Lian Y, Wang X, Men S et al (2024) Toxicity mechanisms of nanoplastics on crop growth, interference of phyllosphere microbes, and evidence for foliar penetration and translocation. Environ Sci Technol 58(2):1010–1021. https://doi.org/10.1021/acs.est.3c03649
172. Zhuang M, Qiao C, Han L, Bi Y, Cao M, et al (2024) Multi-omics analyses reveal the responses of wheat (Triticum aestivum L.) and rhizosphere bacterial community to nano(micro)plastics stress. J Nanobiotechnol 22(1):1–18. https://doi.org/10.1186/s12951-024-02777-x
173. Guo S, Zhang X, Sun H (2023) Transcriptomic mechanism for foliar applied nano-ZnO alleviating phytotoxicity of nanoplastics in corn (Zea mays L.) plants. Sci Total Environ 905:166818. https://doi.org/10.1016/j.scitotenv.2023.166818
174. Liu D, Iqbal S, Gui H, Xu J, An S et al (2023) Nano-iron oxide (Fe_3O_4) mitigates the effects of microplastics on a ryegrass soil-microbe-plant system. ACS Nano 17(24):24867–24882. https://doi.org/10.1021/acsnano.3c05809
175. Karamchandani BM, Dalvi SG, Bagayatkar M, Banat IM, Satpute SK (2024) Prospective applications of chitosan and chitosan-based nanoparticles formulations in sustainable agricultural practices. Biocatal Agric Biotechnol 58:103210. https://doi.org/10.1016/j.bcab.2024.103210
176. Mawale KS, Giridhar P (2024) Chitosan nanoparticles modulate plant growth, and yield, as well as thrips infestation in Capsicum spp. Int J Biol Macromol 254:127682. https://doi.org/10.1016/j.ijbiomac.2023.127682
177. Arshad T, Lihong W, Alawadi HFN, Mahood A, Zia MA, et al (2024) Alleviating adverse effects of environmental stress in plants through chitosan application. Not Bot Horti Agrobot Cluj-Napoca 52(3):13538–13538. https://doi.org/10.15835/NBHA52313538
178. Chouhan D, Dutta P, Dutta D, Dutta A, Kumar A et al (2024) Effect of silver nanochitosan on control of seed-borne pathogens and maintaining seed quality of wheat. Phytopathol Res 6(1):1–17. https://doi.org/10.1186/s42483-024-00260-x

179. Munaro D, Mazo CH, Bauer CM, da Silva Gomes L, Teodoro EB, et al (2024) A novel biostimulant from chitosan nanoparticles and microalgae-based protein hydrolysate: improving crop performance in tomato. Sci Hortic (Amsterdam). 323:112491. https://doi.org/10.1016/j.scienta.2023.112491
180. Rani N, Kumari K, Hooda V (2024) Utilizing chitosan and titanium dioxide nanomaterials and their nanocomposites for improving the growth and biochemical responses of Sorghum bicolor (L.). Plant Sci Today 11(2):512–526. https://doi.org/10.14719/pst.2921
181. Steven S, Islam MS, Ghimire A, Methela NJ, Kwon EH, et al (2024) Chitosan-GSNO nanoparticles and silicon priming enhance the germination and seedling growth of soybean (Glycine max L.). Plants 13(10):1290. https://doi.org/10.3390/plants13101290
182. Sharma A, Kumar S, Singh R (2022) Synthesis and characterization of a novel slow-release nanourea/chitosan nanocomposite and its effect on Vigna radiata L. Environ Sci Nano 9(11):4177–4189. https://doi.org/10.1039/d2en00297c
183. Ali EF, El-Shehawi AM, Ibrahim OHM, Abdul-Hafeez EY, Moussa MM et al (2021) A vital role of chitosan nanoparticles in improvisation the drought stress tolerance in Catharanthus roseus (L.) through biochemical and gene expression modulation. Plant Physiol Biochem 161:166–175. https://doi.org/10.1016/j.plaphy.2021.02.008
184. Mohamed NG, Abdel-Hakeem MA (2023) Chitosan nanoparticles enhance drought tolerance in tomatoes (Solanum lycopersicum L) via gene expression modulation. Plant Gene 34:100406. https://doi.org/10.1016/j.plgene.2023.100406
185. Gad-Allah M, El-khamissi H, Yousef H, Elbakery AM (2024) Chitosan nanoparticles as combat salinity stress to improve biochemical characteristics and seedling vigor in maize (Zea maize L.). Al-Azhar J Agric Res 48(3). https://doi.org/10.21608/ajar.2023.226392.1212
186. Hayati A, Hosseini SM, Rahimi MM, Kelidari A (2022) The effect of humic acid and nanochelate iron on the absorption of some nutrients, grain yield, oil, and essential oil of (Nigella sativa) under drought stress. Commun Soil Sci Plant Anal 53(14):1744–1755. https://doi.org/10.1080/00103624.2022.2063319
187. Mahmoud AWM, Esmail SEA, El-Attar AB, Othman EZ, El-Bahbohy RM (2022) Prospective practice for compound stress tolerance in thyme plants using nanoparticles and biochar for photosynthesis and biochemical ingredient stability. Agronomy 12(5):1069. https://doi.org/10.3390/agronomy12051069
188. Seyed Hajizadeh H, Azizi S, Aghaee A, Karakus S, Kaya O (2023a) Nano-silicone and Ascophyllum nodosum-based biostimulant down-regulates the negative effect of in vitro induced-salinity in Rosa damascena. BMC Plant Biol 23(1):1–17. https://doi.org/10.1186/S12870-023-04584-2/FIGURES/10
189. Afify RRM, El-Nwehy SS (2023) Nano fertilizers with algae extract as biostimulant affecting growth, bulb yield, and quality of onion (Allium cepa L.). Sabrao J Breed Genet 55(6):2128–2139. https://doi.org/10.54910/sabrao2023.55.6.24
190. Hanif S, Sajjad A, Zia M (2023) Synergistic effect of glycine betain-ZnO nanocomposite in vitro for the amelioration of drought stress in coriander. Plant Cell Tissue Organ Cult 155(2):505–519. https://doi.org/10.1007/s11240-023-02476-9
191. Noaema AH, Alabdulla SA, Alhasany AR (2021) Impact of foliar application of seaweed extract and nano humic acid on growth and yield of wheat varieties. Int J Agric Stat Sci 16:1169–1174. https://connectjournals.com/03899.2020.16.1169. Accessed on 7 Oct 2024
192. Selim D (2022) Physiological responses of Phaseolus vulgaris to some nano bio-stimulants under salt stress conditions. Ann Agric Sci Moshtohor 60(3):809–820. https://doi.org/10.21608/assjm.2022.263823
193. Turan M, Ekinci M, Kul R, Kocaman A, Argin S, et al (2022) Foliar applications of humic substances together with Fe/Nano Fe to increase the iron content and growth parameters of Spinach (Spinacia oleracea L.). Agronomy 12(9):2044. https://doi.org/10.3390/agronomy12092044
194. Sayed EG, Desoukey SF, Desouky AF, Farag MF, EL-kholy RI, et al (2024) Synergistic influence of arbuscular mycorrhizal fungi inoculation with nanoparticle foliar application enhances chili (Capsicum annuum L.) antioxidant enzymes, anatomical characteristics, and

productivity under cold-stress conditions. Plants 13(4):517. https://doi.org/10.3390/plants13040517
195. Ogunkunle CO, Ahmed El-Imam AM, Bassey E, Vishwakarma V, Fatoba PO (2020) Co-application of indigenous arbuscular mycorrhizal fungi and nano-TiO_2 reduced Cd uptake and oxidative stress in pre-flowering cowpea plants. Environ Technol Innov 20:101163. https://doi.org/10.1016/j.eti.2020.101163
196. Alharby HF, Ali S (2022) Combined role of Fe nanoparticles (Fe NPs) and Staphylococcus aureus L. in the alleviation of chromium stress in rice plants. Life 12(3):338. https://doi.org/10.3390/life12030338
197. Rady MM, El-Shewy AA, Seif El-Yazal MA, Abd El-Gawwad IFM (2020) Integrative application of soil P-solubilizing bacteria and foliar nano P improves phaseolus vulgaris plant performance and antioxidative defense system components under calcareous soil conditions. J Soil Sci Plant Nutr 20(3):820–839. https://doi.org/10.1007/s42729-019-00168-y
198. Ayman M, Fahmy MA, Elnahal ASM, Alfassam HE, Rudayni HA et al (2024) Enhancing wheat tolerance to salinity using nanomaterials, proline, and biochar-inoculated with Bacillus subtilis. Heliyon 10(17):e37160. https://doi.org/10.1016/j.heliyon.2024.e37160
199. Hamed R, Jodeh S, Alkowni R (2024) Nano bio fertilizer capsules for sustainable agriculture. Sci Rep 14(1):1–17. https://doi.org/10.1038/s41598-024-62973-5
200. Hanif S, Zia M (2023) Glycine betaine capped ZnO NPs eliminate oxidative stress to coriander plants grown under NaCl presence. Plant Physiol Biochem 197:107651. https://doi.org/10.1016/j.plaphy.2023.107651
201. Adibah FSF, Jahan MS, Fatihah HNN, Alias N, Khandaker MM (2020) Effects of betaine-rich nano fertilizer in improving physiology of zea mays var. saccharata and arabidopsis thaliana under salt stress. Bulg J Agric Sci 26(1):177–185
202. Hallaji B, Haghighi M, Abolghasemi R, Mozafarian M (2024) Effect of foliar applications of aminolevulinic acid (bulk and nano-encapsulated) on bell pepper under heat stress. Plant Stress 12:100477. https://doi.org/10.1016/j.stress.2024.100477
203. Guerriero G, Sutera FM, Hoffmann J, Leclercq CC, Planchon S et al (2023) Nanoporous quercetin-loaded silicon-stabilized hybrid lipid nanoparticles alleviate salt stress in tomato plants. ACS Appl Nano Mater 6(5):3647–3660. https://doi.org/10.1021/acsanm.2c05384

Microbial Biosynthesis of Biostimulant Nanomaterials and Nanofertilizers

Joanna Trzcińska-Wencel, Patrycja Golińska, Aniket Gade, Pramod U. Ingle, Sudhir S. Shende, and Mahendra Rai

Abstract Enhancing crop productivity and soil fertility has driven extensive research into novel approaches for fulfilling demand via sustainable agricultural practices. The present chapter explores the emerging field of microbial biosynthesis as a sustainable and eco-friendly strategy for the synthesis of biostimulant nanomaterials (NMs) and nanofertilizers (NFs). Microorganisms, like bacteria and fungi, are studied for their unique ability to synthesize nanoparticles (NPs) with explicit properties valuable for plant growth and nutrient uptake. Present study provides an overview of the current challenges in agriculture, including nutrient deficiencies, soil degradation, and environmental concerns associated with traditional fertilization practices. It also delves into the principles of microbial biosynthesis and its impending applications in the production of biostimulant NMs. The complex processes involved in microbial-mediated nanoparticle synthesis, including extracellular and intracellular

J. Trzcińska-Wencel · P. Golińska · A. Gade · S. S. Shende
Department of Microbiology, Nicolaus Copernicus University in Toruń, Toruń, Poland
e-mail: trzcinska@doktorant.umk.pl

P. Golińska
e-mail: golinska@umk.pl

A. Gade
e-mail: aniket.gade@umk.pl

S. S. Shende
e-mail: sudhir.shende@umk.pl

A. Gade
Department of Biological Science and Biotechnology, Institute of Chemical Technology (Formely UDCT), Nathalal Parekh Marg, Mumbai, Maharashtra, India

P. U. Ingle · M. Rai (✉)
Nanobiotechnology Laboratory, Department of Biotechnology, SGB Amravati University, Amravati, India
e-mail: mahendrarai@sgbau.ac.in

P. U. Ingle
e-mail: pingle23@gmail.com

M. Rai
Department of Chemistry, Federal University of Piaui (UFPI), Teresina, Piaui, Brazil

© The Author(s), under exclusive license to Springer Nature Singapore Pte Ltd. 2025
A. Juárez-Maldonado et al. (eds.), *Plant Biostimulation with Nanomaterials*, Smart Nanomaterials Technology, https://doi.org/10.1007/978-981-96-4648-7_12

mechanisms, are discussed to understand the mechanism associated. The subsequent sections highlight the diverse types of biostimulant NMs and NFs generated through microbial biosynthesis, emphasizing their unique physicochemical characteristics and their impact on plant growth and development. Case studies and experimental findings showcase the efficacy of these microbial-derived NMs in promoting seed germination, root progression, and nutrient assimilation. Environmental sustainability of microbial biosynthesis, highlighting its potential to mitigate the ecological footprint accompanying conventional fertilizer production has also been discussed. The regulatory aspects, safety considerations, and future prospects of integrating microbial biosynthesized NMs into normal agricultural practices are also debated. Keen review of literature concludes the microbial synthesis of NMs is a promising avenue for the fabrication of biostimulant NMs.

Keywords Biostimulant · Growth promoter · Nutrient efficacy · Nutrient uptake · Microbial synthesis

1 Introduction

Globally there are various challenges faced by agriculture including food security for fast-growing populations and climate change. It is estimated that the world population will increase to 9.7 billion by 2050 [181] which will additionally need 60–70% food for survival of the global population. Moreover, climate change is another problem due to the enhanced use of agrochemicals including pesticides, fungicides, and huge amounts of chemical fertilizers. It is estimated that globally 3 billion kg of pesticide is used per year which will further increase at a fast pace [259]. In this context, there is an urgent need for alternative solutions. The use of biological control of the pests and pathogens is another strategy by employing natural products such as plant extracts or essential oils or the use of beneficial microbes including mycorrhizal fungi [83], plant growth promoting rhizobacteria (PGPR), phosphate solubilizing bacteria (PSB) [25, 154]. Although these methods are environment friendly, they suffer from one or another drawback due to changing environmental conditions and types of plant species.

Biostimulants can be defined as natural or synthetic substances suitable for application to soil, seeds, and plants for enhancement of their growth [58]. These are used in small quantities for the growth promotion of the plants. Different nanomaterials such as NPs and their oxides including copper nanoparticles (CuNPs), copper oxide nanoparticles (CuONPs), iron oxide nanoparticles (Fe_3O_4), and magnesium oxide nanoparticles (MgONPs) are used in minute quantities for enhancement of growth of plants. Such materials are used as NFs. Biostimulation by NMs depends upon the type of NPs and also their shape, size, surface charge, and agglomeration. According to Juárez-Maldonado et al. [127], this may take place in two stages. The first phase is a physicochemical phase that depends upon the interaction of surface charge, the second phase includes after entry of NMs and their interaction with cell membranes and proteins.

NMs can be synthesized by physical, chemical, and biological methods. The physical and chemical methods involve high temperature, pressure or toxic chemicals. Taking into consideration these facts, biogenic NMs are required which are green materials and support sustainable production. Recently, green synthesis has garnered the attention of scientists. The biogenic synthesis involves plants and microbes such as fungi, bacteria algae, etc. This type of synthesis can be carried out at ambient temperature and pressure. It is cost-effective, eco-friendly, and without toxicity [302]. The biosynthesized nanobiostimulants are very useful for the growth enhancement of plants leading to sustainable agriculture. Modern agriculture is facing a plethora of challenges, such as unpredictable climate change, environmental contamination, and a significant increase in demand for food, and therefore the need for modern and precise strategies to increase crop production is emerging [100].

Biostimulation is defined as the interaction between organisms and external factors or stimuli resulting in alterations in the organism by modification of cellular and metabolic processes with beneficial effects on plant growth. Biostimulant NMs are materials designed at the nanoscale to improve plant growth, development, and tolerance to abiotic and biotic stresses [127]. While these materials provide assurance in agriculture because of their potential to enhance crop yield and decrease reliance on conventional chemical inputs, they also establish environmental risks that require cautious apprehension.

This chapter will explore the role of different microbes in the biosynthesis of diverse NMs, challenges, their role as biostimulants and NFs, their interaction with the environment, and safety concerns. Moreover, the future perspective and direction of research will also be discussed.

2 Microbial Biosynthesis

NPs are gaining keen attention due to their diverse application in various industries and agriculture. Recently, newly fabricated NMs ranging between 10 and 1000 nm are primarily considered in agriculture and other applications [22]. Most NPs that are below 100 nm are known to interact with biological systems uniquely, which encourages the fabrication and development of multifunctional NMs for the desired application in agriculture [85]. Microorganisms play a vital role in synthesizing NPs in intra- and extracellular modes. Microbial synthesis of NPs is considered a green technology as they do not significantly harm the environment and ecosystem. The production of NMs using microbes is a bottom-up method where most reactions are oxidation and reduction of the precursors [131]. These biologically synthesized NPs are preferred over chemically and physically synthesized NPs because of their stability in colloidal state at room temperature, cost-effectiveness, high surface area, and surface energy [32]. Extracellular microbial synthesis is advantageous due to minimal downstream processing. Successive sections will focus on various microorganisms exploited for the synthesis of NMs, the proposed hypothesis of the mechanism of synthesis, factors affecting the synthesis, and the advantages of microbial synthesis [146].

2.1 Choice of Microorganisms

The synthesis of metal NPs using microorganisms, a process known as biosynthesis, gained significant attention due to the ability of microorganisms to synthesize the NPs at ambient temperature and pressure, its eco-friendly nature, and cost-effective approach. Microorganisms such as bacteria [226], actinomycetes [87], filamentous fungi [78], algae [82], viruses [168], and yeast [148] are commonly employed in this approach. The choice of microorganism plays a crucial role in determining the size, shape and concentration of the synthesized NPs, which in turn affect properties of the synthesized NPs. Some of the factors that influence the selection of microorganisms for nanoparticle synthesis are

(i) Inherent reducing agents and their concentration present in the microorganisms: Many microorganisms possess inherent reducing agents, such as enzymes, peptides, and secondary metabolites that can effectively reduce metal ions to form NPs.
(ii) Metabolism and biomolecules: The metabolic processes of microorganisms can influence the synthesis of NPs. The biomolecules produced during metabolism, such as enzymes, proteins, and polysaccharides, can act as capping or stabilizing agents for the NPs, preventing their agglomeration.
(iii) Tolerance of microorganisms: The growth conditions of microorganisms, including temperature, salt concentration and pH, are important factors. Microorganisms that can thrive under a wide range of these conditions provide flexibility in controlling the reaction parameters during nanoparticle synthesis.
(iv) Ease of cultivation: Microorganisms that are easy to cultivate and maintain in the laboratory are preferred for nanoparticle synthesis.
(v) Scale up potential: Microorganisms that can be easily scaled up for industrial production are preferred in practical applications. The scalability of the synthesis process is essential for commercial viability.
(vi) Non-pathogenic nature: The non-pathogenic microorganisms against humans are preferred for the synthesis of NPs, due to their ease pf handling and safety.

The choice of microorganisms for nanoparticle synthesis depends on various factors, including their inherent reducing capabilities, metabolic processes, tolerance to environmental conditions, ease of cultivation, scalability, and non-pathogenic nature. Researchers carefully consider these factors to optimize the synthesis process and tailor the properties of the NPs for specific applications.

2.2 *Microorganisms Involved in Nanomaterial Synthesis*

Numerous studies have reported the synthesis of a huge number of NMs by different types of microorganisms. These include bacteria, fungi, actinomycetes, algae, yeast, and viruses [248]. All these microbes are rich biomolecule sources responsible for stabilizing synthesized NPs within or outside the cells [151]. NP synthesis by microorganisms is primarily due to their ability to detoxify heavy metals by reducing the metal salts into NMs [12]. Various types of microorganisms applied for the synthesis of NMs are discussed below.

2.2.1 Bacteria

Different bacterial species are used for the NMs synthesis, mainly the extracellular approach, which is time efficient and requires very little downstream processing. Bacterial species used are the key to large-scale production, eco-friendly synthesis, and cost-effectiveness of the NPs. Various bacterial species include *Klebsiella pneumoniae, Enterobacter cloacae, Escherichia coli,* and *Bacillus licheniformis* [216]. Similarly, *Pseudomonas stutzeri* and *P. aeruginosa, Thiobacillus ferrooxidans, T. thiooxidans,* and *Sulfolobus acidocaldarius* are used for the NMs synthesis like silver and iron oxide NPs [119]. *Rhodopseudomonas capsulata* can synthesize gold NPs in different sizes and shape that is controlled by pH [98]. Lengke et al. [153] prepared gold NPs using filamentous cyanobacteria *Plectonema boryanum* UTEX 485, which produced cubic gold NPs with size range of <10–25 nm and octahedral gold platelets of ~1–10 μm size by interacting *P. boryanum* UTEX 485 with aqueous Au $(S_2O_3)_2^{3-}$ and $AuCl_4^-$ solutions and incubating at 25–100 °C for one month followed by one day heating at 200 °C. Other bacteria-mediated NPs are summarized in Table 1.

2.2.2 Fungi

The continuous use of fungi in nanobiotechnology has fascinated the worldwide consideration of producing green metallic NPs [173]. Mycogenic (i.e., fungus mediated) synthesis of NPs is preferred due to various advantages such as ease of handling, less nutritional requirements, greater biomass production, higher yield, greater tolerance, and non-pathogenicity towards human use [151, 173]. Extracellular fungal synthesis of metallic NPs has been considered in the last few decades to a large extent [59]. Cell-free extracts from various species of filamentous fungi such as *Fusarium oxysporium* [64], *Cryptococcus laurentii* and *Rhodotorula glutinis* [72], *Phoma glomerata* [38, 78, 117] are being explored as a nano factory for the synthesis of NPs. *Aspergillus* spp. (like *A. niger*, *A. flavus*, *A. terreus*, and *A. clavatus*) is another group of industrially important fungi capable of nanoparticle synthesis [219]. Some of the mycogenic NPs are listed in Table 1.

Table 1 Various types of nanoparticles synthesized by microorganisms

Sr. no.	Type of NPs	Name of microbe	Size in nm	Shape of NPs	References
Bacteria					
1	Silver	*Aeromonas* sp. SH10	6.4	–	Rai et al. [220]
2	Silver	*Bacillus cereus*	20–40	Spherical	Sunkar and Nachiyar [281]
3	Gold	*Deinococcus radiodurans*	43.75	Face-centered cubic structure of atomic gold crystals, spherical, triangular and irregular shapes	Li et al. [155]
4	Zinc sulfide	*Desulfobacteraceae*	2–5	Spherical	Labrenz et al. [150]
5	Magnetite	*Desulfovibrio magneticus* strain RS-1	Up to 30	Crystalline	Pósfai et al. [206]
6	Cadmium sulfide	*Escherichia coli*	2–5	Wurtzite crystal	Sweeney et al. [283]
7	Gold	*Geobacillus* sp.	5–50	Quasi-hexagonal	Correa-Llantén et al. [52]
8	Silver	*Klebsiella pneumoniae*	40–80	Pseudo-spherical shape	Sayyid and Zghair [250]
9	Cadmium sulfide	*Rhodopseudomonas palustris*	8.01 ± 0.25	Crystalline, face-centered cubic	Bai et al. [27]
10	Selenium	*Shewanella* sp. HN-41	<100	Roughly spherical	Ho et al. [101], Rajput et al. [226]
Fungi					
1	Silver	*Aspergillus niger*	20	Spherical	Gade et al. [77]
2	Silver	*Fusarium solani*	5–35	Spherical	Ingle et al. [116]
3	Silver	*Pleurotus sajor-caju*	5–50	Spherical	Nithya and Ragunathan [184]
4	Silver	*Rhizopus stolonifera*	9.47	Spherical	AbdelRahim et al. [4]
5	Silver	*Penicillium diversum*	10–15	Spherical	Ganachari et al. [80]
6	Cadmium sulfide	*Trichoderma harzianum*	3–8	Spherical	Bhadwal et al. [36]
7	Zinc oxide	*Aspergillus terreus*	28–63	Spherical	Baskar et al. [34]
8	Aluminium oxide	*Colletotrichum* sp.	30–50	Spherical	Suryavanshi et al. [282]

(continued)

Table 1 (continued)

Sr. no.	Type of NPs	Name of microbe	Size in nm	Shape of NPs	References
9	Iron oxide	*Aspergillus niger* BSC-1	20–40	Crystalline	Chatterjee et al. [47]
10	Titania, Silica	*Fusarium oxysporum*	10.2 9.8	–	Bansal et al. [33]
11	Zinc oxide	*Aspergillus fumigatus* TFR-8	1.2–6.8	Spherical and hexagonal	Raliya and Tarafdar [228]
12	Copper oxide	*Phoma glomerata* MTCC-2210	119	Spherical	Ingle et al. [118]
Yeasts					
1	Cadmium sufide	*Schizosaccharomyces pombe*	1–1.5	Wurtzite (Cd(16)S(20))-type hexagonal lattice	Kowshik et al. [143]
2	Lead sulfide	*Torulopsis* sp.	<5	Spherical	Kowshik et al. [144]
3	Cadmium sulfide	*Candida glabrata*	2–3	Hexamer	Dameron et al. [56]
4	Silver	Yeast strain MKY3	2–5	Twinned or multitwinned, some hexagonal	Kowshik et al. [142]
5	Gold	*Pichia jadinii*(*Candida utilis*)	1–1.5	Polymorphic	Sanghi and Verma [247]
6	Gold	*Yarrowia lipolytica* NCIM3589	Variable	Particles and plates	Pimprikar et al. [202]
7	Zirconium	Yeast sp.	Variable	Irregular mesoporous	Tian et al. [289]
Actinomycetes					
1	Gold	*Rhodococcus* sp.	20 ± 8	Spherical	Ahmad et al. [7]
2	Silver	Actinomycete strain SF23	3-36	Spherical	Wypij et al. [302]
Viruses					
1	Cadmium sulfide, Lead sulfide, Silica oxide, and Iron oxide	Tobacco mosaic virus	10–40	Nanotubes	Shenton et al. [265]
2	Zinc sulfide, Cadmium sulfide	M13 bacteriophage	560 × 20	Quantum dot nanowires	Mao et al. [168]
3	Gold	Tobacco mosaic virus	5	Spherical	Kobayashi et al. [139]
4	Lead	Tobacco mosaic virus	3–4	Nanotubes	Fan et al. [69]
5	Nanoconjugates	Hepatitis E virus	27–34	Icosahedral	Chen et al. [48]

(continued)

Table 1 (continued)

Sr. no.	Type of NPs	Name of microbe	Size in nm	Shape of NPs	References
6	Nanocarriers	Potato virus X	13	Helical	Le et al. [152]
Algae					
1	Gold	*Chlorella vulgaris*	2–10	Spherical	Annamalai and Nallamuthu [21]
2	Gold	*Spirogyra varians*	35	Quasi spherical	Salari et al. [241]
3	Lead	*Chlorella vulgaris*	5–20	Spherical	Garole et al. [82]
4	Silver	*Gelidium ammansii*	27–54	Spherical	Pugazhendhi et al. [213]
5	Zinc oxide	*Sargassum muticum*	0–57	Hexagonal	Sanaeimehr et al. [246]
6	Silver	*Caulerpa racemosa*	5–25	Spherical, triangular	Kathiraven et al. [133]
7	Copper oxide	*Bifurcaria bifurcata*	5–45	Crystalline	Abboud et al. [1]
8	Aluminum oxide	*Sargassum ilicifolium*	35	Hexagon	Koopi and Buazar [141]

2.2.3 Yeasts

Besides bacteria and filamentous fungi, the non-filamentous fungi that reproduce by budding are capable of producing NPs, which are usually monodisperse with well-defined morphology [49, 59, 134, 148, 301]. *Saccharomyces cerevisiae* synthesizes titanium dioxide (TiO_2) NPs and mesoporous magnetite (Fe_3O_4) NPs with an approximate size of 20 nm [124, 320]. Some species of yeasts have been reported to synthesize AgNPs, showing antifungal activity [68]. The enzyme from yeasts and fungi that is responsible for the production of NPs is nitrate reductase. Other NPs yeasts synthesize include selenium, lead, gold, silver etc. [59]. Various yeast species that are capable of producing NPs are *Nematospora coryli* [232], *Torulopsis* sp. [144], *Candida utilise* [247], *Yarrowia* sp. [202], etc. [173].

2.2.4 Actinomycetes

Other than bacteria, fungi and yeasts, actinomycetes have recently been explored for the synthesis of various types of NPs. Golinska et al. [87] summarized the various types of NPs synthesized by different species of actinomycetes. *Streptomyces* spp. are the most common among the species studied. NPs are intracellularly synthesized by actinomycetes over the surface of mycelia along with the cytoplasmic membrane. The possible mechanism of intracellular synthesis of silver NPs was proposed by

Sunitha and co-workers [280]. Similarly, several researchers have attempted the elucidation of the mechanism for intracellular synthesis of metal NPs using actinomycetes [31, 208, 279, 293]. The enzyme responsible for nitrate to nitrite conversion, i.e. nitrate reductase, is based on the mechanism of extracellular synthesis of NPs by *Streptomyces* sp. LK3 was hypothesized by Karthik et al. [132].

2.2.5 Other Microorganisms

The biological molecules such as proteins, viruses, and DNA self-assemble in a solution, forming engineering nanostructures to form functional nanoconstructs. These are used in various fields for the targeted delivery of active compounds [300]. Viral capsids can self-assemble around genetic materials under physiological conditions that allow modifications and alterations to form nanoconstruct. Plant viruses and bacteriophages like tobacco mosaic virus (TMV), red clover necrotic mosaic virus (RCNMV), cowpea chlorotic mottle virus (CCMV), bacteriophage MS2, bacteriophage P22 are intensively studied as they are safer to use in humans [59, 300].

Another group of aquatic oxygenic photoautotrophs, i.e., algae, can produce various NPs. *Caularepa racemosa, Nannochloropsis oculata, Dunaliella salina*, and *Chlorella vulgaris* can produce silver NPs [133, 175]. Other members, like *Rhizoclonium fontinale, Spirulina subsalsa* and *Lyngbya majuscule* can produce gold NPs. Table 1 summarizes the algae-facilitated synthesis of NPs [131]. Very little is known about protozoa as a nano factory and their role in NP synthesis and mechanism [59]. *Leishmania* sp. is explored to produce silver and gold NPs [231]. Marine protozoa foraminifera fabricates magnetic nanocomposites [167]. *Tetrahymena thermophila* SB210 can produce nano-Se ranging between 50 and 500 nm [53]. Thus, a wide range of microorganisms is available for the biological synthesis of various types of NMs, from metals and metal oxides to nanocomposites.

2.3 Mechanisms of Microbial Synthesis of NPs

Microorganisms are the novel factories of biomolecules that produce various metabolites intracellularly and secrete a variety of proteins and enzymes extracellularly. These metabolites are used as a reducing agent and capping materials during the synthesis of NMs. Many researchers have attempted and proposed hypothetical illustrations to describe the NP synthesis mechanism using biological methods involving biomolecules from their cell-free extracts. Bacteria secrete various enzymes, e.g. nitrate reductase, which is capable of converting monovalent silver ions (Ag^+) into the zero-valent core of silver NPs (Ag^0) by donating an electron [146]. Another mechanism states that the metal ions are initially captured inside the microbial cell or on the cell surface, where the reduction occurs to form a nanoparticle in the presence of enzymes. The bound enzyme serves as a nucleation site for the provision

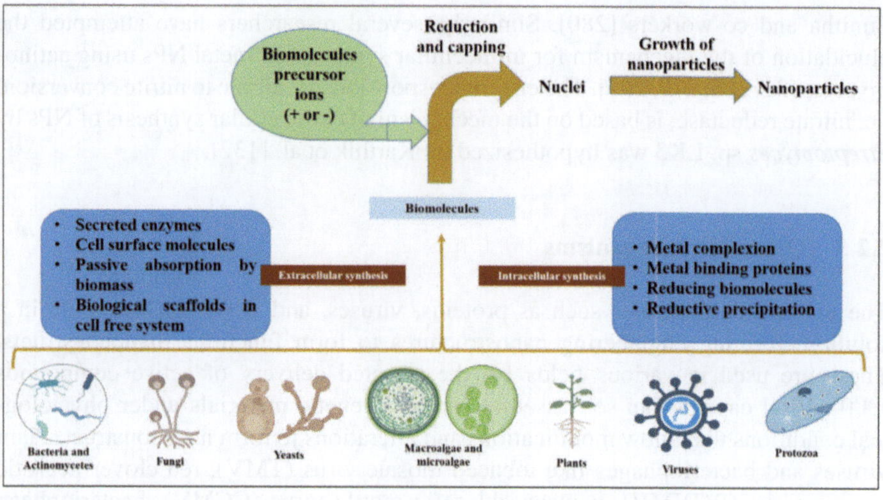

Fig. 1 Schematic representation of the microbial synthesis of NPs (Recreated from Ghosh et al. [85], Dhanker et al. [59], Ahmed et al. [12], Kumar [146])

of an electron for the reduction of the metal ions. Microorganisms also produce many polymeric substances which aid in the stabilization of the NPs [85]. Intracellular mechanism involves various metal complexes, binding proteins, and reductive precipitations associated with the NP synthesis. In contrast, extracellular synthesis involves secretory enzymes, cell surface molecules, passively absorbed biomass, biological scaffolds, etc. [59]. Figure 1 represents the mechanistic representation of nanoparticle synthesis by microbes.

2.4 *Factors Influencing Microbial Biosynthesis*

Although biological synthesis is a simple process, numerous factors affect the course of synthesis. These parameters define the stability, morphology, and physio-chemical properties of NMs. Ahmed et al. [12] have summarized the various factors that impose significant effects on the synthesis of NPs. Factors that influence NP synthesis include process parameters like temperature and concentration, the interactions between the precursor metal ions and the reducing mediators comprising process kinetics, and the adsorption kinetics involving the NMs and stabilizing agents [200]. As far as the physiological factors [120] are concerned, NP synthesis is affected by the type of water or solvent, salinity, organic chemical components that possibly produce material accumulation, and pH [122, 242]. For microbial synthesis selection of microbial species is imparted a scale-up of production methods, eco-friendly, and cost-effective [179] synthesis of NPs [216]. In the case of physical methods for synthesis, the desired shape of NPs (like nanowires, dendrimers, and tubes) is obtained by changing the surfactants and precursor molecules, using the heating technique, etc. [89]. Several important factors that influence the synthesis process are particle shape and size [84], the concentration of precursors and reactants [30, 287], time [255, 271], reaction temperature [30, 297], pressure [70], etc. [24].

2.5 Advantages and Challenges of Microbial Biosynthesis

Biological NPs show novel benefits over those synthesized by physical and chemical methods. As these NPs are cost-effective, less energy is consumed, they have safer synthesis methods, have fewer byproducts, minimal waste materials, and are very versatile to be applied in various fields like biomedical, pharmaceutical including agriculture [12, 94, 114]. Monodisperse NPs can be synthesized biologically by manipulating the pH, temperature, incubation period, and mixing ratio [274]. Capping of NPs by biological materials and biomolecules imparts them a biological activity that attributes a higher pharmacological activity to the biogenic NPs. Extracellular enzymes and bioactive components of actinomycetes make them favorable material for the biosynthesis of NPs [64, 65]. Bacteria-mediated NPs are preferably mono-disperse, highly pure, and small size range but are less tolerant to metal salt stress than fungi. Bacteria produce a number of NMs and need high maintenance and incubation time. Fungi produce polydispersed NPs of various types because of the enormous range of enzymes they produce. They have a higher capacity for metal ion accumulation and biomass production in a short period of incubation, preferring a scale-up process, making them economically feasible. Algae are abundantly available and accessible to cultivate under natural conditions. Large biomass production in small spaces and accumulation of high quantities of metal ions are their beneficial properties for use as biosynthesis materials. However, the incubation time for algal growth is very high, which might constrain production. Actinomycetes also require high maintenance, which may reduce their applicability. In contrast, they produce a range of antibiotic compounds that can impart biologically important properties to NPs. Viruses need a host cell for their growth, multiplication, and self-assembly. This ultimately causes an increase in the cost, limiting their use in the biosynthesis of NMs [12, 85, 317].

3 Diversity in Biostimulant Nanomaterials

The stimulation of plants in agriculture facilitates growth enhancement, improved resistance to adverse biotic or abiotic factors, and consequently increased yields [127]. The use of biostimulants is a frequent procedure in agriculture, as biostimulants, including organic and inorganic materials and/or microorganisms, increase nutrient uptake, stimulate growth, and increase stress tolerance or crop quality.

Currently, studies are focusing on improving products for use as biostimulants. Therefore, nanotechnology encompasses a wide range of scientific disciplines, from engineering and materials sciences, life sciences, as well as information technology, and handles the design, generation, exploration of the properties, application potential, and ecological impact of materials at nanoscale (size ranging between 1 and 100 nm) [46]. More recently, products of nanotechnology, such as nanoparticles (NPs), nanomaterials (NMs) and their nanocomposites, have been broadly explored regarding their potential in medicine, industry and agriculture e.g., as nanopesticides, NFs or biostimulants [88]. Among NMs are those synthesized by physical, chemical and biological methods, while due to simplicity, non-toxicity and cost effectiveness the green and biological methods continues to attract growing interest in recent years [253].

3.1 Types of Biostimulant Nanomaterials

In general, biostimulators in agricultural cultivation include microorganisms and substances (or mixtures of substances), excluding nutrients, plant hormones, or pesticides [236]. Jardin [123] summarized categories of plant biostimulants that include microorganisms (beneficial bacteria/fungi) as well as substances (humic and fluvic acids, protein hydrolysates, seaweed extracts and botanicals, biopolymers and inorganic compounds). NMs applied in agriculture are classified into inorganic, organic and hybrid materials. The inorganic materials mainly cover NPs of metals, metal oxides, non-metals, or carbon-based materials, whereas the organic NMs involve various biomolecules (e.g. proteins, peptides lipids or polysaccharides) in the form of micelles, nanogels, polymeric nanospheres/nanoparticles [172]. Recently, studies have indicated the high potential of biologically synthesized NMs as agents to facilitate agricultural crop improvement [221].

3.1.1 Inorganic Nanoparticles

To date, a large number of inorganic NPs have been successfully synthesized by green approach form a wide range of microorganisms e.g. bacteria, fungi, actinomycetes, yeast [218, 272, 273]. Recent studies on the different types of NPs (silver, gold, copper, magnesium, zinc NPs) biosynthesized by microorganisms are shown in Table 2.

Microorganisms have unique potential to synthesize NPs without using hazardous and expensive chemicals [272, 273]. The microbial-mediated synthesis of NPs is demonstrated by representatives of diverse bacteria (e.g. *Bacillus altitudinis* WM-2/2, *Klebsiella pneumoniae* SN35, *Shigella flexneri* SNT22), actinomycetes (e.g. *Streptomyces* sp. OSIP-1, *Streptomyces pseudogriseolus* Acv-11), fungi (e.g. *Aspergillus carbonarious* D-1, *Fusarium proliferatum* YNS2) or yeast (e.g. *Pichia kudriavzevii*

Table 2 Review of inorganic nanoparticles biosynthesized from microorganisms

Nanoparticles	Microorganism	UV-Vis-max absorbance λ (nm)	Average size (or size range) in nm	Reference
AgNPs	*Streptomyces* sp. OSIP-1	425	TEM: 8 (2.5–13)	Bakhtiari-Sardari et al. [29]
AgNPs	*Streptomyces* sp. OSNP-14	415	TEM: 15 (7–26)	Bakhtiari-Sardari et al. [29]
AgNPs	*Pichia kudriavzevii* HA-NY2	430	TEM: 12.4	Ammar et al. [20]
AgNPs	*Saccharomyces uvarum* HA-NY3	455	TEM: 20.65	Ammar et al. [20]
AgNPs	*Fusarium oxysporum*	465	TEM: 56 (50–100)	Abdel-Hadi et al. [2]
AgNPS	*Aspergillus oryzae* MTCC 3107	420	TEM: 40	Gupta and Saxena [91]
AgNPs	*Alternaria arborescens* MK629314	416; 404	HR-TEM: 6.5; 2.5	Alharbi et al. [14]
AgNPs	*Letendraea* sp. WZ07	390–450	TEM: 33.8 (8–56)	Qiao et al. [215]
AgNPs	*Trichoderma saturnisporum*	430	TEM: 10–70	Soliman et al. [277]
AgNPs	*Trichoderma harzianum*	409	DLS: 57.02	Guilger-Casagrande et al. [90]
AgNPs	*Lysinibacillus fusiformis* sp.	420	TEM: 50	Liu et al. [159]
AgNPs	*Fusarium culmorum* JTW1	430	TEM: 15.56 (3.58–58.63)	Trzcińska-Wencel et al. [291]
AuNPs	*Alternaria chlamydospora*	540	TEM: 12–15	Ameen et al. [17]
AuNPs	*Fusarium solani*	510–560	SEM: 40–45	Clarance et al. [51]
AuNPs	*Trichoderma hamatum* SU136	530	TEM: 5–30	Abdel-Kareem and Zohri [3]
AuNPs	*Trichoderma saturnisporum*	550	TEM: 8–31	Soliman et al. [277]
CuNPs	*Bacillus altitudinis* WM-2/2	330	TEM: 29.16–78.56 SEM: 29.11–51.41	Noman et al. [189]
CuNPs	*Klebsiella pneumoniae* SN35	338	SEM: 19.01–47.47	Noman et al. [188]
CuNPs	*Shigella flexneri* SNT22	334.5	TEM: 17.24–38.03	Noman et al. [186]
CuNPs	*Xenorhabdus* sp.	335	TEM: 5–20	Guru Bharathi et al. [92]
CuONPs	*Streptomyces zaomyceticus* Oc-5	400	TEM: 78	Hassan et al. [95]

(continued)

Table 2 (continued)

Nanoparticles	Microorganism	UV-Vis-max absorbance λ (nm)	Average size (or size range) in nm	Reference
CuONPs	*Streptomyces pseudogriseolus* Acv-11	400	TEM: 80	Hassan et al. [95]
CuONPs	*Fusarium proliferatum* YNS2	575	TEM: 12.5 (10–16) SEM: 12.5–54.87	Kalaimurugan et al. [129]
MgONPs	*Enterobacter* sp. RTN2	307	TEM and SEM: 38–57	Ahmed et al. [10]
MgONPs	*Lactobacillus* sp.	N.a	TEM: < 30	Mohanasrinivasan et al. [174]
MgONPs	*Rhizopus oryaze*	282	TEM: 20.38 (8.0–47.5)	Hassan et al. [96]
MgONPs	*Aspergillus carbonarious* strain D-1	400	TEM: 20.0–86.0	Fouda et al. [73]
ZnO-CuO NPs	*Aspergillus fumigatus*	375	TEM: 54.18 (14.73–72.07)	Gaber et al. [76]
ZnONPs	*Bacillus cereus* RNT6	382	TEM: 21–35	Ahmed et al. [11]
ZnONPs	*Bacillus subtilis* ZBP4	375	TEM and SEM: 22–59	Hamk et al. [93]
ZnONPs	*Penicillium expansum* ATCC 7861	N.a	TEM: 3.50–67.30	Abdelaziz et al. [5]

N.a.; not available

HA-NY2, *Saccharomyces uvarum* HA-NY3). According to the literature, microbial biomolecules (e.g., enzymes, polysaccharides) are involved in the formation of metallic cores and act as coating agents on NPs surface. For instance, Alharbi and coworkers [14] used laccase enzyme from endophytic strain *Alternaria arborescens* MK629314 as a reducing and capping agent for AgNPs biofabrication. FTIR results confirmed the presence of various functional groups on AgNPs surface, including free amino groups of enzyme. The high values of Zeta potential (−58 and −98 mV, depending upon purification of enzyme used for synthesis) indicated the long-term stability and non-agglomeration tendency of AgNPs in the solution [14]. Furthermore, the majority of bio-NPs have a small size (<100 nm), crystalline structure, and a high surface-to-volume ratio, which affects their bioactivity [253]. Given the properties of bio-NPs, increasing attention is being paid to their potential applications, particularly as biostimulants in crops. Therefore, research efforts are aimed at unraveling the mechanisms responsible for their beneficial effects on plant organisms, leading to more efficient growth and increased yields.

Investigations into the effects of metallic nanoparticles (MNPs) on plants mainly have focused on CuNPs [186, 188] or AgNPs [113]. However, the majority of studies endorse their potential to inhibit phytopathogens and/or pests, but also indicate a risk of their phyto- and eco-toxicity [291, 309]. In a study conducted by Ibrahim et al. [113], AgNPs synthetized from *Bacillus siamensis* strain C1, at a concentration of 10 μg mL^{-1}, stimulated the growth of rice (*Oryza sativa* L.) which was evidenced by increased root and shoot length, fresh and dry weight. Similary, application of AgNPs from *Bacillus cereus* SZT1 at concentrations range of 25–100 mg L^{-1} resulted in improved growth parameters and reduced damage in *Xanthhomonas*-infected rice plants [9]. Importantly, there is a concern that higher doses of AgNPs may impair plant development, as observed in tomato seedlings after seed nanopriming with AgNPs from *Trichoderma citrinoviride* at concentrations >25 μg mL^{-1} [278]. In turn, copper (Cu) as an essential trace element is associated with several proteins and metalloenzymes, and is involved in both the metabolism and nutrition of plants. Hence, it has been reported that biosynthesized Cu-based NPs act as fungicides, nanofertilizers as well as stimulators of seed germination and plant growth [95, 263]. For instance, increased leaf yield of tea with primary root disease (*Poria hypolateritia*) was observed after application of CuNPs synthesized from *Streptomyces griseus* isolated from the rhizosphere of tea plants. Studies suggested that improved yielding (compared to plants treated with conventional fungicide or bulk copper) resulted from antimicrobial activity of CuNPs as well as beneficial impact on total nutrient uptake (phosphorus and potassium) [205].

The research on stimulating crop production also extends to metal oxide nanoparticles (MONPs), and covers magnesium oxide (MgONPs), zinc oxide (ZnONPs), iron oxide (Fe_3O_4NPs), and titanium dioxide (TiO_2NPs) [13]. In addition, some of the studies focus on Mn_3O_4 NPs [161] or non-metal and semi-metal NPs, e.g. selenium (SeNPs) or silicon (SiNPs, SiO_2NPs) [28, 294]. Case in point, ZnONPs are recognized as highly important in agricultural application for plant growth and metabolism processes, as zinc is a crucial element for enzymes. They may function as biostimulants, plant fertilizers, pesticides, and most importantly, these NPs are considered safe [145]. Interestingly, Raliya and coworkers [229] reported ZnONPs synthetized by *Aspergillus fumigatus* as promoters of mung bean growth, which resulted in improved P-solubilization by rhizosphere microorganisms in response to ZnONPs application. Moreover, several studies involve the development of NPs that specifically enhance plant resistance to biotic and abiotic stresses. It has been reported that NPs can mitigate a variety of stresses, including salinity, heavy metal pollution, drought, infection or pest implications [225]. Plant exposure to stress factors led to accumulation of reactive oxygen species (ROS), including superoxide radicals (O_2^-), hydrogen peroxide (H_2O_2), and hydroxyl radicals (OH^-). ROS restricts the accessibility of tissues for pathogens, while also being responsible for signal transduction pathways in plant defense systems [99]. In this context, the beneficial role of biosynthesized ZnONPs in modulating plant growth and disease resistance enhancement were presented by many authors [180, 191, 230, 238]. Recently, Attia et al. [23] reported mycosynthesis of ZnONPs, CuONPs, and bimetallic ZnO-CuONPs from *Aspergillus fumigatus* OQ519856. Results showed that foliar application of all types of NPs on *Fusarium*-infected *Vicia faba* resulted in higher soluble protein content, lower MDA content and increased peroxidase (POD) and polyphenol oxidase (PPO) activity, as well as improved plant growth parameters indicating their potential to increase resistance to *Fusarium oxysporum* pathogen. Besides, Abdelfattah and coworkers [6] reported synthesis of small (27 nm) and spherical MgONPs by *Saccharomyces cerevisiae*. Field experiment on the growth and development of cowpea (*Vigna unguiculata* L.) after foliar application of bio-MgONPs (100, 150 and 200 ppm) showed significant increase in plant shoot, root and biomass and improved photosynthesis at concentrations of 150 ppm. In addition, MgONPs exhibited potential to suppress *Callosobruchus maculatus* Fab, what confirmed their insecticidal potential [6]. Similary, the improvement of plant growth parameters after application of MgONPs synthesized by *Aspergillus niger*, *Aspergillus flavus* TRF-12, and *Paenibacillus polymyxa* Sx3 at concentrations of 50–500 ppm, 15 mg L^{-1} and 16 mg L^{-1}, respectively were described [191, 230, 268]. It is suggested that MgONPs support plant defense systems due to extremely high surface areas and crystal structure, with notable edges/corners and reactive sites on the surface [115].

Furthermore, SeNPs synthesized from *Halomonas venusta* strain GUSDM4 at a concentration of 0.8 and 1 mg L^{-1} alleviated arsenate (As (V)) stress in rice crop var. *Jyoti*, by stimulation of antioxidants synthesis, namely proline and phenolic compounds [294]. While, application of SeNPs biofabricated extracellularly by *Lactobacillus acidophilus* ML14 at a concentration of 100 $\mu g\ L^{-1}$ resulted in a 75% lower susceptibility of wheat (*Triticum asetivum* L.) to crown and root rot diseases (CRDs) [66]. Modulation of plant immune response by microbially synthesized NPs have been reported in studies conducted by Noman and coworkers [187], where biogenic FeNPs synthesized by using cell-free supernatant of *Bacillus marisflavi* ZJ-4 were linked with salicylic acid (SA) and their effects on watermelon seedlings were evaluated. Both bio-FeNPs, as well as salicylic acid-loaded bio-FeNPs nanocomposites (SINCs), improved watermelon seedlings growth when compared to controls (untreated or treated with ionic Fe or SA). Moreover, suppressing the potential of bio-FeNPs and SINCs was confirmed by biochemical and molecular analyses, indicating induction of systemic acquired resistance (SAR) in plants [187]. In another study, Cr contaminated soil were supplemented with CuNPs synthesized from *Klebsiella pneumoniae* at concentration of 25 and 50 mg kg^{-1}, which resulted in higher plant growth and cellular antioxidants content, whereas lowered reactive oxygen species and Cr contents in plant tissues were observed [188]. The above research indicates the possibility of using NPs synthesized by microorganisms to support plant growth, especially under stress conditions.

3.1.2 Carbon-Based Nanomaterials

Carbon-based NMs (CNMs) consist of carbon atoms and are categorized by their geometric structure, namely, graphene is arranged in a nanostructure of a hexagonal network consisting of a single layer of carbon atoms, carbon nanotubes (CNTs) are hollow tubes made of rolled graphene sheets with multiple length, diameter, chirality, while fullerenes include a number of atomic clusters in the shape of ellipsoids or spheres. The synthesis methods of CNMs include arc discharge, laser ablation and chemical vapor deposition (CVD) [269]. There are limited investigations into the microbial synthesis of CNMs. For instance, there are some reports on the production of carbon dots with the use of citric acid, which is widely produced by the fungus *Aspergillus niger* [137, 295]. In another study, composite of CNTs and citric acid act as encapsulating agents for water-soluble chemicals and metal ions, thus providing an opportunity to formulate encapsulated bioactive molecules for use in agriculture [249]. In addition, studies focus on incorporation of CNMs into microbial polymer-based nanocomposites, including e.g. polyhydroxyalkanoates [275, 314], bacterial cellulose [310], pullulan [240] or curdlan [284].

According to the literature, carbon-based materials display both positive and negative effects on plant growth. In a study reported by Hatami [97] treatment of *Cucurbita pepo* L. seeds with multiwalled CNTs resulted in lower germination percentage and reduced biomass of seedlings. Recently, Xin et al. [304] reported improved growth of corn (*Zea mays L.*) after foliar and soil application of carbon NPs. Authors suggested that higher plant biomass resulted from increased nutrient uptake (N, P, K, Ca, Mg, Fe, Mn, Cu, Zn) and improved photosynthesis efficiency. Husen and coworkers [112] summarized effect of carbon NMs on crop plants and revealed that these NMs can penetrate seed coats and plant cell walls, depending on their size, concentration, solubility as well as plant species. CNMs have the potential to induce changes in crop growth and development, physiological and metabolic processes, as well as gene expression, which result in improved biomass or grain yield. Meanwhile, the authors pointed out the negative effects of CNMs, and suggested that the applied dose needs to be controlled within an acceptable range to avoid damage to crop species. Consequently, further research is needed due to the uncertain nature of the CNMs effects.

3.1.3 Organic Nanomaterials

The components of organic NMs consist mainly of small molecules, polymers, biomolecules (including polysaccharides or proteins), and their potential for use as nanocarriers of bioactive compounds is determined by their size and structure [44]. Therefore, polymer NPs are colloidal systems that typically range in size from 100 to 500 nm. The term is used to describe both polymer nanospheres/nanoparticles as well as polymer nanocapsules [214]. Polymer nanospheres are matrix particles in which all the mass is polymer, but they can also serve as carriers for biologically active molecules, which can be adsorbed either on the surface of the sphere or encapsulated in particles. In contrast, polymer nanocapsules are vesicular systems in which the bioactive agents are embedded into an aqueous core and surrounded by a protective polymer layer [162]. As a substrate for the preparation of polymeric nanoparticles both synthetic and natural origin polysaccharide might be used, including a wide range of bacterial EPS [162, 237]. Microbial polysaccharides used for formulation of nanocomposites are for instance: alginate, dextran, levan, pullulan, xanthan, curdlan, polyhydroxyalkanoates (PHA) [55, 63, 162, 217, 244]. It has been shown, that some bacteria spontaneously produce nanosized PHA granules under imbalanced conditions of growth, such as overabundant carbon sources or limitation of nutrients (e.g.

nitrogen or phosphorus) [140, 214]. In a study reported by Senthilkumar et al. [254], nanoprecipitation method was used for encapsulation of curcumin in polyhydroxybutyrate (PHB) synthetized from *Pseudomonas aeruginosa* SU-1, the fabricated nanocomposite shows potential as a tool for precise delivery of bioactive compounds in agriculture. In turn, synthesis of bacterial nanocellulose (BCN) from *Gluconacetobacter xylinus* was described by Kuo et al. [149] and by Zhao et al. [318]. Notably, Diaz-Ramirez et al. [60] used white grape pomace for reuse as a source of organic matter and microorganisms for the biosynthesis of bacterial sphere-like cellulose with potential application in agriculture due to high superabsorbent capacity and slow release ability. In addition, biosynthesis of silver and gold NPs with levan was reported by Ahmed et al. [8]. The selection of biostimulants and nanosystems for their delivery must be precise and consider the plant species, time of application as well as environmental conditions [127]. Milenković et al. [171] described the influence of CeO_2NPs coated with glucose, levan, and pullulan on pea and wheat growth. They indicated that the effect of all CeO_2NPs depended on the target plant, with the greatest effect of treatments during seed germination. All of CeO_2NPs tested stimulated shoot elongation, but only in wheat seedlings, while improved root elongation was observed in wheat after application of CeO_2NPs coated with glucose or levan and in pea by CeO_2NPs coated with pullulan.

3.2 Applications and Benefits of Biostimulant Nanomaterials in Agriculture

Application of biostimulant NMs in agriculture results in improved germination of seeds, plant growth, and yield quantity and quality. Some effects of application of microbially biosynthesized NPs on crop plants are summarized in Table 3.

The biostimulatory effect of nanoscale biomaterials is attributed to their extremely small size, high surface area to volume ratio, compared to their bulk counterparts, and natural components of capping agents on their surface [42]. The widespread physicochemical properties affect the biological activity of NMs, and accordingly, some of them support plant defense systems under stress, while those with catalytic activity make them resemble enzymes which are called enzyme NPs or nanozymes [195]. Whereas, further nanotechnology products have the potential to improve conventional biostimulants through nano-encapsulation or as nanocarriers for a wide range of chemicals. Diverse types of biological nanocarriers (e.g. CNMs, polymeric NPs, lipid NPs, nanogels, nanofibers) are being used to improve the effective and sustained release of biostimulants to deliver crop growth stimulants in the most accurate way [45]. The biostimulating effect is achieved through the appropriate delivery in time and dose of elements, and biomolecules that stimulate various processes to promote metabolic activity and improve resistance to various biotic and abiotic stresses, as well as better utilization of available nutrients [260].

Table 3 Biostimulating effect of microbially synthesized nanomaterials on plants

Type of nanomaterial	Microorganism	Applied contentrations	Effect on plants	References
AgNPs	*Bacillus siamensis* strain C1	10 μg mL^{-1}	Increase of rice (*Oryza sativa L.*) root length, shoot length, fresh weight and dry weight	Ibrahim et al. [113]
AgNPs, TiO_2NPs	*Trichoderma citrinoviride*	25 and 50 μg L^{-1}	Positive effect of TiO_2NPs on tomato seed germination and seedling vigor, reduced seed germination, seedling vigor, increase in the activities of antioxidant enzymes (catalase, superoxide dismutase, and peroxidase) in seedlings treated with AgNPs	Sonawane et al. [278]
AgNPs ZnONPs	*Fusarium solani* IOR 825	1–256 μg mL^{-1}	Sterilization effect on maize (*Zea mays*) seeds after treatment with AgNPs at the concentrations range 32–256 μg mL^{-1}; improved seedlings growth after seeds priming with ZnONPs (16–256 μg mL^{-1}) demonstrated by higher fresh and dry biomass	Trzcińska-Wencel et al. [292]
CuNPs	*Klebsiella pneumoniae*	25 and 50 mg kg^{-1} of soil	Increase of wheat (*Triticum aestivum* L.) growth, biomass, and cellular antioxidants contents, decreased the reactive oxygen species (ROS) and Cr translocation from Cr contaminated soil to plant roots and shoots	Noman et al. [188]

(continued)

Table 3 (continued)

Type of nanomaterial	Microorganism	Applied contentrations	Effect on plants	References
CuONPs	*Serratia* sp. ZTB29	100 mg L^{-1}	Improved the growth characteristics of maize (*Zea mays*) plants (shoot and root length, plant biomass, total chlorophyll)	Singh et al. [272]
MgOHNPs	*Aspergillus niger*	50–1000 ppm	Enhancement of maize (*Zea mays*) germination rate (at concentrations in range 200–700 ppm) increase in shoot and root length (<500 pmm), higher photosynthesis efficiency	Shinde et al. [268]
MgONPs	*Aspergillus flavus* TFR-12	15 mg L^{-1}	Improvement of shoot–root growth (18.2 to 49.2%) and chlorophyll photosynthetic pigment (76.1%) in clusterbean (*Cyamopsis tetragonoloba*)	Raliya et al. [230]
MoNPs	*Bacillus* sp. ZH16	100 mg L^{-1}	Improvement of the efficacy of *Bacillus* sp. ZH16—PGPR bacteria—under Asl stress conditions, enhancement of wheat (*Triticum aestivum* L.) growth, increased length, fresh and dry biomass of seedlings under regular and As stress conditions	Ahmed et al. [12]
SeNPs	*Halomonas venusta* GUSDM4	0.8 and 1 mg L^{-1}	Increase in germination and seed growth of rice crop var. *Jyoti* As(V) stress alleviation by increasing proline and phenolic contents	Vaigankar et al. [294]

(continued)

Table 3 (continued)

Type of nanomaterial	Microorganism	Applied contentrations	Effect on plants	References
SeNPs	*Penicillium chrysogenum*	20 ppm	Enhancement of the growth parameters of sunflower (*Helianthus annuus* L.), increase in chlorophylls, carbohydrates, proteins, phenolic compounds and free proline contents	Amin et al. [19]
SeNPs	*Lactobacillus acidophilus* ML14	100 μg L^{-1}	Reduced the incidence by 75% of wheat (*Triticum asetivum* L.) crown and root rot diseases (CRDs), enhanced plant growth, grain quantity and quality by 5–40%, increased photosynthetic pigments and gas exchange parameters, enhanced tolerance with drought and heat stress	El-Saadony et al. [66]
TiO_2NPs	*Aspergillus flavus* TFR 7	10 mg L^{-1}	Enhanced shoot and root growth, chlorophyll content and total soluble leaf protein in mung bean (*Vigna radiata*)	Raliya et al. [227]

(continued)

Table 3 (continued)

Type of nanomaterial	Microorganism	Applied contentrations	Effect on plants	References
ZnONPs	*Aspergillus fumigatus* TFR-8	10 ppm	Improvement in clusterbean (*Cyamopsis tetragonoloba* L.) biomass, shoot and root length, root area, chlorophyll content, total soluble leaf protein, rhizospheric microbial population, enzymatic activity (acid phosphatase, alkaline phosphatase and phytase) in rhizosphere	Raliya and Tarafdar [228]
ZnONPs	*Aspergillus fumigatus* TFR-8	10 mg L^{-1}	Improved mung bean growth (stem height, root volume), increased biochemical indicators (leaf protein and chlorophyll contents)	Raliya et al. [229]
ZnONPs	*Rhizoctonia bataticola* TFR-6	10 mg L^{-1}	Improvement in pearl millet (*Pennisetum americanum* L.) cv. HHB 67 shoot and root length, chlorophyll and total soluble leaf protein contents, dry biomass of plant, and enzyme activities (acid and alkaline phosphatase, phytase and dehydrogenase), improved grain yield	Tarafdar et al. [285]
ZnONPs	*Bacillus subtilis*	2, 4, 8 and 16 mg L^{-1}	Increase in maize (*Zea mays*) growth parameters (shoot and root length, leaf area), higher level of protein (maximum effect at 8 mg L^{-1} of ZnO NPs)	Sabir et al. [238]

(continued)

Table 3 (continued)

Type of nanomaterial	Microorganism	Applied contentrations	Effect on plants	References
ZnONPs, MnO_2NPs, MgONPs	*Paenibacillus polymyxa* Sx3	16 mg L^{-1}	Increased shoot and root length, fresh and dry weight of rice seedlings infected with *Xanthomonas oryzae* pv. *oryzae* (Xoo) strain GZ 0006 after application of all bio-NPs separately	Ogunyemi et al. [191]
Ag/SiO_2NC	*Escherichia coli* D8	40 ppm	Improved resistance of faba bean (*Vicia faba* L.) to *Botrytis cinerea* (higher total phenolic content, peroxidase and polyphenol oxidase activity), without cytotoxic effect to plant cells	Baka and El-Zahed [28]
FeO_x NPs, MnO_x NPs, bimetallic MnO_x/ FeO_xNPs	*Paenibacillus polymyxa*	10 g L^{-1}	Germination rates, root growth and fresh weight in maize (*Zea mays*) plantlets improved after application of bimetallic MnO_x/ FeO_x NPs	de França Bettencourt et al. [57]

4 Nanofertilizers

The fertility of the soil for crop plants is one of the primary determinants of crop abundance. Fertilizers are defined as any kind of material produced naturally or synthetically to supply one or more nutrients critical for plant growth. Based on nutrient delivered by fertilizers, two main types are identified: macronutrient fertilizers and micronutrient fertilizers. Fertilizers provide nutrients to maximize crop productivity and generally are applied in solid or liquid forms by broadcasting the soil surface or along with field irrigation [256]. However, inappropriate fertilization of agricultural land is a potential hazard causing damage to ecosystems [316]. It was estimated that use of the synthetic fertilizers per hectare of cropland increased from 8.54 (N), 7.94 (P), and 6.54 (K) kilograms in 1961 to 72.88 (N), 30.9 (P), and 25.20 (K) in 2020, respectively [111]. The efficiency of fertilization, and simultaneously the minimal adverse effects on the environment, are determined by the type of fertilizer, soil properties, plant adsorption capacities, as well as the time and form of application [185]. Recently, a wide range of studies suggest improved nutrient use efficiency by application of nanoproducts [128, 307]. In this context, this section provides an overview of nanoforms and nanocomposites containing the main macronutrients, with a focus on biological methods of synthesis and their application for the most efficient use and reduction of environmental pollution.

4.1 Chemical Versus Nanofertilizers

The low efficiency of using conventional fertilizers, as well as further contamination of the environment, is becoming a great concern for agriculture [39]. The nitrogen fertilizers in use consist of various substances, such as ammonium nitrates, urea, ammonium sulphate, tetraamines, and sodium/chloride nitrates, and are taken up by crop plants in the form of NH^{4+} cations or NO^{3-} anions. However, these products have some disadvantages and pose risks to humans and the environment [245]. For example, ammonium nitrate-based fertilizers pose a risk of detonation and can act as an oxidizer to fuel a fire. To prevent moisture absorption and reduce the potential risk, it is produced in granular form Marlair and Kordek [170]. Urea is the most concentrated form of solid nitrogen fertilizer and after application to the soil may be hydrolyzed to ammonium carbonate, causing a short-term increase in pH, affecting in multiple ways the properties and functioning of the soil ecosystems and subsequently plant growth [138]. To achieve the most effective use of nitrogen from fertilizers, the method of application, the timing, climatic conditions or soil type are of great significance [245]. Phosphorus fertilizers are generally recognized as low efficient due to their high immobilization in the soil and consequently lower availability to plants. Naturally occurring rock phosphate or phosphorite are commonly used to prepare P fertilizer, but the production process is expensive and its availability to plants is limited due to its low solubility [234]. Microbial solubilization of P is a natural process that occurs in the environment, and some studies have shown the use of microorganisms to convert various P-rich materials, or present in soil insoluble inorganic or organic form into more plant-available forms, among them NPs [16, 207, 313]. Most commonly used potassium fertilizers include two main forms, namely potassium chloride (KCl) or potassium sulphate (K_2SO_4), but formulation depends on target of delivery and may include potassium nitrate (KNO_3), sodium chloride (NaCl) or magnesium sulfate ($MgSO_4$) [251]. The modifications are needed to increase nutrient availability and plant uptake [307].

Nanotechnology approaches in comparison to conventional ones, consist of preparing fertilizers in the form of nanoproducts, by providing the fertilizer ingredient in the form of NPs or nanoemulsions, coating nutrients with a thin layer of polymer or its entrapment in nanoporous materials [190]. In comparison to fertilizers currently in use, nano-sized formulations possess unique characteristics, including a high surface area to volume ratio, sorption capabilities, and slow/controlled-release

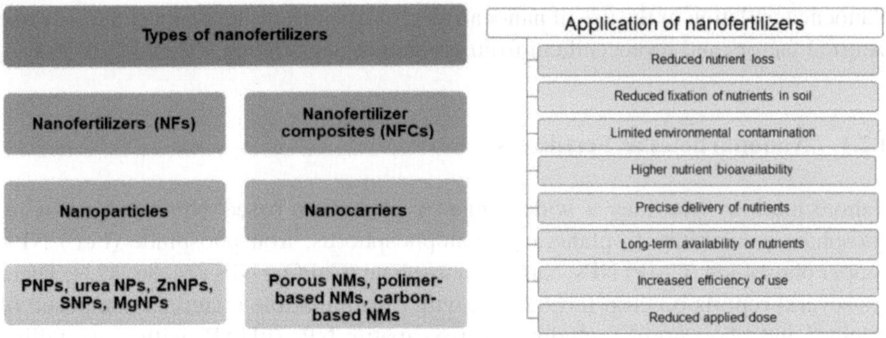

Fig. 2 The benefits of the use of various types of NFs in agriculture

properties [190]. The types of NFs and the benefits of their implementation are illustrated in Fig. 2. The use of NFs facilitates precise regulation and prolongation of nutrient delivery to plants, as well as a decrease in fertilizer outflow. It results in higher application efficiency, a significant reduction of nutrients losses and environmental contamination [192]. The effect of NFs application in crop production is associated with their strong stability, which reduces the degradation rate of the nutrients and enables adequate level of uptake by the plants. NFs formulations exhibit ability for slow or controlled release of fertilizers due to their high surface area and then ensure the long-term availability of bioactive ingredients at the recommended dose. An important issue related to the precise delivery of fertilizers using nanoscale products is also the reduction of environmental pollution, which affects all organisms, from bacteria to animals and humans [239].

4.2 Types of Nanofertilizers

The definition of NFs and their classification is based primarily on two key aspects, such as size and properties. NFs are defined as materials with a size of less than 100 nm, which simultaneously perform the function of a fertilizer by itself [257]. In this category belong NPs, e.g. nanoparticles of phosphorus, urea, magnesium, zinc, sulphur [190]. In contrast, another term is nanofertilizer composite/formulation (NFCs), where the NM is a carrier or encapsulating agent for the nutrient ingredient. These include carriers and encapsulating NMs such as polymer-based, carbon-based, nanogel forms, and porous NMs [257]. The main idea in applying NFs/NFCs is to increase in use efficiency of fertilizers by improving the delivery and uptake of nutrients by plants and reducing losses and environmental contamination. Therefore, the proper selection of nanoproducts and their composites becomes essential. Factors affecting NMs efficiency in nutrient delivery include intrinsic properties of NMs such as size, surface coatings, and nanostructure of NMs and extrinsic properties e.g. soil texture, organic matter, microorganisms, soil pH or exposure route (soil or foliar).

Nanoencapsulation or the use of nanocarriers make nutrients less exposed to adverse external factors and protects them from premature degradation [125].

4.2.1 Nanoparticles as Fertilizers

Nanotechnology provides a wide range of phosphate-based NPs for improving phosphate availability to plants e.g. nanophosphorus, iron phosphide (FeP) NPs, copper phosphide (Cu_3P) NPs, calcium phosphate ($CaPO_4$) NPs [74, 86, 321]. There are several reports on crop fertilization with biocompatible materials composed of calcium and phosphorus including hydroxyapatite NPs (nHAP) with a crystalline structure, or amorphous CaPNPs [43, 204]. While urea-based NPs have been investigated for their potential as a source of nitrogen [79]. Furthermore, there are also reports on the preparation and application of hybrid NPs e.g. nHAP dopped with urea, Mg and Zn [203, 258].

4.2.2 Microbial Biosynthesis of Nanofertilizers

The urea is one of the most widely used conventional nitrogen fertilizers in agriculture, but it has low nutrient use efficiency and a large part of it is dispersed into the environment. Recently, urea nanoparticles (TNU) were biosynthesized from commercial urea by co-cultured *Bacillus* sp. strains C1 and C2. TNU were spherical with size ranging from 2.2 to 45.6 nm and surrounded by the bacterial lipids, which indicated their potential to slow-release and improvement in nitrogen use efficiency [211]. In turn, Tarafdar and coworkers [286] reported synthesis of spherical phosphorus nanoparticles (PNPs) with sizes from 4.6 to 51.8 nm and an average diameter of 28.2 nm by *Aspergillus tubingensis* TFR-5. In another study, rhizospheric fungus *Aspergillus fumigatus* and tricalcium phosphate $Ca_3(PO_4)_2$ were used for the biosynthesis of PNPs with sizes of 10–15 nm and 40–100 nm [312]. While yeast-mediated biosynthesis of iron phosphate NPs with mesoporous structure with pore size of 13.9 and 14.8 nm were reported by Zhou et al. [320]. Priyam and coworkers [209–212] optimized bacterial-mediated method for nanohydroxyapatite (nHAP) (calcium phosphate—based NM) from dipotassium hydrogen phosphate (K_2HPO_4) and calcium chloride ($CaCl_2$) as precursors and culture supernatants of *Bacillus licheniformis.* The final nHAP were diverse in shape and size, namely spherical/platelet ones with average diameters about 35 nm, while the length and width of rod-like particles ranged between 98.9–129.7 nm and 36.6–48.4 nm, respectively. Results from HPLC indicated that organic acids (mainly gluconic acid) were responsible for solubilization of P from precursor and subsequent phases of synthesis involved gelation of P and Ca for nucleation and growth of NPs. In contrast, *Bacillus tequilensis*-mediated synthesis with an additional calcination step resulted in formation of smaller nHAP with needle-like morphology and average size of 30–60 nm (length) and 3–10 nm (width) [197]. Paramasivan et al. [198] used cell surface of Gram-negative bacteria *Serratia marcescens* for needle-like shaped hydroxyapatite NPs

(HANPs) synthesis. Moreover synthesis condition were optimized by substitution of $CaCl_2$ by $SrCl_2$, $MgCl_2$, and $ZnCl_2$, separately (at a concentration in the range 1–3%) and in combination (at a concentration of 1%). The dimensions of biofabricated HANPs ranged between 10–40 nm and 8–10 nm (length and width, respectively). Higher values were observed for Sr-substitution, where average length and width ranged between 20 and 50 nm.

4.2.3 Nanofertilizers Composites

Precise and controlled delivery of nutrients assumes the use of specific carriers e.g., polymeric nanosystems, nanoemulsions or mesoporous NMs [125]. The delivery of agrochemicals in the form of nanocomposites involves different release mechanisms, such as diffusion, dissolution or biodegradation, which are determined by the basic material used, but also by external factors such as the method of application, environmental conditions (e.g. soil microorganisms, temperature or humidity) [75].

Polymer-Based Nanocomposites

Polymer-based nanoformulations use a variety of biodegradable, synthetic or natural polymers. The most frequently used natural-derived ones are chitosan, pectin, and carboxymethylcellulose, while synthetic polymers include polyethylene glycol (PEG), polyvinyl alcohol (PVA), poly(lactic-co-glycolic acid) (PLGA). Among the most widely microbially synthesized polymers are dextran, gellan, kefiran, xanthan, pullulan, and polyhydroxyalkanoates (PHA) [244, 303]. The functionality of polymer-based delivery systems can be achieved by controlling the nanostructure of polymers or incorporating nanofillers. The development and generation process relies on the use of pre-existing polymers or by direct polymerization of monomers, therefore the selection of synthesis conditions contributes to the desired structure [296]. The use of polymer-based nanocomposites in nanocarrier systems is becoming increasingly common due to their high biocompatibility, biodegradability, low toxicity, and most importantly, their ability to release active ingredients in a controlled manner [201].

Nanoemulsions

Nanoemulsions are colloidal systems consistent of two immiscible liquids, where the dispersed phase is kinetically stable and forms droplet (with a limited size of 20–500 nm) in the continuous phase. The formation of a droplet with a small size from 10 to 300 nm, provides superior transparency, translucency, and stability when compared to systems with larger droplets [67]. The microbially derived emulsifiers generally encompass carbohydrates, oligosaccharides, lipids, polysaccharides, amino acids, peptides and proteins or its combinations [169]. Hence, glycolipid biosurfactants such as rhamnolipids were produced by *Pseudomonas aeruginosa* [26, 276], sophorolipids were synthesized by yeast e.g. *Starmerella bombicola* [37], while lipopeptide, surfactin, were fabricated by *Bacillus amyloliquefaciens* [319]. Nanoemulsions are versatile matrices for encapsulation, based on their ability to dissolve large amounts of poorly soluble nutrients as a result of their mutual compatibility. Besides, they represent a protective force for nutrients against enzymatic degradation and hydrolysis during storage or application [160].

Porous Nanomaterials—Based Formulations

Porous materials are classified according to their pore size, and are recognized as microporous (<2 nm), mesoporous (2–50 nm), and macroporous (>50 nm) materials, and among them are zeolites, silica NPs, nanoporous metal–organic frameworks, covalent organic frameworks [257]. Their properties allow for application as substrates for nanoferitilizer composites production. In this sense, for instance, mesoporous silica NPs due to the high entrapment capacity can act as carrier of urea [299]. The synthesis of mesoporous NPs was reported by bacterial strains e.g. *Bacillus subtilis* or *Serratia marcescens* [135, 198, 257, 306, 307].

4.2.4 Nanoscale Nutrient Delivery Systems

The physical and chemical properties of the soil, which include structure, the content of clay minerals, pH, cation exchange capacity (CEC), organic matter as well as soil microbiome considerably determine the fate, behaviour, immobilization, and bioavailability of fertilizers [256]. The premature degradation or undesired transformation impedes or inhibits the uptake of nutrients by plants, for this reason, NMs acting as carriers or encapsulating agents play a protective role and ensure their availability to plants by decelerating their release into the environment or neutralizing the adverse effects of environmental factors (such as soil properties, enzymes or excessive water) on nutrients content. Interestingly, hydroxyapatite plays a dual role, as a source of phosphorus for crops, but also as a carrier of other nutrients [196, 258]. For instance, urea interacts with nHAP through carbonyl groups and amino groups, H-bonding and metal–ligand interactions between N (urea) and Ca (nHAP), allowing slow release and preventing premature decomposition of urea in the soil, which occurs through excess water, volatilization or in the presence of the urease enzyme, prior to being effectively adsorbed by plants [165, 166]. The diversity of microbial metabolic pathways offers the possibility of using microbial cells as biofactories to produce nanostructured and functional agents, such as biopolymers, biomolecules or enzymes for preparation of nanoscale nutrient delivery systems with attractive agricultural applications [298]. In study conducted by Almasi et al. [15], CuNPs were incorporated into bacterial cellulose nanofibers (BCN) or chitosan nanofibers (CHN). The results indicated the potential of both nanohybrids for controlled and slow release, as after 72 h the level of CuO release from the materials was 65.5%

and 47.7% for the BCN and CHN composites, respectively. Recently, Yadav and coworkers [306] proposed use of *Bacillus circulans* MTCC 6811 for biosynthesis of mesoporous short rod-shaped silica NPs from coal fly ash (CFA). In another study, Zhao et al. [318] showed one-step in situ biomineralization by *Acetobacter xylinum* for bacterial cellulose silica nanocomposite production. Mesoporous NMs are a promising system for developing controlled release systems owing up to their attributes such as porosity, versatile surface functionalization, controllable pore size, stability, biocompatibility, high surface area and low toxicity [178].

4.2.5 Controlled-Release Nanofertilizers

In order to avoid excessive and harmful use of conventional fertilizers and assure long-term sustainability of nutrients to plant, nanotechnology provides a group of controlled-release nanofertilizers (CRNFs). To this category belong a wide range of bionanocomposites, including these based on NPs, biopolymer NMs and nanofillers. The various mechanisms of release depend on the inducing factor, e.g. thermal release is determined by reaching the required environment temperature, while moisture release is activated once the water is available, others include pH-dependent release, rapid release after contact with the plant surface, or slow release over a long period [239]. The release of nutrients from a nanocomposite is primarily related to its properties, which depend on the matrix and substrates used, the methods and conditions of preparation, as well as the structure acquired (e.g. the ratio of coating agent to nutrients). For instance, modification of nanohydroxyapatite surface by organic acids, e.g. citric acid, improves P-availability for plants by reducing the P-release rate [243]. Timely and dose-appropriate delivery of agrochemicals requires consideration of modifications occurring in the surrounding plant environment, such as abiotic or biotic stimuli (light, pH, temperature, humidity, enzymes, etc.), which serve as triggers to alter the physicochemical properties of nanocarriers resulting in controlled release [261]. An example is the use of nanocarriers sensitive to changes in pH, which is lowered during seedling development and plant growth due to the release of secretions containing organic acids, sugars or amino acids [305]. In addition, Sigmon and coworkers [270] investigated delivery of phosphorus (P) by biodegradable PHA-based nanocomposite. The results showed that the PHA-Ca-P nanocomposites were biodegradable by microbial activity in the soil, thereby allowing the gradual release of P during the early stages of plant growth. PHA-Ca-P PNCs promoted plant productivity relative to a conventional P source ($CaHPO_4$), while also significantly reducing P loss from the soil by more than 80%. Considering the harmful implications of eutrophication from phosphorus-rich agricultural runoff, those findings demonstrate the feasibility of using this type of composite to reduce nutrient losses and their negative environmental impacts [270].

4.3 Advantages and Challenges of Nanofertilizers Applications

It has been reported that NFs, including those synthesised by green approach showed beneficial effects on germination, growth and development of a wide range of crop plants, among them rice (*Oryza sativa* L.), cowpea (*Vigna unguiculata* L.), maize (*Zea mays*), eggplant (*Solanum melongena*), sunflower (*Helianthus annuus* L.), pearl millet (*Pennisetum glaucum* L.) and many others [5, 6, 10, 19, 57, 180, 292]. The supplementation with NFs results in improved germination percentages, increased root and shoot length, improved rates of photosynthesis, higher biomass and yield, depending on the type of fertilizer used, the application method, environmental conditions or the plant species. The small size and large surface area improve the bioavailability of nutrients to plants, CRNFs provide prolonged exposure to nutrients for plants, as well as less environmental pollution. Furthermore, balanced nutrition ensures better growth and less susceptibility to various biotic and abiotic stresses [307].

However, research on NMs as fertilizers also has to address market needs and regulatory requirements. The process of synthesis of nanocomponents should be optimized (high efficiency, low costs) for large-scale production. The identification of the stability of NMs and their sensitivity to various factors is an important aspect to ensure appropriate storage conditions and the preservation of the physicochemical properties that determine their effective activity [183]. Furthermore, it is essential to assess the performance of nanocomponents through comprehensive full-scale experiments, considering the growth of plants to full maturity and evaluating both the quality and quantity of yields. The results provide a basis for developing products with desirable properties and application strategies customized to crop requirements in comparison with currently used products [62]. Although the use of NFs and other nanoagrochemicals offers many benefits for agriculture, they may exhibit adverse effects associated with excessive or long-term use. The challenges of ensuring safety include testing of fate and behaviour of NFs among them solubility and dispersion, transport and transformations, as well as environmental interactions with microbes, plants, animals and humans. Potential ecotoxicity, immunotoxicity or genotoxicity should be considered not only by in vitro studies but also by in vivo evaluations [235, 271]. The following outcomes might be used for environmental risk assessment (ERA). Recently, the ERA of NMs has been facilitated by the use of machine learning and artificial intelligence, but there is still a lack of adequate and high-quality databases and standard guidelines [252]. Furthermore, there is still urgent need for studies related to the potential transgenerational effect of applied NMs, due to the limited number of available reports [164].

5 Environmental Implications and Safety Considerations

Despite the several roles and advancements in the field of nanobiotechnology, there is even now concern about the presence of NMs in environmental spheres, their fate, and the subsequent toxicity implications [176, 223, 224]. Thus, the focus of this section is to specifically present an overview of the environmental implications and safety considerations of the deliberate emission of NMs because of their enhanced applications, especially as biostimulant NMs in agriculture. There are several risk factors associated with biostimulant NMs applications for instance, (1) their longer persistence in the environment because of their tiny size, possibly leading to inadvertent accumulation in water bodies, soils, and accidental exposure to non-target organisms, (2) their uptake and bioavailability by plants and subsequent translocation into the food chain, raising issues about their bioaccumulation and potential effects on public health [262], (3) there are concerns that the biostimulant NMs could have toxicity influence on soil microbes, useful insects, and other non-target organisms [147, 223]; however, their effect on ecosystem and food chain is yet not completely known, (4) they may have unintentional impacts or altering the pH of soil, which is one of the critical parameters that affects the soil microbiota, its dynamics, nutrient availability and cycling, overall soil health, and plant growth and development [71, 264], as well as ecosystem functions are not still well-known, airborne concerns regarding potential the ecological disturbance to natural ecosystems, (5) there may be insufficient regulation and consideration of the risks pretended by biostimulant NMs, leading to impeding environmental and health risks because of inadequate testing and monitoring [50].

Besides this, biostimulant NMs, as with every promising technology, appear to have both prospects and potential interests, considering safety and environmental implications. Several safety concerns need to be considered, including (1) toxicity: biostimulant NMs could have NPs that may have toxic impacts on plants, animals, humans, or ecosystems unless they are not systematically tested for safety, (2) regulatory surveillance: regulation and surveillance could not be deep-rooted or adequately robust to address the exceptional challenges posed by NMs used in biostimulants, (3) exposure risks: there may be risks associated with exposure during the manufacture, employment or consumption of these stimulants, particularly if they are absorbed via the skin or inhaled by individuals [50, 262], (4) health effects: the long-term health effects of exposure to biostimulant NMs could not be completely known because of their novelty. Furthermore, several environmental impacts need to be considered, such as (1) bioaccumulation: there is a possibility that these NMs may accumulate in living organisms, leading to possible disruptions in food chains and ecosystems, (2) degradation and persistence: biostimulant NMs could persist in the environment for an extended duration, posing a risk of accumulation over time, (3) ecotoxicity: nano-scale materials may have adverse effects on soil, life underwater, and overall ecosystems if released in higher amounts or without appropriate control measures [50, 290]. Although biostimulant NMs hold promise to enhance agricultural yield and sustainable development, safety, and environmental concerns should be considered, and it would be a priority to warrant their sensible and sustainable use.

5.1 Ecotoxicity and Environmental Impact of Biostimulant Nanomaterials

In recent years, there has been an innovative advancement in the field of nanobiotechnology, with novel and diverse applications such as biostimulant NMs developing day by day. However, there is insufficient data about the toxicity of NMs and their fate in living systems. Inhalation, ingestion, and dermal dissemination are the possible exposure routes for the entry of NPs, while particle size, shape, surface area, and chemistry together define the NPs toxicity [262, 308]. The potential for nanobiotechnology is thought to be practically endless, and the potential for benefiting from creating and marketing these developments is the dynamic strength behind an exceptionally quick rush to disperse these applications in the market [136]. Though the manufacture, usage, and disposal of produced NMs lead to discharges into soil, air, and water systems, it is therefore important to investigate their transport into and through the environment, along with their impacts on ecosystems. The arbitrary use of NMs with unknown toxicological properties influences wildlife and human health by creating a variety of hazards for the environment [35]. Our knowledge about the harmful effects of biostimulant NMs is still very insufficient, and at this time, no specific regulations have been developed for the use of these formulations. There is a plausible threat to the exposure of humans and the environment to NMs during their life cycle, starting from production to disposal [315]. Unintended spillages or permitted release of waste from industries engaged in formulating biostimulant NMs into waterways and aquatic systems could result in direct exposure of NMs to humans through skin contact, inhalation of water aerosols, and direct ingestion of contaminated drinking water or particles adsorbed on vegetables or other foodstuffs. The tiny size of NMs facilitates their uptake into cells and translocation across epithelial and endothelial cells. They may also travel to other places in the body and interact with tissues, prolonging their stay in the body [35]. There is a prevalent accumulation of NMs in organs with high phagocytic activity, primarily in the liver, kidney, and spleen. The toxic effects of NMs have been documented at the cutaneous, cardiac, renal, pulmonary, cellular, and reproductive levels [262, 158].

In plant systems, the accumulation of NMs may alter the physiological process by reducing photosynthesis and transpiration rate, disturbing the cell integrity at cellular and sub-cellular organelles, and influencing plant growth and development [222, 233]. Cullen et al. [54] found that the alteration of nanoscale zero-valent iron (nZVI) in the soil caused a significant rise in pH in the soils they studied. Comparable information was claimed by Shi et al. [266], studying CuONPs effects on soil pH, where CuONPs exploit H^+ from the soil to generate Cu ions as well as $Cu(OH)^+$ ionic complex. Moreover, this process is confirmed to be further enhanced in acidic soil [266]. Similarly, Ag, Au, Ti, and Zn NMs have also been reported to influence soil pH, and their occurrences have been concurrent with adverse effects on helpful microbes and nematodes in the soil [81]. The level of harmful influences induced by the manifestation of NMs in soil is affected by their type and concentration, soil type, and the enzymatic activity of soil [267]. Additionally, an elevated concentration of NMs is associated with reduced dehydrogenase activity, which interrupts the nutrient equilibrium and fertility levels of the soil [126, 267]. Moreover, the absorption and incorporation of such biostimulant NMs by microorganisms considerably affect normal cell functioning by damaging the mycelium [18]. As soon as NMs enter water bodies, they act as contaminants and are ingested by lower aquatic biota. In line with this, metal-based NMs were tested by Dietz and Herth [61] in three different animal models signifying various trophic levels, including the *Danio rerio* (zebrafish), *Daphnia pulex* (daphnia), and *Pseudokirchneriella subcapitata* (microalgae), which demonstrated acute toxicity and filter-feeding. The authors noted that the NPs were more toxic to daphnia and microalgae than those of zebrafish, which is reasonable as daphnids are particle filter feeders and, therefore, are more sensitive to being exposed to NMs [61]. Therefore, after entering biological systems, NMs influence biochemical processes at the molecular as well as tissue-organ levels [156, 177]. The most common toxicological effects of NMs involved ROS-induced oxidative stress, DNA damage, disruption in protein folding, and cell death [182, 193, 311]. Thus, before the implementation of NMs as biostimulants, these ecotoxicity and environmental impacts should be considered to alleviate these risks and increase the benefits of biostimulant NMs in ways that promote sustainable applications [50]. In this direction, more research is required to systematically comprehend their activities in the environment, their short- and long-term effects on ecosystems, and the prospective related risks.

5.2 *Regulatory Considerations for Nanomaterial-Based Products*

To ensure the secure and reliable deployment of these nanomaterial-based products in agriculture applications, there should be developed regulatory frameworks. Regulatory considerations and risk assessment controlling biostimulant NMs include evaluating their efficiency, safety, and potential impacts on the environment. Some key considerations need to be taken into account during the development of regulatory frameworks and risk assessment as their hazard identification for determining the potential hazard related to persistence, toxicity, and effects on ecosystems including non-target organisms; evaluation of their exposure to understanding how these biostimulant NMs could enter and transport through different systems such as plants, animals, humans, and the environment by different routes and levels of their exposure; their risk characterization assessment for determining the possibility and significance of adverse impacts by biostimulant NMs on the basis of hazard and exposure information obtained during the evaluation [50, 163]. There is always a critical discussion about whether nanotechnology or nanotechnology-based products are entitled to extraordinary regulatory and administrative guidelines. This discussion is associated with the conditions in which it is necessary and appropriate to evaluate new substances prior to their release into the market, community, and environment [194, 199]. Regulatory bodies, like the United States Environmental Protection Agency (USEPA) and the Food and Drug Administration (FDA) in the U.S., on the other hand, the Directorate-General Healts and Consumer Protection (SANCO) of the European Commission, have originated to manage the potential risks exhibited by NMs. Until now, neither engineered NMs nor the substances and items that have them are subjected to any extraordinary regulations concerning their manufacturing, labeling, and handling. The Material Safety Data Sheet (MSDS) that should be distributed for specific materials frequently does not distinguish between bulk and nanoscale size of the material being mentioned, and nevertheless, when it does, these material safety data sheets are recommended only [41, 194]. In the Consumer Product Safety Commission (CPSC) and Nanotechnology, E. Marla Felcher anticipated that the CPSC, which is accused of confirming the general population in contradiction of the absurd risks of impairment and passing associated with consumer products, is poorly equipped to manage the safety of advanced products that utilize nanotechnology [40].

Besides this, some regulatory frameworks and guidelines should be developed, for example, for policy development by collaborating with regulatory bodies to launch robust and adaptive regulatory frameworks that warrant the safe, and reliable application of biostimulant NMs in agriculture [157]. Furthermore, there is a need for standardization and guidelines related to testing methods, the development of standardized methods, and policies for assessing the safety, efficiency, and environmental impacts of bistimulant NMs to accelerate regulatory sanction and market access. Regulatory considerations relate to the risk management strategies that could implement measures to alleviate identified risks, involving labelling obligations, restrictions on usage, guidelines for their application, and safety measures during their disposal. Moreover, data requirements for establishing the standard testing protocols for biostimulant NMs, such as toxicity studies, evaluations of efficiency, and assessments of environmental fate [50, 290].

Nonetheless, prescribed specific guidelines for using different NMs for marketing and application of nano-derived industrial products have not been prepared, which may be accepted equally. By addressing the challenges in investment, research quality, and participation of the private sector during the development of a robust regulatory framework for nanotechnology, it will be necessary particularly for India to completely leverage the potential of this emerging scientific area. Taking the above discussion into account, it can be suggested that only a few regulations with specific provisions for NMs have been framed, which opens up new prospects for debate about nanomaterial-specific regulatory guidelines and risk management. Such regulatory guidelines are immediately required in current circumstances for the appropriate and cautious application of nanotechnology products in daily life.

5.3 Risk Assessment and Management Strategies

Because of their unique physicochemical properties, NMs with biostimulant characteristics are becoming important. However, sometimes, because of additional interference, NMs could also have negative impacts, leading to major issues for plants, public health, and the environment. Thus, international regulatory bodies around the world have realized the risks associated with the applications of NMs. It is a priority to strike a balance between the NMs' efficiency and toxicity before applying such materials [130, 288]. International collaborations must be established for harmonization of standards and guidelines for the assessment and regulation of biostimulant NMs to confirm consistency and accelerate trade. Proper monitoring and post-market surveillance should be implemented to track the presence and behaviour of biostimulant NMs in the environment, involving their effects on ecosystems and community health. The post-market surveillance of approved biostimulant NMs after entry into the market must be continuously accessed for safety and efficacy to detect and report any unexpected risks or adverse impacts [50]. Community involvement and stakeholder engagement are also important to ensure transparent communication between industry stakeholders, regulatory organizations, researchers and scientists, and the general public to foster trust, share information, and address concerns. Industrial engagements in safety analysis and regulatory processes while confirming compliance with verified standards should be encouraged. In a review by Jain et al. [121], it is argued that NMs that are anticipated to be employed in food products and medicine should be investigated carefully for their toxicity by in vitro, in silico, and in vivo analysis. Many government agencies, such as the Royal Academy of Engineering, USEPA, Royal Society, EU's Scientific Committee, and several other non-governmental organizations, like Friends of the Earth and Xpert Arena, are actively operational on NMs' potential toxicity and risk assessment [121].

In India, the regulatory framework for managing the potential risks associated with nanotechnology is still not fully developed. However, there are guidelines in place for the secure handling of NMs in research and industry; there are no specific regulations for nanotechnology [105]. Nevertheless, existing waste and environmental management laws, like the Environment Protection Act of 1986 [103] and the Hazardous Waste (Management, Handling, and Transboundary Movement) Rules of 2008 [108], provide a foundation for regulating nanotechnology waste and categorizing NMs as harmful materials. The prevailing chemical regulations, like the Manufacture, Storage, and Import of Hazardous Chemical Rules, 1989 [104], and the Chemical Accidents (Emergency Planning, Preparedness, and Response) Rules, 1996 [106], might also be relevant to NMs; nonetheless, there is a necessity for further transparency and definition in this field. The application of different nano-based products could fall under various regulations, like the Drug and Cosmetics Act of 1940 [110], the Insecticide Act of 1968 [102], and the Food Safety and Standards (Contaminants, Toxins, and Residues) Regulations of 2011 [109]. But these regulations do not address the current potential risks established by NMs. The recent institutional mechanisms for examining and executing environmental laws are supervised by the

Central Pollution Control Board and respective State Pollution Control Boards, but they are not prepared to control the risks associated with NMs. Safety regulations and occupational health, like the 'Guidelines and Best Practices for Safe Handling of Nanomaterials in Research Laboratories and Industries,' are critical for the protective handling and disposal of NMs in research and industrial settings [107]. Some important regulations have been established by the EU for supporting exclusive requirements for NMs' direct or indirect usage, as given in Table 4 [121, 130].

Addressing the regulatory considerations and conducting a comprehensive risk assessment for their management are crucial steps in confirming the safe and responsible application of biostimulant NMs in agriculture while diminishing the potential negative impacts on community health and the environment. The bio-based NMs and biostimulants from the waste biomass have the potential to counteract several problems that plague agriculture, which could be helpful in managing the risks associated with the technology [290]. Overall, there is a necessity for a comprehensive regulatory framework that explicitly addresses the potential risks associated with nanotechnology in India and at the global level. Furthermore, continued research, collaborative activities, and adaptive monitoring frameworks will be necessary to manage the risks associated with this novel, evolving technologies.

Table 4 List of important regulations recognized by "European Union" for providing specific provisions for nanomaterials' direct or indirect usage [121, 130]

Name of regulation and specific features	Particulars
Regulation (EC) no 1333/2008 For food additives	States that a food additive already authorized but obtained by nanotechnology requires an evaluation before marketing
Regulation (EC) no 1332/2008 For food enzymes	States that a food enzyme already included in the community list but prepared by different methods or using different starting materials from those included in the risk assessment of authority should be submitted for re-evaluation
Regulation (EC) 450/2009[a] Active and intelligent materials and articles intended to contact with food	Though NMs are not directly stated, there is a direct reference to "substances deliberately engineered to particle size that exhibit functional physical and chemical properties that significantly differ from those at a larger scale"; therefore, a case-by-case analysis must be followed for active and intelligent materials and articles containing NMs
Regulation (EU) no 10/2011[a] Plastic materials and articles intended to be used to contact with food material	States that substances with NMs must only be used if stated in Annex I of the regulation

[a] It can be noted that both regulations 450/2009 and 10/2011 state the functional barrier concept which means that it is not directly applicable to NMs

6 Conclusion and Future Perspectives

The synthesis of NMs using diverse microorganisms such as bacteria, fungi, and algae offers significant advantages as a low-cost, rapid, and environmentally friendly process. These green synthesized NMs are required in minimum quantity for plant growth. A comprehensive understanding of the mechanisms behind green synthesis, the influence of various environmental factors, and the challenges involved are important for nurturing sustainable agriculture.

The application of nano-based biostimulants and NFs holds promise in mitigating the environmental impact of greenhouse gases and other detrimental effects. The chapter delves into different types of NM-based biostimulants, including metallic NPs, metal oxide NPs, carbon-based NPs, and organic NMs.

The discussion on the application of NMs and NFs as biostimulants and their role in promoting plant growth is particularly relevant for reducing the pollution generated by the extensive use of chemicals in modern agriculture. Addressing environmental hazards and safety is pivotal in the development and application of biostimulant NMs. Detailed consideration of ecotoxicity and regulatory issues underlines the importance of balancing innovation with environmental accountability.

Given their eco-friendly nature, the future of biosynthesized NMs for application as biostimulants and NFs appears highly promising. Current trends in biosynthesis indicate a need for deeper understanding through thorough studies in this area. Further studies on novel microorganisms and the optimization of biosynthetic pathways have the potential to lead to breakthroughs in the development of more sustainable and eco-friendly biostimulant products. The ongoing exploration of these avenues holds the key to advancing environmentally responsible agriculture practices.

Acknowledgements J.T.W. acknowledges grant No. 2022/45/N/NZ9/01483 from National Science Centre, Poland. Mahendra Rai thanks to CNPq Brazil for financial assistance (Processo: 100125/2021-2) and Federal University of Piaui (UFPI), Teresina, Brazil. SSS and PG acknowledge the YUFE4Postdocs (Y4P) programme, financially supported by European Union funds as a part of the Marie Sklodowska-Curie Actions Cofund programme (Agreement No. 101081327), Ministry of Science and Higher Education under the "International Co-financed Projects" programme (Agreement No. 5678/HE/2023/2024/2). AG acknowledges the National Science Centre and the European Union Framework Programme (Project No. UMO-2022/45/P/NZ9/01571), co-funded by Research and Innovation Horizon 2020 under the Marie Skłodowska-Curie grant agreement No.945339.

References

1. Abboud Y, Saffaj T, Chagraoui A, El Bouari A, Brouzi K, Tanane O et al (2014) Biosynthesis, characterization and antimicrobial activity of copper oxide nanoparticles produced using brown alga extract (*Bifurcaria bifurcata*). App Nanosci 4:571–576. https://doi.org/10.1007/s13204-013-0233-x
2. Abdel-Hadi A, Iqbal D, Alharbi R, Jahan S, Darwish O, Alshehri B, Fatima F (2023) Mycosynthesis of silver nanoparticles and their bioactive role against pathogenic microbes. Biology 12(5):661. https://doi.org/10.3390/biology12050661

3. Abdel-Kareem MM, Zohri AA (2018) Extracellular mycosynthesis of gold nanoparticles using *Trichoderma hamatum*: optimization, characterization and antimicrobial activity. Lett App Microbiol 67(5):465–475. https://doi.org/10.1111/lam.13055
4. AbdelRahim K, Mahmoud SY, Ali AM, Almaary KS, Mustafa AMA, Husseiny SM (2017) Extracellular biosynthesis of silver nanoparticles using *Rhizopus stolonifer*. Saudi J Biol Sci 24:208–216. https://doi.org/10.1016/j.sjbs.2016.02.025
5. Abdelaziz AM, Salem SS, Khalil AM, El-Wakil DA, Fouda HM, Hashem AH (2022) Potential of biosynthesized zinc oxide nanoparticles to control *Fusarium* wilt disease in eggplant (*Solanum melongena*) and promote plant growth. Biometals 35:601–616. https://doi.org/10.1007/s10534-022-00391-8
6. Abdelfattah NA, Yousef MA, Badawy AA, Salem SS (2023) Influence of biosynthesized magnesium oxide nanoparticles on growth and physiological aspects of cowpea (*Vigna unguiculata* L.) plant, cowpea beetle, and cytotoxicity. Biotechnol J 18:2300301. https://doi.org/10.1002/biot.202300301
7. Ahmad A, Senapati S, Khan MI, Kumar R, Ramani R, Srinivas V et al (2003) Intracellular synthesis of gold nanoparticles by a novel alkalo tolerant actinomycete, *Rhodococcus* species. Nanotechnology 14:824–828. https://doi.org/10.1088/0957-4484/14/7/323
8. Ahmed KBA, Kalla D, Uppuluri KB, Anbazhagan V (2014) Green synthesis of silver and gold nanoparticles employing levan, a biopolymer from *Acetobacter xylinum* NCIM 2526, as a reducing agent and capping agent. Carbohydr Polym 112:539–545. https://doi.org/10.1016/j.carbpol.2014.06.033
9. Ahmed T, Shahid M, Noman M, Niazi MBK, Mahmood F, Manzoor I, Chen J (2020) Silver nanoparticles synthesized by using *Bacillus cereus* SZT1 ameliorated the damage of bacterial leaf blight pathogen in rice. Pathogens 9(3):160. https://doi.org/10.3390/pathogens9030160
10. Ahmed T, Noman M, Manzoor N, Shahid M, Hussaini KM, Rizwan M, Li B (2021) Green magnesium oxide nanoparticles-based modulation of cellular oxidative repair mechanisms to reduce arsenic uptake and translocation in rice (*Oryza sativa* L.) plants. Environ Pollut 288:117785. https://doi.org/10.1016/j.envpol.2021.117785
11. Ahmed T, Wu Z, Jiang H, Luo J, Noman M, Shahid M, Li B (2021) Bioinspired green synthesis of zinc oxide nanoparticles from a native *Bacillus cereus* strain RNT6: characterization and antibacterial activity against rice panicle blight pathogens *Burkholderia glumae* and *B. gladioli*. Nanomaterials 11(4):884. https://doi.org/10.3390/nano11040884
12. Ahmed SF, Mofijur M, Rafa N, Chowdhury AT, Chowdhury S, Nahrin M, Ong HC et al (2022) Green approaches in synthesising nanomaterials for environmental nanobioremediation: technological advancements, applications, benefits and challenges. Env Res 204:111967. https://doi.org/10.1016/j.envres.2021.111967
13. Alabdallah NM, Hasan MM, Hammami I, Alghamdi AI, Alshehri D, Alatawi HA (2021) Green synthesized metal oxide nanoparticles mediate growth regulation and physiology of crop plants under drought stress. Plants 10(8):1730. https://doi.org/10.3390/plants10081730
14. Alharbi RM, Alshammari SO, Abd El Aty AA (2023) Statically improved fungal laccase-mediated biogenesis of silver nanoparticles with antimicrobial applications. J Appl Pharm Sci 13(1):241–253. https://doi.org/10.7324/JAPS.2023.130105
15. Almasi H, Jafarzadeh P, Mehryar L (2018) Fabrication of novel nanohybrids by impregnation of CuO nanoparticles into bacterial cellulose and chitosan nanofibers: characterization, antimicrobial and release properties. Carbohydr Polym 186:273–281. https://doi.org/10.1016/j.carbpol.2018.01.067
16. Alori ET, Glick BR, Babalola OO (2017) Microbial phosphorus solubilization and its potential for use in sustainable agriculture. Front Microbiol 8:971. https://doi.org/10.3389/fmicb.2017.00971
17. Ameen F, Al-Maary KS, Almansob A, AlNadhari S (2023) Antioxidant, antibacterial and anticancer efficacy of *Alternaria chlamydospora*-mediated gold nanoparticles. Appl Nanosci 13(3):2233–2240. https://doi.org/10.1007/s13204-021-02047-4
18. Ameen F, Alsamhary K, Alabdullatif JA, ALNadhari S (2021) A review on metal-based nanoparticles and their toxicity to beneficial soil bacteria and fungi. Ecotoxicol Environ Saf 213:112027

19. Amin MA, Ismail MA, Badawy AA, Awad MA, Hamza MF, Awad MF, Fouda A (2021) The Potency of fungal-fabricated selenium nanoparticles to improve the growth performance of *Helianthus annuus* L. and control of cutworm *Agrotis ipsilon*. Catalysts 11(12):1551. https://doi.org/10.3390/catal11121551
20. Ammar HA, El Aty AAA, El Awdan SA (2021) Extracellular myco-synthesis of nano-silver using the fermentable yeasts *Pichia kudriavzevii* HA-NY2 and *Saccharomyces uvarum* HA-NY3, and their effective biomedical applications. Bioprocess Biosyst Eng 44:841–854. https://doi.org/10.1007/s00449-020-02494-3
21. Annamalai J, Nallamuthu T (2015) Characterization of biosynthesized gold nanoparticles from aqueous extract of *Chlorella vulgaris* and their antipathogenic properties. Appl Nanosci 5:603–607. https://doi.org/10.1007/s13204-014-0353-y
22. Arshad A (2017) Bacterial synthesis and applications of nanoparticles. Nano Sci Nano Technol 11:119
23. Attia MS, El-Sayyad GS, Abdelaziz AM, Gaber SE, Khalil AMA, Saleh AM, Hashem AH (2023) Protective role of mycosynthesized bimetallic ZnO–CuO nanoparticles as therapeutic nutrients to enhance the resistance of *Vicia faba* against *Fusarium* wilt disease. Agronomy 13(11):2725. https://doi.org/10.3390/agronomy13112725
24. Azad A, Zafar H, Raza F, Sulaiman M (2023) Factors influencing the green synthesis of metallic nanoparticles using plant extracts: a comprehensive review. Pharm Fronts 5:e117–e131. https://doi.org/10.1055/s-0043-1774289
25. Backer R, Rokem JS, Ilangumaran G et al (2018) Plant growth-promoting rhizobacteria: context, mechanisms of action, and roadmap to commercialization of biostimulants for sustainable agriculture. Front Plant Sci 9:1473. https://doi.org/10.3389/fpls.2018.01473
26. Bai L, McClements DJ (2016) Formation and stabilization of nanoemulsions using biosurfactants: rhamnolipids. J Colloid Interface Sci 479:71–79. https://doi.org/10.1016/j.jcis.2016.06.047
27. Bai HJ, Zhang ZM, Guo Y, Yang GE (2009) iosynthesis of cadmium sulfide nanoparticles by photosynthetic bacteria *Rhodopseudomonas palustris*. Colloids Surf B Biointerfaces 70(1):142–146
28. Baka ZA, El-Zahed MM (2022) Antifungal activity of silver/silicon dioxide nanocomposite on the response of faba bean plants (*Vicia faba* L.) infected by *Botrytis cinerea*. Bioresour Bioprocess 9(1):102. https://doi.org/10.1186/s40643-022-00591-7
29. Bakhtiari-Sardari A, Mashreghi M, Eshghi H, Behnam-Rasouli F, Lashani E, Shahnavaz B (2020) Comparative evaluation of silver nanoparticles biosynthesis by two cold-tolerant *Streptomyces* strains and their biological activities. Biotechnol Lett 42:1985–1999. https://doi.org/10.1007/s10529-020-02921-1
30. Bala A, Rani G (2020) A review on phytosynthesis, affecting factors and characterization techniques of silver nanoparticles designed by green approach. Int Nano Lett 10:159–176. https://doi.org/10.1007/s40089-020-00309-7
31. Balagurunathan R, Radhakrishnan M, Rajendran RB, Velmurugan D (2011) Biosynthesis of gold nanoparticles by actinomycete *Streptomyces viridogens* strain HM10. J Biochem Biophys 48:331–335. http://nopr.niscpr.res.in/handle/123456789/12941
32. Balakrishnan S, Mukherjee S, Das S Bhat FA, Raja SP, Patra CR et al (2017) Gold nanoparticles-conjugated quercetin induces apoptosis via inhibition of EGFR/PI3K/Akt-mediated pathway in breast cancer cell lines (MCF-7 and MDA-MB-231). Cell Biochem Funct 35:217–231. https://doi.org/10.1002/cbf.3266
33. Bansal V, Rautaray D, Bharde A, Ahire K, Sanyal A, Ahmad A, Sastry M (2005) Fungus-mediated biosynthesis of silica and titania particles. J Mater Chem 15:2583–2589. https://doi.org/10.1039/b503008k
34. Baskar G, Chandhuru J, Sheraz Fahad K, Praveen AS, Chamundeeswari M, Muthukumar T (2015) Anticancer activity of fungal l-asparaginase conjugated with zinc oxide nanoparticles. J Mater Sci Mater Med 26:43. https://doi.org/10.1007/s10856-015-5380-z
35. Bawa R, Bawa SR, Mehra RN (2016) The translational challenge in medicine at the Nanoscale. In: Bawa R, Audette GF, Reese BE (eds) Handbook of clinical nanomedicine: law, business, regulation, safety, and risk, pp. 1291–1346. 978-981-4669-23-8 (eBook)

36. Bhadwal AS, Tripathi RM, Gupta RK, Kumar N, Singh RP, Shrivastav A (2014) Biogenic synthesis and photocatalytic activity of CdS nanoparticles. RSC Adv 4:9484–9490. https://doi.org/10.1039/c3ra46221h
37. Bhangale A, Wadekar S, Kale S, Bhowmick D, Pratap A (2014) Production of sophorolipids synthesized on castor oil with glucose and glycerol by using *Starmerella bombicola* (ATCC 22214). Eur J Lipid Sci Technol 116(3):336–343. https://doi.org/10.1002/ejlt.201300236
38. Birla SS, Tiwari VV, Gade AK, Ingle AP, Yadav AP, Rai MK (2009) Fabrication of silver nanoparticles by *Phoma glomerata* and its combined effect against *Escherichia coli*, *Pseudomonas aeruginosa* and *Staphylococcus aureus*. Lett Appl Microbiol 48(2):173–179. https://doi.org/10.1111/j.1472-765X.2008.02510.x. Epub 2008 Jan 3. PMID: 19141039
39. Bisht N, Chauhan PS (2020) Excessive and disproportionate use of chemicals cause soil contamination and nutritional stress. In: Larramendy L, Soloneski S. (eds.) Soil contamination-threats and sustainable solutions, IntechOpen, London. https://doi.org/10.5772/intechopen.94593
40. Bowman D, Hodge G (2006) Nanotechnology: mapping the wild regulatory frontier. Futures 38(9):1060–1073
41. Bowman D, Hodge G (2007) A small matter of regulation: an international review of nanotechnology regulation. C Sci Technol Law Rev 8:1–32
42. Bulgari R, Franzoni G, Ferrante A (2019) Biostimulants application in horticultural crops under abiotic stress conditions. Agronomy 9(6):306. https://doi.org/10.3390/agronomy9060306
43. Carmona FJ, Dal Sasso G, Bertolotti F, Ramírez-Rodríguez GB, Delgado-López JM, Pedersen JS, Guagliardi A (2020) The role of nanoparticle structure and morphology in the dissolution kinetics and nutrient release of nitrate-doped calcium phosphate nanofertilizers. Sci Rep 10(1):12396. https://doi.org/10.1038/s41598-020-69279-2
44. Castro PM, Sarmento B, Madureira AR, Pintado ME (2019) Organic nanocomposites for the delivery of bioactive molecules. In: Hasnain MS, Nayak AK (eds) Natural polysaccharides in drug delivery and biomedical applications. Academic Press, Cambridge https://doi.org/10.1016/B978-0-12-817055-7.00020-0
45. Chadha S (2021) Recent advances in nano-encapsulation technologies for controlled release of biostimulants and antimicrobial agents. In: Jogaiah S, Singh HB, Fraceto LF, De Lima (eds) Advances in nano-fertilizers and nano-pesticides in agriculture. Elsevier, Amsterdam https://doi.org/10.1016/B978-0-12-820092-6.00002-1
46. Chaikittisilp W, Yamauchi Y, Ariga K (2022) Material evolution with nanotechnology, nanoarchitectonics, and materials informatics: what will be the next paradigm shift in nanoporous materials? Adv Mater 34(7):2107212. https://doi.org/10.1002/adma.202107212
47. Chatterjee S, Mahanty S, Das P, Chaudhuri P, Das S (2019) Biofabrication of iron oxide nanoparticles using manglicolous fungus *Aspergillus niger* BSC-1 and removal of Cr(VI) from aqueous solution. Chem Engg J 123790. https://doi.org/10.1016/j.cej.2019.123790
48. Chen CC, Stark M, Baikoghli M, Cheng RH (2018) Surface functionalization of hepatitis E virus nanoparticles using chemical conjugation methods. J Vis Exp e57020. https://doi.org/10.3791/57020, PMID: 29806824
49. Chin HS, Cheong KY, Razak KA (2010) Review on oxides of antimony nanoparticles: synthesis, properties, and applications. J Mater Sci 45(22):5993–6008. https://doi.org/10.1007/s10853-010-4849-x
50. Chittibomma K, Sai D V, Singh H B, Reddy M G, Sai Vaishnav C (2023) Environmental impacts of nano-biostimulants in agriculture. Pharma Innov J SP-12(12):609–615
51. Clarance P, Luvankar B, Sales J, Khusro A, Agastian P, Tack JC, Kim HJ (2020) Green synthesis and characterization of gold nanoparticles using endophytic fungi *Fusarium solani* and its in-vitro anticancer and biomedical applications. Saudi J Biol Sci 27(2):706–712. https://doi.org/10.1016/j.sjbs.2019.12.026
52. Correa-Llantén DN, Muñoz-Ibacache SA, Castro ME, Muñoz PA, Blamey JM (2013) Gold nanoparticles synthesized by *Geobacillus* sp. strain ID17 a thermophilic bacterium isolated from Deception Island, Antarctica. Microb Cell Factories 12(1):75

53. Cui YH, Li LL, Zhou NQ, Liu JH, Huang Q, Wang HJ, Tian J, Yu HQ (2016) In vivo synthesis of nano-selenium by *Tetrahymena thermophila* SB210. Enzyme and microbial technology 95:185–191. https://doi.org/10.1016/j.enzmictec.2016.08.017
54. Cullen LG, Tilston EL, Mitchell GR, Collins CD, Shaw LJ (2011) Assessing the impact of nano-and micro-scale zerovalent iron particles on soil microbial activities: particle reactivity interferes with assay conditions and interpretation of genuine microbial effects. Chemosphere 82:1675–1682
55. Daba GM, Elnahas MO, Elkhateeb WA (2021) Contributions of exopolysaccharides from lactic acid bacteria as biotechnological tools in food, pharmaceutical, and medical applications. Int J Biol Macromol 173:79–89. https://doi.org/10.1016/j.ijbiomac.2021.01.110
56. Dameron CT, Reese RN, Mehra RK, Kortan AR, Carroll PJ, Steigerwald ML, Brus LE, Winge DR (1989) Biosynthesis of cadmium sulphide quantum semiconductor crystallites. Nature 338:596–597
57. de França Bettencourt GM, Degenhardt J, Torres LAZ, de Andrade Tanobe VO, Soccol CR (2020) Green biosynthesis of single and bimetallic nanoparticles of iron and manganese using bacterial auxin complex to act as plant bio-fertilizer. Biocatal Agric Biotechnol 30:101822. https://doi.org/10.1016/j.bcab.2020.101822
58. De Vasconcelos ACF, Chaves LHG (2020) Biostimulants and their role in improving plant growth under abiotic stresses. In: Mirmajlessi SM, Radhakrishnan R (eds) Biostimulants in plant science, IntechOpen. https://doi.org/10.5772/intechopen.88829
59. Dhanker R, Hussain T, Tyagi P, Singh KJ, Kamble SS (2021) The emerging trend of bioengineering approaches for microbial nanomaterial synthesis and its applications. Front Microbiol 12:638003. https://doi.org/10.3389/fmicb.2021.638003
60. Diaz-Ramirez J, Urbina L, Eceiza A, Retegi A, Gabilondo N (2021) Superabsorbent bacterial cellulose spheres biosynthesized from winery by-products as natural carriers for fertilizers. Int J Biol Macromol 191:1212–1220. https://doi.org/10.1016/j.ijbiomac.2021.09.203
61. Dietz KJ, Herth S (2011) Plant nanotoxicology. Trends Plant Sci 16:582–589
62. Dimkpa CO, Bindraban PS (2017) Nanofertilizers: new products for the industry? J Agric Food Chem 66(26):6462–6473. https://doi.org/10.1021/acs.jafc.7b02150
63. Domżał-Kędzia M, Ostrowska M, Lewińska A, Łukaszewicz M (2023) Recent developments and applications of microbial Levan, a versatile polysaccharide-based biopolymer. Molecules 28(14):5407. https://doi.org/10.3390/molecules28145407
64. Durán N, Marcato PD, Alves OL, De Souza GI, Esposito E (2005) Mechanistic aspects of biosynthesis of silver nanoparticles by several *Fusarium oxysporum* strains. J Nanobiotechnol 3:8. https://doi.org/10.1186/1477-3155-3-8
65. Durán N, Marcato PD, De Souza GI, Alves OL, Esposito E (2007) Antibacterial effect of silver nanoparticles produced by fungal process on textile fabrics and their effluent treatment. J Biomed Nanotechnol 3:203–208. https://doi.org/10.1166/jbn.2007.022
66. El-Saadony MT, Saad AM, Najjar AA, Alzahrani SO, Alkhatib FM, Shafi ME, Hassan MA (2021) The use of biological selenium nanoparticles to suppress *Triticum aestivum* L. crown and root rot diseases induced by Fusarium species and improve yield under drought and heat stress. Saudi J Biol Sci 28(8):4461–4471. https://doi.org/10.1016/j.sjbs.2021.04.043
67. Espitia PJ, Fuenmayor CA, Otoni CG (2019) Nanoemulsions: synthesis, characterization, and application in bio-based active food packaging. CRFSFS 18(1):264–285. https://doi.org/10.1111/1541-4337.12405
68. Eugenio M, Müller N, Frases S, Almeida-Paes R, Lima LM, Lemgruber L, Farina M, de Souza W, Sant'Anna C (2016) Yeast-derived biosynthesis of silver/silver chloride nanoparticles and their antiproliferative activity against bacteria. Rsc Adv 6(12):9893–9904. https://doi.org/10.1039/C5RA22727E
69. Fan XZ, Pomerantseva E, Gnerlich M, Brown A, Gerasopoulos K, McCarthy M et al (2013) Tobacco mosaic virus: a biological building block for micro/nano/bio systems. J Vac Sci Technol A 31:050815. https://doi.org/10.1116/1.4816584
70. Feng L, Li X, Xu C, Sadeghzadeh SM (2020) Green synthesis of Dy2Ce2O7 nanoparticles immobilized on fibrous nano-silica for synthesis of 3-Aryl-2-oxazolidinones from alkenes,

amines, and carbon dioxide. Catal Lett 150:1729–1740. https://doi.org/10.1007/s10562-019-03073-2
71. Fernández FG, Hoeft RG (2009) Managing soil pH and crop nutrients. Ill Agron Handb 24:91–112
72. Fernández JG, Fernández-Baldo MA, Berni E, Camí G, Durán N, Raba J, Sanz MI (2016) Production of silver nanoparticles using yeasts and evaluation of their antifungal activity against phytopathogenic fungi. Process Biochem 51(9):1306–1313. https://doi.org/10.1016/j.procbio.2016.05.021
73. Fouda A, Hassan SED, Abdel-Rahman MA, Farag MM, Shehal-deen A, Mohamed AA, Azab MS (2021) Catalytic degradation of wastewater from the textile and tannery industries by green synthesized hematite (α-Fe_2O_3) and magnesium oxide (MgO) nanoparticles. Curr Res Biotechnol 3:29–41. https://doi.org/10.1016/j.crbiot.2021.01.004
74. Franco RT, Silva AL, Licea YE, Serna JD, Alzamora M, Sánchez DR, Carvalho NM (2021) Green synthesis of iron oxides and phosphates via thermal treatment of iron polyphenols synthesized by a *Camellia sinensis* extract. Inorg Chem 60(8):5734–5746. https://doi.org/10.1021/acs.inorgchem.0c03794
75. França D, Messa LL, Souza CF, Faez R (2019) Nano and microencapsulated nutrients for enhanced efficiency fertilizer In: Gutiérrez T (eds) Polymers for agri-food applications. Springer, Cham. https://doi.org/10.1007/978-3-030-19416-1_3
76. Gaber SE, Hashem AH, El-Sayyad GS, Attia MS (2023) Antifungal activity of myco-synthesized bimetallic ZnO–CuO nanoparticles against fungal plant pathogen *Fusarium oxysporum*. Biomass Convers Biorefin 1-15. https://doi.org/10.1007/s13399-023-04550-w
77. Gade A, Bonde P, Ingle A, Marcato P, Duran N, Rai M (2008) Exploitation of Aspergillus niger for synthesis of silver nanoparticles. J Biobased Mater Biol 2:243–247
78. Gade A, Shende S, Rai M (2022) Potential role of *Phoma* spp. for mycogenic synthesis of silver nanoparticles. In: Rai M, Zimowska B, Kövics GJ (eds) Phoma: diversity, taxonomy, bioactivities, and nanotechnology. Springer, Cham. https://doi.org/10.1007/978-3-030-81218-8_17
79. Gaiotti F, Lucchetta M, Rodegher G, Lorenzoni D, Longo E, Boselli E, Pii Y (2021) Urea-doped calcium phosphate nanoparticles as sustainable nitrogen nanofertilizers for viticulture: implications on yield and quality of *Pinot gris* grapevines. Agronomy 11(6):1026. https://doi.org/10.3390/agronomy11061026
80. Ganachari SV, Bhat R, Deshpande R, Venkataraman A (2012) Extracellular biosynthesis of silver nanoparticles using fungi *Penicillium diversum* and their antimicrobial activity studies. Bionanoscience 2:316–321. https://doi.org/10.1007/s12668-012-0046-5
81. García-Gómez C, Fernández MD, García S, Obrador AF, Letón M, Babín M (2018) Soil pH effects on the toxicity of zinc oxide nanoparticles to soil microbial community. Environ Sci Pollut Res 25:28140–28152
82. Garole VJ, Choudhary BC, Tetgure SR, Garole DJ, Borse AU (2019) Palladium nanocatalyst: green synthesis, characterization and catalytic application. Int J Environ Sci Technol 16:7885–7892. https://doi.org/10.1007/s13762-018-2173-1
83. Gelvez-Pardo I, Lobo-Berbesi L, Santos-Díaz A (2023) Biological efficacy of plant growth-promoting bacteria and arbuscular mycorrhizae fungi: assessments in laboratory and greenhouse conditions. Curr Prot 3(4):e732. https://doi.org/10.1002/cpz1.732
84. Ghahremanzadeh R, Yazdi Samadi F, Yousefi M (2019) Green synthesis of gold nanoparticles using three medicinal plant extracts as efficient reducing agents. Iran J Chem Chem Eng 38:1–10
85. Ghosh S, Ahmad R, Zeyaullah M, Khare SK (2021) Microbial nano-factories: synthesis and biomedical applications. Front Chem 9:626834. https://doi.org/10.3389/fchem.2021.626834
86. Gizzie N, Mayne R, Patton D, Kendrick P, Adamatzky A (2016) On hybridising lettuce seedlings with nanoparticles and the resultant effects on the organisms' electrical characteristics. Biosystems 147:28–34. https://doi.org/10.1016/j.biosystems.2016.07.002
87. Golinska P, Wypij M, Ingle AP, Gupta I, Dahm H, Rai M (2014) Biogenic synthesis of metal nanoparticles from actinomycetes: biomedical applications and cytotoxicity. Appl Microbiol Biotechnol 98:8083–8097. https://doi.org/10.1007/s00253-014-5953-7

88. González-Morales S, Cárdenas-Atayde PA, Garza-Alonso CA, Robledo-Olivo A, Benavides-Mendoza A (2022) Plant biostimulation with nanomaterials: a physiological and molecular standpoint. In: Fernandes Fraceto L, Pereira de Carvalho HW, de Lima R, Ghoshal S, Santaella C (eds) Inorganic nanopesticides and nanofertilizers. Springer, Cham. https://doi.org/10.1007/978-3-030-94155-0_5
89. Gour A, Jain NK (2019) Advances in green synthesis of nanoparticles. Artif Cells Nanomed Biotechnol 41:844–851. https://doi.org/10.1080/21691401.2019.1577878
90. Guilger-Casagrande M, Germano-Costa T, Pasquoto-Stigliani T, Fraceto LF, Lima RD (2019) Biosynthesis of silver nanoparticles employing *Trichoderma harzianum* with enzymatic stimulation for the control of *Sclerotinia sclerotiorum*. Sci Rep 9(1):14351. https://doi.org/10.1038/s41598-019-50871-0
91. Gupta T, Saxena J (2023) Biogenic synthesis of silver nanoparticles from *Aspergillus oryzae* mtcc 3107 against plant pathogenic fungi *Sclerotinia sclerotiorum* MTCC 8785. J Microbiol Biotechnol Food Sci 12(4):e9387–e9387. https://doi.org/10.55251/jmbfs.9387
92. Guru Bharathi B, Lalitha K, Shivakumar MS (2022) Biosynthesis of copper nanoparticles using symbiotic bacterium *Xenorhabdus* sp, isolated from entomopathogenic nematode and its antimicrobial and insecticidal activity against *Spodoptera litura*. Inorg Nano-Metal Chem 1–13. https://doi.org/10.1080/24701556.2022.2078359
93. Hamk M, Akçay FA, Avcı A (2023) Green synthesis of zinc oxide nanoparticles using *Bacillus subtilis* ZBP4 and their antibacterial potential against foodborne pathogens. Prep Biochem Biotechnol 53(3):255–264. https://doi.org/10.1080/10826068.2022.2076243
94. Hassaan MA, Hosny S (2018) Green synthesis of Ag and Au nanoparticles from micro and macro algae—review. Int J Atmos Ocean Sci 2:10–22. https://doi.org/10.11648/j.ijaos.20180201.12
95. Hassan SED, Fouda A, Radwan AA, Salem SS, Barghoth MG, Awad MA, El-Gamal MS (2019) Endophytic actinomycetes *Streptomyces* spp mediated biosynthesis of copper oxide nanoparticles as a promising tool for biotechnological applications. J Biol Inorg Chem 24:377–393. https://doi.org/10.1007/s00775-019-01654-5
96. Hassan SED, Fouda A, Saied E, Farag MM, Eid AM, Barghoth MG, Awad MF (2021) *Rhizopus oryzae*-mediated green synthesis of magnesium oxide nanoparticles (MgO-NPs): a promising tool for antimicrobial, mosquitocidal action, and tanning effluent treatment. J Fungi 7(5):372. https://doi.org/10.3390/jof7050372
97. Hatami M (2017) Toxicity assessment of multi-walled carbon nanotubes on *Cucurbita pepo* L. under well-watered and water-stressed conditions. Ecotoxicol Environ Saf 142:274–283. https://doi.org/10.1016/j.ecoenv.2017.04.018
98. He S, Guo Z, Zhang Y, Zhang S, Wang J, Gu N (2007) Biosynthesis of gold nanoparticles using the bacteria *Rhodopseudomonas capsulate*. Mater Lett 61(18):3984–3987
99. Henry E, Yadeta KA, Coaker G (2013) Recognition of bacterial plant pathogens: local, systemic and transgenerational immunity. New Phytol 199(4):908–915. https://doi.org/10.1111/nph.12214
100. Hickey LTN, Hafeez A, Robinson H, Jackson SA, Leal-Bertioli SC, Tester M, Wulff BB (2019) Breeding crops to feed 10 billion. Nat Biotechnol 37(7):744–754. https://doi.org/10.1038/s41587-019-0152-9
101. Ho CT, Nguyen TH, Lam TT, Le DQ, Nguyen CX, Lee J, Hur HG (2021) Biogenic synthesis of selenium nanoparticles by *Shewanella* sp. HN-41 using a modified bioelectrochemical system. Elect J Biotech 54:1–7. https://doi.org/10.1016/j.ejbt.2021.07.004
102. https://agricoop.gov.in/Documents/46%20of%201968.pdf. Accessed 10 Jan 2024
103. https://cpcb.nic.in/displaypdf.php?id=aG9tZS9lcGEvZXByb3RlY3RfYWN0XzE5ODYucGRm. Accessed 10 Jan 2024
104. https://dish.tn.gov.in/assets/pdf/HAZARDOUS%20CHEMICAL%20RULES.pdf. Accessed 10 Jan 2024
105. https://dst.gov.in/sites/default/files/Draft-Guidelines%20.pdf. Accessed 08 Jan 2024
106. https://dste.py.gov.in/ppcc/pdf/act/THE%20CHEMICAL%20ACCIDENTS%20RULES-1996.pdf. Accessed 10 Jan 2024

107. https://ksandk.com/information-technology/nanotechnology-regulations-in-india/#comprehensive-regulatory-framework-to-manage-potential-risks-of-nanotechnology. Accessed 08 Jan 2024
108. https://thc.nic.in/Central%20Governmental%20Rules/Hazardous%20Waste%20(%20Management,%20Handling%20and%20Transboundary%20Movement)%20Rules,%202008.pdf. Accessed 10 Jan 2024
109. https://www.fssai.gov.in/upload/uploadfiles/files/Compendium_Contaminants_Regulations_20_08_2020.pdf Accessed 10 Jan 2024
110. https://cdsco.gov.in/opencms/export/sites/CDSCO_WEB/Pdf-documents/acts_rules/2016DrugsandCosmeticsAct1940Rules1945.pdf. Accessed 10 Jan 2024
111. https://ourworldindata.org/fertilizers. Accessed 08 Jan 2024
112. Husen A (2020) Carbon-based nanomaterials and their interactions with agricultural crops. In: Husen A, Jawaid M (eds) Nanomaterials for agriculture and forestry applications. Elsevier, Amsterdam https://doi.org/10.1016/B978-0-12-817852-2.00008-1
113. Ibrahim E, Fouad H, Zhang M, Zhang Y, Qiu W, Yan C, Chen J (2019) Biosynthesis of silver nanoparticles using endophytic bacteria and their role in inhibition of rice pathogenic bacteria and plant growth promotion. RSC Adv 9(50):29293–29299. https://doi.org/10.1039/C9RA04246F
114. Ijaz I, Gilani E, Nazir A, Bukhari A (2020) Detail review on chemical, physical and green synthesis, classification, characterizations and applications of nanoparticles. Green Chem Lett Rev 13: 59–81. https://doi.org/10.1080/17518253.2020.1802517
115. Imada K, Sakai S, Kajihara H, Tanaka S, Ito S (2016) Magnesium oxide nanoparticles induce systemic resistance in tomato against bacterial wilt disease. Plant Pathol 65(4):551–560. https://doi.org/10.1111/ppa.12443
116. Ingle A, Rai M, Gade A, Bawaskar M (2009) *Fusarium solani*: a novel biological agent for the extracellular synthesis of silver nanoparticles. J Nanopart Res 11:2079. https://doi.org/10.1007/s11051-008-9573-y
117. Ingle PU, Shende SS, Hande D, Rai M, Golinska P, Gade AK (2023) Mycogenic copper oxide nanoparticles for fungal infection management in agricultural crop plants. BioNanoScience, 1–9. https://doi.org/10.1007/s12668-023-01255-2
118. Ingle PU, Shende SS, Hande D, Rai M, Golinska P, Gade AK (2024) Mycogenic copper oxide nanoparticles for fungal infection management in agricultural crop plants. BioNanoScience 14(1):359–367. https://doi.org/10.1007/s12668-023-01255-2
119. Iravani S (2014) Bacteria in nanoparticle synthesis: current status and future prospects. Int Sch Res Not 18:2014. Article ID 359316. https://doi.org/10.1155/2014/359316
120. Jahan S, Yusoff I, Bin Alias YB, Bakar AFBA (2017) Reviews of the toxicity behaviour of five potential engineered nanomaterials (ENMs) into the aquatic ecosystem. Toxicol Reports 4:211–220. https://doi.org/10.1016/j.toxrep.2017.04.001
121. Jain A, Ranjan S, Dasgupta N, Ramalingam C (2016) Nanomaterials in food and agriculture: an overview on their safety concerns and regulatory issues. Crit Rev Food Sci Nutr. https://doi.org/10.1080/10408398.2016.1160363
122. Jameel MS, Aziz AA, Dheyab MA (2020) Green synthesis: proposed mechanism and factors influencing the synthesis of platinum nanoparticles. Green Process Synth 9:386–398. https://doi.org/10.1515/gps-2020-0041
123. Jardin P (2015) Plant biostimulants: definition, concept, main categories and regulation. Sci Hortic 196:3–14. https://doi.org/10.1016/j.scienta.2015.09.021
124. Jha A, Prasad K, Kulkarni AR (2009) Synthesis of TiO_2 nanoparticles using microorganisms. Colloids Surf B Biointerfaces 71:226–229
125. Jiménez-Arias D, Morales-Sierra S, Borges AA, Díaz Díaz D (2020) Biostimulant nanoencapsulation: the new keystone to fight hunger. J Agric Food Chem 68:7083–7085. https://doi.org/10.1021/acs.jafc.0c02872
126. Joško I, Oleszczuk P, Futa B (2014) The effect of inorganic nanoparticles (ZnO, Cr_2O_3, CuO and Ni) and their bulk counterparts on enzyme activities in different soils. Geoderma 232:528–537

127. Juárez-Maldonado A, Ortega-Ortíz H, Morales-Díaz AB, González-Morales S, Morelos-Moreno Á, Cabrera-De la Fuente M, Benavides-Mendoza A (2019) Nanoparticles and nanomaterials as plant biostimulants. IJMS 20(1):162. https://doi.org/10.3390/ijms20010162
128. Kah M, Kookana RS, Gogos A, Bucheli TD (2018) A critical evaluation of nanopesticides and nanofertilizers against their conventional analogues. Nature Nanotechnol 13(8):677–684. https://doi.org/10.1038/s41565-018-0131-1
129. Kalaimurugan D, Sivasankar P, Lavanya K, Shivakumar MS, Venkatesan S (2019) Antibacterial and larvicidal activity of *Fusarium proliferatum* (YNS2) whole cell biomass mediated copper nanoparticles. J Cluster Sci 30:1071–1080. https://doi.org/10.1007/s10876-019-01568-x
130. Kaphle A, Navya PN, Umapathi A, Daima HK (2018) Nanomaterials for agriculture, food and environment: applications, toxicity and regulation. Environ Chem Lett 16:43–58. https://doi.org/10.1007/s10311-017-0062-y
131. Kapoor RT, Salvadori MR, Rafatullah M, Siddiqui MR, Khan MA, Alshareef SA (2021) Exploration of microbial factories for synthesis of nanoparticles–a sustainable approach for bioremediation of environmental contaminants. Front Microbiol 4(12):658294. https://doi.org/10.3389/fmicb.2021.658294
132. Karthik L, Kumar G, Vishnu Kirthi A, Rahuman AA, Bhaskara Rao KV (2014) Streptomyces sp. LK3 mediated synthesis of silver nanoparticles and its biomedical application. Bioprocess Biosyst Eng 37:261–267
133. Kathiraven T, Sundaramanickam A, Shanmugam N, Balasubramanian T (2014) Green synthesis of silver nanoparticles using marine algae *Caulerpa racemosa* and their antibacterial activity against some human pathogens. Appl Nanosci 5:499–504. https://doi.org/10.1007/s13204-014-0341-2
134. Kato Y, Suzuki M (2020) Synthesis of metal nanoparticles by microorganisms. SSRN: https://ssrn.com/abstract=3560625 or https://doi.org/10.2139/ssrn.3560625
135. Khan SA, Uddin I, Moeez S, Ahmad A (2014) Fungus-mediated preferential bioleaching of waste material such as fly-ash as a means of producing extracellular, protein capped, fluorescent and water soluble silica nanoparticles. PLoS ONE 9(9):e107597. https://doi.org/10.1371/journal.pone.0107597
136. Khan HA, Shanker R (2015) Toxicity of nanomaterials. BioMed Res Inter 1–3. Article ID 521014
137. Kirimura K, Yoshioka I (2019) Citric acid. In: Moo-Young M (eds) Comprehensive biotechnology, 3rd edn. Pergamon, Oxford. https://doi.org/10.1016/B978-0-444-64046-8.00157-9
138. Klimczyk M, Siczek A, Schimmelpfennig L (2021) Improving the efficiency of urea-based fertilization leading to reduction in ammonia emission. Sci Total Environ 771:145483. https://doi.org/10.1016/j.scitotenv.2021.145483
139. Kobayashi M, Tomita S, Sawada K, Shiba K, Yanagi H, Yamashita I et al (2012) Chiral meta-molecules consisting of gold nanoparticles and genetically engineered tobacco mosaic virus. Opt Express 20:24856–24863. https://doi.org/10.1364/OE.20.024856
140. Koller M (2018) Biodegradable and biocompatible polyhydroxy-alkanoates (PHA): auspicious microbial macromolecules for pharmaceutical and therapeutic applications. Molecules 23(2):362. https://doi.org/10.3390/molecules23020362
141. Koopi H, Buazar F (2018) A novel one-pot biosynthesis of pure alpha aluminum oxide nanoparticles using the macroalgae *Sargassum ilicifolium*: a green marine approach. Ceram Int 44:8940–8945. https://doi.org/10.1016/j.ceramint.2018.02.091
142. Kowshik M, Ashtaputre S, Kulkani SK, Parknikar KMM (2003) Extracellular synthesis of silver nanoparticles by a silver-tolerant yeast strain MKY3. Nanotechnology 14:95–100
143. Kowshik M, Deshmukh N, Vogel W, Urban J, Kulkarni SK, Paknikar K (2002) Microbial synthesis of semiconductor CdS nanoparticles, their characterization, and their use in the fabrication of an ideal diode. Biotechnol Bioeng 78:583–588. https://doi.org/10.1002/bit.10233

144. Kowshik M, Vogel W, Urban J, Kulkarni SK, Paknikar KM (2002) Microbial synthesis of semiconductor PbS nanocrystallites. Adv Mater 14:815–818. https://doi.org/10.1002/1521-4095(20020605)14:11%3c815::AID-ADMA815%3e3.0.CO;2-K
145. Kumar A, Singh IK, Mishra R, Singh A, Ramawat N, Singh A (2021) The role of zinc oxide nanoparticles in plants: a critical appraisal. In: Sharma N, Sahi S (eds) Nanomaterial biointeractions at the cellular, organismal and system levels. Nanotechnology in the life sciences. Springer, Cham. https://doi.org/10.1007/978-3-030-65792-5_10
146. Kumar P (2022) Nanomaterial's synthesis, types and their use in Bioremediation and Agriculture. Natr Resour Human Health 2(3):349–365 https://doi.org/10.53365/nrfhh/144289
147. Kumari A, Kumari P, Rajput VD, Sushkova SN, Minkina T (2021) Metal(loid) nanosorbents in restoration of polluted soils: geochemical, ecotoxicological, and remediation perspectives. Environ Geochem Health 44:235–246
148. Kumari KA, Mangatayaru KG, Reddy GB (2023) Fungal and yeast-mediated biosynthesis of metal nanoparticles: characterization and bio applications. In: Abd-Elsalam KA (ed) Fungal cell factories for sustainable nanomaterials productions and agricultural applications. A volume in nanobiotechnology for plant protection, pp 309–336. https://doi.org/10.1016/B978-0-323-99922-9.00018-0
149. Kuo CH, Huang CY, Shieh CJ, Wang HMD, Tseng CY (2019) Hydrolysis of orange peel with cellulase and pectinase to produce bacterial cellulose using *Gluconacetobacter xylinus*. Waste Biomass Valorization 10:85–93. https://doi.org/10.1007/s12649-017-0034-7
150. Labrenz M, Druschel GK, Thomsen-Ebert T et al (2000) Formation of sphalerite (ZnS) deposits in natural biofilms of sulfate-reducing bacteria. Science 290(5497):1744–1747
151. Lahiri D, Nag M, Sheikh HI, Sarkar T, Edinur HA, Pati S, Ray RR (2021) Microbiologically-synthesized nanoparticles and their role in silencing the biofilm signaling cascade. Front Microbiol 12:636588. https://doi.org/10.3389/fmicb.2021.636588
152. Le DHT, Lee KL, Shukla S, Commandeur U, Steinmetz NF (2017) Potato virus X, a filamentous plant viral nanoparticle for doxorubicin delivery in cancer therapy. Nanoscale 9:2348–2357. https://doi.org/10.1039/C6NR09099K
153. Lengke MF, Fleet ME, Southam G (2006) Morphology of gold nanoparticles synthesized by filamentous cyanobacteria from gold(I)-Thiosulfate and gold(III)-chloride complexes. Langmuir 22(6):2780–2787
154. Li HP, Han QQ, Liu QM et al (2023) Roles of phosphate-solubilizing bacteria in mediating soil legacy phosphorus availability. Micro Res 272:127375. https://doi.org/10.1016/j.micres.2023.127375
155. Li J, Li Q, Ma X, Tian B, Li T, Yu J, Dai S, Weng Y, Hua Y (2016) Biosynthesis of gold nanoparticles by the extreme bacterium *Deinococcus radiodurans* and an evaluation of their antibacterial properties. Int J Nanomed 11:5931–5944. https://doi.org/10.2147/IJN.S119618
156. Liu J, Fan D, Wang L, Shi L, Ding J, Chen Y, Shen S (2014) Effects of ZnO, CuO, Au, and TiO_2 nanoparticles on *Daphnia magna* and early life stages of zebrafish *Danio rerio*. Environ Prot Eng 40:139–149
157. Liu R, Lal R, Potdar N (2020) Nanotechnology interventions in agri-food systems: a critical review. Sustainability 12(1):134
158. Liu Y, Zhu S, Gu Z, Chen C, Zhao Y (2022) Toxicity of manufactured nanomaterials. Particuology 69:31–48. https://doi.org/10.1016/j.partic.2021.11.007
159. Liu P, Zhou R, Yin T, Wang Q, Guo Z, Qiwen T, Li X (2020) Novel bio-fabrication of silver nanoparticles using the cell-free extract of *Lysinibacillus fusiformis* sp. and their potent activity against pathogenic fungi. Material Res Express 6(12):1250f2. https://doi.org/10.1088/2053-1591/ab664a
160. Lohith Kumar DH, Sarkar P (2018) Encapsulation of bioactive compounds using nanoemulsions. Environ Chem Lett 16:59–70. https://doi.org/10.1007/s10311-017-0663-x
161. Lu L, Huang M, Huang Y, Corvini PFX, Ji R, Zhao L (2020) Mn_3O_4 nanozymes boost endogenous antioxidant metabolites in cucumber (*Cucumis sativus*) plant and enhance resistance to salinity stress. Environ Sci Nano 7(6):1692–1703. https://doi.org/10.1039/d0en00214c

162. Lu XY, Wu DC, Li ZJ, Chen GQ (2011) Polymer nanoparticles. Prog Mol Biol Transl Sci 104:299–323. https://doi.org/10.1016/B978-0-12-416020-0.00007-3
163. Ma X, Geiser-Lee J, Deng Y, Kolmakov A, Li Z (2010) Interactions between engineered nanoparticles (ENPs) and plants: phytotoxicity, uptake and accumulation. Sci Total Environ 408(16):3053–3061
164. Ma X, Wang X (2022) Multigenerational exposure of plants to engineered nanoparticles (ENPs). In: Rico MC (ed) Plant exposure to engineered nanoparticles. Academic Press, Cambridge https://doi.org/10.1016/B978-0-323-85032-2.00002-6
165. Madusanka N, Sandaruwan C, Kottegoda N, Sirisena D, Munaweera I, De Alwis A, Amaratunga GA (2017) Urea–hydroxyapatite-montmorillonite nanohybrid composites as slow release nitrogen compositions. Appl Clay Sci 150:303–308. https://doi.org/10.1016/j.clay.2017.09.039
166. Maghsoodi MR, Najafi N, Reyhanitabar A, Oustan S (2020) Hydroxyapatite nanorods, hydrochar, biochar, and zeolite for controlled-release urea fertilizers. Geoderma 379:114644. https://doi.org/10.1016/j.geoderma.2020.114644
167. Magnabosco G, Hauzer H, Fermani S, Calvaresi M, Corticelli F, Christian M, Albonetti C, Morandi V, Erez J, Falini G (2019) Bionic synthesis of a magnetic calcite skeletal structure through living foraminifera. Mate Horiz 6(9):1862–1867. https://doi.org/10.1039/C9MH00495E
168. Mao C, Flynn CE, Hayhurst A, Sweeney R, Qi J (2003) Viral assembly of oriented quantum dot nanowires. Proc Natl Acad Sci USA 100:6946–6951. https://doi.org/10.1073/pnas.0832310100
169. Markande AR, Patel D, Varjani S (2021) A review on biosurfactants: properties, applications and current developments. Bioresour Technol 330:124963. https://doi.org/10.1016/j.biortech.2021.124963
170. Marlair G, Kordek MA (2005) Safety and security issues relating to low capacity storage of AN-based fertilizers. J Hazard Material 123:13–28. https://doi.org/10.1016/j.jhazmat.2005.03.028
171. Milenković I, Mitrović A, Algarra M, Lázaro-Martínez JM, Rodríguez-Castellón E, Maksimović V, Radotić K (2019) Interaction of carbohydrate coated cerium-oxide nanoparticles with wheat and pea: stress induction potential and effect on development. Plants 8(11):478. https://doi.org/10.3390/plants8110478
172. Mittal D, Kaur G, Singh P, Yadav K, Ali SA (2020) Nanoparticle-based sustainable agriculture and food science: recent advances and future outlook. Front Nanotechnol 2:579954. https://doi.org/10.3389/fnano.2020.579954
173. Moghaddam AB, Namvar F, Moniri M, Md Tahir P, Azizi S, Mohamad R (2015) Nanoparticles biosynthesized by fungi and yeast: a review of their preparation, properties, and medical applications. Molecules 20(9):16540–16565. https://doi.org/10.3390/molecules200916540
174. Mohanasrinivasan V, Subathra Devi C, Mehra A, Prakash S, Agarwal A, Selvarajan E, Jemimah Naine S (2018) Biosynthesis of MgO nanoparticles using *Lactobacillus* sp. and its activity against human leukemia cell lines HL-60. BioNanoScience 8:249–253. https://doi.org/10.1007/s12668-017-0480-5
175. Mohseniazar M, Barin M, Zarredar H, Alizadeh S, Shanehbandi D (2011) Potential of microalgae and lactobacilli in biosynthesis of silver nanoparticles. BioImpacts: BI 1(3):149–152. https://doi.org/10.5681/bi.2011.020
176. Mollaei H, Safaie N, Aliasgharzad N, Behmanesh M (2021) Engineered nanoparticles in agriculture: a new paradigm for plant fertilization and protection. Environ Technol Innov 23:101799
177. Monteiro-Riviere NA, Inman AO, Zhang LW (2009) Limitations and relative utility of screening assays to assess engineered nanoparticle toxicity in a human cell line. Toxicol Appl Pharmacol 234:222–235
178. Naaz H, Rawat K, Saffeullah P, Umar S (2023) Silica nanoparticles synthesis and applications in agriculture for plant fertilization and protection: a review. Environ Chem Lett 21(1):539–559. https://doi.org/10.1007/s10311-022-01515-9

179. Nagababu P, Umamaheswara Rao V (2016) Cost–effective green synthesis and characterization of silver nanoparticles from *Avicennia alba blume* leaves and their antibacterial activity. Asian J Pharmaceut Clin Res 9:301–303
180. Nandhini M, Rajini SB, Udayashankar AC, Niranjana SR, Lund OS, Shetty HS, Prakash HS (2019) Biofabricated zinc oxide nanoparticles as an eco-friendly alternative for growth promotion and management of downy mildew of pearl millet. Crop Prot 121:103–112. https://doi.org/10.1016/j.cropro.2019.03.015
181. Nejat N, Ramalingam A, Mantri N (2018) Advances in transcriptomics of plants. Adv Biochem Engg/Biotech 164:161–185. https://doi.org/10.1007/10_2017_52
182. Nel A, Xia T, Mädler L, Li N (2006) Toxic potential of materials at the nanolevel. Science 311:622–627
183. Nielsen MB, Baun A, Mackevica A, Thit A, Wallinder IO, Gallego JA, Hansen SF (2021) Nanomaterials in the European chemicals legislation–methodological challenges for registration and environmental safety assessment. Environ Sci Nano 8(3):731–747. https://doi.org/10.1039/D0EN01123A
184. Nithya R, Ragunathan R (2009) Synthesis of silver nanoparticles using *Pleurotus sajor caju* and its antimicrobial study. Digest J Nanomat Biostru 4:623–629
185. Niu J, Liu C, Huang M, Liu K, Yan D (2021) Effects of foliar fertilization: a review of current status and future perspectives. J Soil Sci Plant Nutr 21:104–118. https://doi.org/10.1007/s42729-020-00346-3
186. Noman M, Ahmed T, Hussain S, Niazi MBK, Shahid M, Song F (2020) Biogenic copper nanoparticles synthesized by using a copper-resistant strain *Shigella flexneri* SNT22 reduced the translocation of cadmium from soil to wheat plants. J Hazard Mater 398:123175. https://doi.org/10.1016/j.jhazmat.2020.123175
187. Noman M, Ahmed T, Shahid M, Nazir MM, Li D, Song F (2023) Salicylic acid-doped iron nano-biostimulants potentiate defense responses and suppress *Fusarium* wilt in watermelon. J Advan Res. https://doi.org/10.1016/j.jare.2023.06.011
188. Noman M, Shahid M, Ahmed T, Tahir M, Naqqash T, Muhammad S, Aslam Z (2020) Green copper nanoparticles from a native *Klebsiella pneumoniae* strain alleviated oxidative stress impairment of wheat plants by reducing the chromium bioavailability and increasing the growth. Ecotoxicol Environ Saf 192:110303. https://doi.org/10.1016/j.ecoenv.2020.110303
189. Noman M, Ahmed T, White JC, Nazir MM, Azizullah Li D, Song F (2023) *Bacillus altitudinis*-stabilized multifarious copper nanoparticles prevent bacterial fruit blotch in watermelon (*Citrullus lanatus* L.): direct pathogen inhibition, in planta particles accumulation, and host stomatal immunity modulation. Small 19(15):2207136. https://doi.org/10.1002/smll.202207136
190. Nongbet A, Mishra AK, Mohanta YK, Mahanta S, Ray MK, Khan M, Chakrabartty I (2022) Nanofertilizers: a smart and sustainable attribute to modern agriculture. Plants 11(19):2587. https://doi.org/10.3390/plants11192587
191. Ogunyemi SO, Zhang M, Abdallah Y, Ahmed T, Qiu W, Ali MA, Li B (2020) The biosynthesis of three metal oxide nanoparticles (ZnO, MnO_2, and MgO) and their antibacterial activity against the bacterial leaf blight pathogen. Front Microbiol 11:588326. https://doi.org/10.3389/fmicb.2020.588326
192. Okey-Onyesolu CF, Hassanisaadi M, Bilal M, Barani M, Rahdar A, Iqbal J, Kyzas GZ (2021) Nanomaterials as nanofertilizers and nanopesticides: an overview. ChemistrySelect 6(33):8645–8663. https://doi.org/10.1002/slct.202102379
193. Olaru D, Olaru A, Hussein Kassem G, Popescu-Drigă MV, Pinosanu LR, Dumitrascu DI, Popescu EL, Hermann DM, Popa-Wagner A (2019) Toxicity and health impact of nanoparticles. Basic biology and clinical perspective. Rom J Morphol Embryol 60:787–792
194. Osman EM (2019) Environmental and health safety considerations of nanotechnology: nano safety. Biomed J Sci Tech Res 19(4):14501–14515. https://doi.org/10.26717/BJSTR.2019.19.003346 BJSTR. MS.ID.003346

195. Pal RS, Mondal T, Dev R, Parihar M, Kumar A, Sharma D, Kant L (2023) Role of enzyme-mimicking nanoparticles in crop plants. In: Mishra KK, Santosh K (eds) Biotic stress management of crop plants using nanomaterials. CRC Press, Boca Raton https://doi.org/10.1201/9781003322122
196. Paramasivan M, Kumar TSS, Chandra TS (2022) Microbial synthesis of hydroxyapatite-nanocellulose nanocomposites from symbiotic culture of bacteria and yeast pellicle of fermented kombucha tea. Sustainability 14(13):8144. https://doi.org/10.3390/su14138144
197. Paramasivan M, Kumar TS, Kanniyappan H, Muthuvijayan V, Chandra TS (2023) Microbial biomineralization of hydroxyapatite nanocrystals using *Bacillus tequilensis*. Ceram Int 49(4):5621–5629. https://doi.org/10.1016/j.ceramint.2022.10.138
198. Paramasivan M, Sampath Kumar TS, Kanniyappan H, Muthuvijayan V, Chandra TS (2023) Biomimetic ion substituted and Co-substituted hydroxyapatite nanoparticle synthesis using *Serratia marcescens*. Sci Rep 13(1):4513. https://doi.org/10.1038/s41598-023-30996-z
199. Park HG, Yeo MK (2016) Review paper: nanomaterial regulatory policy for human health and environment. Mol Cell Toxicol 12:223–236
200. Patra JK, Baek KH (2014) Green nanobiotechnology: factors affecting synthesis and characterization techniques. J Nanomat 12:2014. Article ID 417305. https://doi.org/10.1155/2014/417305
201. Peponi L, Puglia D, Torre L, Valentini L, Kenny JM (2014) Processing of nanostructured polymers and advanced polymeric based nanocomposites. Mater Sci Eng: R Rep 85:1–46. https://doi.org/10.1016/j.mser.2014.08.002
202. Pimprikar PS, Joshi SS, Kumar AR, Zinjarde SS, Kulkarni SK (2009) Influence of biomass and gold salt concentration on nanoparticle synthesis by the tropical marine yeast *Yarrowia lipolytica* NCIM 3589. Colloids Surf B Biointerfaces 74:309–316
203. Pohshna C, Mailapalli DR (2021) Engineered urea-doped hydroxyapatite nanomaterials as nitrogen and phosphorus fertilizers for rice. ACS Agric Sci Technol 2(1):100–112. https://doi.org/10.1021/acsagscitech.1c00191
204. Pokale P, Shende S, Gade A, Rai M (2014) Biofabrication of calcium phosphate nanoparticles using the plant *Mimusops elengi*. Environ Chem Lett 12:393–399. https://doi.org/10.1007/s10311-014-0460-8
205. Ponmurugan P, Manjukarunambika K, Elango V, Gnanamangai BM (2016) Antifungal activity of biosynthesised copper nanoparticles evaluated against red root-rot disease in tea plants. J Exp Nanosci 11(13):1019–1031. https://doi.org/10.1080/17458080.2016.1184766
206. Pósfai M, Moskowitz BM, Arató B et al (2006) Properties of intracellular magnetite crystals produced by *Desulfovibrio magneticus* strain RS-1. Earth Planet Sci Lett 249(3–4):444–455
207. Prabhu N, Borkar S, Garg S (2019) Phosphate solubilization by microorganisms: overview, mechanisms, applications and advances. In Meena SN, Naik MM (eds), Advances in biological science research. Academic Press, Cambridge https://doi.org/10.1016/B978-0-12-817497-5.00011-2
208. Prakasham RS, Buddana SK, Yannam SK, Guntuku GS (2012) Characterization of silver nanoparticles synthesized by using marineisolate *Streptomyces albidoflavus*. J Microbiol Biotechnol 22:614–621. https://doi.org/10.4014/jmb.1107.07013
209. Priyam A, Yadav N, Reddy PM, Afonso LO, Schultz AG, Singh PP (2022) Uptake and benefits of biogenic phosphorus nanomaterials applied via fertigation to Japonica Rice (Taipei 309) in low-and high-calcareous soil conditions. ACS Agric Sci Technol 2(3):462–476. https://doi.org/10.1021/acsagscitech.1c00244
210. Priyam A, Das RK, Schultz A, Singh PP (2019) A new method for biological synthesis of agriculturally relevant nanohydroxyapatite with elucidated effects on soil bacteria. Sci Rep 9(1):15083. https://doi.org/10.1038/s41598-019-51514-0
211. Priyam A, Seth P, Mishra J, Manna PK, Singh PP (2023) Occupational safety assessment of biogenic urea nanofertilisers using in vitro pulmonary, and in vivo ocular models. Heliyon 9(11). https://doi.org/10.1016/j.heliyon.2023.e21623
212. Priyam A, Yadav N, Reddy PM, Afonso LO, Schultz AG, Singh PP (2022) Fertilizing benefits of biogenic phosphorous nanonutrients on *Solanum lycopersicum* in soils with variable pH. Heliyon 8(3). https://doi.org/10.1016/j.heliyon.2022.e09144

213. Pugazhendhi A, Prabakar D, Jacob JM, Karuppusamy I, Saratale RG (2018) Synthesis and characterization of silver nanoparticles using *Gelidium amansii* and its antimicrobial property against various pathogenic bacteria. Microb Pathog 114:41–45. https://doi.org/10.1016/j.micpath.2017.11.013
214. Pulingam T, Foroozandeh P, Chuah JA, Sudesh K (2022) Exploring various techniques for the chemical and biological synthesis of polymeric nanoparticles. Nanomaterials 12(3):576. https://doi.org/10.3390/nano12030576
215. Qiao ZP, Wang MY, Liu JF, Wang QZ (2022) Green synthesis of silver nanoparticles using a novel endophytic fungus *Letendraea* sp. WZ07: characterization and evaluation of antioxidant, antibacterial and catalytic activities (3-in-1 system). Inorg Chem Commun 138:109301. https://doi.org/10.1016/j.inoche.2022.109301
216. Rafique M, Sadaf I, Rafique MS, Tahir MB (2017) A review on green synthesis of silver nanoparticles and their applications. Artif cells nanomedicine Biotechnol 45:1272–1291
217. Rai M, Wypij M, Ingle AP, Trzcińska-Wencel J, Golińska P (2021) Emerging trends in pullulan-based antimicrobial systems for various applications. IJMS 22(24):13596. https://doi.org/10.3390/ijms222413596
218. Rai M, Golinska P (Eds) (2023) Myconanotechnology: emerging trends and applications. CRC Press, Boca Raton. https://doi.org/10.1201/9781003327356
219. Rai M, Gupta I, Bonde S, Ingle P, Shende S, Gaikwad S, Razzaghi-Abyaneh M, Gade A (2022) Industrial applications of nanomaterials produced from aspergillus species. IntechOpen. https://doi.org/10.5772/intechopen.98780
220. Rai A, Singh A, Ahmad A, Sastry M (2006). Role of halide ions and temperature on the morphology of biologically synthesized gold nanotriangles. Langmuir 22(2):736–741. https://doi.org/10.1021/la052055q
221. Raj NB, Swamy MK, Purushotham B, Sukrutha SK (2021) Applications of microbe-based nanoparticles in agriculture: present state and future challenges. In: Lateef A, Gueguim-Kana EB., Dasgupta N, Ranjan S (eds) Microbial nanobiotechnology. Materials horizons: from nature to nanomaterials. Springer, Singapore. https://doi.org/10.1007/978-981-33-4777-9_12
222. Rajput V, Minkina T, Fedorenko A, Sushkova S, Mandzhieva S, Lysenko V, Duplii N, Fedorenko G, Dvadnenko K, Ghazaryan K (2018) Toxicity of copper oxide nanoparticles on spring barley (*Hordeum sativum* distichum). Sci Total Environ 645:1103–1113
223. Rajput V, Minkina T, Mazarji M, Shende S, Sushkova S, Mandzhieva S, Burachevskaya M, Chaplygin V, Singh A, Jatav H (2020) Accumulation of nanoparticles in the soil-plant systems and their effects on human health. Ann Agric Sci 65:137–143
224. Rajput VD, Minkina T, Upadhyay SK, Kumari A, Ranjan A, Mandzhieva S, Sushkova S, Singh RK, Verma KK (2022) Nanotechnology in the restoration of polluted soil. Nanomaterials 12:769. https://doi.org/10.3390/nano12050769
225. Rajput VD, Singh A, Minkina T, Rawat S, Mandzhieva S, Sushkova S, Upadhyay SK (2021) Nano-enabled products: challenges and opportunities for sustainable agriculture. Plants 10(12):2727. https://doi.org/10.3390/plants10122727
226. Rajput VD, Minkina T, Kimber RL, Singh VK, Shende S, Behal A, Sushkova S, Mandzhieva S, Lloyd JR (2021) Insights into the Biosynthesis of Nanoparticles by the Genus *Shewanella*. Appl Environ Microbiol 87(22):e0139021. https://doi.org/10.1128/AEM.01390-21. PMID: 34495739; PMCID: PMC8552890
227. Raliya R, Biswas P, Tarafdar JC (2015) TiO_2 nanoparticle biosynthesis and its physiological effect on mung bean (*Vigna radiata* L.). Biotechnol Rep 5:22–26. https://doi.org/10.1016/j.btre.2014.10.009
228. Raliya R, Tarafdar JC (2013) ZnO nanoparticle biosynthesis and its effect on phosphorous-mobilizing enzyme secretion and gum contents in clusterbean (*Cyamopsis tetragonoloba* L.). Agric Res 2:48–57. https://doi.org/10.1007/s40003-012-0049-z
229. Raliya R, Tarafdar JC, Biswas P (2016) Enhancing the mobilization of native phosphorus in the mung bean rhizosphere using ZnO nanoparticles synthesized by soil fungi. J Agric Food Chem 64(16):3111–3118. https://doi.org/10.1021/acs.jafc.5b05224

230. Raliya R, Tarafdar JC, Singh SK, Gautam R, Choudhary K, Maurino VG, Saharan V (2014) MgO nanoparticles biosynthesis and its effect on chlorophyll contents in the leaves of clusterbean (*Cyamopsis tetragonoloba* L.). Adv Sci Eng Med 6(5):538–545. https://doi.org/10.1166/asem.2014.1540
231. Ramezani F, Jebali A, Kazemi B (2012) A green approach for synthesis of gold and silver nanoparticles by *Leishmania* sp. Appl Biochem Biotechnol 168:1549–1555. https://doi.org/10.1007/s12010-012-9877-3
232. Rasouli M (2019) Biosynthesis of Selenium Nanoparticles using yeast Nematospora coryli and examination of their anti-candida and anti-oxidant activities. IET Nanobiotech 13:214–218. https://doi.org/10.1049/iet-nbt.2018.5187
233. Rico CM, Majumdar S, Duarte-Gardea M, Peralta-Videa JR, Gardea-Torresdey JL (2011) Interaction of nanoparticles with edible plants and their possible implications in the food chain. J Agric Food Chem 59:3485–3498
234. Roberts TL, Johnston AE (2015) Phosphorus use efficiency and management in agriculture. Resour Conserv Recycl 105:275–281. https://doi.org/10.1016/j.resconrec.2015.09.013
235. Romeo D, Salieri B, Hischier R, Nowack B, Wick P (2020) An integrated pathway based on in vitro data for the human hazard assessment of nanomaterials. Environ Int 137:105505. https://doi.org/10.1016/j.envint.2020.105505
236. Rouphael Y, Colla G (2020) Biostimulants in agriculture. Front Plant Sci 11:40. https://doi.org/10.3389/fpls.2020.00040
237. Saadat YR, Khosroushahi AY, Gargari BP (2019) A comprehensive review of anticancer, immunomodulatory and health beneficial effects of the lactic acid bacteria exopolysaccharides. Carbohydr Polym 217:79–89. https://doi.org/10.1016/j.carbpol.2019.04.025
238. Sabir S, Zahoor MA, Waseem M, Siddique MH, Shafique M, Imran M, Muzammil S (2020) Biosynthesis of ZnO nanoparticles using *Bacillus subtilis*: characterization and nutritive significance for promoting plant growth in *Zea mays* L. Dose-Response 18(3):1559325820958911. https://doi.org/10.1177/1559325820958911
239. Sadhukhan R, Sharma LD, Sen S, Karmakar S, Banerjee K, Baral K (2021) Enhancing the productivity of field crops through nano-fertilizer. In: Asaduzzaman M, Afroz M (eds) Agricultural development in Asia: potential use of nano-materials and nano-technology. IntechOpen, London. https://doi.org/10.5772/intechopen.101146
240. Saeaeh K, Thummarungsan N, Paradee N, Choeichom P, Phasuksom K, Lerdwijitjarud W, Sirivat A (2019) Soft and highly responsive multi-walled carbon nanotube/pullulan hydrogel composites as electroactive materials. Eur Polym J 120:109231. https://doi.org/10.1016/j.eurpolymj.2019.109231
241. Salari Z, Danafar F, Dabaghi S, Ataei SA (2016) Sustainable synthesis of silver nanoparticles using macroalgae *Spirogyra varians* and analysis of their antibacterial activity. J Saudi Chem Soc 20:459–464. https://doi.org/10.1016/j.jscs.2014.10.004
242. Saleh TA (2020) Nanomaterials: classification, properties, and environmental toxicities. Environ Technol Innov 20:101067. https://doi.org/10.1016/j.eti.2020.101067
243. Samavini R, Sandaruwan C, De Silva M, Priyadarshana G, Kottegoda N, Karunaratne V (2018) Effect of citric acid surface modification on solubility of hydroxyapatite nanoparticles. J Agric Food Chem 66(13):3330–3337. https://doi.org/10.1021/acs.jafc.7b05544
244. Sampathkumar K, Tan KX, Loo SCJ (2020) Developing nano-delivery systems for agriculture and food applications with nature-derived polymers. Iscience 23(5). https://doi.org/10.1016/j.isci.2020.101055
245. Samuel AM, Dines L, Finch HJS, Lane GPF (eds) (2022) In Samuel A, Dines L (eds) Lockhart and Wiseman's crop husbandry including grassland, 10th edn. Woodhead publishing series in food science, technology and nutrition. Elsevier, Amsterdam https://doi.org/10.1016/B978-0-323-85702-4.00014-5
246. Sanaeimehr Z, Javadi I, Namvar F (2018) Antiangiogenic and antiapoptotic effects of green-synthesized zinc oxide nanoparticles using *Sargassum muticum* algae extraction. In: Tiwari A, Turner APF (eds) Cancer nanotechnology, vol 9. Scrivener Publishing, pp 1–16. https://doi.org/10.1186/s12645-018-0037-5

247. Sanghi R, Verma P (2010) PH dependant fungal proteins in the "green" synthesis of gold nanoparticles. Adv Mater Lett 1:193–199
248. Saratale RG, Karuppusamy I, Saratale GD, Pugazhendhi A, Kumar G, Park Y, Ghodake GS, Bharagava RN, Banu JR, Shin HS (2018) A comprehensive review on green nanomaterials using biological systems: recent perception and their future applications. Colloids Surf B Biointerfaces 170:20–35. https://doi.org/10.1016/j.colsurfb.2018.05.045
249. Sarlak N, Taherifar A, Salehi F (2014) Synthesis of nanopesticides by encapsulating pesticide nanoparticles using functionalized carbon nanotubes and application of new nanocomposite for plant disease treatment. J Agric Food Chem 62(21):4833–4838. https://doi.org/10.1021/jf404720d
250. Sayyid NH, Zghair ZR (2021) Biosynthesis of silver nanoparticles produced by *Klebsiella pneumoniae*. Mater Today: Proc 42:2045–2049. https://doi.org/10.1016/j.matpr.2020.12.257
251. Scherer HW (2005) Fertilizers and fertilization. In: Hillel D (ed) Encyclopedia of soils in the environment. Elsevier, Amsterdam https://doi.org/10.1016/B0-12-348530-4/00229-0
252. Scott-Fordsmand JJ, Amorim MJ (2023) Using machine learning to make nanomaterials sustainable. Sci Total Environ 859:160303. https://doi.org/10.1016/j.scitotenv.2022.160303
253. Selmani A, Kovačević D, Bohinc K (2022) Nanoparticles: from synthesis to applications and beyond. Adv Coll Interf Sci 303:102640. https://doi.org/10.1016/j.cis.2022.102640
254. Senthilkumar P, Dawn SS, Samanvitha KS, Kumar SS, Kumar GN, Samrot AV (2017) Optimization and characterization of poly [R] hydroxyalkanoate of *Pseudomonas aeruginosa* SU-1 to utilize in nanoparticle synthesis for curcumin delivery. Biocatal Agric Biotechnol 12:292–298. https://doi.org/10.1016/j.bcab.2017.10.019
255. Shah M, Fawcett D, Sharma S, Tripathy SK, Poinern GE (2015) Green synthesis of metallic nanoparticles via biological entities. Materials 8:7278–7308. https://doi.org/10.3390/ma8115377
256. Shahena S, Rajan M, Chandran V, Mathew L (2021) Conventional methods of fertilizer release. In: Lewu FB, Volova T, Sabu T, Rakhimol KR (eds) Controlled release fertilizers for sustainable agriculture. Academic Press, Cambridge. https://doi.org/10.1016/B978-0-12-819555-0.00001-7
257. Sharma RK, Wang SC, Maity JP, Banerjee P, Dey G, Huang YH, Chen CY (2021) A novel BMSN (biologically synthesized mesoporous silica nanoparticles) material: synthesis using a bacteria-mediated biosurfactant and characterization. RSC Adv 11(52):32906–32916. https://doi.org/10.1039/D1RA05852E
258. Sharma B, Shrivastava M, Afonso LO, Soni U, Cahill DM (2022) Zinc-and magnesium-doped hydroxyapatite nanoparticles modified with urea as smart nitrogen fertilizers. ACS Appl Nano Mater 5(5):7288–7299. https://doi.org/10.1038/s41598-022-20772-w
259. Sharma A, Shukla A, Attri K et al (2020) Global trends in pesticides: a looming threat and viable alternatives. Ecotoxi Env Safety 201:110812. https://doi.org/10.1016/j.ecoenv.2020.110812
260. Shelke DB, Islam NF, Chambhare MR, Sonawane HB, Patowary R, Prasad R, Sarma H (2023) Enhancing secondary metabolites and alleviating environmental stress in crops with mycogenic nanoparticles: a comprehensive review. Biocatal Agric Biotechnol 102805. https://doi.org/10.1016/j.bcab.2023.102805
261. Shen M, Liu S, Jiang C, Zhang T, Chen W (2023) Recent advances in stimuli-response mechanisms of nano-enabled controlled-release fertilizers and pesticides. Eco-Environ Health. https://doi.org/10.1016/j.eehl.2023.07.005
262. Shende S, Rajput V, Gade A, Minkina T, Rai M, Singh SP (2020) Impact of nanomaterials on human health through/via food chain. In: Keswani C (ed) Book intellectual property issues in nanotechnology, 1st edn. CRC Press, p 36. eBook ISBN9781003052104
263. Shende S, Rathod D, Gade A, Rai M (2017) Biogenic copper nanoparticles promote the growth of pigeon pea (*Cajanus cajan* L.). IET Nanobiotechnol 11(7):773–781. https://doi.org/10.1049/iet-nbt.2016.0179

264. Shende SS et al (2021) Interaction of nanoparticles with microbes. In: Singh P, Singh R, Verma P, Bhadouria R, Kumar A, Kaushik M (eds) Plant-microbes-engineered nano-particles (PM-ENPs) nexus in agro-ecosystems. Advances in science, technology & innovation. Springer, Cham. https://doi.org/10.1007/978-3-030-66956-0_12
265. Shenton W, Douglas T, Young M, Stubbs G, Mann S (1999) Inorganicorganic nanotube composites from template mineralization of tobacco mosaic virus. Adv Mater 11:253–256. https://doi.org/10.1002/(sici)1521-4095(199903)11:33.0.co;2-7
266. Shi J, Ye J, Fang H, Zhang S, Xu C (2018) Effects of copper oxide nanoparticles on paddy soil properties and components. Nanomaterials 8:839
267. Shin YJ, Kwak JI, An YJ (2012) Evidence for the inhibitory effects of silver nanoparticles on the activities of soil exoenzymes. Chemosphere 88:524–529
268. Shinde S, Paralikar P, Ingle AP, Rai M (2020) Promotion of seed germination and seedling growth of *Zea mays* by magnesium hydroxide nanoparticles synthesized by the filtrate from *Aspergillus niger*. Arab J Chem 13(1):3172–3182. https://doi.org/10.1016/j.arabjc.2018.10.001
269. Shoukat R, Khan MI (2021) Carbon nanotubes: a review on properties, synthesis methods and applications in micro and nanotechnology. Microsys Technol, 1–10. https://doi.org/10.1007/s00542-021-05211-6
270. Sigmon LR, Adisa IO, Liu B, Elmer WH, White JC, Dimkpa CO, Fairbrother DH (2021) Biodegradable polymer nanocomposites provide effective delivery and reduce phosphorus loss during plant growth. ACS Agric Sci Technol 1(5):529–539. https://doi.org/10.1021/acsagscitech.1c00149
271. Singh AV, Ansari MHD, Rosenkranz D, Maharjan RS, Kriegel FL, Gandhi K, Luch A (2020) Artificial intelligence and machine learning in computational nanotoxicology: unlocking and empowering nanomedicine. Adv Healthc Mater 9(17):1901862. https://doi.org/10.1002/adhm.201901862
272. Singh D, Jain D, Rajpurohit D, Jat G, Kushwaha HS, Singh A, Upadhyay SK (2023) Bacteria assisted green synthesis of copper oxide nanoparticles and their potential applications as antimicrobial agents and plant growth stimulants. Front Chem 11:1154128. https://doi.org/10.3389/fchem.2023.1154128
273. Singh NA, Narang J, Garg D, Jain V, Payasi D, Suleman S, Swami RK (2023) Nanoparticles synthesis via microorganisms and their prospective applications in agriculture. Plant Nano Biol 100047. https://doi.org/10.1016/j.plana.2023.100047
274. Singh A, Gautam PK, Verma A, Singh V, Shivapriya PM, Shivalkar S, Sahoo AK, Samanta SK (2020) Green synthesis of metallic nanoparticles as effective alternatives to treat antibiotics resistant bacterial infections: a review. Biotechnol Rep 25:e00427. https://doi.org/10.1016/j.btre.2020.e00427
275. Sirajudeen AAO, Annuar MSM, Subramaniam R (2021) Composite of medium-chain-length polyhydroxyalkanoates-co-methyl acrylate and carbon nanotubes as innovative electrodes modifier in microbial fuel cell. Biotechnol Appl Biochem 68(2):307–318. https://doi.org/10.1002/bab.1928
276. Soberón-Chávez G, González-Valdez A, Soto-Aceves MP, Cocotl-Yañez M (2021) Rhamnolipids produced by *Pseudomonas*: from molecular genetics to the market. Microb Biotechnol 14(1):136–146. https://doi.org/10.1111/1751-7915.13700
277. Soliman MKY, Salem SS, Abu-Elghait M (2023) Biosynthesis of silver and gold nanoparticles and their efficacy towards antibacterial, antibiofilm, cytotoxicity, and antioxidant activities. Appl Biochem Biotechnol 195:1158–1183. https://doi.org/10.1007/s12010-022-04199-7
278. Sonawane H, Arya S, Math S, Shelke D (2021) Myco-synthesized silver and titanium oxide nanoparticles as seed priming agents to promote seed germination and seedling growth of *Solanum lycopersicum*: a comparative study. Int Nano Lett 11(4):371–379. https://doi.org/10.1007/s40089-021-00346-w
279. Sukanya MK, Saju KA, Praseetha PK, Sakthivel G (2013) Potential of biologically reduced silver nanoparticles from actinomycete cultures. J Nanosci 2013:940719. https://doi.org/10.1155/2013/940719

280. Sunitha A, Geo S, Sukanya S, Praseetha PK, Dhanya RP (2014) Biosynthesis of silver nanoparticles from actinomycetes for therapeutic applications. Int J Nano Dimens 5:155–162
281. Sunkar S, Nachiyar CV (2012) Biogenesis of antibacterial silver nanoparticles using the endophytic bacterium *Bacillus cereus* isolated from *Garcinia xanthochymus*. Asian Pacific J Trop Biomed 2(12):953–959
282. Suryavanshi P, Pandit R, Gade A, Derita M, Zachino S, Rai M (2017) *Colletotrichum* sp.-mediated synthesis of sulphur and aluminium oxide nanoparticles and its in vitro activity against selected food-borne pathogens. LWT 81:188–194. https://doi.org/10.1016/j.lwt.2017.03.038
283. Sweeney RY, Mao C, Gao X et al (2004) Bacterial biosynthesis of cadmium sulfide nanocrystals. Chem Biol 11(11):1553–1559
284. Tamesue S, Takashima Y, Yamaguchi H, Shinkai S, Harada A (2011) Photochemically controlled supramolecular curdlan/single-walled carbon nanotube composite gel: preparation of molecular distaff by cyclodextrin modified curdlan and phase transition control. Eur J Org Chem 2011:2801–2806. https://doi.org/10.1002/ejoc.201100077
285. Tarafdar JC, Raliya R, Mahawar H, Rathore I (2014) Development of zinc nanofertilizer to enhance crop production in pearl millet (*Pennisetum americanum*). Agric Res 3:257–262. https://doi.org/10.1007/s40003-014-0113-y
286. Tarafdar JC, Raliya R, Rathore I (2012) Microbial synthesis of phosphorous nanoparticle from tri-calcium phosphate using *Aspergillus tubingensis* TFR-5. J Bionanosci 6(2):84–89. https://doi.org/10.1166/jbns.2012.1077
287. Tarannum N, Gautam YK (2019) Facile green synthesis and applications of silver nanoparticles: a state-of-the-art review. RSC Adv 9:34926–34948. https://doi.org/10.1039/c9ra04164hrsc.li/rsc-advances
288. Thorley AJ, Tetley TD (2013) New perspectives in nanomedicine. Pharmacol Ther 140:176–185. https://doi.org/10.1016/j.pharmthera.2013.06.008
289. Tian X, He W, Cui J, Zhang X, Zhou W, Yan S, Sun X, Han X, Han S, Yue Y (2010) Mesoporous zirconium phosphate from yeast biotemplate. J Colloid Interface Sci 343:344–349
290. Tolisano C, Buono DD (2023) 2023 Biobased: biostimulants and biogenic nanoparticles enter the scene. Sci Total Environ 885:163912. https://doi.org/10.1016/j.scitotenv.2023.163912
291. Trzcińska-Wencel J, Wypij M, Rai M, Golińska P (2023) Biogenic nanosilver bearing antimicrobial and antibiofilm activities and its potential for application in agriculture and industry. Front Microbiol 14:1125685. https://doi.org/10.3389/fmicb.2023.1125685
292. Trzcińska-Wencel J, Wypij M, Terzyk AP, Rai M, Golińska P (2023) Biofabrication of novel silver and zinc oxide nanoparticles from *Fusarium solani* IOR 825 and their potential application in agriculture as biocontrol agents of phytopathogens, and seed germination and seedling growth promoters. Front Chem 11. https://doi.org/10.3389/fchem.2023.1235437
293. Usha R, Prabu E, Palaniswamy M, Venil CK, Rajendran R (2010) Synthesis of metal oxide nanoparticles by *Streptomyces* sp. for development of antimicrobial textiles. Glob J Biotechnol Biochem 5:153–160
294. Vaigankar DC, Mujawar SY, Mohanty AK, Dubey SK (2022) Biogenesis of selenium nanospheres using *Halomonas venusta* strain GUSDM4 exhibiting potent environmental applications. Archives Microbiol 204(7):372. https://doi.org/10.1007/s00203-022-02977-9
295. Vallan L, Imahori H (2022) Citric acid-based carbon dots and their application in energy conversion. ACS Appl Electron Mater 4(9):4231–4257. https://doi.org/10.1021/acsaelm.2c01021
296. Vemula M, Reddy A (2023) Polymeric nanoparticles as effective delivery systems in agriculture sustainability. Nanotechnol Environ Eng 8:805–814. https://doi.org/10.1007/s41204-023-00319-8
297. Vijayaraghavan K, Ashokkumar T (2017) Plant-mediated biosynthesis of metallic nanoparticles: a review of literature, factors affecting synthesis, characterization techniques and applications. J Environ Chem Eng 5:4866–4883. https://doi.org/10.1016/j.jece.2017.09.026
298. Vinzant K, Rashid M, Khodakovskaya MV (2023) Advanced applications of sustainable and biological nano-polymers in agricultural production. Front Plant Sci 13:1081165. https://doi.org/10.3389/fpls.2022.1081165

299. Wanyika H, Gatebe E, Kioni P, Tang Z, Gao Y (2012) Mesoporous silica nanoparticles carrier for urea: potential applications in agrochemical delivery systems. J Nanosci Nanotechnol 12(3):2221–2228. https://doi.org/10.1166/jnn.2012.5801
300. Wu Y, Yang H, Shin HJ (2013) Viruses as self-assembled nanocontainers for encapsulation of functional cargoes. Korean J Chem Eng 30:1359–1367. https://doi.org/10.1007/s11814-013-0083-y
301. Wu Z, Ren Y, Liang Y, Huang L, Yang Y, Zafar A, Hasan M, Yang F, Shu X (2021) Synthesis, characterization, immune regulation, and antioxidative assessment of yeast-derived selenium nanoparticles in cyclophosphamide-induced rats. ACS Omega 6(38):24585–24594. https://doi.org/10.1021/acsomega.1c03205. PMID: 34604640; PMCID: PMC8482516
302. Wypij M, Jędrzejewski T, Trzcińska-Wencel J et al (2021) Green synthesized silver nanoparticles: antibacterial and anticancer activities, biocompatibility, and analyses of surface-attached proteins. Front Microbiol 12:632505. https://doi.org/10.3389/fmicb.2021.632505
303. Wypij M, Rai M, Zemljič LF, Bračič M, Hribernik S, Golińska P (2023) Pullulan-based films impregnated with silver nanoparticles from the Fusarium culmorum strain JTW1 for potential applications in the food industry and medicine. Front Bioeng Biotechnol 11. https://doi.org/10.3389/fbioe.2023.1241739
304. Xin X, Nepal J, Wright AL, Yang X, He Z (2022) Carbon nanoparticles improve corn (*Zea mays* L.) growth and soil quality: comparison of foliar spray and soil drench application. J Clean Prod 363:132630. https://doi.org/10.1016/j.jclepro.2022.132630
305. Xu T, Wang Y, Aytac Z, Zuverza-Mena N, Zhao Z, Hu X, Demokritou P (2022) Enhancing agrichemical delivery and plant development with biopolymer-based stimuli responsive core–shell nanostructures. ACS Nano 16(4):6034–6048. https://doi.org/10.1021/acsnano.1c11490
306. Yadav VK, Amari A, Mahdhi N, Elkhaleefa AM, Fulekar MH, Patel A (2023) A novel and economical approach for the synthesis of short rod-shaped mesoporous silica nanoparticles from coal fly ash waste by *Bacillus circulans* MTCC 6811. World J Microbiol Biotechnol 39(11):289. https://doi.org/10.1007/s11274-023-03734-w
307. Yadav A, Yadav K, Abd-Elsalam KA (2023) Exploring the potential of nanofertilizers for a sustainable agriculture. Plant Nano Biol 100044. https://doi.org/10.1016/j.plana.2023.100044
308. Yah CS, Iyuke SE, Simate GS (2011) A review of nanoparticles toxicity and their routes of exposures. Iran J Pharma Sci 8(1):299–314
309. Yan A, Chen Z (2019) Impacts of silver nanoparticles on plants: a focus on the phytotoxicity and underlying mechanism. Int J Molec Sci 20(5):1003. https://doi.org/10.3390/ijms20051003
310. Yan Z, Chen S, Wang H, Wang B, Jiang J (2008) Biosynthesis of bacterial cellulose/multi-walled carbon nanotubes in agitated culture. Carbohydr Polym 74(3):659–665. https://doi.org/10.1016/j.carbpol.2008.04.028
311. Yang W, Wang L, Mettenbrink EM, DeAngelis PL, Wilhelm S (2021) Nanoparticle toxicology. Annu Rev Pharmacol Toxicol 61:269–289
312. Yaseen R, Amin BH (2021) The use of phosphorus nanoparticles synthesized by rhizospheric fungus *Aspergillius fumigatus* as a nanofertilizer for flax plant. Malays J Microbiol 17(3):244–253. https://doi.org/10.21161/mjm.200852
313. Yong P, Macaskie LE, Sammons RL, Marquis PM (2004) Synthesis of nanophase hydroxyapatite by a *Serratia* sp. from waste-water containing inorganic phosphate. Biotechnol Lett 26:1723–1730. https://doi.org/10.1007/s10529-004-3744-4
314. Yun SI, Gadd GE, Latella BA, Lo V, Russell RA, Holden PJ (2008) Mechanical properties of biodegradable polyhydroxyalkanoates/single wall carbon nanotube nanocomposite films. Polym Bull 61:267–275. https://doi.org/10.1007/s00289-008-0936-4
315. Zawrah MF, Snousy MG (2017) Environmental challenges of nanotechnology in subsurface. In: Zhang TC, Gurjar BR, Govil JN (eds) Environmental science and engineering. Industrial processing & nanotechnology, vol 10, no 16. Studium Press LLC, USA, pp 379–416
316. Zhang W, Tang X, Feng X, Wang E, Li H, Shen J, Zhang F (2019) Management strategies to optimize soil phosphorus utilization and alleviate environmental risk in China. J Environ Qual 48(5):1167–1175. https://doi.org/10.2134/jeq2019.02.0054

317. Zhang C, Shu Z, Sun H, Yan L, Peng C, Dai Z, Yang L, Fan L, Chu Y (2023) Cu (II)@ ZIF-8 nanoparticles with dual enzyme-like activity bound to bacteria specifically for efficient and durable bacterial inhibition. Appl Surf Sci 611:155599. https://doi.org/10.1016/j.apsusc.2022.155599
318. Zhao Y, Sun B, Wang T, Yang L, Xu X, Chen C, Sun D (2020) Synthesis of cellulose–silica nanocomposites by in situ biomineralization during fermentation. Cellulose 27:703–712. https://doi.org/10.1007/s10570-019-02824-z
319. Zhi Y, Wu Q, Xu Y (2017) Genome and transcriptome analysis of surfactin biosynthesis in *Bacillus amyloliquefaciens* MT45. Sci Rep 7(1):40976. https://doi.org/10.1038/srep40976
320. Zhou W, He W, Zhang X, Zhao H, Li Z, Yan S, Han X (2009) Biosynthesis and characterization of mesoporous organic–inorganic hybrid iron phosphate. Mater Chem Phys 116:319–322. https://doi.org/10.1016/j.matchemphys.2009.04.013
321. Zhu G, Sun Y, Shakoor N, Zhao W, Wang Q, Wang Q, Rui Y (2023) Phosphorus-based nanomaterials as a potential phosphate fertilizer for sustainable agricultural development. Plant Physiol Biochem 108172. https://doi.org/10.1016/j.plaphy.2023.108172

Plant Biosynthesis of Biostimulant Nanomaterials for Application as Nanofertilizers

Luis Alejandro Martínez-Chávez, Alejandra Jimenez-Hernandez, Jesús Antonio Angole-Tierrablanca, Ramón G. Guevara-González, Ana A. Feregrino-Pérez, and Karen Esquivel Escalante

Abstract Nanotechnology involves synthesizing and applying diverse materials and devices by managing and controlling their shape and size at the nanoscale. In this sense, in the past two decades, the use of nanomaterials in the agriculture sector has been focused on the increment of crop yield by nanofertilizers, which can facilitate efficient nutrient uptake, soil fertility, water and nutrient absorption, increased photosynthesis, reduced soil toxicity, decreased application frequency, plant health improvement, and reduced environmental pollution. Nevertheless, some risks must be addressed before using these "smart tools" in food production. That is why one way to start on this path is the generation of green synthesis processes for nanofertilizers by using extracts from biological sources such as microorganisms and plants, obtaining more bio-compatible nanomaterials, boosting the advantages of its application and reducing the toxic effects over the environment.

Keywords Biosynthesis · Nanomaterials · Biostimulant · Nanofertilizer · Green synthesis

1 Introduction

According to how the population rate increases over time, new agricultural processes are needed to keep the food supply all over the globe. In this matter, fertilizers have been spread in various crops' cultivation for augmenting crop yields and

L. A. Martínez-Chávez · A. A. Feregrino-Pérez · K. Esquivel Escalante (✉)
División de Investigación y Posgrado, Facultad de Ingeniería, Universidad Autónoma de Querétaro, Santiago de Querétaro, México
e-mail: karen.esquivel@uaq.mx
URL: http://www.uaq.mx

A. Jimenez-Hernandez · J. A. Angole-Tierrablanca · R. G. Guevara-González
Biosystems Engineering Group, Center of Applied Research in Biosystems (CARB-CIAB), School of Engineering, Autonomous University of Querétaro, Santiago de Querétaro, México

© The Author(s), under exclusive license to Springer Nature Singapore Pte Ltd. 2025

A. Juárez-Maldonado et al. (eds.), *Plant Biostimulation with Nanomaterials*, Smart Nanomaterials Technology, https://doi.org/10.1007/978-981-96-4648-7_13

preserving soil fertility. Conventional fertilizers (urea, nitrogen, phosphorous, potassium, monoammonium phosphate, and diammonium phosphate) are used to supplement essential nutrients in the soil. However, this conventional practice has diverse disadvantages, such as low nutrient utilization efficiency due to leaching, leading to substantial economic losses and decreased soil fertility [171, 179].

That is how the nanofertilizers (NFs) have found an opportunity for application as a promising solution to offer higher efficiency and reduced environmental impacts. NFs can be manufactured from conventional fertilizers and bulk materials or extracted from other plants by encapsulating/coating them with nanomaterials (NMs). Also, due to the large surface area of the nanoparticles (NPs), they can hold plenty of nutrients while slowly and steadily releasing them to the plants and soil [55]. With this, it is possible to reduce the ecotoxicity, loss of nutrients to the soil, and prevention of groundwater pollution [88].

The classification of NFs has been discussed by several scientific groups, leading to a general classification according to (i) action-based, i.e., controlled release, targeted delivery, plant growth stimulating, and water and nutrient loss controlling; (ii) nutrient based, i.e., inorganic, organic, hybrid, and nutrient loads; (iii) consistency based, i.e., surface coated, synthetic polymer coated, biological product coated, and nanocarrier based [13, 89, 109, 154]. Among the different NFs, it is crucial to seek those obtained with the less invasive methods in processing and disposal, such as the green synthesis methods (bio-synthesis methods) [120] in accordance of reducing the impact on the environment and toxicity over plants, soil, and even human beings [103]. In this chapter, a general overview of what is a Nanomaterial (NM), its classification, synthesis methods, and how it can be applied as biostimulant NMs for further use as NFs in crops, fruits, and vegetables is presented, along with some ethical implications and perspectives for the future years.

2 Nanomaterials and Their Classification

A nanometer (nm) is quantitatively defined as one-billionth of a meter. Consequently, nanomaterials refer to materials that possess at least one dimension—be it length, width, or depth—falling within the nanometric scale, specifically ranging between 1 and 100 nm, as delineated in the field of nanoscience. These materials can further be categorized (Fig. 1) into various structures, contingent upon the particular dimensions manifested at the nanoscale [15, 151].

Nanomaterials can be systematically classified by size based on the number of dimensions within the nanoscale range of 1–100 nm (Table 1).

Each classification denotes a different subset of nanomaterials, affording a distinct set of properties and potential applications reflective of their dimensional characteristics at the nanoscale. Apart from dimensional classification, nanomaterials can also be categorized based on their specific shapes (Table 2).

Nanomaterials can be classified into four distinct groupings according to their chemical composition: carbon-based, organic, inorganic, and composite (Table 3).

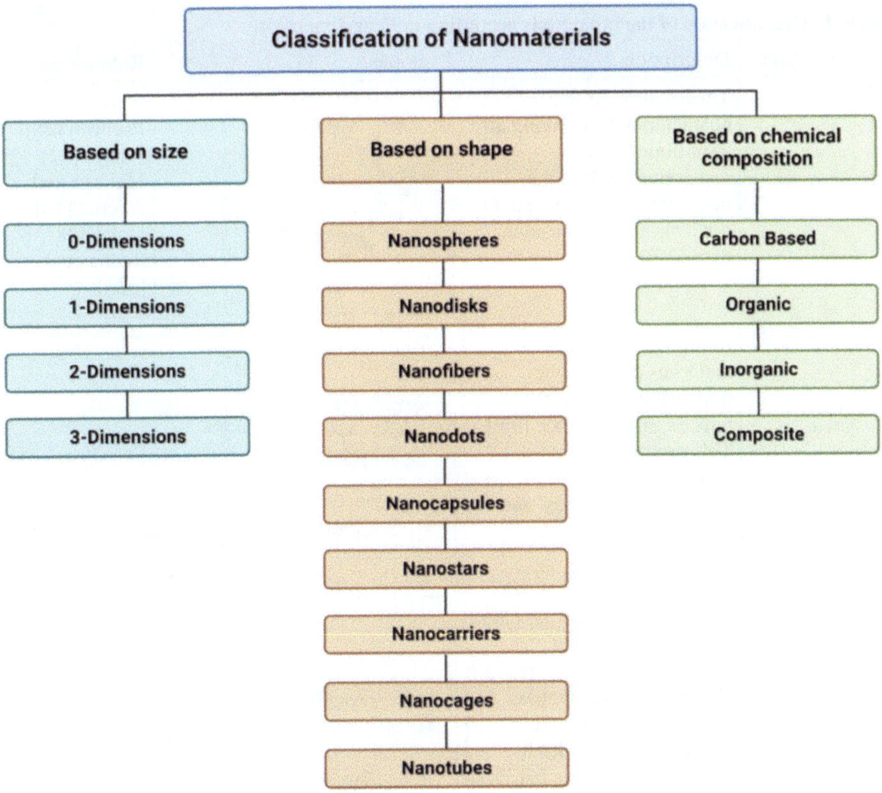

Fig. 1 Classification of Nanomaterials based on size, shape, and chemical composition

3 Nanomaterials Synthesis Methods

The realm of nanomaterial synthesis embraces two primary methodologies. One methodology, known as the top-down approach, involves the reduction of larger molecular entities to achieve nanoscale dimensions, typically ranging from 1 to 100 nm. On the contrasting front, the bottom-up approach entails the amalgamation of diminutive molecules, particles, and even individual atoms, culminating in the creation of nanomaterials [140]. Owing to nanomaterials' revolutionary and captivating potential in advancing the next industrial era, nanotechnology has garnered substantial scientific attention and intrigue throughout its evolution [141, 151]. Table 4 provides a comprehensive overview of the various synthesis techniques employed in *bottom-up* and *top-down* approaches for nanomaterials synthesis, focusing on more specialized methods in some instances [3, 42].

These approaches categorize distinct synthesis methods, emphasizing various classifications, including chemical, physical, and biological techniques.

Table 1 Classification of nanomaterials according to their dimension

Classification	Description	Example	References
0D	These materials are characterized by having all their dimensions constrained within the nanoscale range, meaning none of their dimensions exceed 100 nm	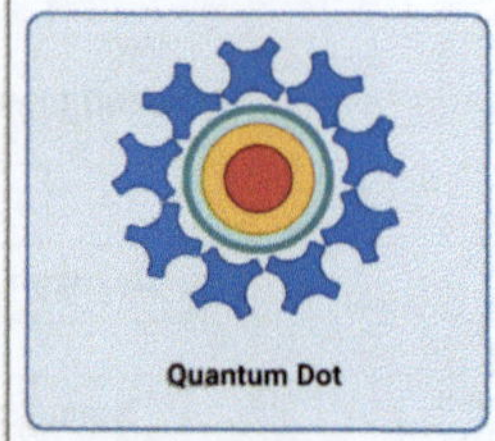	Buzea and Pacheco [28], Findik [52], Mekuye and Abera [110], Dolez [38], Sudha et al. [166]
1D	In this category, the materials possess one dimension that extends beyond the nanoscale limit. This group includes structures such as nanotubes, nanorods, and nanowires	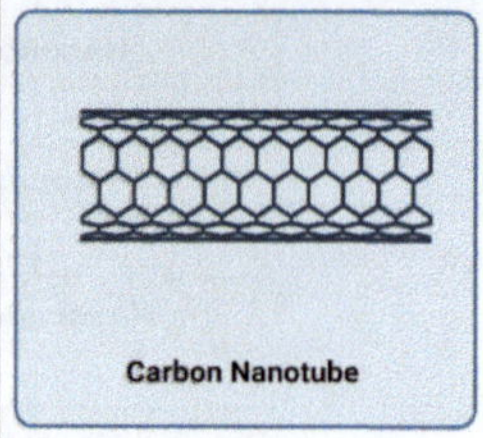	
2D	These materials have two dimensions that surpass the nanoscale threshold. Typical structures in this classification encompass sheets, thin films, and coatings, which exhibit nanoscale thickness but have larger lateral dimensions		
3D	These materials are characterized by having all three dimensions exceeding the 100 nm boundary. Nanocomposites, powders, assemblages of nanotubes or multilayers, and Micro-Electro-Mechanical Systems (MEMS) are examples of 3D nanomaterials with a complex three-dimensional architecture involving nanoscale components	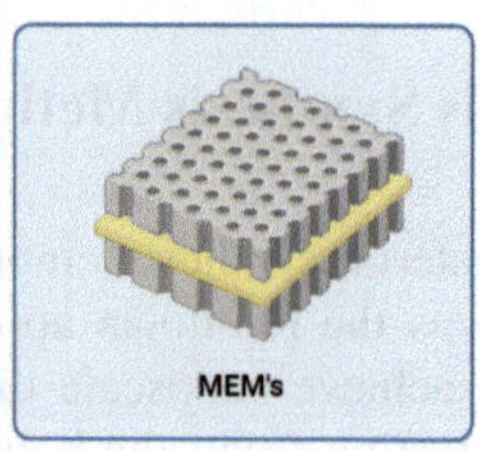	

Table 2 Classification of nanomaterials based on their shape

Nanomaterial	Description	Representative Image	References
Nanospheres	Materials with a spherical shape	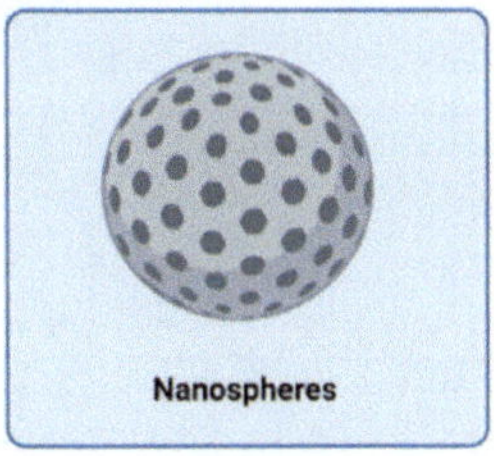	Mekuye and Abera [110], Pathak et al. [131], Liu et al. [97], Pokropivny et al. [133], Al Mashary et al. [9], Saira et al. [150], Singh [163], Guerrero-Martínez et al. [57], Fery and Weinkamer [51]
Nanodisks	Nanomaterials with disk-like or plate-like morphology	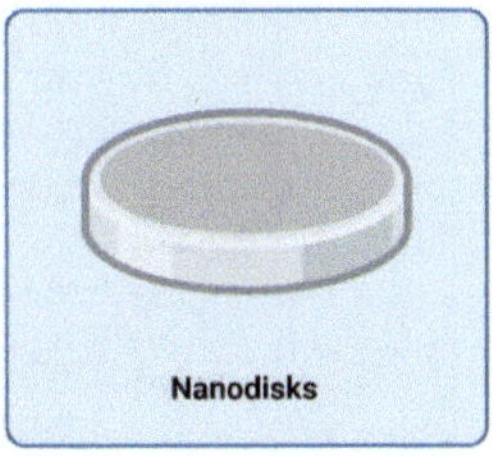	
Nanofibers	Materials with fiber-like shapes	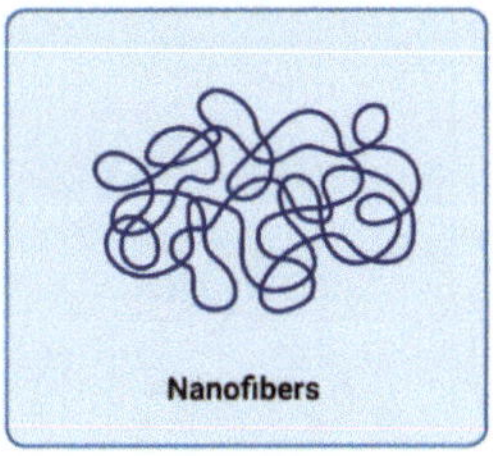	
Nanodots	Dot-like nanostructures are often utilized in quantum dot technology	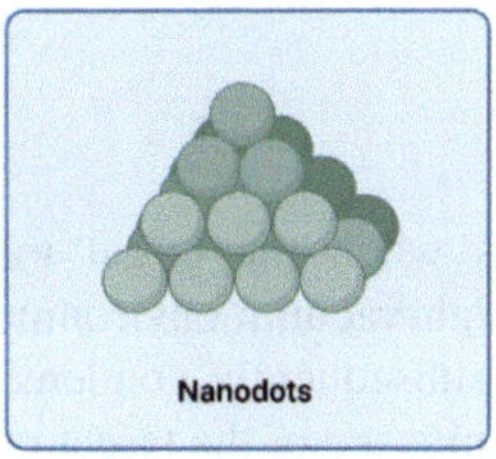	
Nanocapsules	Nanomaterials are shaped like capsules, having a core–shell structure	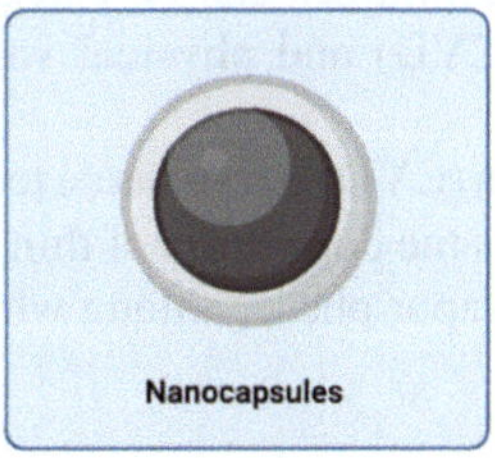	

(continued)

Table 2 (continued)

Nanomaterial	Description	Representative Image	References
Nanostars	Materials with star-like shapes are often used in sensing and imaging applications	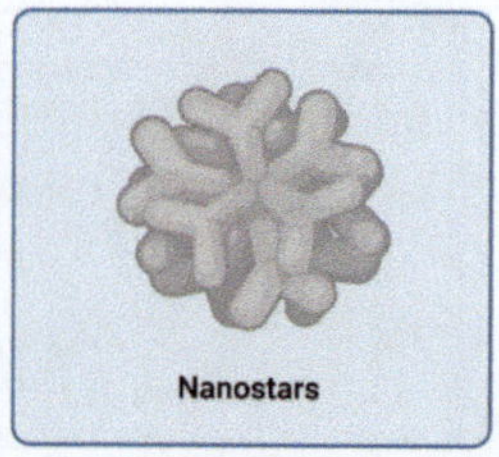	
Nanocarriers	Materials designed to carry other substances are often used in drug delivery systems	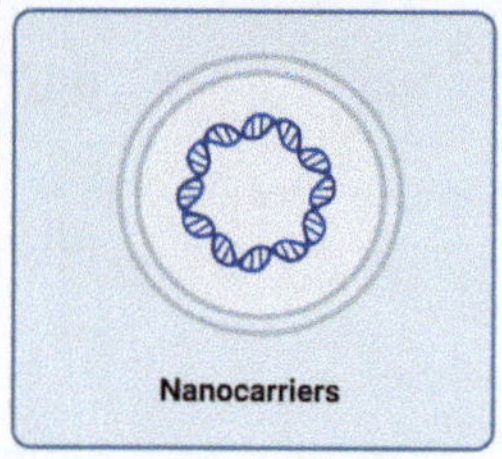	
Nanocages	Materials with cage-like structures are useful in various applications, including drug delivery and catalysis	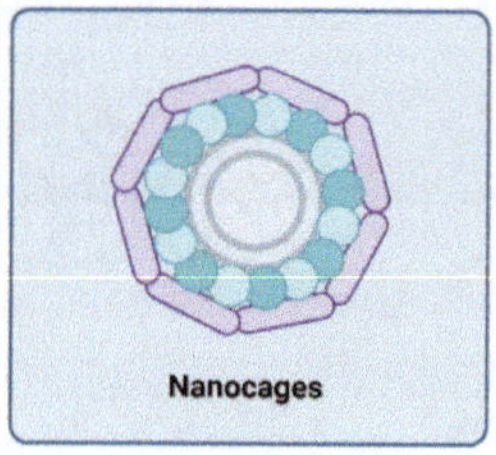	

3.1 Physical Processes

Physical vapor deposition is when the target material evaporates into atoms, molecules, or ions within a high-vacuum environment [159]. This resulting vapor, comprised of atoms or ions, subsequently condenses onto the substrate's surface, forming a thin film coating composed of the target material. Based on the governing principle, vapor deposition techniques can be categorized into two major classes: chemical vapor deposition (CVD) and physical vapor deposition (PVD) (Fig. 2) [36].

Chemical Vapor Deposition (CVD) is a versatile technique for synthesizing nanostructured materials. It involves the deposition of thin films or coatings on a substrate by chemical reactions in the vapor phase, among which are processes such as:

Table 3 Classification of nanomaterials based on their chemical composition

Classification	Description	Example	References
Carbon-based nanomaterials	Carbon-based nanomaterials are composed primarily of carbon atoms. They encompass various structures, such as fullerenes, carbon nanotubes, and graphene, each with unique properties and applications. Fullerenes are spherical carbon molecules, carbon nanotubes exhibit cylindrical structures, and graphene is a single layer of carbon atoms arranged in a hexagonal lattice. These materials are known for their exceptional mechanical, electrical, and thermal properties, making them crucial in nanoelectronics, composites, and energy storage applications	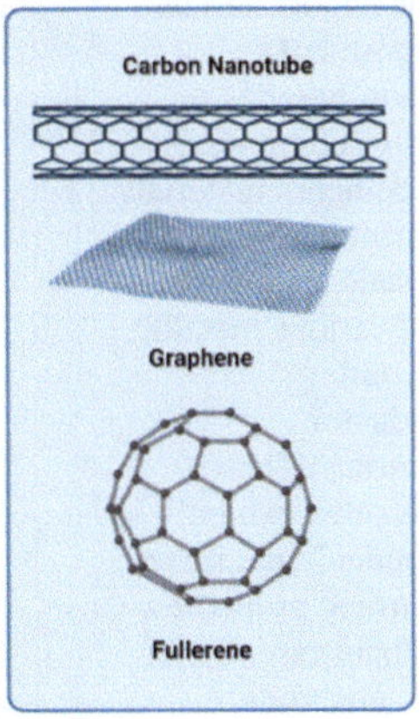	Ealias and Saravanakumar [40], Hola et al. [66], Liu et al. [98], Bhattacharya et al. [22]
Organic nanomaterials	Organic nanomaterials are derived from organic compounds and can include structures like organic nanoparticles, micelles, and dendrimers. They are known for their versatility and biocompatibility, making them useful in drug delivery and sensing and as building blocks for nanoscale assembly. Organic nanomaterials are vital in pharmaceuticals, biotechnology, and materials science	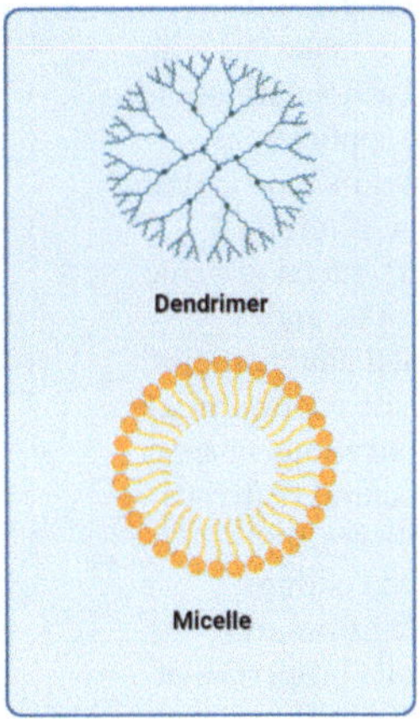	Armarego [15], Alshammari et al. [14], Ali et al. [12], Khalid et al. [79]

(continued)

Table 3 (continued)

Classification	Description	Example	References
Inorganic nanomaterials	Inorganic nanomaterials are composed of non-carbon-based elements or compounds. They encompass materials such as nanoparticles, nanocrystals, quantum dots, and various metallic, ceramic, and semiconductor nanomaterials. These materials often exhibit unique optical, magnetic, and electronic properties, making them essential in catalysis, electronics, optics, and materials science	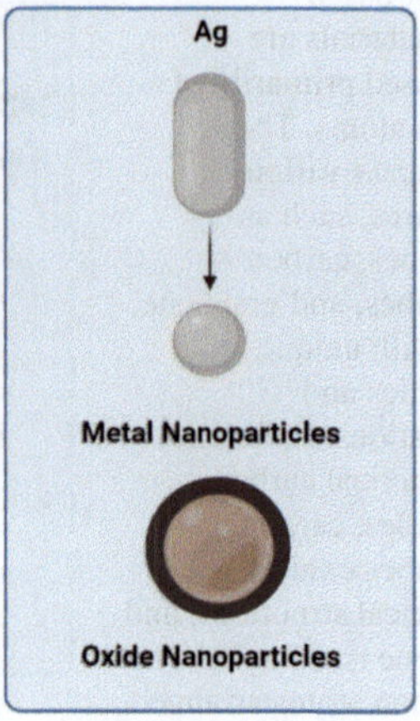	Khalid et al. [79], Saravanan et al. [155], Jeevanandam et al. [69], Saleh [151], Armarego [15]
Composite nanomaterials	Composite nanomaterials are hybrid materials composed of two or more distinct components, with at least one component at the nanoscale. These composites combine the properties of their individual constituents, leading to materials with tailored characteristics for specific applications. Examples include carbon nanotube-reinforced polymers, metal–organic frameworks, and core–shell nanoparticles. Composite nanomaterials find applications in areas like structural materials, drug delivery, and advanced coatings, benefiting from the synergistic properties of their components	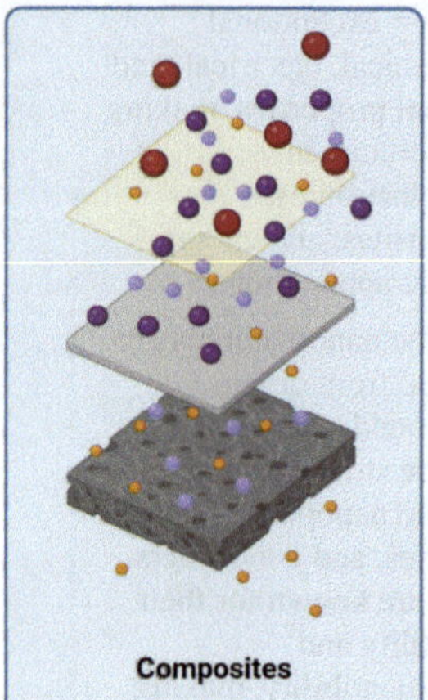	Baig et al. [19], Reduwan Billah [143], Alarifi [11], Liu et al. [99]

Table 4 Different synthesis techniques of nanomaterials from the bottom-up and top-down processes

Method	Technique	References
Top-down	1. Sol–Gel Method 2. Chemical Vapor Deposition (CVD) 3. Hydrothermal Synthesis 4. Solvothermal Synthesis 5. Co-Precipitation 6. Chemical Reduction 7. Microemulsion 8. Electrochemical Deposition 9. Green Synthesis 10. Biochemical Synthesis	Baig et al. [19], Abid et al. [3], El-Khawaga et al. [42], Biswas et al. [24]
Bottom-up	1. Ball Milling 2. Lithography (e.g., E-beam lithography, Nanoimprint lithography) 3. Plasma Arc Discharge 4. Laser Ablation 5. Mechanical Exfoliation (e.g., Scotch tape method) 6. Template-Based Synthesis 7. Chemical Etching 8. Atomic Layer Deposition (ALD) 9. Electrospinning 10. Jet Milling	

- Metal–Organic Chemical Vapor Deposition (MOCVD)

MOCVD is used for depositing thin films of various materials, including semiconductors and superconductors. It utilizes volatile metal–organic precursors that decompose at elevated temperatures to deposit the desired material. The metal–organic precursor is vaporized and carried to the substrate in a controlled atmosphere. At the substrate surface, the precursor decomposes, releasing the metal and organic components. The metal atoms then react to form the desired material on the substrate [29, 92, 177].

- Atomic Layer Deposition (ALD)

ALD is a precise technique for depositing conformal, uniform layers of materials at the atomic scale. It is commonly used for nanoscale coatings and thin films. ALD is based on a self-limiting, alternating sequence of precursor gases. A precursor is introduced in each cycle and chemisorbs onto the substrate's surface. Any excess precursor is purged, and a second precursor is introduced, which reacts with the first layer. This cycle is repeated until the desired film thickness is achieved, ensuring precise control over film thickness and composition [93, 149].

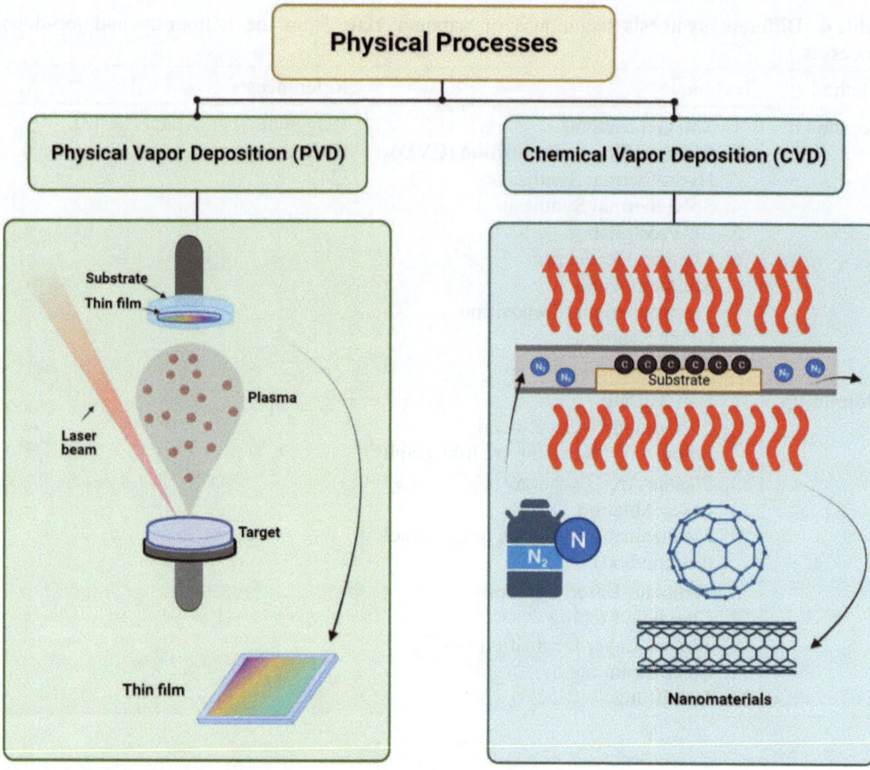

Fig. 2 Classification of nanomaterial synthesis processes by physical processes

- Plasma-Enhanced Chemical Vapor Deposition (PECVD)

PECVD is used for the deposition of thin films, primarily in microelectronics and photovoltaics. It involves the use of plasma to enhance the chemical reactions in the vapor phase. A low-pressure plasma is created in the reactor using a radiofrequency or microwave source. Precursor gases are introduced into the plasma, where they are dissociated and ionized. The highly reactive species in the plasma then interact with the substrate surface, leading to the deposition of thin films. PECVD allows for precise control of film properties and can be used for various materials, including amorphous and crystalline films [39, 106, 182].

Physical Vapor Deposition (PVD) is another versatile technique for synthesizing nanostructured materials, which involves the deposition of thin films or coatings through physical processes in a vacuum environment. Physical vapor deposition is subcategorized as sputtering, laser ablation, and vacuum arc deposition [100].

- Sputtering

Sputter deposition is a PVD technique that uses high-energy ions to dislodge material from a target and deposit it onto a substrate. A target material is placed near a substrate in a vacuum chamber, and a plasma of high-energy ions bombards both. These ions impact the target surface, causing atoms or molecules to be ejected. These ejected species then condense onto the substrate, forming a thin film. The direction and energy of the sputtered species can be controlled, allowing for precise film thickness and composition control. Sputter deposition is widely used in microelectronics, optics, and material science to produce high-quality, uniform, thin films [5, 69].

- Laser Ablation

Laser ablation is a PVD technique that employs a high-energy laser to remove material from a target and deposit it on a substrate. A high-energy laser beam is focused on a target material in a vacuum chamber. The intense energy of the laser rapidly heats and vaporizes the target, causing it to ablate and generate a plasma plume. The material in the plasma plume is then directed toward a substrate and deposits to form a thin film. Laser ablation offers precise control over the ablated material, making it suitable for creating complex and customized materials, such as high-temperature superconductors, thin films for optics, and thin film electronics [16, 45, 142].

- Vacuum arc deposition

Vacuum arc deposition is a PVD technique that uses an electrical discharge to vaporize and ionize a target material and deposit it on a substrate. A high-current electrical discharge or vacuum arc is generated in a vacuum chamber between a cathode made of the target material and an anode. This discharge vaporizes and ionizes the target material, creating a highly ionized plasma. The ions from the plasma are then accelerated and deposited onto a substrate, forming a thin film. Vacuum arc deposition is known for creating durable and adherent coatings, making it suitable for applications like wear-resistant coatings, decorative films, and thin film solar cells [26, 174].

3.2 Chemical Processes

Chemical synthesis is a method that involves a series of chemical and physical processes, encompassing various chemical reactions that can be controlled and replicated to obtain the desired primary product. In the synthesis of nanomaterials, it is customary to employ liquid-phase precursors. Achieving superior material properties, precise morphology, and consistent reproducibility necessitates a meticulous selection of precursors their concentrations, understanding of reaction dynamics, and careful control of synthesis parameters [46].

These methods have garnered more popularity when contrasted with preceding physical synthesis techniques. They offer proficient stoichiometry control, thanks to their capacity to function under ambient conditions and lower temperatures and with basic equipment. This results in cost reduction and renders them more appealing than physical methods. The synthesis category encompasses chemical precipitation, hydrothermal processes, sol–gel methods, microwave-coupled sol–gel approaches, and sonochemistry-coupled sol–gel methods (Fig. 3) [54, 104].

- Chemical precipitation

Among the various techniques, chemical precipitation is one of the most popular and straightforward methods, offering high performance when synthesizing metal-chalcogenides with precise composition control. The primary reaction involved in this process is co-precipitation, which encompasses simultaneous nucleation, growth, coarsening, and agglomeration processes [104]. During a co-precipitation experiment, inorganic salts are dissolved in a suitable medium to create a uniform solution. Preparing a basic precursor solution by gradually introducing the salts into solvents is crucial to maintaining optimal pH control since most materials only precipitate in alkaline solvents [128]. This technique is characterized by its simplicity, enabling rapid preparation and control over particle size, composition, and homogeneity while operating at low temperatures and consuming minimal energy and resources [30].

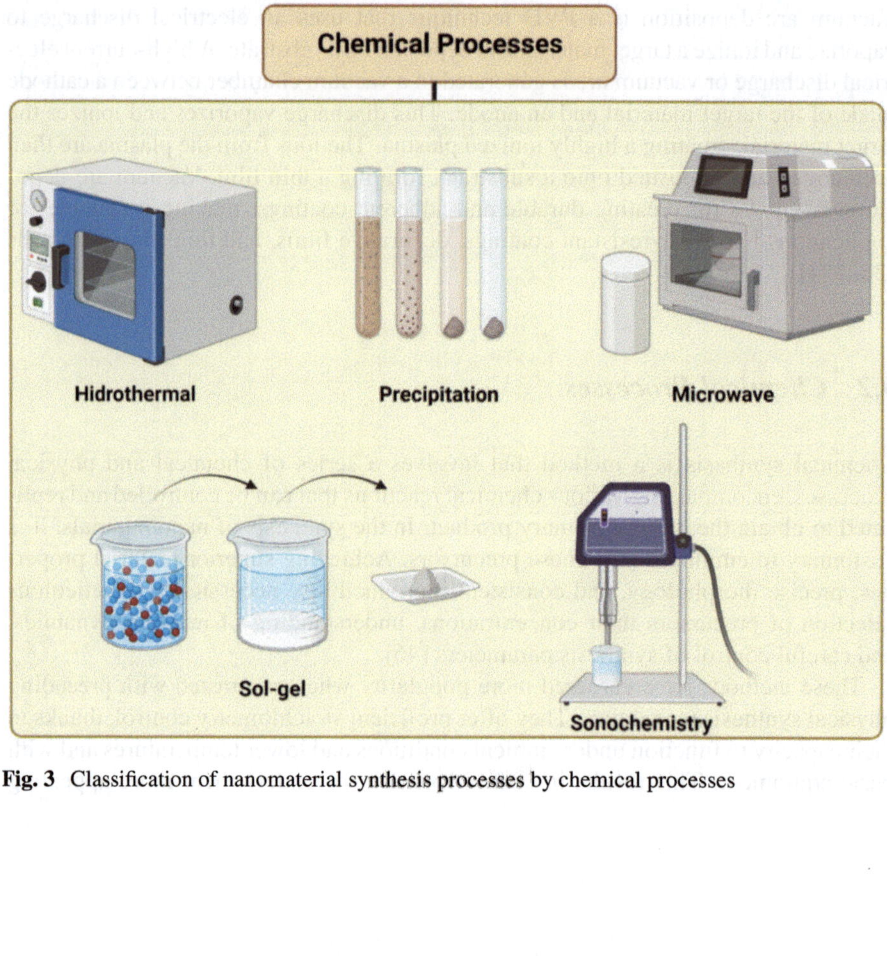

Fig. 3 Classification of nanomaterial synthesis processes by chemical processes

- Hydrothermal processes

The hydrothermal technique involves subjecting chemical reactions to high temperatures and pressures that exceed typical ambient conditions. This method capitalizes on the unique property of water and aqueous solutions to dissolve nearly insoluble substances under normal circumstances, particularly at elevated temperatures and pressures of up to 300 MPa. In the hydrothermal technique, water serves as the primary solvent, hence the name of the approach [140]. There are two main categories within this method. The first is the batch hydrothermal system, where the process occurs with carefully controlled phase ratios, and the second is the continuous hydrothermal system, designed for high-speed reactions conducted in a shorter timeframe [118]. The hydrothermal technique is valued for its energy efficiency and waste reduction, making it a favored choice. Its notable advantages include low costs and the ability to maintain controlled and reproducible conditions for nucleation control. However, it necessitates using autoclaves, operates at high temperatures, and lacks real-time observation during the growth process [183].

- Sol–gel method

The sol–gel technique is the prevailing and well-established method for creating colloidal nanoparticles through liquid-phase manipulation of nanocomposite powders. Its chief advantage lies in its ability to operate at low temperatures, conserve energy during the synthesis process, and offer superior precision [104]. This process can be broken down into five distinct steps [2, 72, 130]: (i) Hydrolysis: The initial step involves the breakdown of chemical compounds through a reaction with water, typically resulting in the formation of various precursors. (ii) Polycondensation: In this stage, these precursor compounds undergo a process known as polycondensation, where they link together to create more extensive and more complex molecules or networks. (iii) Aging: During this phase, the synthesized material is left to age or mature, allowing further reactions and structural changes to occur, which can have a significant impact on the final properties of the material. (iv) Drying: The material is then dried, which removes excess moisture or solvents and helps stabilize the structure. (v) Thermal Decomposition: Finally, the material undergoes thermal decomposition, which can form the desired nanomaterial by releasing gases and rearranging atoms or molecules to shape texture and surface attributes.

- Microwave-coupled sol–gel

Microwave-assisted methodologies have gained widespread popularity across diverse fields, from biochemical processes to nanotechnology [67]. Microwave-assisted heating offers a unique advantage in precisely activating nanomaterial precursor substances. Microwaves, classified as electromagnetic waves with frequencies ranging between 300 MHz and 300 GHz, correspond to wavelengths spanning from 1 to 1 mm respectively [47]. The efficacy of microwave heating becomes evident when polar and charged molecules are exposed to these intensely agitated electromagnetic waves. This exposure leads to the dissipation of radiation energy and a subsequent rise in the internal temperature of the subject exposed to microwaves. Remarkably, the interaction of microwaves with matter results in an inverted temperature gradient profile compared to conventional heating methods [32].

However, it is important to note that uneven heating can disrupt product or reagent stability, reduce reaction yields, encourage the formation of undesired by-products, and necessitate purification stages. Nevertheless, microwave-assisted techniques have numerous advantages, including shortened reaction times, increased reaction yields, selective heating, enhanced crystallinity, and uniform morphological characteristics [181].

- Sonochemistry

Sonochemistry, in contrast to alternative methods, represents an environmentally friendly process harnessed through the influence of ultrasound within a liquid medium [180]. This technique generates unique hotspots capable of achieving temperatures as high as 5000 Kelvin, pressures of up to 1000 atmospheres, and exceptionally rapid cooling rates exceeding 10^{10} Ks^{-1}. The intriguing chemical and physical effects of ultrasound do not stem directly from the interaction between chemical species and sound waves. Instead, they arise from the intriguing physical phenomenon known as acoustic cavitation, characterized by bubble formation, expansion, and sudden collapse [180].

Sonochemical methodology encompasses two distinct processes: primary sonochemistry, which involves volatile precursors formed within the confines of bubbles, and secondary sonochemistry, which utilizes non-volatile precursors situated outside the bubbles. These two processes act synergistically to yield unique results. The manipulation of parameters such as temperature, sonication duration, and ultrasonication power plays a pivotal role in shaping the properties of the resulting nanomaterials [34, 63, 146].

3.3 *Biosynthesis*

Biofabrication, or bioprinting or biosynthesis, is an innovative approach to creating nanostructured materials or tissue constructs using biological components, such as cells, biomass, and biomaterials (Fig. 4). It is a multidisciplinary field that merges biology, materials science, and engineering [85].

A variety of synthesis methods offer numerous processing advantages, yet they often come with environmental drawbacks. This is where the concept of "green chemistry" (GC) emerges as a response to the need for environmentally friendly alternatives [6]. Green chemistry involves using environmentally benign materials as a sustainable approach to produce various materials, particularly nanomaterials (NMs). Its primary goal is to tackle pollution and enable low-cost production, aligning

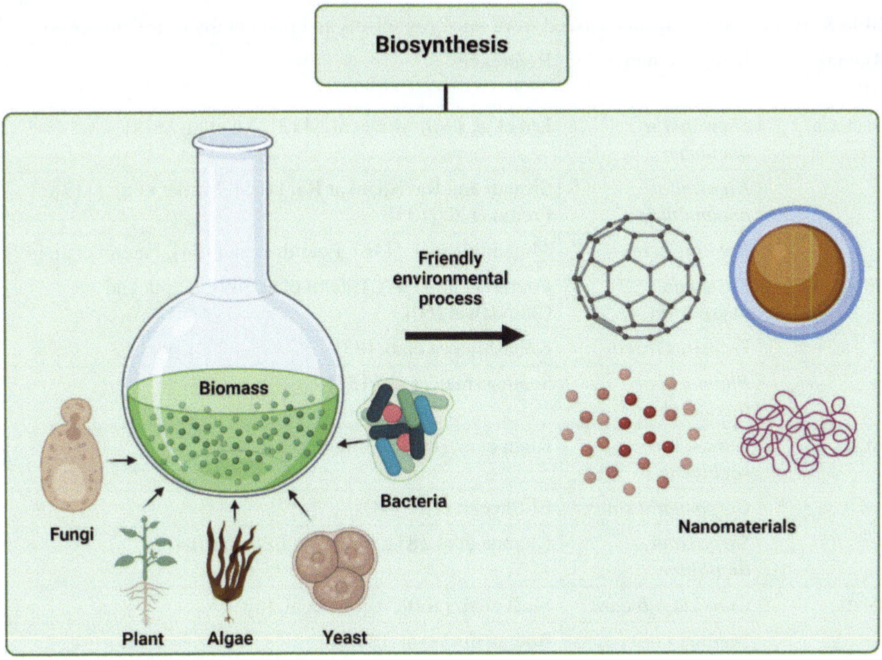

Fig. 4 Nanomaterial synthesis from a biosynthesis process

with the broader objective of creating a more sustainable and eco-friendly approach to material synthesis [21].

The utilization of chemical compounds derived from microorganisms or plant extracts is attributed to the abundant presence of oxidoreductive proteins, alkaloids, flavonoids, saponins, steroids, tannins, and phenols [87]. These components serve as both reducing agents and stabilizers, obviating the need for the use of potentially hazardous reagents like sodium borohydride ($NaBH_4$). This approach enhances safety and aligns with sustainability and eco-friendliness in chemical processes [175].

Numerous factors influence the ultimate properties of NMs obtained through biosynthesis [31]. These factors encompass the type of precursors employed, the reaction's duration, the reaction medium's pH, the temperature, the concentration of extracts used, and more [85]. Various microbes contribute to the process of biosynthesis, including bacteria, fungi, algae, viruses, and yeast [33]. When it comes to plant extracts, the diversity is even more extensive, primarily because the sole criterion for their application is their antioxidant activity. This diversity in microbial and plant resources allows for a versatile and adaptable approach to NM production tailored to specific desired properties [152]. Table 5 showcases some of the most significant microorganisms and plants in recent literature, as identified by their contributions to various applications and research outcomes.

Table 5 Bio-reduction agents derived from microorganisms and plant extracts are employed

Biomass type	Scientific name	References
Bacteria	*Shewanella oneidensi*	Lin et al. [96], Min et al. [112], Xu et al. [178]
	Klebsiella pneumoniae	Srinath and Ravishankar Rai [165], Müller et al. [117], Prema et al. [135]
	Bacillus cereus	Mujaddidi et al. [116], Pourali et al. [134], Fatemi et al. [49]
Fungi	*Fusarium oxysporum*	Pourali et al. [134], Fatemi et al. [49], Gupta and Chundawat [60]
	Trametes trogii	Kobashigawa et al. [83]
	Phomopsis liquidambaris	Seetharaman et al. [158]
Alga	*Tetraselmis kochinensis*	Khan et al. [80], Kalimuthu et al. [74]
	Colpomenia sinusa	El-Sheekh et al. [44]
	Sargassum ilicifolium	Khanna et al. [81], Koopi and Buazar [84]
Yeast	*Candida albicans*	Shah et al. [160], Ağçeli et al. [6]
	Yarrowia lipolytica	Bolbanabad et al. [25]
	Colletotrichum sp.	Suryavanshi et al. [168]
Plant	*Cinnamomum camphora*	Li et al. [95], Ramya et al. [139]
	Eucalyptus sp.	Salgado et al. [153], Sawalha et al. [157], Syukri et al. [169]
	Theobroma cacao	Thatikayala et al. [170], Mellinas et al. [111]

4 Nanomaterials and Their Application for Biostimulation and Plant Fertilization

The rapid increase in global population has been an essential issue for agriculture due to the necessity of food production to ensure food safety, especially in cereal crops. Therefore, technologies and advanced tools are necessary to prompt the radical shift to precision farming [120]. Nanotechnology is a multidisciplinary field of research that has shown promise in many sciences through the incorporation of nanoparticles (NPs), becoming a promising strategy to enhance plant growth and productivity due to its properties such as small size, high surface área, high adsorption, high catalytic activity, and high chemical stability [147, 120].

A strategy to enhance crop production sustainably and healthily is the proper application of fertilizers; since the rising use of conventional fertilizers has resulted in significant environmental risks, nanotechnology represents an emerging alternative to cope with it through the use of nano-fertilizers (NFs) [88], which are fertilizers made at nanoscale (1–100 nm) that increase production due to its features such as slow/controlled release or plant response-based release improving the fertilizer use efficiency and reducing pollution [86, 173].

Some oxide nanomaterials made of metallic polymer have demonstrated positive effects on plant growth [1]. The NPs of SiO_2 and TiO_2 showed a beneficial effect on maize crops by elevating the action of nitrate and increasing plant absorption [121]. Nevertheless, using metal oxide nanomaterials and the synthesis process carried out by some chemical compounds in agriculture has been reported as phytotoxic to the environment [1, 90].

Nowadays, nanotechnology has made some techniques aimed at environmentally sustainable methods for synthesizing NPs using plant extracts, microorganisms, and enzymes [90]. Using these biological agents to synthesize NPs generates a nonhazardous, eco-friendly strategy for humans and ecosystems [23]. Bioactive molecules in plants such as proteins, carbohydrates, vitamins, alkaloids, organic acids, and secondary metabolites have been found to act as bioreductive, capping, and stabilizing agents for precursors of chloride and nitrate in the biosynthesis of NPs [64].

4.1 Nanofertilizers in Cereal Crops

The most significant application of nanotechnology in agriculture is using nanofertilizers (NFs), which can feed plants progressively, improving nutrient use efficiency [20]. Furthermore, NFs offer the best stress tolerance to plants and reduce groundwater pollution,among NFs, biological NFs have been used to enhance the growth of crops and have been proven to be sustainable and non-toxic [41]. It has been reported that using plant extracts and microorganisms to synthesize NFs is an effective and environmentally friendly way [88]. Also, the biological mechanism of NFs in plants has a tremendous impact on crop productivity, increasing crop yields [90, 115].

Cereal crops are considered basic food that satisfies the main food demands of human health; using biological NFs in this crop type has improved crop growth and yield parameters [73], as shown in Table 6.

However, using biological NFs targets precision farming by involving small amounts of fertilizers in submicroscopic capsules and slow nutrient release [41, 132]. This slow release is controlled by different environmental fluctuations such as temperature, moisture, and soil acidity to enhance nutrient use and improve plant growth [8].

Table 6 Biological NFs used globally in cereal crops

Nanofertilizer	Content	Crops	Yield increment (%)	Manufacturer	References
Nano Chitosan Organic Fertilizer	Organic acids, chitosan, salicylic acids, and phenolic compounds	Wheat	14.6	Pannaraj Intertrade, Thailand	Pirzadah et al. [132], Muhammad Aamir [115]
Nano Max NPK Fertilizer	Organic acids chelated with major nutrients, amino acids, organic carbon, organic micronutrients, vitamins, and probiotics	Maize	9.71	JU Agri Sciences Pvt. Ltd., Janakpuri, New Delhi, India	Ahmed et al. [8], Lakzian et al. [90], Pirzadah et al. [132]
Nano-Ag Answer®	Microorganisms, sea kelp, and mineral electrolyte	Cereals	4–8.5	Urth Agriculture, CA, United States	Muhammad Aamir [115], Pirzadah et al. [132]
Nano Ultra-Fertilizer + Urea	Organic matter, Macronutrients (NPK), and urea	Wheat	6.5–7.3	SMTET Eco-technologies Co., Ltd., Taiwan	Lakzian et al. [90], Muhammad Aamir [115]
Nano-encapsulated phosphorous	N, 0.5%; P2O5, 0.7%; K2O, 3.9%; Ca, 2.0%; Mg, 0.2%; S, 0.8%; Fe, 2.0%; Mn, 0.004%; Cu, 0.007%; Zn, 0.004%	Maize Wheat	10.9 28.8	The Best International Network Co., Ltd., Thailand	

4.2 Biosythesized Nanomaterials as Fertilizer in Fruits and Vegetables

Using NMs and NPs properties, different nano fertilizers have been developed whose size allows get into the wall cell; plants roots uptake efficiently when particle size is found in 20–50 nm [53, 184]. Scientists found nano fertilizers an opportunity to increase productivity, being more effective than those formed with carbon-based and minerals [65] in horticulture have been studied yield increment in different vegetal models and nano fertilizers, shown in Table 7.

Regarding horticultural applications in fruits and vegetable production using nano-fertilizers, scientists have found answers to use NMs and NPs in crops, for example, in pepper (*Capsicum annum* L.) using ZnO, FeO, MgO increased solubility in micronutrients as Cu, B, Mn, also was obtained plant growth and higher

Table 7 Nanofertilizer impact on yield increment in vegetables (modified from Muhammad Aamir [115])

Nanofertilizer	Crop	Yield increment (%)
Nano-encapsulated phosphorous	Vegetables	12.0–19.7
Nano chitosan	Tomato	20.0
Nano chitosan	Cucumber	9.3
Nano chitosan	Capsicum	11.5
Nanopowder of cotton seed and ammonium fertilizer	Sweet potato	16.0
Nanoparticles of ZnO	Cucumber	6.3
Nanoparticles of ZnO	Cabbage	9.1
Nanoparticles of ZnO	Cauliflower	8.3
Rare earth oxide nanoparticles	Vegetables	7.0–45.0

crop yield [102]. CeO_2 and ZnO Nps were used in cucumber (*Cucumis sativus*), where these NPs decreased Cu accumulation without influencing the phenols and flavonoids quantity and micronutrients [185]. In date palm (*Phoenix dacttylifera* L.) using common fertilizer combinate with nano fertilizers applied foliar and compound by manganese (Mn) 0.7%, Mg 6%, calcium (Ca) 6%, iron (Fe) 4.5%, zinc (Zn) 8%, K 3%, P 3%, N 5%, molybdenum (Mo) 0.1%, boron (B) 0.65%, and copper (Cu) 0.65%,got increase in fruit ripening rate, dry mass, total soluble solids, activity of the enzymes peroxidase and superoxide dismutase, and abscisic acid content [161]. Nanotechnology promises a change in crops, using nanofertilizers to promote germination and increase biomass, impacting enzymatic activity and uptake of nutrients [50]. However, there are not all impacts on horticulture crops. Table 8 shows another effect of nanofertilizers based on NMs and NPs [129].

Table 8 Impact in horticulture crops using nano fertilizers (modified from Muhammad Aamir [115])

Nano fertilizers	Crop	Impact in crop
Colloidal silica + NPK fertilizers	Tomato	Increased resistance to pathogens
Polyethylene + indium oxide	Vegetables	Increased sunlight absorption
Polypropylene + indium–tin oxide	Vegetables	Increased sunlight utilization
Gold nanoparticles + sulfur	Grapes	Increased antioxidants and other human health benefits
Kaolin + SiO_2	Vegetables	Improved water retention
Stevia extract + nanoparticles of Se + organo-Ca + rare-earth elements + chitosan	Vegetables	Enhanced root networking and root diameter

5 Biosynthesized Nanomaterials as Biostimulant

As mentioned in previous paragraphs, nanotechnology has applications in various fields, such as medicine, cosmetics, electronics, environmental remediation, and agriculture [119]. The diversity of applications of nanoparticles has encouraged the development of various techniques to obtain them, among which the so-called green routes or green synthesis stand out, which use natural sources such as algae, fungi, microorganisms, and plant extracts, developing more processes. Economical and environmentally friendly compared to traditional techniques, such as physical and chemical, are expensive and harmful to health and the environment [126].

The green synthesis of NPs has advantages over its counterparts, such as variation in size and shape, reduction in synthesis times, and low production cost and toxicity [17, 126], additionally provides biological characteristics such as antibacterial, antiviral, antioxidant, biocompatibility, among others [127]. Plant extracts are the most used resources for the green synthesis of NPs due to the presence of secondary metabolites (phenols, amides, carboxylic acids, terpenoids, ketones, flavonoids, etc.), which play the role of reducing and stabilizing agents during the synthesis process [105], in addition to providing antioxidant power due to its high content of the −OH functional group, which helps in the neutralization of free radicals by donating its hydrogen atoms or electrons, promoting a reaction type redox [27].

Microorganisms (bacteria, fungi, yeasts) are another resource used for the green synthesis of NPs. Bacteria have unique strategies to ensure their survival and facilitate the absorption of nutrients necessary for their multiplication and propagation. Bacteria develop defense mechanisms, such as sequestration (intracellular), variation in metal ion concentrations, and efflux pump [58, 124] that are part of extracellular biosynthesis and intracellular of living beings. The functional groups of biomolecules in living beings promote biomineralization, which is necessary to convert metal ions into NPs [59]. On the other hand, fungi can synthesize NPs extracellularly or intracellularly [75]. During the synthesis of NPs by fungi, the interaction between metal ions and secreted biomolecules (sugars, carbohydrates, enzymes, and proteins) undergoes oxidation or reduction reactions, which induces the production of NPs [43]. The microorganisms can use various synthesis routes, which allows obtaining various types of NPs with different properties but also makes it difficult to understand the synthesis.

The green synthesis of nanoparticles is a recent practice; however, it has received significant attention in agriculture given its low toxicity, biocompatibility, ecological approach, and diversity of products with biological activity that provides extra value to NPs [18]. Additionally, NPs can be used as growth promoters, stimulators of the germination process, and facilitators of nutrition, in addition to generating tolerance to abiotic stress, and are considered biostimulators [123].

According to the European Biostimulant Industry Council (EBIC), a biostimulant is a bioactive substance that falls under the classification of fertilizer products but is used in minimal quantities and has the sole function of improving nutrition, the profile of quality, tolerance to abiotic stress of crops and the availability of nutrients in the soil or rhizosphere—considering organic materials from by-products and waste from industrial activities as a potential source of biostimulants [94].

As mentioned, nanoparticles from green routes or pathways can use natural resources, including microorganisms and plant extracts. This type of green synthesis allows the use of organic waste derived from industrial activities complying with the term biostimulants, according to the EBIC. Although various studies demonstrate the effect of biostimulants on plants, there are still questionable areas regarding NPs with biostimulant properties regarding their use and application, such as the interaction with silver and microorganisms in the environment, as well as doses at employ and toxicological aspects [59]. Tables 9 and 10 describe some works where NPs obtained by green synthesis with biostimulant effects on cereals, fruits, and vegetables are used.

The response obtained from the interaction of NPs with different crops is diverse and will depend on the type, size, dosage, concentration, and crop type. However, the studies indicated that NPs could promote seed germination and seedling development, stimulate cell division and some biochemical pathways [122], and promote the production of reactive oxygen species (ROS). Also, the NPs might mobilize protein reserves in the plant to increase phytohormone levels and enzymatic activity [37]. Additionally, it increases the flow of electrons and light absorption, influencing the photosynthetic machinery [176]. These changes can generate morphological, physical, and biochemical changes in the study plant.

6 Nanomaterials as Elicitors in Horticulture, Producing Bioactive Compounds

Usually, plants produce hormones that regulate essential processes to survive in the environment, such as the formation and growth of roots, buds, shoots, flowers, and fruits [164]. Nevertheless, any compounds such as NMs and NPs can be biostimulants/elicitors activating the immunological system, increasing the number of roots helping assimilate nutrients [82, 101]. This process occurs because the presence of NMs and NPs induces production of free radicals such as ^{-}OH, $^{-}O_{2,}$ and $H_2O_{2,}$ which affects production important biomolecules as DNA, proteins, and cell membrane [71], generating toxicity, but the most vital plants that can resist in these conditions [7], inducing the expression of genes involved in the biosynthesis of secondary metabolites [10]. This indicates a novel opportunity to study nanomaterials like elicitors in agriculture and principally in vegetable and fruit production as the form of bioactive compounds (Table 11).

Table 9 Biosynthesis of nanomaterials and its influence on cereal crops

Nanoparticle	Biological source	Crops	Biological effects	References
Ag	*Cappaaris spinosa* L. Extract	Seeds of *Triticum aestivum* L.	Stimulate the germination and growth of wheat grains. In addition, it affects the plant phytohormones balance by stimulating indole-3-butyric acid (IBA), 1-naphthalene acetic acid (NAA), 6-benzylaminopurine (BAP) contents and reducing abscicic acid (ABA) content. The AgNP priming might have a role in improving plant tolerance against environmental stresses such as salinity	Abou-Zeid and Ismail [4]
Ag	*Pseudomonas sp., Achromobacter sp., Trichoderma sp., Cephalosporium sp.*	Seeds *Cicer arietinum* L.	High germinability up to 98% and no negative impact was observed on the soil community	Kaur et al. [76]
Cu	*Mentha longifolia* L. Extract	Seeds of four wheat (*Tritium aestivum* L.) varieties like 18-Elite Line 1, 18-Elite Line 3, 18-Elite Line 5, and 18-Elite Line 6	The concentration influences germination and growth depending on the variety and enhances the contents of chlorophyll, carotenoids, phenols, and soluble sugar	Kausar et al. [77]
Ag	*Bacillus subtilis* spizizenni	Seeds *Pennisetum glaucum* L.	48% increase in germination and fresh weight	Sable et al. [148]
Cu	*Allium cepa* Extract	Wheat (*Tritium aestivum* L.)	88% Increased leaf area to 0.4 ppm 139 and 150% improvement in fresh and root dry weights, respectively	Hafeez et al. [61]

(continued)

Table 9 (continued)

Nanoparticle	Biological source	Crops	Biological effects	References
ZnO	*Catharathus roseus* L. leaves	*Eleusine coracana* L.	Increased seed germination and seedling growth	Mishra et al. [113]
Ag-chitosan	Crop residues of *Tritticum vulgare* L	*Triticum vulgare* L.	Increased seed germination and chlorophyll content	Mondéjar-Lόpez et al. [114]
Mn	*Eucalyptus robusta* L. and *Corymbia citriodora* L. Leaves	*Zea mays* L.	Increased seed germination and seedling growth	Gonçalves et al. [56]
ZnO	*Lemna minor* L. Plant tissues	*Zea mays* L.	Increased plant growth, photosynthetic pigments, carotenoid and anthocyanin contents, and antioxidant activity	Del Buono et al. [35]
Cu	*Avicennia marina* L. Leaves	*Triticum aestivum* L.	Increased seed germination and seedling growth, antioxidant activity, and chlorophyll content	Essa et al. [48]
Cu	*Nerium oleander* L. Leaves	*Vigna mungo* L.	Improvement of seed germination, quality, and seedling vigor	Raja et al. [138]
TiO_2	*Melia azedarach* L. Leaves	*Pennisetum glaucum* L.	It increased seed germination	Rafique et al. [137]
Ag	*Capparis spinosa* L. Plant tissues	*Triticum aestivum* L.	Increased seed germination, seedling growth, chlorophyll content, and photosynthetic efficiency	Abou-Zeid and Ismail [4]

7 Conclusion and Perspectives

It seems that using NMs and NPs as nanofertilizers is an adequate idea to sustainably increase crop yield. However, many times, it is not considered the bioaccumulation in humans due to some crops are fertilized with NMs and NPs associated with being phytotoxic and could be transferred to humans [70, 167]. Another limitation is the standardization in the process because NF concentrations vary depending on the plant and climatic conditions [115]. The idea and efficiency of using NFs are out of the question because several studies have proven their effectiveness. Nevertheless, some control in the dosage of the NFs on the soil, groundwater, and air monitoring should

Table 10 Biosynthesis of nanomaterials and its influence on fruits and vegetables

Nanoparticle	Biological source	Crops	Biological effect	References
Ag	Fungal extracts of *Trichoderma harzianum* and *Aspergillus fumigaatus*	*Solanum lycopersicum* L.	Improved growth traits (plant height, fresh and dry biomass, yield) and potentially suppressed bacterial canker disease	Noshad et al. [125]
CuO, ZnO	Lemon peels extract	Citrus	Antifungal property on *Alternaria citri*	Sardar et al. [156]
CuO	Lead tissue of lavender and green tea	*Solanum lycopersicum* L. and *Lactuca sativa* L.	Recorded two folds have more significant germination indices than that of chemically synthesized NPs	Khaldari et al. [78]
Ag	*Berberis lycium*	*Pisum sativum* L.	Increase seeds per pod. No pods per plant and green pod yield and increase protein and carbohydrate of seeds	Mehmood and Murtaza [108]
NiO	*Berberiks balochistanica* L. stem	Radish (*Raphanus sativus* L.)	Increase seedling length	Uddin et al. [172]
Ag	*Moringa oleifera* L. leaves	Kinnow mandarine (*Citrus retifulata* L.)	Increase fruit weight	Hussain et al. [68]
ZnO	*Citrus aurantium* L. Peels	*Lycopersicon esculentum* L.	Increased seed germination and fruit production	Punitha et al. [136]
MgS	Olive mill wastewater (*Olea europea* L.)	*Pisum sativum* L.	Increased seed germination, seedling growth, physiological traits, and antioxidant activity	Hamimed et al. [62]
ZnO	*Lemna minor* L. Plant tissues	*Olea europaea* cv. Moraiolo	Beneficial effects in vitro on physiological and biochemical traits, and antioxidant activity	Regni et al. [144]
Ag	Crop residues (shell) *Pistacia vera*	*Solanum melongena* L.	Increased plant growth, chlorophyll, and carotenoid contents	Noshad et al. [125]
TiO_2	*Cuminum cyminum* L. Seeds	*Vigna radiata* L.	Increase seed germination and seedling growth (root and shoot systems)	Mathew et al. [107]
Cu	*Citrus medica* L. Juice	*Cajanus*	Improved seedling growth (increased biomass and root length)	Shende et al. [162]

Table 11 Nanomaterials associated as elicitor effect in horticulture crops (modified from Lala [91], Rivero-Montejo et al. [145])

Nanomaterial	Treatment	Crop	Impact in crop
Cu absorbed on chitosan hydrogel	One time only prior to transplantation	Saladette tomato	An increment of lycopene, titratable acidity, and catalase activity
Cu absorbed on chitosan hydrogel	One time only prior to transplantation	Jalapeño pepper	Rise the content of antioxidants ABTS and DPPH, total phenols and flavonoids, titratable acidity, and reduce weight loss in fruits post-harvest
Cu	Soil pre-treated	Cucumber	Augmentation of valine, leucine, isoleucine, threonine, tyrosine, xylose, fructose, glycine, proline, benzoic acid, glutaric acid, caprylic acid, linolenic acid, and imidazole. Decrease of lysine and methionine
Cerium oxide (CeO_2) and Citric acid-coated cerium oxide (CeO_2 +CA)	Soil pre-treated	Roma tomato	It was increased the lycopene. CeO_2 + CA decreases, reducing sugar and reduce starch. CeO_2 reduces copper, manganese, magnesium, calcium, and iron
Au	Foliar	Cucumber	Increase in phenolics. Enhancement of antioxidant capacity
Se	Foliar	Tomato	Increase in fruits' vitamin C, glutathione, phenol, and flavonoid content
ZnO	Synthetic matrix	Pepper	Accumulation of phenolics, flavonoids, and tannins while enhancing the seedlings' antioxidant capacity and seed germination
ZnO	Foliar	Radish	It enhanced the concentration of anthocyanins, phenols, tannins, and flavonoids, as well as crude protein and carbohydrates in roots
CuO	Sterile filter paper	Eggplant	Enhancement of anthocyanin, flavonoids, and phenolics in a concentration

be done around the farmland. Rigorous normative must be proposed and developed around the production, distribution, application, and disposal of the NFs, taking into consideration the different characters involved, such as government, business people, the scientific community, farmers, and the final food consumer. In that way, it could be said that the NFs' development and application are under a circular economy state of mind. Further studies in this matter must be carried out, and here is the next stage of the NFs development for the scientific groups.

References

1. Aacharya R, Chhipa H (2020) 15—Nanocarbon fertilizers: implications of carbon nanomaterials in sustainable agriculture production. In: Abd-Elsalam KA (ed) Carbon nanomaterials for agri-food and environmental applications. Elsevier
2. Abdollahi A, Roghani-Mamaqani H, Salami-Kalajahi M, Mousavi A, Razavi B, Shahi S (2018) Preparation of organic-inorganic hybrid nanocomposites from chemically modified epoxy and novolac resins and silica-attached carbon nanotubes by sol-gel process: investigation of thermal degradation and stability. Prog Org Coat 117:154–165
3. Abid N, Khan AM, Shujait S, Chaudhary K, Ikram M, Imran M, Haider J, Khan M, Khan Q, Maqbool M (2022) Synthesis of nanomaterials using various top-down and bottom-up approaches, influencing factors, advantages, and disadvantages: a review. Adv Coll Interface Sci 300:102597–102597
4. Abou-Zeid H, Ismail G (2018) The role of priming with biosynthesized silver nanoparticles in the response of triticum aestivum L to salt stress. Egypt J Bot 58:73–85
5. Adachi H, Wasa K, Kanno I, Kotera H (2012) Handbook of sputter deposition technology: fundamentals and applications for functional thin films, nano-materials and MEMS, William Andrew
6. Ağçeli GK, Hammachi H, Kodal SP, Cihangir N, Aksu Z (2020) A novel approach to synthesize TiO_2 nanoparticles: biosynthesis by using streptomyces sp. HC1. J Inorg Organomet Polym Mater 30:3221–3229
7. Aguirre-Becerra H, Feregrino-Pérez AA, Esquivel K, Perez-Garcia CE, Vazquez-Hernandez MC, Mariana-Alvarado A (2022) Nanomaterials as an alternative to increase plant resistance to abiotic stresses. Front Plant Sci 13:1023636
8. Ahmed NR, Sheha AM, Hamd-Alla WA (2019) Effect of intercropping of some legume forage crops with maize under levels of mineral Npk and nano Npk fertilizer. J Plant Prod 10:833–842
9. Al Mashary FS, De Castro S, Da Silva AF, Felix JF, Piton MR, Galeti HVA, De Giovanni Rodrigues A, Gobato YG, Al Saqri N, Henini M, Al Huwayz MM, Albadri AM, Alyamani AY, Albrathen HA, Alhusaini SA, Aljaber KM, Alanazi AZ, Alghamdi FS (2018) Effect of growth techniques on the structural, optical and electrical properties of indium doped TiO_2 thin films. J Alloy Compd 766:194–203
10. Al-Theyab NS, Abuelizz HA, Al-Hamoud GA, Aldossary A, Liang M (2023) Priestia megaterium metabolism: isolation, identification of naringenin analogues and genes elevated associated with nanoparticle intervention. Curr Issues Mol Biol 45:6704–6716
11. Alarifi IM (2022) Composite nanomaterials. Synth Eng Mater Nanotechnol, 253–275. https://doi.org/10.1016/B978-0-12-824001-4.00014-4
12. Ali I, Basheer AA, Mbianda XY, Burakov A, Galunin E, Burakova I, Mkrtchyan E, Tkachev A, Grachev V (2019) Graphene based adsorbents for remediation of noxious pollutants from wastewater. Environ Int 127:160–180
13. Alrbaihat MR (2023) Agricultural nano fertilizers: macronutrient types and applications review. Singapore. Springer Nature Singapore, pp 306–316
14. Alshammari BH, Lashin MMA, Mahmood MA, Al-Mubaddel FS, Ilyas N, Rahman N, Sohail M, Khan A, Abdullaev SS, Khan R (2023) Organic and inorganic nanomaterials: fabrication, properties and applications. RSC Adv 13:13735–13785
15. Armarego WLF (2022) Nanomaterials. Purif Lab Chem, 586–630. https://doi.org/10.1016/B978-0-323-90968-6.50005-9
16. Ashfold MNR, Claeyssens F, Fuge GM, Henley SJ (2004) Pulsed laser ablation and deposition of thin films. Chem Soc Rev 33:23–31
17. Awan A, Baig A, Zubair M, Rahdar A, Nazar MF, Farooqi AS, Shalan AE, Lanceros-Méndez S, Zafar MN (2022) Green synthesis of molybdenum-based nanoparticles and their applications in energy conversion and storage: a review. Int J Hydrogen Energy 47:31014–31057

18. Bahrulolum H, Nooraei S, Javanshir N, Tarrahimofrad H, Mirbagheri VS, Easton AJ, Ahmadian G (2021) Green synthesis of metal nanoparticles using microorganisms and their application in the agrifood sector. J Nanobiotechnol 19:86
19. Baig N, Kammakakam I, Falath W, Kammakakam I (2021) Nanomaterials: a review of synthesis methods, properties, recent progress, and challenges. Mater Adv 2:1821–1871
20. Bairwa P, Kumar N, Devra V, Abd-Elsalam KA (2023) Nano-biofertilizers synthesis and applications in agroecosystems. Agrochemicals, 2
21. Bhati M (2023) Biogenic synthesis of metallic nanoparticles: Principles and applications. Mater Today: Proc 81:882–887
22. Bhattacharya K, Mukherjee SP, Gallud A, Burkert SC, Bistarelli S, Bellucci S, Bottini M, Star A, Fadeel B (2016) Biological interactions of carbon-based nanomaterials: from coronation to degradation. Nanomed: Nanotechnol Biol Med 12:333–351
23. Bipin DL, Arti SS (2020) Phytonanofabrication: methodology and factors affecting biosynthesis of nanoparticles. In: Tatyana S, Vladimir B (eds) Smart nanosystems for biomedicine, optoelectronics and catalysis. IntechOpen, Rijeka
24. Biswas A, Bayer IS, Biris AS, Wang T, Dervishi E, Faupel F (2012) Advances in top–down and bottom–up surface nanofabrication: techniques, applications & future prospects. Adv Coll Interface Sci 170:2–27
25. Bolbanabad EM, Ashengroph M, Darvishi F (2020) Development and evaluation of different strategies for the clean synthesis of silver nanoparticles using Yarrowia lipolytica and their antibacterial activity. Process Biochem 94:319–328
26. Brown IG (2003) Cathodic arc deposition of films 28:243–269. https://doi.org/10.1146/annurev.matsci.28.1.243
27. Burlacu E, Tanase C, Coman NA, Berta L (2019) A review of bark-extract-mediated green synthesis of metallic nanoparticles and their applications. Molecules, 24.
28. Buzea C, Pacheco I (2017) Nanomaterials and their classification. Springer
29. Chen Z, Prud'homme N, Wang B, Ji V (2011) Residual stress gradient analysis with GIXRD on ZrO_2 thin films deposited by MOCVD. Surf Coat Technol 206:405–410
30. Cui D, Zhang P, Ma Y, He X, Li Y, Zhang J, Zhao Y, Zhang Z (2014) Effect of cerium oxide nanoparticles on asparagus lettuce cultured in an agar medium. Environ Sci Nano 1:459–465
31. Cuong HN, Pansambal S, Ghotekar S, Oza R, Thanh Hai NT, Viet NM, Nguyen VH (2022) New frontiers in the plant extract mediated biosynthesis of copper oxide (CuO) nanoparticles and their potential applications: a review. Environ Res, 203
32. De Greñu BD, De Los Reyes R, Costero AM, Amorós P, Ros-Lis JV (2020) Recent progress of microwave-assisted synthesis of silica materials. Nanomaterials, 10
33. De Jesus RA, De Assis GC, De Oliveira RJ, Costa JAS, Da Silva CMP, Bilal M, Iqbal HMN, Ferreira LFR, Figueiredo RT (2021) Environmental remediation potentialities of metal and metal oxide nanoparticles: mechanistic biosynthesis, influencing factors, and application standpoint. Environ Technol Innov 24:101851–101851
34. De Santiago Colín DM, Martínez-Chávez LA, Cuán Á, Elizalde-Peña EA, Rivera JA, Guzmán C, Escobar-Alarcón L, Esquivel K (2018) Sonochemical coupled synthesis of Cr-TiO_2 supported on Fe_3O_4 structures and chemical simulation of the degradation mechanism of Malachite Green dye. J Photochem Photobiol A: Chem, 364
35. Del Buono D, Di Michele A, Costantino F, Trevisan M, Lucini L (2021) Biogenic ZnO nanoparticles synthesized using a novel plant extract: application to enhance physiological and biochemical traits in maize. Nanomaterials 11:1270
36. Deng Y, Chen W, Li B, Wang C, Kuang T, Li Y (2020) Physical vapor deposition technology for coated cutting tools: a review. Ceram Int 46:18373–18390
37. Do Espirito Santo Pereira A, Caixeta Oliveira H, Fernandes Fraceto L, Santaella C (2021) Nanotechnology potential in seed priming for sustainable agriculture. Nanomaterials 11:267
38. Dolez PI (2015) Nanomaterials definitions, classifications, and applications. Nanoeng: Glob Approaches Health Saf Issues, 3–40. https://doi.org/10.1016/B978-0-444-62747-6.00001-4
39. Droes SR, Kodas TT, Hampden-Smith MJ (1997) Plasma-Enhanced Chemical Vapor Deposition (PECVD). Carbide Nitride Boride Mater Synth Process, 579–603. https://doi.org/10.1007/978-94-009-0071-4_23

40. Ealias AM, Saravanakumar MP (2017) A review on the classification, characterisation, synthesis of nanoparticles and their application. In: IOP conference series: materials science and engineering, vol 263, p 032019
41. El-Bialy SM, El-Mahrouk ME, Elesawy T, Omara AE, Elbehiry F, El-Ramady H, Áron B, Prokisch J, Brevik EC, Solberg SØ (2023) Biological nanofertilizers to enhance growth potential of strawberry seedlings by boosting photosynthetic pigments, plant enzymatic antioxidants, and nutritional status. Plants, 12
42. El-Khawaga AM, Zidan A, El-Mageed AIAA (2023) Preparation methods of different nanomaterials for various potential applications: a review. J Mol Struct 1281:135148–135148
43. El-Seedi HR, El-Shabasy RM, Khalifa SAM, Saeed A, Shah A, Shah R, Iftikhar FJ, Abdel-Daim MM, Omri A, Hajrahand NH, Sabir JSM, Zou X, Halabi MF, Sarhan W, Guo W (2019) Metal nanoparticles fabricated by green chemistry using natural extracts: biosynthesis, mechanisms, and applications. RSC Adv 9:24539–24559
44. El-Sheekh MM, El-Kassas HY, Shams El-Din NG, Eissa DI, El-Sherbiny BA (2021) Green synthesis, characterization applications of iron oxide nanoparticles for antialgal and wastewater bioremediation using three brown algae. Int J Phytorem 23:1538–1552
45. Escobar-Alarcón L, Solis-Casados DA, Romero S, Haro-Poniatowski E (2020) Thin films prepared by a hybrid deposition configuration combining two laser ablation plasmas with one sputtering plasma. Appl Phys A Mater Sci Process 126:1–8
46. Esquivel K, Martínez-Chávez LA, Rosales-Pérez A, Hernández-Rangel R (2023) Chapter 1 nanomaterials: classification, synthesis methods, and physicochemical characterization. Eng Nanoparticles Agric, 1–58. https://doi.org/10.1515/9781501523229-001
47. Esquivel K, Nava R, Zamudio-Méndez A, González MV, Jaime-Acuña OE, Escobar-Alarcón L, Peralta-Hernández JM, Pawelec B, Fierro JLG (2013) Microwave-assisted synthesis of (S)Fe/TiO_2 systems: effects of synthesis conditions and dopant concentration on photoactivity. Appl Catal B: Environ 140:213–224
48. Essa HL, Abdelfattah MS, Marzouk AS, Shedeed Z, Guirguis HA, El-Sayed MMH (2021) Biogenic copper nanoparticles from Avicennia marina leaves: Impact on seed germination, detoxification enzymes, chlorophyll content and uptake by wheat seedlings. PLoS ONE 16:e0249764
49. Fatemi M, Mollania N, Momeni-Moghaddam M, Sadeghifar F (2018) Extracellular biosynthesis of magnetic iron oxide nanoparticles by Bacillus cereus strain HMH1: characterization and in vitro cytotoxicity analysis on MCF-7 and 3T3 cell lines. J Biotechnol 270:1–11
50. Feregrino-Perez AA, Magaña-López E, Guzmán C, Esquivel K (2018) A general overview of the benefits and possible negative effects of the nanotechnology in horticulture. Sci Hortic 238:126–137
51. Fery A, Weinkamer R (2007) Mechanical properties of micro- and nanocapsules: single-capsule measurements. Polymer 48:7221–7235
52. Findik F (2021) Nanomaterials and their applications. Period Eng Nat Sci 9:62–75
53. Fleischer A, O'neill MA, Ehwald R (1999) The pore size of non-graminaceous plant cell walls is rapidly decreased by borate ester cross-linking of the pectic polysaccharide rhamnogalacturonan II. Plant Physiol 121:829–838
54. Freitas DDFDS, Rangel Cruz Da Silva PS, Amantes BDP, Cestari SP, Albitres GAV, Mariano DDM, Mendes LC, Tavares MIB (2021) Sustainable routes and green synthesis for nanomaterials and nanocomposites production. In: Handbook of greener synthesis of nanomaterials and compounds: volume 2: synthesis at the macroscale and nanoscale, pp 637–650. https://doi.org/10.1016/B978-0-12-822446-5.00027-7
55. Gade A, Ingle P, Nimbalkar U, Rai M, Raut R, Vedpathak M, Jagtap P, Abd-Elsalam KA (2023) Nanofertilizers: the next generation of agrochemicals for long-term impact on sustainability in farming systems. Agrochemicals 2:257–278
56. Gonçalves JPZ, Seraglio J, Macuvele DLP, Padoin N, Soares C, Riella HG (2022) Green synthesis of manganese based nanoparticles mediated by Eucalyptus robusta and Corymbia citriodora for agricultural applications. Colloids Surf A: Physicochem Eng Asp 636:128180

57. Guerrero-Martínez A, Barbosa S, Pastoriza-Santos I, Liz-Marzán LM (2011) Nanostars shine bright for you: colloidal synthesis, properties and applications of branched metallic nanoparticles. Curr Opin Colloid Interface Sci 16:118–127
58. Gupta A, Dubey P, Kumar M, Roy A, Sharma D, Khan MM, Bajpai AB, Shukla RP, Pathak N, Hasanuzzaman M (2022) Consequences of arsenic contamination on plants and mycoremediation-mediated arsenic stress tolerance for sustainable agriculture. Plants 11:3220
59. Gupta A, Rayeen F, Mishra R, Tripathi M, Pathak N (2023) Nanotechnology applications in sustainable agriculture: an emerging eco-friendly approach. Plant Nano Biol 4:100033
60. Gupta K, Chundawat TS (2019) Bio-inspired synthesis of platinum nanoparticles from fungus Fusarium oxysporum: its characteristics, potential antimicrobial, antioxidant and photocatalytic activities. Mater Res Express 6:1050d6
61. Hafeez A, Razzaq A, Mahmood T, Jhanzab HM (2015) Potential of copper nanoparticles to increase growth and yield of wheat. J Nanosci Adv Technol 1:6–11
62. Hamimed S, Chamekh A, Slimi H, Chatti A (2022) How olive mill wastewater could turn into valuable bionanoparticles in improving germination and soil bacteria. Ind Crops Prod 188:115682
63. Hangxun X, Zeiger BW, Suslick KS (2013) Sonochemical synthesis of nanomaterials. Chem Soc Rev 42:2555–2567
64. Hassanisaadi M, Bonjar GH, Rahdar A, Pandey S, Hosseinipour A, Abdolshahi R (2021) Environmentally safe biosynthesis of gold nanoparticles using plant water extracts. Nanomaterials, 11
65. Hayatu NG, Liu Y-R, Han T-F, Daba NA, Zhang L, Shen Z, Li J-W, Muazu H, Lamlom SF, Zhang H-M (2023) Carbon sequestration rate, nitrogen use efficiency and rice yield responses to long-term substitution of chemical fertilizer by organic manure in a rice–rice cropping system. J Integr Agric 22:2848–2864
66. Hola K, Zhang Y, Wang Y, Giannelis EP, Zboril R, Rogach AL (2014) Carbon dots—emerging light emitters for bioimaging, cancer therapy and optoelectronics. Nano Today 9:590–603
67. Horikoshi S, Serpone N (2013) Microwaves in nanoparticle synthesis: fundamentals and applications. Wiley
68. Hussain M, Iqbal Raja N, Mashwani ZU, Iqbal M, Ejaz M, Aslam S (2019) Green synthesis and evaluation of silver nanoparticles for antimicrobial and biochemical profiling in Kinnow (Citrus reticulata L.) to enhance fruit quality and productivity under biotic stress. IET Nanobiotechnol 13:250–256
69. Jeevanandam J, Barhoum A, Chan YS, Dufresne A, Danquah MK (2018) Review on nanoparticles and nanostructured materials: history, sources, toxicity and regulations. Beilstein J Nanotechnol 9:1050–1050
70. Jha S, Pudake RN (2016) Molecular mechanism of plant–nanoparticle interactions. In Kole C, Kumar DS, Khodakovskaya MV (eds) Plant nanotechnology: principles and practices. Springer International Publishing, Cham
71. Juárez-Maldonado A, Tortella G, Rubilar O, Fincheira P, Benavides-Mendoza A (2021) Biostimulation and toxicity: the magnitude of the impact of nanomaterials in microorganisms and plants. J Adv Res 31:113–126
72. Kajihara K (2013) Recent advances in sol–gel synthesis of monolithic silica and silica-based glasses. J Asian Ceram Soc 1:121–133
73. Kalia A, Kaur H (2019) Nano-biofertilizers: harnessing dual benefits of nano-nutrient and bio-fertilizers for enhanced nutrient use efficiency and sustainable productivity. In: Pudake RN, Chauhan N, Kole C (eds) Nanoscience for sustainable agriculture. Springer International Publishing, Cham
74. Kalimuthu K, Cha BS, Kim S, Park KS (2020) Eco-friendly synthesis and biomedical applications of gold nanoparticles: a review. Microchem J 152:104296–104296
75. Kapoor RT, Salvadori MR, Rafatullah M, Siddiqui MR, Khan MA, Alshareef SA (2021) Exploration of microbial factories for synthesis of nanoparticles—a sustainable approach for bioremediation of environmental contaminants. Front Microbiol 12:658294

76. Kaur P, Thakur R, Duhan JS, Chaudhury A (2018) Management of wilt disease of chickpea in vivo by silver nanoparticles biosynthesized by rhizospheric microflora of chickpea (Cicer arietinum). J Chem Technol Biotechnol 93:3233–3243
77. Kausar H, Mehmood A, Khan RT, Ahmad KS, Hussain S, Nawaz F, Iqbal MS, Nasir M, Ullah TS (2022) Green synthesis and characterization of copper nanoparticles for investigating their effect on germination and growth of wheat. PLoS One, 17
78. Khaldari I, Naghavi MR, Motamedi E (2021) Synthesis of green and pure copper oxide nanoparticles using two plant resources via solid-state route and their phytotoxicity assessment. RSC Adv 11:3346–3353
79. Khalid K, Tan X, Mohd Zaid HF, Tao Y, Lye Chew C, Chu DT, Lam MK, Ho YC, Lim JW, Chin Wei L (2020) Advanced in developmental organic and inorganic nanomaterial: a review. Bioengineered 11:328–328
80. Khan AU, Khan M, Malik N, Cho MH, Khan MM (2019) Recent progress of algae and blue–green algae-assisted synthesis of gold nanoparticles for various applications. Bioprocess Biosyst Eng 42:1–15
81. Khanna P, Kaur A, Goyal D (2019) Algae-based metallic nanoparticles: synthesis, characterization and applications. J Microbiol Methods 163:105656
82. Khodakovskaya MV, Kim BS, Kim JN, Alimohammadi M, Dervishi E, Mustafa T, Cernigla CE (2013) Carbon nanotubes as plant growth regulators: effects on tomato growth, reproductive system, and soil microbial community. Small 9:115–123
83. Kobashigawa JM, Robles CA, Martínez Ricci ML, Carmarán CC (2019) Influence of strong bases on the synthesis of silver nanoparticles (AgNPs) using the ligninolytic fungi Trametes trogii. Saudi J Biol Sci 26:1331–1337
84. Koopi H, Buazar F (2018) A novel one-pot biosynthesis of pure alpha aluminum oxide nanoparticles using the macroalgae Sargassum ilicifolium: a green marine approach. Ceram Int 44:8940–8945
85. Kulkarni N, Muddapur U (2014) Biosynthesis of metal nanoparticles: a review. J Nanotechnol, 2014
86. Kumar A, Singh K, Verma P, Singh O, Panwar A, Singh T, Kumar Y, Raliya R (2022) Effect of nitrogen and zinc nanofertilizer with the organic farming practices on cereal and oil seed crops. Sci Rep 12:6938
87. Kumar JA, Krithiga T, Manigandan S, Sathish S, Renita AA, Prakash P, Prasad BSN, Kumar TRP, Rajasimman M, Hosseini-Bandegharaei A, Prabu D, Crispin S (2021) A focus to green synthesis of metal/metal based oxide nanoparticles: various mechanisms and applications towards ecological approach. J Clean Prod 324:129198–129198
88. Kumar N, Samota SR, Venkatesh K, Tripathi SC (2023) Global trends in use of nano-fertilizers for crop production: advantages and constraints—a review. Soil Tillage Res 228:105645
89. Kumari A, Sharma B, Singh BN, Hidangmayum A, Jatav HS, Chandra K, Singhal RK, Sathyanarayana E, Patra A, Mohapatra KK (2022) Chapter 10—physiological mechanisms and adaptation strategies of plants under nutrient deficiency and toxicity conditions. In: Aftab T, Roychoudhury A (eds) Plant perspectives to global climate changes. Academic Press
90. Lakzian A, Bayat M, Gadzhikurbanov A, Zargar M (2019) The role of nanotechnology for improving crop production. RUDN: AA 14:297–305
91. Lala S (2021) Nanoparticles as elicitors and harvesters of economically important secondary metabolites in higher plants: a review. IET Nanobiotechnol 15:28–57
92. Lee W, Jeong MC, Myoung JM (2004) Catalyst-free growth of ZnO nanowires by metal-organic chemical vapour deposition (MOCVD) and thermal evaporation. Acta Mater 52:3949–3957
93. Leskelä M, Ritala M (2002) Atomic layer deposition (ALD): from precursors to thin film structures. Thin Solid Films 409:138–146
94. Li J, Trinh HK, Mirmajlessi SM, Haesaert G, Xhaferi R, Delaere I, Höfte M, Raymaekers K, Cammue BPA, Jonckheere W, Van Leeuwen T, De Smet L, Kyndt T, Bernaert N, Van Poucke C, Van Droogenbroeck B, Pequeur E, De Schamphelaere K, Callebaut G, De Cleyn T, Xu L, Ramirez A, Van Gerrewey T, Geelen D (2023) Biopesticide and plant growth-promoting activity in maize distillers' dried grains with solubles. Ind Crops Prod 193:116175

95. Li W, Qu F, Chen Y, Sun Y, Zhang J, Xie G, You Q, Xu H (2021) Antimicrobial activity of sliver nanoparticles synthesized by the leaf extract of Cinnamomum camphora. Biochem Eng J 172:108050–108050
96. Lin T, Ding W, Sun L, Wang L, Liu CG, Song H (2018) Engineered Shewanella oneidensis-reduced graphene oxide biohybrid with enhanced biosynthesis and transport of flavins enabled a highest bioelectricity output in microbial fuel cells. Nano Energy 50:639–648
97. Liu J, Zhao C, Chen WR, Zhou B (2022) Recent progress in two-dimensional nanomaterials for cancer theranostics. Coord Chem Rev 469:214654–214654
98. Liu Y, Li X, Fan L, Li S, Maleki Kheimeh Sari H, Qin J (2019) A review of carbon-based materials for safe lithium metal anodes. Front Chem 7:721–721
99. Liu Y, Liu H, Guo S, Zhao Y, Qi J, Zhang R, Ren J, Cheng H, Zong M, Wu X, Li B (2024) A review of carbon nanomaterials/bacterial cellulose composites for nanomedicine applications. Carbohyd Polym 323:121445–121445
100. Lukaszkowicz K (2011) Review of nanocomposite thin films and coatings deposited by PVD and CVD technology. Nanomaterials. https://doi.org/10.5772/25799
101. Magnabosco P, Masi A, Shukla R, Bansal V, Carletti P (2023) Advancing the impact of plant biostimulants to sustainable agriculture through nanotechnologies. Chem Biol Technol Agric 10:117
102. Mahajan S, Singh S, Dwivedi H (2023) Efficacy of nanomaterials for sustainable crop productivity of Capsicum (Capsicum annuum L.) var. Rani under naturally ventilated polyhouse. J Appl Nat Sci 15:1282–1291
103. Maity D, Gupta U, Saha S (2022) Biosynthesized metal oxide nanoparticles for sustainable agriculture: next-generation nanotechnology for crop production, protection and management. Nanoscale 14:13950–13989
104. Majid A, Bibi M (2018) Wet chemical synthesis methods. Top Min, Metall Mater Eng, 43–101. https://doi.org/10.1007/978-3-319-68753-7_3/COVER
105. Marslin G, Siram K, Maqbool Q, Selvakesavan RK, Kruszka D, Kachlicki P, Franklin G (2018) Secondary metabolites in the green synthesis of metallic nanoparticles. Materials (Basel), 11
106. Martinu L, Zabeida O, Klemberg-Sapieha JE (2010) Plasma-enhanced chemical vapor deposition of functional coatings. In: Handbook of deposition technologies for films and coatings: science, applications and technology, 392–465. https://doi.org/10.1016/B978-0-8155-2031-3.00009-0
107. Mathew SS, Sunny NE, Shanmugam V (2021) Green synthesis of anatase titanium dioxide nanoparticles using Cuminum cyminum seed extract; effect on Mung bean (Vigna radiata) seed germination. Inorg Chem Commun 126:108485
108. Mehmood A, Murtaza G (2017) Application of SNPs to improve yield of Pisum sativum L. (pea). IET Nanobiotechnol 11:390–394
109. Mejias JH, Salazar F, Pérez Amaro L, Hube S, Rodriguez M, Alfaro M (2021) Nanofertilizers: a cutting-edge approach to increase nitrogen use efficiency in grasslands. Front Environ Sci, 9
110. Mekuye B, Abera B (2023) Nanomaterials: an overview of synthesis, classification, characterization, and applications. Nano Select 4:486–501
111. Mellinas C, Jiménez A, Del Carmen Garrigós M (2019) Microwave-assisted green synthesis and antioxidant activity of selenium nanoparticles using Theobroma cacao L. bean shell extract. Molecules, 24
112. Min D, Cheng L, Zhang F, Huang XN, Li DB, Liu DF, Lau TC, Mu Y, Yu HQ (2017) Enhancing extracellular electron transfer of shewanella oneidensis MR-1 through coupling improved Flavin synthesis and metal-reducing conduit for pollutant degradation. Environ Sci Technol 51:5082–5089
113. Mishra D, Chitara MK, Negi S, Pal Singh J, Kumar R, Chaturvedi P (2023) Biosynthesis of zinc oxide nanoparticles via leaf extracts of *Catharanthus roseus* (L.) G. Don and their application in improving seed germination potential and seedling vigor of *Eleusine coracana* (L.) Gaertn. Adv Agric 2023:7412714

114. Mondéjar-López M, López-Jimenez AJ, Ahrazem O, Gómez-Gómez L, Niza E (2023) Chitosan coated—biogenic silver nanoparticles from wheat residues as green antifungal and nanoprimig in wheat seeds. Int J Biol Macromol 225:964–973
115. Muhammad Aamir I (2019) Nano-fertilizers for sustainable crop production under changing climate: a global perspective. In: Mirza H, Marcelo Carvalho Minhoto Teixeira F, Masayuki F, Thiago Assis Rodrigues N (eds) Sustainable crop production. IntechOpen, Rijeka
116. Mujaddidi N, Nisa S, Al Ayoubi S, Bibi Y, Khan S, Sabir M, Zia M, Ahmad S, Qayyum A (2021) Pharmacological properties of biogenically synthesized silver nanoparticles using endophyte Bacillus cereus extract of Berberis lyceum against oxidative stress and pathogenic multidrug-resistant bacteria. Saudi J Biol Sci 28:6432–6440
117. Müller A, Behsnilian D, Walz E, Gräf V, Hogekamp L, Greiner R (2016) Effect of culture medium on the extracellular synthesis of silver nanoparticles using Klebsiella pneumoniae, Escherichia coli and Pseudomonas jessinii. Biocatal Agric Biotechnol 6:107–115
118. Mustafa MK, Iqbal Y, Majeed U, Sahdan MZ (2017) Effect of precursor's concentration on structure and morphology of ZnO nanorods synthesized through hydrothermal method on gold surface. AIP Conf Proc 1788:30120–30120
119. Nasrollahzadeh M, Sajadi SM, Sajjadi M, Issaabadi Z (2019) Chapter 4—applications of nanotechnology in daily life. In: Nasrollahzadeh M, Sajadi SM, Sajjadi M, Issaabadi Z, Atarod M (eds) Interface science and technology. Elsevier
120. Ndaba B, Roopnarain A, Rama H, Maaza M (2022) Biosynthesized metallic nanoparticles as fertilizers: an emerging precision agriculture strategy. J Integr Agric 21:1225–1242
121. Neme K, Nafady A, Uddin S, Tola YB (2021) Application of nanotechnology in agriculture, post-harvest loss reduction and food processing: food security implication and challenges. Heliyon 7:e08539
122. Nguyen NTT, Nguyen LM, Nguyen TTT, Liew RK, Nguyen DTC, Tran TV (2022) Recent advances on botanical biosynthesis of nanoparticles for catalytic, water treatment and agricultural applications: a review. Sci Total Environ 827:154160
123. Nile SH, Thiruvengadam M, Wang Y, Samynathan R, Shariati MA, Rebezov M, Nile A, Sun M, Venkidasamy B, Xiao J, Kai G (2022) Nano-priming as emerging seed priming technology for sustainable agriculture—recent developments and future perspectives. J Nanobiotechnol 20:254
124. Niño-Martínez N, Salas Orozco MF, Martínez-Castañón GA, Torres Méndez F, Ruiz F (2019) Molecular mechanisms of bacterial resistance to metal and metal oxide nanoparticles. Int J Mol Sci, 20
125. Noshad A, Iqbal M, Folkers L, Hetherington C, Khan A, Numan M, Ullah S (2019) Antibacterial effect of silver nanoparticles (AgNPs) synthesized from trichoderma harzianum against clavibacter michiganensis. J Nano Res 58:10–19
126. Palit S, Hussain CM (2018) Recent advances in green nanotechnology and the vision for the future. Green Met Nanoparticles
127. Pandit C, Roy A, Ghotekar S, Khusro A, Islam MN, Emran TB, Lam SE, Khandaker MU, Bradley DA (2022) Biological agents for synthesis of nanoparticles and their applications. J King Saud Univ-Sci 34:101869
128. Panova G, Nikolaev A, Kovalenko A, Sinelnikov A, Kopitsa G, Udalova O, Artem'eva A, Kornyuchin D, Anikina L, Zhuravleva A, Khomyakov Y, Vertebnyi V, Dubovitskaya V, Shilova O, Khamova T, Baranchikov A (2021) Aqueous chemical co-precipitation of iron oxide magnetic nanoparticles for use in agricultural technologies. Lett Appl NanoBioSci 10:2215–2239
129. Paramo LA, Feregrino-Pérez AA, Guevara R, Mendoza S, Esquivel K (2020) Nanoparticles in agroindustry: applications, toxicity, challenges, and trends. Nanomaterials 10:1654
130. Parashar M, Shukla VK, Singh R (2020) Metal oxides nanoparticles via sol–gel method: a review on synthesis, characterization and applications. J Mater Sci: Mater Electron 31:3729–3749
131. Pathak C, Vaidya FU, Pandey SM (2019) Mechanism for development of nanobased drug delivery system. Appl Target Nano Drugs Deliv Syst: Nanosci Nanotechnol Drug Deliv, 35–67. https://doi.org/10.1016/B978-0-12-814029-1.00003-X

132. Pirzadah TB, Malik B, Maqbool T, Rehman RU (2019) Development of nano-bioformulations of nutrients for sustainable agriculture. In: Prasad R, Kumar V, Kumar M, Choudhary D (eds) Nanobiotechnology in bioformulations. Springer International Publishing, Cham
133. Pokropivny V, Lohmus R, Hussainova I, Pokropivny A, Vlassov S (2007) Introduction to nanomaterials and nanotechnology. Tartu University Press Ukraine
134. Pourali P, Badiee SH, Manafi S, Noorani T, Rezaei A, Yahyaei B (2017) Biosynthesis of gold nanoparticles by two bacterial and fungal strains, Bacillus cereus and Fusarium oxysporum, and assessment and comparison of their nanotoxicity in vitro by direct and indirect assays. Electron J Biotechnol 29:86–93
135. Prema P, Iniya PA, Immanuel G (2016) Microbial mediated synthesis, characterization, antibacterial and synergistic effect of gold nanoparticles using Klebsiella pneumoniae (MTCC-4030). RSC Adv 6:4601–4607
136. Punitha VN, Vijayakumar S, Alsalhi MS, Devanesan S, Nilavukkarasi M, Vidhya E, Kumar SP, Kim W (2023) Biofabricated ZnO nanoparticles as vital components for agriculture revolutionization–a green approach. Appl Nanosci 13:5821–5827
137. Rafique M, Jahangir J, Amin BAZ, Bilal Tahir M, Nabi G, Isa Khan M, Khalid NR, Gillani SSA, Sadaf I (2019) Investigation of photocatalytic and seed germination effects of TiO_2 nanoparticles synthesized by Melia azedarach L. leaf extract. J Inorg Organomet Polym Mater 29:2133–2144
138. Raja K, Sowmya R, Sudhagar R, Moorthy PS, Govindaraju K, Subramanian KS (2019) Biogenic ZnO and Cu nanoparticles to improve seed germination quality in blackgram (Vigna mungo). Mater Lett 235:164–167
139. Ramya AV, Joseph N, Balachandran M (2021) Facile synthesis of few-layer graphene oxide from cinnamomum camphora. Nanobiotechnol Rep 16:183–187
140. Rane AV, Kanny K, Abitha VK, Thomas S (2018) Methods for synthesis of nanoparticles and fabrication of nanocomposites. Synth Inorg Nanomater, 121–139. https://doi.org/10.1016/B978-0-08-101975-7.00005-1
141. Rani A, Reddy R, Sharma U, Mukherjee P, Mishra P, Kuila A, Sim LC, Saravanan P (2018) A review on the progress of nanostructure materials for energy harnessing and environmental remediation. J Nanostruct Chem 8:255–291
142. Ravi-Kumar S, Lies B, Lyu H, Qin H (2019) Laser ablation of polymers: a review. Procedia Manuf 34:316–327
143. Reduwan Billah SM (2019) Composites and nanocomposites, 1–67. https://doi.org/10.1007/978-3-319-92067-2_15-1
144. Regni L, Del Buono D, Micheli M, Facchin SL, Tolisano C, Proietti P (2022) Effects of biogenic ZnO nanoparticles on growth, physiological, biochemical traits and antioxidants on olive tree in vitro. Horticulturae 8:161
145. Rivero-Montejo SDJ, Vargas-Hernandez M, Torres-Pacheco I (2021) Nanoparticles as novel elicitors to improve bioactive compounds in plants. Agriculture 11:134
146. Rosales A, Maury-Ramírez A, Gutiérrez RMD, Guzmán C, Esquivel K (2018) $SiO_2@TiO_2$ coating: synthesis, physical characterization and photocatalytic evaluation. Coatings 8:120–120
147. Sabir S, Zahoor MA, Waseem M, Siddique MH, Shafique M, Imran M, Hayat S, Malik IR, Muzammil S (2020) Biosynthesis of ZnO nanoparticles using bacillus subtilis: characterization and nutritive significance for promoting plant growth in Zea mays L. Dose-Response 18:1559325820958911
148. Sable SV, Ranade S, Joshi S (2018) Role of AgNPs in the enhancement of seed germination and its effect on plumule and radicle length of Pennisetum glaucum. IET Nanobiotechnol 12:922–926
149. Saha D, Ajimsha RS, Rajiv K, Mukherjee C, Gupta M, Misra P, Kukreja LM (2014) Spectroscopic ellipsometry characterization of amorphous and crystalline TiO_2 thin films grown by atomic layer deposition at different temperatures. Appl Surf Sci 315:116–123
150. Saira F, Yaqub A, Razzaq H, Sohail MG, Saleemi S, Mumtaz M, Rafiq MA, Qaisar S (2022) Hollow nanocages for electrochemical glucose sensing: a comprehensive review. J Mol Struct 1268:133646–133646

151. Saleh TA (2020) Nanomaterials: classification, properties, and environmental toxicities. Environ Technol Innov 20:101067–101067
152. Salem MZM, El-Hefny M, Ali HM, Abdel-Megeed A, El-Settawy AAA, Böhm M, Mansour MMA, Salem AZM (2021) Plants-derived bioactives: novel utilization as antimicrobial, antioxidant and phytoreducing agents for the biosynthesis of metallic nanoparticles. Microb Pathog, 158
153. Salgado P, Mártire DO, Vidal G (2019) Eucalyptus extracts-mediated synthesis of metallic and metal oxide nanoparticles: current status and perspectives. Mater Res Express 6:82006–82006
154. Saraiva R, Ferreira Q, Rodrigues GC, Oliveira M (2022) Phosphorous nanofertilizers for precise application in rice cultivation as an adaptation to climate change. Climate 10:183
155. Saravanan A, Kumar PS, Karishma S, Vo DVN, Jeevanantham S, Yaashikaa PR, George CS (2021) A review on biosynthesis of metal nanoparticles and its environmental applications. Chemosphere 264:128580–128580
156. Sardar M, Ahmed W, Al Ayoubi S, Nisa S, Bibi Y, Sabir M, Khan MM, Ahmed W, Qayyum A (2022) Fungicidal synergistic effect of biogenically synthesized zinc oxide and copper oxide nanoparticles against Alternaria citri causing citrus black rot disease. Saudi J Biol Sci 29:88–95
157. Sawalha H, Abiri R, Sanusi R, Shaharuddin NA, Noor AA, Ab Shukor NA, Abdul-Hamid H, Ahmad SA (2021) Toward a better understanding of metal nanoparticles, a novel strategy from eucalyptus plants, 10
158. Seetharaman PK, Chandrasekaran R, Gnanasekar S, Chandrakasan G, Gupta M, Manikandan DB, Sivaperumal S (2018) Antimicrobial and larvicidal activity of eco-friendly silver nanoparticles synthesized from endophytic fungi Phomopsis liquidambaris. Biocatal Agric Biotechnol 16:22–30
159. Selvakumar N, Barshilia HC (2012) Review of physical vapor deposited (PVD) spectrally selective coatings for mid- and high-temperature solar thermal applications. Sol Energy Mater Sol Cells 98:1–23
160. Shah Z, Gul T, Ali Khan S, Shaheen K, Anwar Y, Suo H, Ismail M, Alghamdi KM, Salman SM (2021) Synthesis of high surface area AgNPs from Dodonaea viscosa plant for the removal of pathogenic microbes and persistent organic pollutants. Mater Sci Eng B 263:114770–114770
161. Shareef HJ, Al-Yahyai RA, Omar AE-DK, Barus WA (2021) Foliar nano-fertilization enhances fruit growth, maturity, and biochemical responses of date palm. Can J Plant Sci 101:299–306
162. Shende S, Rathod D, Gade A, Rai M (2017) Biogenic copper nanoparticles promote the growth of pigeon pea (Cajanus cajan L.). IET Nanobiotechnol 11:773–781
163. Singh SB (2020) Nanocapsules: an engineered nanomaterial for smart self-healing coatings and catalysis. Handb Nanomater Manuf Appl, 175–189. https://doi.org/10.1016/B978-0-12-821381-0.00007-7
164. Small CC, Degenhardt D (2018) Plant growth regulators for enhancing revegetation success in reclamation: a review. Ecol Eng 118:43–51
165. Srinath BS, Ravishankar Rai V (2015) Biosynthesis of highly monodispersed, spherical gold nanoparticles of size 4–10 nm from spent cultures of Klebsiella pneumoniae. 3 Biotech 5:671–671
166. Sudha PN, Sangeetha K, Vijayalakshmi K, Barhoum A (2018) Nanomaterials history, classification, unique properties, production and market. Emerg Appl Nanoparticles Arch Nanostruct: Curr Prospect Futur Trends, 341–384. https://doi.org/10.1016/B978-0-323-51254-1.00012-9
167. Surendhiran D, Cui H, Lin L (2020) Mode of transfer, toxicity and negative impacts of engineered nanoparticles on environment, human and animal health. ELSI Handb Nanotechnol
168. Suryavanshi P, Pandit R, Gade A, Derita M, Zachino S, Rai M (2017) Colletotrichum sp.-mediated synthesis of sulphur and aluminium oxide nanoparticles and its in vitro activity against selected food-borne pathogens. LWT Food Sci Technol 81:188–194
169. Syukri DM, Nwabor OF, Singh S, Ontong JC, Wunnoo S, Paosen S, Munah S, Voravuthikunchai SP (2020) Antibacterial-coated silk surgical sutures by ex situ deposition of silver

nanoparticles synthesized with Eucalyptus camaldulensis eradicates infections. J Microbiol Methods 174:105955–105955
170. Thatikayala D, Jayarambabu N, Banothu V, Ballipalli CB, Park J, Rao KV (2019) Biogenic synthesis of silver nanoparticles mediated by Theobroma cacao extract: enhanced antibacterial and photocatalytic activities. J Mater Sci Mater Electron 30:17303–17313
171. Tyagi J, Ahmad S, Malik M (2022) Nitrogenous fertilizers: impact on environment sustainability, mitigation strategies, and challenges. Int J Environ Sci Technol 19:11649–11672
172. Uddin S, Safdar LB, Anwar S, Iqbal J, Laila S, Abbasi BA, Saif MS, Ali M, Rehman A, Basit A, Wang Y, Quraishi UM (2021) Green synthesis of nickel oxide nanoparticles from berberis balochistanica stem for investigating bioactivities. Molecules, 26
173. Usman M, Farooq M, Wakeel A, Nawaz A, Cheema SA, Rehman HU, Ashraf I, Sanaullah M (2020) Nanotechnology in agriculture: current status, challenges and future opportunities. Sci Total Environ 721:137778
174. Usmanov RA, Amirov RK, Gavrikov AV, Liziakin GD, Melnikov AD, Polistchook VP, Samoylov IS, Smirnov VP, Vorona NA, Yartsev IM (2020) Diffuse vacuum arc with heated cathode made of ceramic (CeO_2) and metal (Cr) mixture. Plasma Sources Sci Technol 29:015004–015004
175. Vijayaraghavan K, Ashokkumar T (2017) Plant-mediated biosynthesis of metallic nanoparticles: a review of literature, factors affecting synthesis, characterization techniques and applications. J Environ Chem Eng 5:4866–4883
176. Wahab A, Munir A, Saleem MH, Abdulraheem MI, Aziz H, Mfarrej MFB, Abdi G (2023) Interactions of metal-based engineered nanoparticles with plants: an overview of the state of current knowledge, research progress, and prospects. J Plant Growth Regul 42:5396–5416
177. Williams JO (1989) Metal Organic Chemical Vapor Deposition (MOCVD) perspectives and prospects. Angew Chem Int Ed Engl 28:1110–1120
178. Xu H, Xiao Y, Xu M, Cui H, Tan L, Feng N, Liu X, Qiu G, Dong H, Xie J (2019) Microbial synthesis of Pd-Pt alloy nanoparticles using Shewanella oneidensis MR-1 with enhanced catalytic activity for nitrophenol and azo dyes reduction. Nanotechnology, 30
179. Yadav A, Yadav K, Abd-Elsalam KA (2023) Nanofertilizers: types, delivery and advantages in agricultural sustainability. Agrochemicals 2:296–336
180. Yadav VK, Ali D, Khan SH, Gnanamoorthy G, Choudhary N, Yadav KK, Thai VN, Hussain SA, Manhrdas S (2020) Synthesis and characterization of amorphous iron oxide nanoparticles by the sonochemical method and their application for the remediation of heavy metals from wastewater. Nanomaterials 2020 10:1551–1551
181. Yang G, Park SJ (2019) Conventional and microwave hydrothermal synthesis and application of functional materials: a review. Materials 12:1177–1177
182. Yi K, Liu D, Chen X, Yang J, Wei D, Liu Y, Wei D (2021) Plasma-enhanced chemical vapor deposition of two-dimensional materials for applications. Acc Chem Res 54:1011–1022
183. Younes NA, Hassan HS, Elkady MF, Hamed AM, Dawood MFA (2020) Impact of synthesized metal oxide nanomaterials on seedlings production of three Solanaceae crops. Heliyon 6:e03188–e03188
184. Zambryski P (2004) Cell-to-cell transport of proteins and fluorescent tracers via plasmodesmata during plant development. J Cell Biol 164:165–168
185. Zhao L, Sun Y, Hernandez-Viezcas JA, Servin AD, Hong J, Niu G, Peralta-Videa JR, Duarte-Gardea M, Gardea-Torresdey JL (2013) Influence of CeO_2 and ZnO nanoparticles on cucumber physiological markers and bioaccumulation of Ce and Zn: a life cycle study. J Agric Food Chem 61:11945–11951

The Flow and Transformations of Nanomaterials in Agricultural Ecosystems

Hermes Pérez-Hernández, Edgar Vázquez-Núñez, Ileana Vera-Reyes, Alma Hortensia Serafín-Muñoz, Diego Iván Escobar-Hernández, and Esperanza Loera-Alvarado

Abstract Nanotechnology (NT), using nanometric materials, has impressed many scientists, technologists, and students due to the results observed when applied in crop fields and the environment. The effectiveness of nanometric-scale materials depends on the unique properties of Nanomaterials (NMs) or Nanoparticles (NPs). Size, shape, surface charge, surface/volume ratio, and catalytic activity, among other aspects, favor positive changes in plant growth or the decontamination of soils contaminated with organic, inorganic, or heavy metal contaminants. Despite the efficiency of nanometric products, at the laboratory and field level, it has been suggested that the indiscriminate use of NMs and NPs may represent risks for plants, soil organisms, and humans. The extensive literature mentions that plants and some soil animals can accumulate high concentrations of different nanometric products. Also, it is suggested that within cells, nanomaterials or nanoparticles can transformed when they come into contact with organic components or with the physical, chemical,

H. Pérez-Hernández (✉)
Instituto Nacional de Investigaciones Forestales, Agrícolas y Pecuarias (INIFAP), Campeche, Mexico
e-mail: hermesph@hotmail.com

E. Vázquez-Núñez
Grupo de Investigación Sobre Aplicaciones Nano y Biotecnológicas para la sostenibilidad (NanoBioTs), Departamento de Ingenierías Química, Electrónica y Biomédica, División de Ciencias e Ingenierías, Universidad de Guanajuato, León, Guanajuato, Mexico

I. Vera-Reyes
Departamento de Biociencias y Agrotecnología, Centro de Investigación en Química Aplicada (CIQA), Saltillo, Coahuila, Mexico

A. H. Serafín-Muñoz
Departamento de Ingeniería Civil, Universidad de Guanajuato, Guanajuato, Mexico

D. I. Escobar-Hernández
Departamento de Botánica, Universidad Autónoma Agraria Antonio Narro, Buenavista, Saltillo, Coahuila de Zaragoza, Mexico

E. Loera-Alvarado
CONAHCYT-Universidad Autónoma Chapingo, Centro Regional Universitario Centro Occidente (CRUCO), Morelia, Michoacán, Mexico

© The Author(s), under exclusive license to Springer Nature Singapore Pte Ltd. 2025
A. Juárez-Maldonado et al. (eds.), *Plant Biostimulation with Nanomaterials*, Smart Nanomaterials Technology, https://doi.org/10.1007/978-981-96-4648-7_14

and biological properties of the soil. Therefore, in this section, the implications of nanomaterial transformations in edible plants and soil organisms are discussed.

Keywords Transformation of nanomaterials · Biodegradation of nanomaterials · Soil and nanomaterials · Nanomaterials in the food chain · Engineered Nanomaterials

1 Introduction

Nanotechnology (NT) has driven the development of many areas of science. Worldwide, more and more companies are generating products based on nanomaterials (NMs) and nanoparticles (NPs) for the agriculture sector. As commented in other chapters, nanotechnology in agriculture aims to reduce chemical products, minimize nutrient losses in fertilization, improve the morphological and biochemical characteristics of plants, and consequently improve the quality and yield of crops. On the other hand, nano pesticides, nanosensors, and others have been tested to detect organic and inorganic substances in the agri-food sector. Nevertheless, the latest reports that suggest NT as an advanced technology in the agri-food sector to promote sustainability have also been the cause of controversy [23, 33].

The concern is heightened as some studies predict that due to the indiscriminate use of NPs in agriculture and other industries that employ them, nanomaterials can bioaccumulate in the environment, in plants and consequently, with the possibility of entering the food chain, which can trigger unknown effects on the health of consumers [29].

Regarding the movement of materials in the food chain, the date, final destination, and behavior of NMs and NPs are unknown. There is no information on the physical and biochemical phenomena at each trophic level, mainly on the reactions involved during interactions at the cellular level, whether in plants, soil organisms, or humans. In addition, controversial results caused by size, shape, evaluated concentration, exposure time, surface/volume ratio, surface chemistry, potential surface charge, and catalytic activity, among other aspects, must be clarified. Therefore, it is necessary to monitor the life cycle of NMs, from their plant-soil application to food, humans, and domestic animals. One study highlighted the collateral effects that nanomaterials can have on human and environmental health. In detail, of all the analyzed studies on toxicity reported between 2012 and 2021, approximately 24% of the analyzed manuscripts did not report ecological or human risks, while 54% of the analyzed manuscripts reported unfavorable effects of nanomaterials on ecosystems, human health, or the environment [55].

This section addresses the topic of the flow and transformations of NMs in agricultural systems. We discuss how soil factors (physical, chemical, and biological) enhance the mobility and transformations of NPs in the soil. In addition, we dedicate a section to discuss the implications of NM transformations in edible plants and soil organisms.

2 Soil and Biologic Factors that Determine the Transformation of Nanomaterials

2.1 Soil Properties Affecting NM Transformations

Soil provides 95% of food by supplying essential plant nutrients, water, oxygen, and root support; is a reservoir of nutrients from the decomposition of organic matter and erosion of minerals; cleans, filters, and stores water; regulates climate and flooding; removes carbon dioxide from the atmosphere; acts as a filter and remediator of pollutants; and is a habitat for micro and macroorganisms [66]. However, agricultural soils face challenges due to erosion, loss of organic carbon, inadequate nutrient management, salinization and sodification, soil and groundwater contamination, compaction and the accumulation of nanomaterials and nanoparticles as a result of new technologies [75].

The transformation and behavior of NMs in soil environments are governed by multiple soil characteristics, including pH, redox potential, organic matter content, and ionic strength [19]. These factors play critical roles in determining the extent and nature of nanomaterial transformations, influencing their mobility, stability, and eventual ecological impacts. As engineered nanomaterials (ENMs) increasingly enter agricultural systems, understanding how these soil factors control NM transformations is essential for predicting their environmental fate.

Figure 1 illustrates the interactions between soil properties and ENMs. It is represented how factors such as pH, organic matter, and ionic strength contribute to processes like aggregation, dissolution, or surface modification of nanomaterials. These transformations directly affect the mobility and bioavailability of ENMs in soils, ultimately determining their impact on the environment and living organisms. For example, high ionic strength may promote aggregation of NMs, reducing their mobility, whereas acidic conditions can enhance dissolution, increasing their potential toxicity. Understanding these interactions is vital for predicting the long-term behavior of ENMs in agricultural soils.

A detailed explanation of how these factors affect the ENMs characteristics and behavior in the environment involves understanding how pH, redox potential, organic matter, and ionic strength influence the transformations and interactions of nanomaterials, ultimately determining their environmental fate and impact. The impact of soil microbial communities is also described.

pH and Redox Potential

The pH of soil is a fundamental factor in NM stability and reactivity. In alkaline or neutral soils, NMs such as Ag, Cu, and Zn tend to aggregate due to enhanced electrostatic interactions, which reduces their mobility and bioavailability [42]. Aggregation under these conditions occurs through the van der Waals and electrostatic forces, leading to larger clusters of nanoparticles that are more likely to be retained in the soil matrix [1, 2]. For example, studies have demonstrated that ZnO and CuO

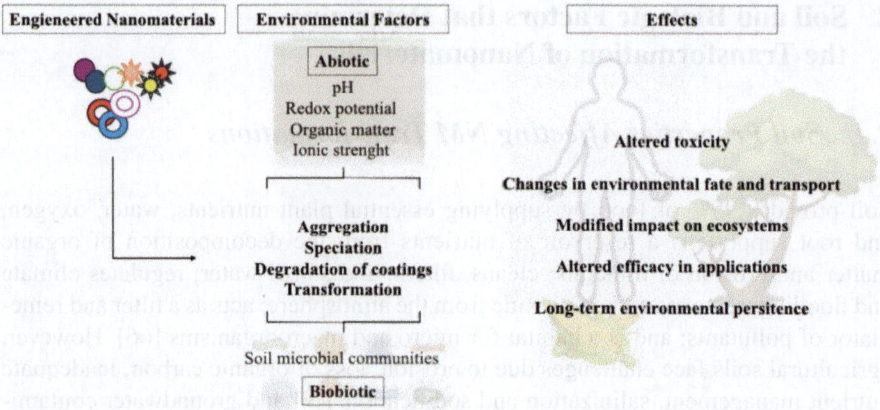

Fig. 1 Influence of soil properties on the transformation and behavior of Engineered Nanomaterials (ENMs)

nanoparticles dissolve more readily in acidic soils, increasing the release of metal ions, which can become toxic to soil organisms [32].

Redox conditions also significantly influence NM transformations, particularly for metal-based nanomaterials. In anaerobic or reducing conditions, redox-sensitive NMs can undergo chemical speciation, as observed with Ag nanoparticles (AgNPs) transforming into less reactive silver sulfide (Ag_2S) under sulfate-reducing conditions [34, 53]. Similarly, iron oxide nanoparticles are subject to redox-driven transformations that lead to the formation of stable iron sulfides, which are less mobile and less reactive compared to their pristine forms [36]. These transformations ultimately reduce the toxicity and environmental impact of NMs in reducing environments.

Organic Matter and NM Interactions

Soil organic matter (OM) plays a crucial role in modulating NM behavior by forming an 'eco-corona' on NM surfaces. This eco-corona can influence the interaction between NMs and soil organisms, modify the mobility and bioavailability of the particles, and impact their aggregation and stability in the soil matrix. By interacting with OM, NMs acquire a biological identity that can either mitigate or enhance their toxicological effects in the environment, making it essential to consider the eco-corona in studies of NM behavior in soil ecosystems [52]. This organic coating consists of humic and fulvic acids that alter the surface reactivity of NMs, often reducing their toxicity and promoting stability. The presence of dissolved organic matter (DOM) can further enhance the dispersion of NMs, preventing aggregation and increasing their mobility through soil pore spaces [19].

The interaction between organic matter and NMs is dynamic, with varying effects depending on the concentration of OM. At low concentrations, NMs tend to adsorb onto soil particles, reducing their mobility. However, at higher OM concentrations, organic molecules may fully coat the NM surfaces, thereby stabilizing the nanoparticles in suspension and facilitating their transport [34]. The effect of OM on NM mobility is particularly significant for carbon-based NMs, such as carbon nanotubes (CNTs), which exhibit increased dispersion in the presence of humic substances [31].

Ionic Strength and Aggregation

The ionic strength of the soil solution strongly influences NM aggregation and transport. High ionic strength increases the likelihood of NM aggregation by compressing the electrical double layer around the nanoparticles, reducing the repulsive forces that keep NMs dispersed [37]. As ionic strength rises, the nanoparticles tend to aggregate into larger particles, which are more likely to be retained in soil micropores. For instance, increased ionic strength due to the presence of divalent cations like calcium can lead to enhanced retention of ZnO nanoparticles in soil, as observed by Kang et al. [34].

This process of aggregation and retention reduces the mobility of NMs in soil environments, but in some cases, NMs can still move through macropores, especially if they form stable aggregates with natural colloids such as clays and organic matter [14, 61]. The extent of NM retention and mobility depends not only on ionic strength but also on the composition of the soil colloidal fraction, with clays like smectite exhibiting different interactions with NMs compared to less negatively charged minerals like kaolinite [38].

2.2 *Additional Factors Influencing NM Transformations*

Several other factors, such as soil porosity and hydraulic properties, also affect NM behavior. As highlighted by recent studies, nanomaterials can alter soil physical properties, such as hydraulic conductivity and pore structure. For example, the deposition of CuO and Al_2O_3 nanoparticles has been shown to reduce soil hydraulic conductivity by clogging soil pores, leading to decreased water flow and potentially affecting soil aeration [47, 68]. Additionally, changes in soil structure due to NM retention can influence the retention of other contaminants, as observed with the retention of arsenic in soils treated with iron-based nanoparticles [7, 28].

Soil Microbial Influence on NM Transformations

Microorganisms within the soil ecosystem are crucial mediators of nanomaterial (NM) transformation processes. Soil microbial communities engage in complex interactions with NMs, driving a variety of biotransformations that can affect both the environmental fate and the potential toxicity of these materials [77]. These microbial transformations include redox reactions, degradation of NM coatings, dissolution of metal-based NMs, and the formation of new NM species due to microbial activity [62]. The biotic influence of microorganisms on NMs plays an essential role in soil health and the broader ecological consequences of NM use in agriculture.

Microbial Redox Reactions and NM Transformations

Microorganisms can induce redox transformations in NMs, particularly for metal-based nanoparticles [3]. For instance, microbial respiration under anaerobic conditions can lead to the reduction of metallic nanoparticles such as Ag nanoparticles (AgNPs) to silver sulfide (Ag_2S), a transformation that results in a less bioavailable and less toxic form of the material [10]. This process typically involves a two-step reaction in which the initial oxidation of silver to Ag^+ is followed by its reduction to Ag_2S. Sulfidation of AgNPs, catalyzed by microbial activity, is one of the most common transformation processes in the soil environment. Sulfidation has been shown to significantly decrease the bioavailability of silver ions, reducing the potential toxicity of AgNPs to plants and soil organisms [43].

Similarly, iron-based nanoparticles (e.g., Fe_3O_4) can undergo microbial-driven redox transformations, where soil bacteria reduce iron oxides to iron sulfides or iron(II) ions. These biotic transformations stabilize the particles in the soil matrix, limiting their mobility and environmental impact. Such microbial-driven processes are important for understanding how NMs evolve in soil environments over time, particularly in the context of environmental safety and sustainability.

Microbial Degradation of NM Coatings and Organic Transformations

The transformation of NMs in soil is not limited to redox reactions. Microbial communities can also degrade the organic coatings of NMs, which are often applied to enhance their stability and dispersibility [42]. As soil microbes metabolize organic coatings, the core NM is exposed, altering its reactivity, surface charge, and overall environmental behavior. For example, carbon-based nanomaterials, such as carbon nanotubes (CNTs), can undergo microbial degradation, where the breakdown of organic surface coatings facilitates further transformation or aggregation of the core NM [46].

In addition, extracellular enzymes secreted by microbes can degrade the polymeric coatings of nanoparticles, leading to changes in their chemical composition and potential interactions with other soil components. These transformations can modify the solubility and reactivity of metal oxides like ZnO and CuO, affecting the release of metal ions and their subsequent uptake by plants and microbes [3].

Aggregation, EPS Secretion, and NM Retention

Soil microbes can also influence the aggregation state of NMs through the secretion of extracellular polymeric substances (EPS). EPS consists of polysaccharides, proteins, and nucleic acids that form a protective matrix around microbial communities [27]. This matrix can trap NMs, promoting the formation of heteroagglomerates, which reduce NM mobility by increasing particle size. The retention of NMs within these biofilm-like structures limits their transport through soil and decreases the potential for broader environmental exposure [21].

EPS and microbial exudates not only promote aggregation but can also alter the surface charge and hydrophobicity of NMs, thereby influencing their interaction with soil particles and plant roots [89]. These modifications can either enhance or reduce

NM mobility, depending on the specific environmental conditions and microbial species involved.

Nanomaterial Aging in Soil Environments

As NMs age in soil, their physical and chemical properties undergo significant changes due to interactions with soil components and microbial activity. This aging process leads to a series of transformations that alter the long-term behavior, bioavailability, and effects of NMs in soil ecosystems [54]. Aging typically involves processes such as aggregation, dissolution, and the formation of new chemical species. For example, sulfidation of AgNPs or phosphatization of ZnO and CuO nanoparticles results in the formation of less toxic species like Ag_2S, ZnS, and Cu_2S [64].

The transformation of NMs due to aging can also result in their incorporation into soil colloids, which immobilizes them and reduces their bioavailability [1, 2]. However, this process is highly dependent on soil type, organic matter content, and microbial activity. Soils rich in organic carbon and with active microbial communities tend to facilitate faster NM transformations, as microbes and organic ligands play a key role in altering NM surfaces and enhancing their incorporation into stable soil aggregates [88].

In soils where microbial activity is limited, the aging process may proceed more slowly, allowing NMs to remain in more reactive, pristine forms for extended periods. This variability in NM aging rates across different soil types underscores the importance of site-specific assessments when evaluating the long-term environmental risks of nanomaterials [60, 68].

Bioavailability and Toxicity of Transformed NMs

The transformation of NMs has a direct impact on their bioavailability and toxicity in soil ecosystems. Pristine NMs are often highly reactive and can exhibit significant toxicity to soil organisms, including plants, microorganisms, and invertebrates.

Certain transformed NMs can still pose environmental risks under specific conditions. For example, oxidative dissolution of Ag_2S in the rhizosphere, where redox conditions fluctuate, can lead to the release of Ag^+ ions, which may be taken up by plants and result in localized toxicity [57]. Similarly, while ZnO and CuO nanoparticles may form less reactive sulfide species, under certain conditions, these forms may dissolve and re-release toxic metal ions [6, 24].

In summary, soil microorganisms are key players in NM transformations, driving biotic processes that influence the environmental fate, bioavailability, and toxicity of nanomaterials. These transformations are crucial for understanding the long-term implications of NM use in agriculture and their potential impact on soil health and ecosystem functioning.

The transformation of NMs in soil environments is a complex, multifaceted process that is strongly influenced by soil properties such as pH, redox potential, organic matter content, and ionic strength. These factors determine the behavior, mobility, and ecological impact of NMs. For instance, soil pH can drive nanoparticle aggregation, while redox conditions can induce chemical changes that stabilize or detoxify NMs. Organic matter plays a dual role in either stabilizing or facilitating the mobility of NMs, with the formation of an ‘eco-corona’ significantly affecting their interactions with soil biota.

Moreover, soil microorganisms actively mediate NM transformations through processes such as redox reactions, coating degradation, and aggregation, all of which influence the long-term fate and toxicity of these materials. Microbial interactions, particularly the secretion of extracellular polymeric substances, enhance NM retention in the soil matrix and reduce mobility. Additionally, the aging of NMs through interactions with soil and microbial activity alters their chemical composition and reduces their bioavailability.

The environmental risks associated with nanomaterials depend on their transformation processes and the specific soil conditions they encounter. While many transformations reduce NM toxicity, some transformed species can still pose hazards under certain conditions. As nanomaterials become more integrated into agricultural practices, understanding these soil and biological interactions is essential for predicting their long-term impacts on soil health and ecosystem functioning.

3 Implications of Biotransformation or Biodegradation of Nanomaterials Within the Plant System

As mentioned earlier, NT is expected to play an important role in agriculture, particularly in the delivery of fertilizers and pesticides that impact consumer plants. Plants interact directly with soil, water, and the atmosphere, thus constituting a potential pathway of exposure to NMs for higher species through the food chain. Plants absorb nanoparticles mainly by foliar spray or root uptake, Fig. 2. Roots can absorb nanoparticles through their hairy roots or root tips. Nanoparticles must translocate through either the symplastic or apoplastic pathways. In the apoplastic pathway, nanoparticles must move out of cells through the cell wall and intercellular spaces. In the other pathway, they must penetrate the cell and through plasmodesmata and translocate between cells [69]. In the case of foliar uptake, nanoparticles must pass through the cell wall, which is a permeable barrier against nanoparticle translocation [30]. The biodegradation of nanomaterials within plant systems is a complex process that involves the interaction of various biological and environmental factors. It refers to the breakdown of nanomaterials into smaller, less harmful components by organisms or their metabolic processes [30]. The composition, size, shape, and surface chemistry of nanomaterials can significantly influence their biodegradability. For instance, metal nanoparticles may be more susceptible to biodegradation than carbon-based nanomaterials.

Plant species, growth stage, and physiological conditions can affect the plant’s ability to interact with and degrade nanomaterials. The presence and activity of microorganisms in the rhizosphere (the region of soil immediately adjacent to plant roots) play a crucial role in the biodegradation process [56]. Microorganisms can degrade nanomaterials through various mechanisms, such as oxidation, reduction,

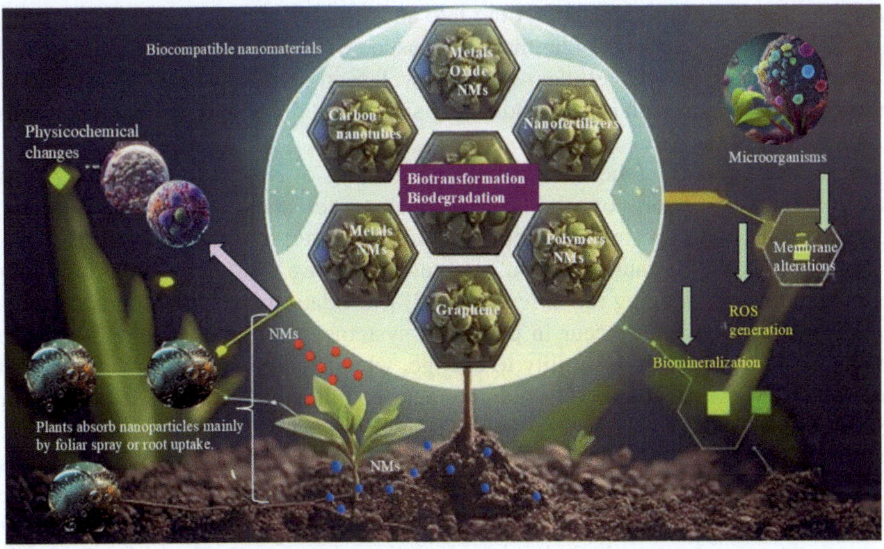

Fig. 2 Schematics of the interaction, biotransformation, and biodegradation of nanomaterials in a plant system. Many types of NMs transformations occur within plants, in the rhizosphere, and at leaf interfaces. Microorganisms can degrade nanomaterials through various mechanisms, including bioremediation; metal nanoparticles and ions can generate reactive oxygen species (ROS) that damage cellular structures; and metal nanoparticles can accumulate and dissolve in the bacterial membrane, altering its permeability

and complexation. Factors like soil pH, temperature, moisture content, and nutrient availability can influence the rate and extent of nanomaterial biodegradation. Under this context, studies have shown that some nanomaterials, such as carbon nanotubes, are relatively resistant to biodegradation, but some studies have shown that microorganisms can degrade them under certain conditions [5]. Understanding the biodegradation of nanomaterials can inform the development of sustainable nanomaterials and their applications in agriculture, environmental remediation, and other fields. For example, biodegradable nanomaterials could be used as controlled-release fertilizers or to deliver pesticides or other agents to plants [51, 56].

In plant systems, NMs can undergo various types of physicochemical changes by plant compounds, Fig. 2. The root can produce large amounts of secretions such as mineral ions, organic matter of minute molecules (such as phenols, aldehydes, amino acids, and organic acids), and high molecular weight pectins (such as polysaccharides and fatty acids) that create a thin environment around the root [22]. Thus, NMs can undergo physicochemical deformations by interaction with root secretions since, in several cases, nanomaterials have direct contact with roots. Most of the nanomaterials are easily absorbed and pile up on the root surface [71, 73]. This physical deformation limits the adsorption of nanomaterials by roots and their subsequent displacement. On the other hand, the suction of nanomaterials on the root surface causes them to connect directly with root secretions and thus increases the likelihood of deformation.

These types of deformations affect the fate and toxicity of nanomaterials to plants. In one study, large amounts of lanthanum phosphate ($LaPO_4$) clusters were observed in cucumber roots in intercellular regions as well as in vacuoles and cytoplasm under a five-day treatment with La_2O_3, indicating significant deformation of nanoparticles in plants [30, 44].

Root exudates affect the transformation behaviors of NPs. Roots of living plants secrete complex mixtures of soluble organic substances, such as sugars, amino acids, organic acids, and enzymes, among others. Root exudates modify rhizosphere chemistry and affect NPs transformation, such as increasing soluble species, inhibiting aggregation, and inducing redox reactions [22]. Oxidation and reduction are important reactions that often occur in plant-soil systems. Many nanomaterials contain metal components with the ability to change, which can be reduced, oxidized, and subsequently biotransformed through interaction with reducing agents in plants. For example, one study observed the transformation of these nanoparticles in plants, which showed that silver nanoparticles are oxidized as Ag_2S or Ag_2O in the roots of *Lolium multiflorum* [80, 87]. There have been reports of studies in which the following have been observed that copper oxide nanoparticles can return from the stem to the root, during which they are partially reduced to Cu_2S and Cu_2O [9, 12, 17]. Cerium oxide nanoparticles are among the nanoparticles whose transformation has been most studied in plants [4, 40]. CeO_2 nanoparticles are very stable nanoparticles in their environment and are used as model nanoparticles compared to other nanoparticles that are easily dissolved and unstable (such as ZnO, Ag, etc.).

Other studies found that $nCeO_2$ is not very firm and is likely to be degraded and transformed into Ce(III) species [18, 26, 41]. Using transmission electron microscopy, large amounts of needle-like clusters were found to be present in the intercellular and epidermal space of cucumber root cells under 21-day treatment with $nCeO_2$. Combined EDS analysis showed that these clusters were formed by Ce and P in an approximate atomic ratio of 1:1 and might be in the form of $CePO_4$. This hypothesis was further confirmed by XANES and STXM analysis, which provided a two-dimensional distribution and cluster formation. XANES mass studies showed that Ce is mainly present as $CePO_4$ and CeO_2 in roots, while Ce and CeO_2 are present as carboxylates in stems and leaves. The combination of the above results and subsequent simulations revealed to some extent the deformation and displacement mechanism of $nCeO_2$ in cucumber. CeO_2 releases Ce^{3+} ions with the help of organic acids and reducing agents in root secretions, which are subsequently converted to $CePO_4$ and Ce carboxylates. Some of the released Ce^{3+} ions are immobilized by phosphates, which are abundantly found in nutrient solutions and plant tissues.

Nanomaterials can induce stress in plants and subsequently cause the generation of reactive oxygen species (ROS), which in the next step affects plant organelles, structures, DNA, proteins, lipids, carbohydrates, and secondary metabolites [16, 74]. Symplastic penetration of nanoparticles through the plant cell membrane occurs through endocytosis, formation of pores, membrane proteins, plasmodesmata, and ion channels. Endocytosis happens when the plasma membrane folds inside out (invagination), and the membrane vesicles that are made can move to different parts of the cell.

Roots secrete organic acids, which play an important role in their biotransformation by enhancing the dissolution of nanoparticles. Another important route of NP uptake in plant cells is endocytosis. Invagination of the plasma membrane of plant cells introduces NPs into cells. For example, studies have shown that protoplasts can internalize particles of size less than 1 μm through endocytosis, so nanoparticles taken up by endocytosis do not have particle size selectivity [25]. In addition to this, plants can take up nanoparticles by combining them with carrier proteins in the outer epidermis. In one study, they pretreated the cells with wortmannin (an inhibitor of plant cell endocytosis), the cell fluorescence of single-walled carbon nanotubes stained with FITC decreased significantly; thus, it was observed that endocytosis functions as a main pathway for carbon nanotubes to enter the plant cell. In another study, they proposed that carbon nanoparticles and carbon nanotubes were taken up by *Catharanthus roseus* root cells via endocytosis [63, 70].

In the foliar biotransformation of nanomaterials in leaves varies according to their growth phases and the life cycle of the plants. Studies have shown that the presence of leaf hair and cuticular wax reduce the penetration of nanomaterials, and this will depend on NP size, leaf surface, epidermal structure and growth stage [72]. Cell walls and waxes also hinder the absorption of nanomaterials. However, the lipophilic nature of epidermal wax favors the penetration of hydrophobic NMs while hindering the absorption of hydrophilic nanosuspensions. The hydrophobic waxy cuticle contains very small pores (<5.0 nm), preventing the uptake of all NPs ranging in size from 0.6 to 4.8 nm but allowing it. The stability of the nanoparticles is related to the retention of a primary, non-aggregate state of the particles. The uptake of NPs through the stomata and its translocation are particularly dependent on the size of the NPs [83]. The size of the pores of the stomata is about 25 μm long and 3–10 μm wide. Therefore, NPS having a size less than 50 nm can penetrate the stomatal pores. A study in *Arabidopsis thaliana* showed that NPS having a size of 43 nm can be captured through the stomatal route, but NPS of size 1.1 μm cannot be captured [49, 67]. On the other hand, metal NMs can be transformed into organic forms in plants. For example, after foliar spray application *to A. thaliana*, Se NPs were transformed into selenoamino acids, including (SeCys2) and methylselenocysteine (MeSeCys) in shoots or roots [81]. In addition, two selenium-binding proteins were found in plant tissues. GSH-coated CdSe/ZnS QDs were biotransformed into volatile alkyl selenides, including dimethyl selenide (DMSe), dimethyl diselenide (DMDSe), and sulfur-containing alkyl selenides, through chalcogenide exchange, methylation, and demethylation reactions mediated by rice plants and rhizosphere microorganisms. Sulfur from MoS_2 NPs exposed to soil was assimilated to thiol compounds.

The biotransformation and biodegradation of nanomaterials in plants represent a complex interaction between nanomaterial properties and plant biological processes. These processes may have significant implications for the fate and effects of nanomaterials in the environment and agricultural systems. Within a sustainable impact perspective, studies should be conducted to assess the long-term environmental implications of the biotransformation and biodegradation of nanomaterials, including potential risks to ecosystems and human health. As well as the development of predictive models, it can help to assess the fate and effects of nanomaterials in different plant systems and environmental conditions.

4 Implications of Nanomaterial Transformations on Beneficial Soil Organisms and Long-Term Effects

The increasing use of nanomaterials in agroecosystems has led to a growing interest in understanding their impact on soil health and the organisms within it. Research extensively examines how these materials interact with soil, microbial communities, and plant systems, revealing potential benefits and risks. Relationship between soil and nanomaterials shows that interactions can occur in both water and solid fractions. The role of soil pores as sites for nanomaterial sorption and biotransformation emphasizes how soil mineral structures can be affected by the presence of nanomaterials. Additionally, the dissolution of nanomaterials depends on factors such as surface area and charge, ultimately influencing their fate in the environment. These findings suggest that the quick dissolution of specific nanomaterials, like zinc oxide, could limit their persistence in soil, raising concerns about their long-term effects on soil microbial communities [58].

Expanding on this understanding, El-Ramady et al. [15] address the urgent need for soil restoration in the face of global degradation. They introduce the concept of nano-restoration, emphasizing the potential of nanomaterials in remediating polluted soils. However, the excessive use of these materials could lead to nano-toxicity, affecting both plants and the food chain.

Nanomaterials have the potential to enhance agricultural productivity while also considering their implications for soil health and microbial dynamics, which is crucial for optimizing their use in sustainable agriculture [48, 50].

The role of soil pores as sites for the sorption and biotransformation of nanomaterials is particularly significant. This suggests that the physical structure of soil can influence the fate of nanomaterials, potentially affecting their bioavailability to soil organisms. The biopores created by organisms such as earthworms can facilitate the transport and discharge of nanoparticles, thereby altering the dynamics of nanomaterial interactions within the soil ecosystem. Additionally, the impact of soil mineral structures on nanomaterial behavior is noted, with distortions and substitutions within the crystal structure potentially leading to electric imbalances. This phenomenon could affect the interactions between nanomaterials and microbial communities, as positively charged soil components may engage in hetero aggregation with negatively charged nanomaterials. The variability in soil types, particularly in terms of organic matter and clay fractions, is highlighted as a factor that can influence these interactions. This suggests that the effects of nanomaterials may not be uniform across different soil environments [58].

The role of soil microbes is evolving, and they are no longer seen solely in a negative light due to their association with soil degradation. Instead, they can play a crucial role in transforming degraded soils, particularly in the pharmaceutical sector. This change in perspective suggests a promising area for future research, focusing on their role alongside nanoparticles and plants in improving soil fertility [15]. However, it's important to consider the potential toxicity of metal-based nanoparticles to soil invertebrates and beneficial microorganisms. These interactions could lead to changes in soil health and biodiversity, which are essential for maintaining ecosystem functions. The long-term effects of these interactions on soil organisms are not yet fully understood, so further research is necessary to assess the ecological risks involved.

Since a couple of years ago, Garner et al. [20] demonstrated that under a dynamic simulation model of the accumulation of metal oxide NPs (CeO_2, CuO, TiO_2, and ZnO) over ten years, soluble NPs can accumulate in the environment, reaching concentrations that exceed the minimum toxic threshold in fresh water and certain soils. Lin et al. [39] predicts that specific NMs such as Ag, Cu, and even C, could disfavor the enzymatic interaction of the soil, particularly in soils with little organic matter and/or neutral to alkaline pH, when applied in high and prolonged doses. The use of metallic NMs in agricultural soils is greater than in urban soils since in agriculture there are direct applications of biosolids, irrigation water (with NM residues) and use of nanopesticides and nanofertilizers [35].

Carbon nanotubes and graphene induce genetic modification by inducing abiotic resistance, such as drought and salinity [11]. As the exposure time increases, the impact of CNMs on soil microbial communities tends to gradually stabilize [82]. CNMs doses <1 mg kg^{-1} promote increased soil respiration, but at doses >500 mg kg^{-1} this is inhibited and disrupts the soil carbon cycle [39]. A 29% decrease in soil respiration has been shown due to the poisoning of bacteria affected by Ag^{+} NPs. In addition, Cu and Ti NPs also reduce respiration by 7%. CuO NPs, their solubility allows the release of Cu ions under various environmental conditions [8, 39, 76]. The effects of ZnO NPs (50, 250, and 500 mg kg^{-1} and Cd (5 mg kg^{-1}), with and without mycorrhizal fungus (AM) inoculation, on *Sorghum bicolor* L. were evaluated. High doses of ZnO NPs (250 and 500 mg kg^{-1} inhibited growth and caused synergistic toxicity with Cd, while the low dose of 50 mg kg^{-1} was not phytotoxic, showing antagonistic interactions [78, 79].

Furthermore, the breakdown of nanomaterials depends on various physicochemical properties such as surface area, charge, and shape. This is important because it affects how nanomaterials move and become available in the soil. This understanding highlights the complexity of interactions at the interface of the nano-bio-ecosystem and the need for more research to understand these processes. The slow clumping together of nanomaterials in most soils is due to the level of ions and the pH, which usually stay below the point where clumping happens quickly. This discovery suggests that although nanomaterials may stay in the soil for a long time, their behavior is influenced by many factors that need to be carefully considered when evaluating their potential effects on helpful soil organisms.

Soil degradation affects approximately 25% of global soils, which impairs their ability to support human needs, especially in food production. Nanomaterials have the potential to interact with soil microbes and cultivated plants, which could enhance soil fertility and restore degraded soils. However, it's important to note that while nanomaterials can provide benefits through remediation and improved nutrient uptake, there is also a significant risk of nano-toxicity.

For the above, it is necessary to establish a regulatory framework to govern the use of nanomaterials in agricultural practices. We must approach nano-restoration cautiously and advocate for thorough assessments of the potential health implications for both humans and the environment. Further investigation is needed to fully understand the implications of nano-restoration strategies, particularly the interactions between nanoparticles, plants, and microbes in various environmental conditions, especially under the influence of climate change.

The multifaceted implications of nanomaterial transformations on beneficial soil organisms. While there is potential for nanomaterials to enhance soil health and agricultural productivity, the risks associated with their use cannot be overlooked. The interactions between nanomaterials and soil ecosystems are complex and warrant further research to develop sustainable agricultural practices that protect soil health and the organisms that inhabit it.

5 A Brief Review of Nanomaterials in the Food Chain

The nanomaterial residues present in the soil and in edible crops increase with the routes of exposure, bioaccumulation in the environment, and consequently, with the possibility of entering the food chain [29]. In this sense, a hundred articles have been published in the last decade on beneficial and harmful effects of which the absorption, translocation, and accumulation of NPs in plants and organisms play a critical role in toxicity with possible consequences for human food safety [23].

The possible effects of NMs in the food chain deserve careful observation. There are several studies that, from the title, claim that nanometric materials accumulate in us humans. In a search carried out on the web of science (research articles), when used the words "nanomaterials", "nanoparticles", "food chain", "terrestrial ecosystems", "plants", and "humans", we found that in the titles, there are at least 258 reports that mention collateral damage in animals and humans caused by NMs. However, independent of the quality of the studies, this does not include a direct evaluation on humans. With the above, we do not mean or rule out that humans are not currently consuming nanometric materials. Therefore, despite the fact that the physicochemical properties of NMs have been studied, synthesis methods for use in various sectors of industry, in environmental and agricultural applications, little is known about their possible adverse environmental impacts, once they are released into terrestrial ecosystems and, of course, the possible effects on human health [23]. In fact, recent reports evaluating the movement of NMs in organisms (soil-plant-terrestrial arthropods) show interesting results of NMs bioaccumulation in at least one trophic level [13].

It is evident that we humans, being consumers of fruits and vegetables treated with NMs, are undoubtedly part of the trophic chain, and in many cases, we are the first trophic level. However, there are gaps in knowledge about living organisms. In the first instance, little is known about the movement, processes, and chemical reactions that occur in cells in plants, animals, and humans [45]. On the other hand, investigations argue that commercial edible products, such as foods with additives, contain nanoparticles, which can cause harmful effects on the human body. However, reports have shown that the intestinal epithelial barrier protects the human body from several foreign particles. Although, the malfunction of this set of cells may allow the translocation of synthetic NMs and consequently magnify the damage [59]. Nevertheless, to date, even when in vitro, in vivo, and ex vivo analysis have been performed, the potential negative impacts of NMs on human health have been speculative and unsubstantiated [65]. In addition to the above, little is known about the trophic transfer and possible damage of NPs from plants to humans.

Recent findings reported the trophic transfer of Ti and Ag NPs from lettuce plants (*Lactuca sativa* L.) to snails (*Cornu asperum*). For the researchers, it was interesting to learn that the Ag NPs ingested by the mollusks were excreted through the feces, while the Ti NPs accumulated in the digestive glands. It was even more interesting to learn that Ti and Ag NPs inhibited the snail's growth. In the conclusions, the authors suggest that there is a possibility of the transfer of NPs to higher levels of the food chain [84]. Two years later, the same authors reported that using the same model organisms confirmed that individual exposure of TiO_2 NPs and mixed Ag^+ TiO_2 NPs induced biomagnification of NPs from lettuce to snails [85].

As well as the examples of research that evaluates the trophic transfer of NMs from plants to terrestrial animals, there is diverse information from studies for aquatic environments. However, although studies show evidence of nanotoxicity, we do not rule out that we humans are ingesting NPs both from products from agriculture and rivers, lakes, and the sea.

We argue that with the desire to increase the production and quality of food, on the one hand, many researchers, technicians, and students try to propose and evaluate nanomaterials, nanocomposites that, to a certain extent, some of these nanometric materials are not essential in the metabolism in plants, animals, and humans. Although research groups want to demonstrate new findings with sophisticated methodologies and tools, it is worth being careful with research proposals. On the other hand, it calls our attention that various investigations conclude that the use of non-essential elements for plants and animals can replace other essential nutrients. These arguments are worth careful observation.

6 Conclusions

The biotransformation of NMs in the food chain appears to be a complex process, initiated and influenced by the soil matrix and prolonged by plant biological processes. Depending on the particular soil conditions, plant species, and NMs characteristics, nanometric materials may or may not dictate a risky impact on the food chain. Therefore, to understand the long-term impacts of NMs on soil health and possibly on humans, an immense understanding of NM-soil-plant or animal interactions is required.

To get a sustainable agri-food system, researchers and technicians must continue with research and thus be able to clarify gaps in knowledge. Our future research must be judicious with high scientific and health rigor. On the one hand, we must be judicious with the conclusions of laboratory, greenhouse, or field experiments. From now on, we cannot recommend the use of NMs as an ally of agriculture, when the experiments were carried out in controlled conditions and not in the field.

Governments around the world must also create tougher laws to regulate the release of nanometric products and thus minimize or eliminate the risks caused by NMs. The reality is that, at an international level, there are no clear guidelines for the labeling of products based on nanometric materials. Although there are guidelines in the cosmetics and drug industry, as far as we know, little progress has been made in the agri-food sector. Therefore, NMs are undoubtedly present and play a role in the agri-food industry, where it seems that this innovative technology is expected to allow for the production of sufficient and healthy products. However, it is necessary to continue research and clarify doubts to achieve a sustainable agri-food system.

References

1. Abbas Q, Yousaf B, Ali MU, Munir MAM, El-Naggar A, Rinklebe J, Naushad M (2020) Transformation pathways and fate of engineered nanoparticles (ENPs) in distinct interactive environmental compartments: a review. Environ Int 138:105646
2. Abbas Q, Yousaf B, Ullah H, Ali MU, Ok YS, Rinklebe J (2020) Environmental transformation and nano-toxicity of engineered nano-particles (ENPs) in aquatic and terrestrial organisms. Crit Rev Environ Sci Technol 50(23):2523–2581
3. Ameen F, Alsamhary K, Alabdullatif JA, ALNadhari S (2021) A review on metal-based nanoparticles and their toxicity to beneficial soil bacteria and fungi. Ecotoxicol Environ Saf 213:112027
4. Annu AA, Gadkari R, Sheikh JN, Ahmed S (2019) Phytomediated synthesis of cerium oxide nanoparticles and their applications. In: Nanomaterials and plant potential. https://doi.org/10.1007/978-3-030-05569-1_10
5. Bele S, Samanidou V, Deliyanni E, Orth ES, Ferreira JGL, Fonsaca JES, Blaskievicz SF, Domingues SH, Dasgupta A, Terrones M, Zarbin AJG, Quast K, Pat-Espadas AM, Field JAJAJA, Otero-Gonzalez L, Razo-Flores E, Cervantes FJFJJ, Sierra-Alvarez R, Toral-Sánchez E, Hull MS et al (2014) Chemical engineering research and design effect of the reduction degree of graphene oxide on the adsorption of bisphenol A. J Hazard Mater 9(1)
6. Bolan S, Sharma S, Mukherjee S, Zhou P, Mandal J, Srivastava P, Hou D, Edussuriya R, Vithanage M, Truong VK (2024) The distribution, fate, and environmental impacts of food additive nanomaterials in soil and aquatic ecosystems. Sci Total Environ 170013
7. Bolaños-Guerrón D, Galarza I, Llumiquinga E, Toulkeridis T, Cumbal L (2023) Design and construction of a prototype for arsenic retention in mining-contaminated waters by application of nanoparticles-based technosols. Sustainability 15:1286

8. Bondarenko O, Ivask A, Käkinen A, Kahru A (2012) Sub-toxic effects of CuO nanoparticles on bacteria: kinetics, role of Cu ions and possible mechanisms of action, 169. https://doi.org/10.1016/j.envpol.2012.05.009
9. Borgatta JR, Lochbaum CA, Elmer WH, White JC, Pedersen JA, Hamers RJ (2021) Biomolecular corona formation on CuO nanoparticles in plant xylem fluid. Environ Sci Nano 8(4). https://doi.org/10.1039/d1en00140j
10. Cao C, Ding ZH, Huang J, Yan CN (2024) Comprehensive response of microbes to Ag and Ag_2S nanoparticles and silver spatial distribution in constructed wetlands. Sci Total Environ 906:167683
11. Cherati SR, Shanmugam S, Pandey K, Khodakovskaya MV (2021) Whole-transcriptome responses to environmental stresses in agricultural crops treated with carbon-based nanomaterials. https://doi.org/10.1021/acsabm.1c00108
12. Dai Y, Zhao J, Liu X, Yu X, Jiang Z, Bu Y, Xu Z, Wang Z, Zhu X, Xing B (2019) Transformation and species identification of CuO nanoparticles in plant cells (*Nicotiana tabacum*). Environ Sci Nano 6(9). https://doi.org/10.1039/c9en00781d
13. Dang F, Huang Y, Wang Y, Zhou D, Xing B (2021) Transfer and toxicity of silver nanoparticles in the food chain. Environ Sci Nano 8(6):1519–1535. https://doi.org/10.1039/D0EN01190H
14. Das K, Dey D, Mondal P (2024) The headway of nanotechnologymediated approaches toward environmental clean-up. In: Environmental contaminants. Apple Academic Press, pp 277–313
15. El-Ramady H, Brevik EC, Fawzy ZF, Elsakhawy T, El-Dein Omara A, Amer M, Faizy SE-D, Abowaly M, El-Henawy A, Kiss A, Törős G, Prokisch J, Ling W (2022) Nano-restoration for sustaining soil fertility: a pictorial and diagrammatic review article
16. Farooq A, Khan I, Shehzad J, Hasan M, Mustafa G (2024) Proteomic insights to decipher nanoparticle uptake, translocation, and intercellular mechanisms in plants. Environ Sci Pollut Res 31:12. https://doi.org/10.1007/s11356-024-32121-7
17. Feigl G (2023) The impact of copper oxide nanoparticles on plant growth: a comprehensive review. J Plant Interact 18:1. https://doi.org/10.1080/17429145.2023.2243098
18. Feng Y, Wang C, Chen F, Cao X, Wang J, Yue L, Wang Z (2023) Cerium oxide nanomaterials improve cucumber flowering, fruit yield and quality: the rhizosphere effect. Environ Sci Nano 10(8). https://doi.org/10.1039/d3en00213f
19. Fernández-Luqueño F, Vázquez-Núñez E, Pérez-Hernandez H (2023) Biophysicochemical transformations of ENMs in soil. In: Physicochemical interactions of engineered nanoparticles and plants. Elsevier, pp 89–114
20. Garner KL, Suh S, Keller AA (2017) Assessing the risk of engineered nanomaterials in the environment: development and application of the nanofate model. Environ Sci Technol 51(10):5541–5551. https://doi.org/10.1021/acs.est.6b05279
21. Ge Z, Lu X (2023) Impacts of extracellular polymeric substances on the behaviors of micro/nanoplastics in the water environment. Environ Pollut, 122691
22. Gonçalves SPC, Delite FDS, Côa F, Neto LLR, da Silva GH, de Sá Bortolozzo L, Ferreira AG, de Medeiros AMZ, Strauss M, Martinez DST (2019) Biotransformation of nanomaterials in the soil environment: nanoecotoxicology and nanosafety implications. In: Nanomaterials applications for environmental matrices: water, soil and air. https://doi.org/10.1016/B978-0-12-814829-7.00007-0
23. Grillo R, Fraceto LF (2022) Impacts of magnetic iron oxide nanoparticles in terrestrial and aquatic environments. In: Williams MA (ed) Toxicology of nanoparticles and nanomaterials in human, terrestrial and aquatic systems, 1st ed. Wiley, pp 147–164. https://doi.org/10.1002/9781119316329.ch7
24. Guo M, Tong H, Cai D, Zhang W, Yuan P, Shen Y, Peng C (2022) Effect of wetting-drying cycles on the Cu bioavailability in the paddy soil amended with CuO nanoparticles. J Hazard Mater 436:129119
25. Holghoomi R, Colagar AH (2024) Applications of biocompatible nanoparticles in plant biotechnology for enhanced secondary metabolite biosynthesis. Inorg Chem Commun, 112753
26. Hong J, Wang L, Sun Y, Zhao L, Niu G, Tan W, Rico CM, Peralta-Videa JR, Gardea-Torresdey JL (2016) Foliar applied nanoscale and microscale CeO_2 and CuO alter cucumber (*Cucumis*

sativus) fruit quality. Sci Total Environ, 563–564. https://doi.org/10.1016/j.scitotenv.2015.08.029
27. Huang L, Jin Y, Zhou D, Liu L, Huang S, Zhao Y, Chen Y (2022) A review of the role of extracellular polymeric substances (EPS) in wastewater treatment systems. Int J Environ Res Public Health 19:12191
28. Hui C, Zhang Y, Ni X, Cheng Q, Zhao Y, Zhao Y, Du L, Jiang H (2021) Interactions of iron-based nanoparticles with soil dissolved organic matter: adsorption, aging, and effects on hexavalent chromium removal. J Hazard Mater 406:124650
29. Iavicoli I, Leso V, Beezhold DH, Shvedova AA (2017) Nanotechnology in agriculture: opportunities, toxicological implications, and occupational risks. Toxicol Appl Pharmacol 329:96–111. https://doi.org/10.1016/j.taap.2017.05.025
30. Jasim SA, Kzar HH, Sivaraman R, Jweeg MJ, Zaidi M, Alkadir OKA, Fahim FS, Aldulaim AKO, Kianfar E (2022) Engineered nanomaterials, plants, plant toxicity and biotransformation: a review. Egypt J Chem 65(12). https://doi.org/10.21608/EJCHEM.2022.131166.5775
31. Jorepalli S, Annavaram V, Reddy RRN, Boya VKN, Koduru JR, Somala AR (2024) Fabrication of carbon-based nanomaterials for wastewater treatment. In: Water treatment using engineered carbon nanotubes. Elsevier, pp 97–129
32. Jośko I, Dobrzyńska J, Dobrowolski R, Kusiak M, Terpiłowski K (2020) The effect of pH and ageing on the fate of CuO and ZnO nanoparticles in soils. Sci Total Environ 721:137771
33. Kah M, Tufenkji N, White JC (2019) Nano-enabled strategies to enhance crop nutrition and protection. Nat Nanotechnol 14(6):532–540
34. Kang J, Zhou N, Zhang Y, Wang Y, Song C, Gao X, Song G, Guo J, Huang L, Ma T (2023) Synthesis, multi-site transformation fate and biological toxicity of silver nanoparticles in aquatic environment: a review. Environ Technol Innov 32:103295
35. Khan ST, Adil SF, Shaik MR, Alkhathlan HZ, Khan M, Khan M (2022) Engineered nanomaterials in soil: their impact on soil microbiome and plant health. Plants 11(1):1–25. https://doi.org/10.3390/plants11010109
36. Klaes B, Thiele-Bruhn S, Wörner G, Höschen C, Mueller CW, Marx P, Arz HW, Breuer S, Kilian R (2023) Iron (hydr) oxide formation in Andosols under extreme climate conditions. Sci Rep 13:2818
37. Li S, Wang B, Zhang X, Wang H, Yi Y, Huang X, Gao X, Zhu P, Han W (2023) Soil particle aggregation and aggregate stability associated with ion specificity and organic matter content. Geoderma 429:116285
38. Li X, Li J, Wang Z, Bol R, Zou H (2024) Heteroaggregation of carbon nanomaterials with mineral-based nanomaterials: a review. J Environ Chem Eng, 113594
39. Lin J, Ma K, Chen H, Chen Z, Xing B (2022). Influence of different types of nanomaterials on soil enzyme activity: a global meta-analysis. Nano Today 42. https://doi.org/10.1016/j.nantod.2021.101345
40. Liu JW, Li YH, Yang TX, Qi J, Chen LL, Wu HH (2023) CeO_2 nanoparticles regulate the homeostasis of reactive oxygen species and nitric oxide levels to improve rice drought tolerance. Chin J Biochem Mol Biol 39(7). https://doi.org/10.13865/j.cnki.cjbmb.2023.05.1006
41. Liu M, Feng S, Ma Y, Xie C, He X, Ding Y, Zhang J, Luo W, Zheng L, Chen D (2019) Influence of surface charge on the phytotoxicity, transformation, and translocation of CeO_2 nanoparticles in cucumber plants. ACS Appl Mater Interfaces 11(18):16905–16913
42. López-Moreno ML, Cedeño-Mattei Y, Bailón-Ruiz SJ, Vazquez-Nuñez E, Hernandez-Viezcas JA, Perales-Pérez OJ, De la Rosa G, Peralta-Videa JR, Gardea-Torresdey JL (2018) Environmental behavior of coated NMs: physicochemical aspects and plant interactions. J Hazard Mater 347:196–217
43. Lowry GV, Espinasse BP, Badireddy AR, Richardson CJ, Reinsch BC, Bryant LD, Bone AJ, Deonarine A, Chae S, Therezien M (2012) Long-term transformation and fate of manufactured Ag nanoparticles in a simulated large scale freshwater emergent wetland. Environ Sci Technol 46:7027–7036
44. Ma Y, He X, Zhang P, Zhang Z, Guo Z, Tai R, Xu Z, Zhang L, Ding Y, Zhao Y (2011) Phytotoxicity and biotransformation of La_2O_3 nanoparticles in a terrestrial plant cucumber (*Cucumis sativus*). Nanotoxicology 5(4):743–753

45. Maharramov AM, Hasanova UA, Suleymanova IA, Osmanova GE, Hajiyeva NE (2019) The engineered nanoparticles in food chain: potential toxicity and effects. SN Appl Sci 1(11):1362. https://doi.org/10.1007/s42452-019-1412-5
46. Mallakpour S, Hussain CM (2021) Environmental applications of carbon nanomaterials-based devices. Wiley Online Library
47. Medhi S, Chowdhury S, Dehury R, Khaklari GH, Puzari S, Bharadwaj J, Talukdar P, Sangwai JS (2024) Comprehensive review on the recent advancements in nanoparticle-based drilling fluids: properties, performance, and perspectives. Energy Fuels 38:13455–13513
48. Mgadi K, Ndaba B, Roopnarain A, Rama H, Adeleke R (2024) Nanoparticle applications in agriculture: overview and response of plant-associated microorganisms
49. Milewska-Hendel A, Zubko M, Stróż D, Kurczyńska EU (2019) Effect of nanoparticles surface charge on the Arabidopsis thaliana (L.) roots development and their movement into the root cells and protoplasts. Int J Mol Sci 20(7). https://doi.org/10.3390/ijms20071650
50. Mirbakhsh M (2023) Role of nano-fertilizer in plants nutrient use efficiency (NUE)-a mini-review
51. Mukhtar M, Zeeshan M, Saba M, Saghir A, Ayub R (2023) Functionalization of green synthesized bionanomaterials. In: Synthesis of bionanomaterials for biomedical applications. https://doi.org/10.1016/B978-0-323-91195-5.00019-2
52. Nasser F, Constantinou J, Lynch I (2020) Nanomaterials in the environment acquire an "Eco-Corona" impacting their toxicity to *Daphnia magna*-a call for updating toxicity testing policies. Proteomics 20:1800412
53. Ogunkunle CO, Awe AA, Orosun MM, Ite AE, Okoro HK, Komolafe ET, Fatoba PO (2024) Nanoremediation of inorganic contaminants in soils: applications, mechanisms of action, and environmental implications. In: Nanotechnology to monitor, remedy, and prevent pollution. Elsevier, pp 569–596
54. Perea Velez YS, Carrillo-González R, Gonzalez-Chavez MDCA (2021) Interaction of metal nanoparticles–plants–microorganisms in agriculture and soil remediation. J Nanopart Res 23(9):206
55. Pérez-Hernández H, Pérez-Moreno A, Sarabia-Castillo CR, García-Mayagoitia S, Medina-Pérez G, López-Valdez F, Campos-Montiel RG, Jayanta-Kumar P, Fernández-Luqueño F (2021) Ecological drawbacks of nanomaterials produced on an industrial scale: collateral effect on human and environmental health. Water Air Soil Pollut 232(10):435. https://doi.org/10.1007/s11270-021-05370-2
56. Rajput VD, Kumari S, Minkina T, Sushkova S, Mandzhieva S (2023) Nano-enhanced microbial remediation of PAHs contaminated soil. Air Soil Water Res 16:11786221231170100. https://doi.org/10.1177/11786221231170099
57. Rani S, Kumari N, Sharma V (2023) Uptake, translocation, transformation and physiological effects of nanoparticles in plants. Arch Agron Soil Sci 69:1579–1599
58. Saleem H, Javaid Zaidi S (2020) Recent developments in the application of nanomaterials in agroecosystems
59. Schwarzfischer M, Rogler G (2022) The intestinal barrier—shielding the body from nano- and microparticles in our diet. Metabolites 12(3):223. https://doi.org/10.3390/metabo12030223
60. Sharma PK, Dhiman A, Anand S, Rai PK (2023) Fate of TiO_2 nanoparticles in the environment: a review on the transport and retention behavior in the soil compartment. New J Chem 47:4145–4165
61. Sharma, R., & Kumar, V. (2024). Nano enabled agriculture for sustainable soil. Waste Management Bulletin.
62. Shende SS, Rajput VD, Gorovtsov AV, Harish, Saxena P, Minkina TM, Kizilkaya R (2021). Interaction of nanoparticles with microbes. Plant-Microbes-Engineered Nano-particles (PM-ENPs) nexus in agro-ecosystems: understanding the interaction of plant, microbes and Engineered Nano-particles (ENPS), 175–188
63. Siddiqui MH, Al-Whaibi MH, Mohammad F (2015) Nanotechnology and plant sciences: nanoparticles and their impact on plants. In: Nanotechnology and plant sciences: nanoparticles and their impact on plants. https://doi.org/10.1007/978-3-319-14502-0

64. Spurgeon DJ, Lahive E, Schultz CL (2020) Nanomaterial transformations in the environment: effects of changing exposure forms on bioaccumulation and toxicity. Small 16:2000618
65. Staroń A, Długosz O, Pulit-Prociak J, Banach M (2020) Analysis of the exposure of organisms to the action of nanomaterials. Materials 13(2):349. https://doi.org/10.3390/ma13020349
66. Stewart ZP, Pierzynski GM, Middendorf BJ, Prasad PV (2020) Approaches to improve soil fertility in sub-Saharan Africa. J Exp Bot 71(2):632–641
67. Su Y, Ashworth V, Kim C, Adeleye AS, Rolshausen P, Roper C, White J, Jassby D (2019) Delivery, uptake, fate, and transport of engineered nanoparticles in plants: a critical review and data analysis. Environ Sci Nano 6(8):2311–2331
68. Suazo-Hernández J, Arancibia-Miranda N, Mlih R, Cáceres-Jensen L, Bolan N, Mora MDLL (2023) Impact on some soil physical and chemical properties caused by metal and metallic oxide engineered nanoparticles: a review. Nanomaterials 13:572
69. Talebi SM (2024) Uptake of nanomaterials by plants and translocation within plants. Microbiome Nano-Cross-Talk, 19–41. https://doi.org/10.1016/B978-0-443-18822-0.00002-5
70. Tun Nisa R, Bashir M, Bahar F, Anwar Bhat M, Singh L, Mahdi S, Palmo T, Rehman U (2022). Nanotechnology in agriculture: prospects and problems. Researchgate.Net 14(1)
71. Uddin MN, Desai F, Asmatulu E (2020) Engineered nanomaterials in the environment: bioaccumulation, biomagnification and biotransformation. Environ Chem Lett 18(4). https://doi.org/10.1007/s10311-019-00947-0
72. Ullah A, Lim SI (2022) Plant extract-based synthesis of metallic nanomaterials, their applications, and safety concerns. Biotechnol Bioeng 119(9):2273–2304
73. Ur Rahim H, Qaswar M, Uddin M, Giannini C, Herrera ML, Rea G (2021) Nano-enable materials promoting sustainability and resilience in modern agriculture. Nanomaterials 11(8). https://doi.org/10.3390/nano11082068
74. Usman M, Farooq M, Wakeel A, Nawaz A, Cheema SA, Rehman HU, Ashraf I, Sanaullah M (2020) Nanotechnology in agriculture: current status, challenges and future opportunities. Sci Total Environ 721. https://doi.org/10.1016/j.scitotenv.2020.137778
75. Vázquez-Núñez E (2023) Uso de nanomateriales en la agricultura y sus implicaciones ecológicas y ambientales. Mundo nano. Revista interdisciplinaria en nanociencias y nanotecnología 16(30)
76. Vencalek BE, Laughton SN, Spielman-sun ER, Rodrigues M, Unrine JM, Lowry GV, Gregory KB (2016) In situ measurement of CuO and Cu $(OH)_2$ nanoparticle dissolution rates in quiescent freshwater mesocosms. https://doi.org/10.1021/acs.estlett.6b00252
77. Vera-Reyes I, Vázquez-Núñez E, Castellano LE, Bautista DIA, Soto JHV, Valle-García JD (2023) Biointeractions of plants–microbes–engineered nanomaterials. Physicochemical interactions of engineered nanoparticles and plants, 201–231
78. Wang F, Adams CA, Shi Z, Sun Y (2018) AC soil with Cd. ECSN. https://doi.org/10.1016/j.chemosphere.2018.06.099
79. Wang M, Gao B, Tang D, Sun H, Yin X, Yu C (2018) Effects of temperature on aggregation kinetics of graphene oxide in aqueous solutions. Colloids Surf A Physicochem Eng Asp 538:63–72
80. Wang X, Luo H, Zheng W, Wang X, Xiao H, Zheng Z (2021) Effects of polydopamine microspheres loaded with silver nanoparticles on lolium multiflorum: bigger size, less toxic. Toxics 9(7). https://doi.org/10.3390/toxics9070151
81. Wei L, Liu J, Jiang G (2024) Nanoparticle-specific transformations dictate nanoparticle effects associated with plants and implications for nanotechnology use in agriculture. Nat Commun 15(1):7389
82. Wu F, You Y, Werner D, Jiao S, Hu J, Zhang X, Wan Y, Liu J, Wang B, Wang X (2020) Carbon nanomaterials affect carbon cycle-related functions of the soil microbial community and the coupling of nutrient cycles. J Hazard Mater 390:122–144
83. Wu H, Li Z (2022) Nano-enabled agriculture: how do nanoparticles cross barriers in plants? Plant Commun 3(6). https://doi.org/10.1016/j.xplc.2022.100346
84. Wu J, Bosker T, Vijver MG, Peijnenburg WJGM (2021) Trophic transfer and toxicity of (mixtures of) Ag and TiO_2 nanoparticles in the lettuce-terrestrial snail food chain. Environ Sci Technol 55(24):16563–16572. https://doi.org/10.1021/acs.est.1c05006

85. Wu J, Sun J, Bosker T, Vijver MG, Peijnenburg WJGM (2023) Toxicokinetics and particle number-based trophic transfer of a metallic nanoparticle mixture in a terrestrial food chain. environmental science & technology, acs.est.2c07660. https://doi.org/10.1021/acs.est.2c07660
86. Wu S, Gaillard JF, Gray KA (2021) The impacts of metal-based engineered nanomaterial mixtures on microbial systems: a review. Sci Total Environ
87. Yin L, Cheng Y, Espinasse B, Colman BP, Auffan M, Wiesner M, Rose J, Liu J, Bernhardt ES (2011) More than the ions: the effects of silver nanoparticles on *Lolium multiflorum.* Environ Sci Technol 45(6). https://doi.org/10.1021/es103995x
88. Zhang M, Tao S, Wang X (2020) Interactions between organic pollutants and carbon nanomaterials and the associated impact on microbial availability and degradation in soil: a review. Environ Sci Nano 7:2486–2508
89. Zhang P, Dai W, Dong C, Zhang Z (2022) Nanomaterial transformation in root–soil interface: a function of root exudate or microbial activity? In: Plant exposure to engineered nanoparticles. Elsevier, pp 209–237

85. Wu J, Sun J, [illegible] (2023) Toxicokinetics and particle number-based trophic transfer of a metallic nanoparticle mixture in a terrestrial food chain. Environmental science & technology, acs.est.2c07660. https://doi.org/10.1021/acs.est.2c07660
86. Wu S, Gaillard JF, Gray KA (2021) The impacts of metal-based engineered nanomaterial mixtures on microbial systems: a review. Sci Total Environ [illegible]
87. Yin L, Cheng Y, Espinasse B, Colman BP, Auffan M, Wiesner M, Rose J, Liu J, Bernhardt ES (2011) More than the ions: the effects of silver nanoparticles on Lolium multiflorum. Environ Sci Technol 45(6). https://doi.org/10.1021/es103995x
88. Zhang M, Zhu S, Wang X (2020) Interactions between organic pollutants and carbon nanomaterials and the associated impact on microbial availability and degradation in soil: a review. Environ Sci Nano 7:2486–2508
89. Zhang P, Du W, Guo Z, Zhang Z (2022) Nanomaterial transformation in rhizosphere: interaction function of root exudate or microbial activity? In: Plant exposure to engineered nanoparticles. Elsevier, pp 200–217

Nanomaterials in Food Chain

Edgar Vázquez-Núñez, Gabriela Medina-Pérez, Hermes Pérez-Hernández, Alma H. Serafín-Muñoz, and Ileana Vera-Hernández

Abstract Nanotechnology is transforming the food industry by offering innovative solutions to enhance production, preservation, safety, and sustainability. Key materials such as nanoparticles, nanocomposites, nanoemulsions, and carbon-based structures drive advancements in intelligent packaging, nutrient delivery, and quality monitoring. Biodegradable and renewable nanocomposites provide eco-friendly alternatives to conventional packaging, while nanosensors enable real-time monitoring of freshness and contamination, ensuring longer shelf life and enhanced safety. Despite its immense potential, nanotechnology raises concerns about nanotoxicity, bioaccumulation, and environmental impacts. Robust regulatory frameworks and thorough risk assessments are essential to address these challenges and support responsible innovation. Public education and transparent labeling play a pivotal role in fostering trust and acceptance, promoting informed decision-making. Future directions emphasize the development of nanosensors for precise quality assessments, sustainable nanomaterials for packaging, and targeted nutrient delivery systems for personalized nutrition. These innovations hold the promise of optimizing food systems while addressing critical health and environmental challenges. Finally, this chapter underscores the dual promise and challenge of integrating nanotechnology

E. Vázquez-Núñez (✉)
Grupo de Investigación Sobre Aplicaciones Nano y Biotecnológicas Para La Sostenibilidad (NanoBioTs), Departamento de Ingenierías Química, Electrónica y Biomédica, División de Ciencias E Ingenierías, Universidad de Guanajuato, León, Mexico
e-mail: edgar.vazquez@ugto.mx; ed.vazquezn@gmail.com

G. Medina-Pérez
ICAP—Instituto de Ciencias Agropecuarias, Universidad Autónoma del Estado de Hidalgo, Tulancingo, México

H. Pérez-Hernández
Departamento de Botánica, Universidad Autónoma Agraria Antonio Narro, Saltillo, México

A. H. Serafín-Muñoz
Department of Civil Engineering, University of Guanajuato, Guanajuato, México

I. Vera-Hernández
Centro de Investigación en Química Aplicada (CIQA), Departamento de Biociencias y Agrotecnología, Saltillo, México

© The Author(s), under exclusive license to Springer Nature Singapore Pte Ltd. 2025
A. Juárez-Maldonado et al. (eds.), *Plant Biostimulation with Nanomaterials*, Smart Nanomaterials Technology, https://doi.org/10.1007/978-981-96-4648-7_15

into the food sector. While it highlights the transformative potential of nanomaterials to enhance food quality, sustainability, and safety, it also emphasizes the need for a balanced approach that considers environmental and health impacts. By fostering interdisciplinary collaboration, advancing regulatory frameworks, and prioritizing transparent communication, nanotechnology can bridge critical gaps in global food systems. This synthesis provides a compelling roadmap for researchers, policymakers, and industry leaders to navigate the opportunities and complexities of nanotechnology, paving the way for a more innovative, sustainable, and equitable future in food science.

Keywords Environmental risks · Food packaging · Food quality · Human health · Nanotoxicity

1 Introduction

In recent decades, nanotechnology has emerged as a transformative force that transcends traditional boundaries and has revolutionized diverse industries. One such field that has witnessed the profound impact of nanotechnology is the realm of food systems. This chapter delves into the intersection of nanotechnology and food, exploring myriad ways in which nanoscale science and engineering are reshaping the landscape of food production, processing, and consumption.

Nanotechnology, at its core, deals with materials and structures at the nanoscale, typically less than 100 nm. In the context of food, this translates into manipulating and controlling matter in dimensions that are on par with fundamental building blocks of life. In this chapter, we will unravel the potential of nanotechnology to enhance the quality, safety, and nutritional value of food products. From novel delivery systems for nutrients to intelligent packaging that monitors freshness, nanotechnology applications in the food industry are as diverse as they are promising.

To address the opportunities afforded by nanotechnology in food, it is imperative to acknowledge the associated challenges and ethical considerations. Concerns regarding the potential toxicity of nanomaterials and their impact on human health and the environment are significant. Striking a balance between harnessing the benefits of nanotechnology and ensuring its responsible and sustainable integration into food systems is a critical aspect that requires careful examination.

This chapter provides a comprehensive overview of the current state of nanotechnology in food systems. It begins by laying the foundation for exploring the basic principles of nanotechnology and its relevance to food science. Subsequent sections will delve into specific applications, ranging from nanoencapsulation of bioactive compounds to the development of smart sensors in food packaging. Through this exploration, we aim to foster a deeper understanding of the transformative potential of nanotechnology while also fostering a nuanced perspective on the challenges and ethical considerations associated with its implementation.

As we start this analysis through the nanoscale intricacies of our food, the chapter aims to provide researchers, industry professionals, and enthusiasts alike with a comprehensive and insightful exploration of the dynamic and evolving field of nanotechnology in food systems. The narrative unfolds against the backdrop of an ever-changing global food landscape, where innovation and responsible science are intertwined to shape the future of what we eat and how we produce it.

2 Types of Nanomaterials Used in Food Industry

The integration of nanotechnology into the food industry has ushered in a new era of innovation, offering a spectrum of nanomaterials that contribute to the enhancement of various aspects of food production, preservation, and packaging. Understanding these diverse nanomaterials is pivotal for understanding their impact on the food landscape. The following are the key types of nanomaterials commonly employed in the food industry.

2.1 Nanoparticles

Nanoparticles are a class of nanomaterials with dimensions ranging from 1 to 100 nm. Metal nanoparticles, such as silver and gold, are known for their antimicrobial properties. In food packaging, they serve as agents that inhibit the growth of bacteria and fungi, thereby effectively extending the shelf life of perishable goods. Metal oxide nanoparticles, including zinc oxide and titanium dioxide, have been utilized to enhance packaging by providing UV protection and acting as barriers against oxygen and moisture [47].

2.2 Nanocomposites

Nanocomposites involve the incorporation of nanomaterials such as clay nanoparticles into packaging materials [87]. This integration enhances mechanical strength, barrier properties, and thermal stability, contributing to improved packaging performance and reduced environmental impact.

2.3 Nanoemulsions

Nanoemulsions consist of droplets of one liquid dispersed within another that are stabilized by surfactants. In the food industry, nanoemulsions play a crucial role in the encapsulation of bioactive compounds, vitamins, and flavors [21]. This encapsulation improves solubility and bioavailability and enhances the sensory attributes of food products.

2.4 *Nanocapsules*

Nanocapsules are nano-sized carriers designed to encapsulate active ingredients such as vitamins, antioxidants, and flavors. These capsules protect encapsulated substances, ensuring controlled release and targeted delivery within food products [4].

2.5 *Carbon-Based Nanomaterials*

Carbon nanotubes and graphene are prominent carbon-based nanomaterials with applications in the food industry [74]. Carbon nanotubes enhance the mechanical strength and barrier properties of food packaging materials [109]. Graphene, with its excellent thermal and electrical conductivities, holds promise for applications in food sensors, facilitating the precise monitoring of food quality parameters [94].

2.6 *Nanostructured Lipids*

Nanostructured lipids, including nanostructured lipid carriers (NLCs) and solid lipid nanoparticles (SLNs), play a significant role in encapsulating lipophilic bioactive compounds. These lipid-based nanoparticles enhance the stability and bioavailability of encapsulated compounds in food formulations [99].

2.7 *Nanoencapsulation Materials*

Liposomes and cyclodextrins are essential nanoencapsulating materials. Liposomes, which consist of lipid bilayers, serve as carriers of both hydrophilic and lipophilic bioactive compounds. Cyclodextrins form inclusion complexes with specific molecules, providing stability and the controlled release of encapsulated substances [66].

2.8 *Quantum Dots*

Quantum dots (QDs), which are semiconductor nanocrystals, are employed for food safety and traceability. They serve as labels for tracking and detecting contaminants or pathogens, contributing to improved monitoring and quality control in the food industry [10].

In utilizing these diverse nanomaterials, the food industry aims not only to enhance product quality and safety but also to address challenges related to food preservation and sustainability. It is imperative to approach their application with a comprehensive understanding of the potential implications and adhere to regulatory guidelines to ensure the responsible and safe integration of nanotechnology into the realm of food production. Relevant examples of the use of nanomaterials in diverse aspects in the food sector are shown in Table 1.

3 Nanomaterials in Food Production

Research on NMs in food production continues to increase. In fact, practical applications in protected agricultural systems and open fields are already a reality, a situation that was not believed to be possible a decade ago. Currently, simple nanomaterials and nanoformulations have been produced for applications such as fertilizers, biostimulants, pesticides, herbicides, and other products to improve the physicochemical characteristics of plants and the quality of edible parts [72].

In general, NMs are classified into three categories: inorganic NMs, nacompounds (i.e., a combination of inorganic NMs with organic NMs), and organic NMs [17]. In the following paragraphs, we describe the use of various inorganic NPs, including metallic NPs and metal oxides, for food production.

3.1 *Zinc Nanoparticles*

Zn is an essential micronutrient for plants. In general, it participates in the synthesis of pigments, increases plant biomass, and activates antioxidant enzymes. Zn mainly promotes the repair of oxidative stress in plants grown under abiotic stress [86].

ZnO NPs are one of the most studied and applied ways to improve crop performance. The application of ZnO NPs (0.12 g pot^{-1}) in wheat crops (*Triticum aestivum* L.) under salt stress improved the chlorophyll content, shoot length, fresh and dry weight of the spike, and increased yield compared to plants treated with conventional ZnO [3]. Regarding foliar application, treatments with ZnO NPs at a concentration of 100 ppm improved the yield of the tomato crop compared to the control [5].

In general, the effects on crop yields indicate that low concentrations have provided better results, as higher concentrations have a toxic or inhibitory effect.

Table 1 Nanomaterials in food sector: types, functions, and results

Nanomaterials type	Function	Results	References
Nanoparticles			
Ag/TiO2 mixed nanoparticles	Antimicrobial activity	The activity index (%) of the antibacterial activity at sample blend + (0.3%) Ag + (0.8%) TiO_2 of the *E. coli*, *S. aureus*, C. Albicans, *A. niger* was 32, 45.8, 77.8, and 92 (%) respectively	[1]
Dextran-coated silver NPS	Antimicrobial activity and Minimization of oxygen penetration and wettability	It was significantly reduced Oxygen Transmission Rates (from 2.07 to 1.40–0.78 $cm^3\ m^{-2} d^{-1}$) and hydrophilicity (from 20.8° to 52.4° for MilliQ water, and from 35–37° to 62–74° for 3% acetic acid and 0.9% NaCl simulant solutions), yielding a 99.9% inhibition of *Escherichia coli* after five repeated cycles of 24 h exposure	[53]
Hydroxypropyl starch/ polyvinyl alcohol loaded with Zn NPs	Bio-based active food package	The prepared bioactive packaging film loaded with ZnONPs (4H) exhibited superior antibacterial activity among other films against *Escherichia coli* ATCC 25,922 and *Staphylococcus aureus* ATCC 25,923 with inhibition zones 15.1 ± 0.76 and 12.1 ± 0.71 mm respectively. Correspondingly, packing film 4H exhibited potential antifungal activity toward *Aspergillus niger* RCMB 02,724, *A. flavus* RCMB 02,782, *Penicillium expansum* IMI 89,372 and *Fusarium oxysporum* RCMB 001,004 with inhibition zones (16 ± 1.0, 22 ± 0.90, 18.0 ± 1.1 and12.3 ± 0.57 mm respectively)	[40]

(continued)

Table 1 (continued)

Nanomaterials type	Function	Results	References
Polymer nanocomposite of sodium alginate (NaAlg)/polyethylene oxide) (PEO) doped by Ni/ZnO nanohybrid	Bioactive food packaging	The antibacterial activity of nanocomposites films was examined against Gram-negative (*Escherichia coli*), Gram-positive (*Staphylococcus aureus*), *Aspergillus niger* (*fungus*) and *Candida albicans* (*yeast*). Though the tested NaAlg/PEO blend did not indicate any activity against the microbial strains, when Ni/ZnO NPs were added, activity against *E. coli, S. aureus, fungus* and *yeast* was enhanced	[7]
Fish oil as bioactive compound encapsulated in lecithin sunflower oil	Improvement of the nutritional value	The technological and sensory evaluation indicated that enriched bread with nano-liposomal fish oil had a higher loaf volume of bread than other samples. The addition of nano-liposomal fish oil had no negative effect on bread textural quality and sensory acceptability. Oxidative stability against lipid oxidation. Positive effects on technological and sensory properties	[70]
α-Tocopherol nanoencapsulated in canola oil and twin 80	Antioxidant agent	The addition of 250 and 500 mg/kg tocopherol NEs not only improved fish sausages' quality during cold storage but also effectively retarded lipid oxidation, as evidenced by the significantly lower peroxide value and higher polyunsaturated fatty acid content. Omega-6 fatty acids of tocopherol NEs added sausages were maintained at 31.01% for NE250 and 29.59% for NE500 on day 16	[30]

(continued)

Table 1 (continued)

Nanomaterials type	Function	Results	References
Zein/chitosan complex particles stabilized emulsion to encapsulate vitamin D_3	Release of vitamin D_3	Emulsions with oil fractions 330 and 500 g kg^{-1} exhibited significant stability during storage, particularly the ones with 500 g kg^{-1} oil fractions which were stable against all the tested parameters. In vitro drug release profile confirmed sustained and controlled release of the encapsulated vitamin D_3	[92]
Solid lipid nanoparticles and nanostructured lipid carriers containing zeaxanthin	Carry zeaxanthin by using stable nanocarriers	The particle size (PS) of NLCs did not exceed 130 nm, but SLNs did show larger sizes. In addition, particle dispersion index (PDI) showed 0.23 ± 0.01 to 0.34 ± 0.01 and 0.23 ± 0.01 to 0.30 ± 0.01 for SLN and NLC, respectively. Glycerol mono-stearate containing zeaxanthin showed phase separation after 60 days at 4 and 25 °C, while this was not observed in nanocarriers produced by glycerol distearate	[73]
Fluorescent carbon dots (CDs) for determination of riboflavin in multivitamin supplements	Method to quantify the riboflavin in supplements	An efficient fluorescent resonance energy transfer (FRET) between the CDs and riboflavin was achieved with CDs acting as donor and riboflavin as acceptor. A linear relationship between FRET and the riboflavin concentration from 0.10 to 3.0 μg/mL was observed, allowing the development of an accurate and fast analytical method to determine this vitamin in multivitamin/mineral supplements	[63]

(continued)

Table 1 (continued)

Nanomaterials type	Function	Results	References
Sensor made of anthocyanin and TiO_2-doped carbon dots derived from sweet potato peels (SPP)	pH sensor film was fabricated using anthocyanin and carbon dots	The composite film added with SPP and 3 wt% Ti-CD showed 100% UV protection, superb antioxidant (100% DPPH and ABTS scavenging assay), and potent antibacterial activity (complete eradication of foodborne *L. monocytogenes* and *E. coli* strains after 3 h incubation)	[83]

Positive effects on yield have been reported for crops such as *Arachis hypogaea* L., *Cicer arietinum* L., *Brassica napus* L., *Lupinus termis* L., Brassica oleracea L., *and Solanum lycopersicum* L. [91].

3.2 *Iron Nanoparticles*

Iron is a microelement that participates in seed germination and plant growth, improves stress tolerance, is involved in the photosynthetic process, and is an enzyme cofactor [89]. The most commonly used forms in food production are magnetite (Fe_3O_4), maghemite (γ-Fe_2O_3), and hematite (Fe_2O_3). Zero-valent iron NPs have been successfully used for crop improvement. In soybean plants grown under drought and well-irrigated conditions, foliar spray of iron NPs (Fe_3O_4) at 200 ppm improved plant growth, yield, and seed quality compared to controls [26]. In flax plants (*Linum usitatissimum* L.), the application of iron NPs (Fe_2O_3) at 75 and 100 ppm increased agronomic parameters, such as stem diameter, stem length, fresh weight, and dry weight of the plant, as well as the yield and seeds per capsule [108]. Numerous studies have summarized the effects of iron NPs on crop production [39].

3.3 *Cooper Nanoparticles*

Copper is a micronutrient that participates in the regulation of plant growth and development, and influences the formation of chlorophyll and seed production [79]. It also participates in the regulation of enzymatic processes, respiration, and metabolism of proteins and carbohydrates. In addition, it helps plants tolerate drought stress. [9]. In its simple form, the most used NPs are Cu and CuO [75]. In the nanocomposites, the forms evaluated in the cultures were $CuCl_2$, $CuSO_4$, and $CuNO_3$.

In tomato plants (*S. lycopersicum*) conditioned under saline stress, foliar application of 250 mg L^{-1} Cu NPs mitigated the effect of stress. These results demonstrated an increase in antioxidant enzymes. Additionally, they promote growth, crop yield, and higher amounts of vitamin C and lycopene [77]. In contrast, in corn plants subjected to drought stress, Cu NPs mitigated stress. When a concentration of 450 mL of each pot of 69.4 μM of Cu NPs was applied in maize plants, the crop presented an increase in the number of seeds and grain yield. Furthermore, under drought stress, an increase in the content of chlorophyll and carotenoids and a decrease in reactive oxygen species (ROS) were observed compared to plants not treated with NPs [68].

3.4 Silicon Nanoparticles

Silicon is a non-metallic element and a micronutrient that participates in the growth and development of plants, particularly in the lignification of the cell wall and the production of antioxidant enzymes and, therefore, in the mitigation of damage caused by biotic and abiotic stresses. In addition, they improve the photosynthetic rate and promote nutrient absorption [113]. SiO_2 is the most commonly used form of foliar application for crops. For example, in *T. aestivum*, foliar and soil applications improved the photosynthetic rate and plant biomass, improved the quality of grains in *O. sativa*, increased chlorophyll levels and length of shoots and roots in *G. max*, increased the germination percentage in *P. vulgaris*, and improved the germination rate, plant vigor, and antioxidant defense system in *Z. mays*. Many other examples can be found in a study by [113].

However, research has reported that combinations of organic NPs with nanomaterials play a role in crop performance. An example of this is the chitosan-silicon (CS–Si) nanocomposite. [51] reported an increase in the antioxidant enzyme system and an increase in corn production. More information on the effects of Cu, Fe, Si, Zn NPs, and other NMs on crop production and protection can be found in other studies [4, 104].

It is important to note that not all NPs are essential elements for plants. However, numerous studies have shown that Ag, Ti, Au, and other NPs positively affect the growth and development of crops, mainly used as nanopesticides because of their antimicrobial properties.

3.5 Ag Nanoparticles

Although Ag NPs have been used to counteract disease problems, low concentrations of nanometric Ag have also shown improvements in seed germination, plant growth, mainly in the shoot and root growth, increased plant height, and an increase in biomass. Changes in chlorophyll levels, antioxidant systems, and other biochemical parameters have also been reported. The crops that these NPs have mostly been tested are *Vigna radiata* L, *Glycine max* L, *Zea mays* L, *Phaseolus vulgaris* L, *Triticum aestivum* L, *Vigna sinensis* L, *Brassica juncea* L, *Sorgo bicolor* L, *Lolium multiflorum* L, *and Pisum sativum* L. [60, 95].

3.6 Titanium Nanoparticles

Titanium is not an essential element for plants but has been shown to the soil or foliage increase plant length, dry weight, and seed germination [29]. They have also been documented to increase the content of chlorophyll and antioxidant enzymes. Studies have reported that they help mitigate salt stress in corn and wheat crops [65]. Although the supply of Ti NPs has had important effects on plants, low concentrations are usually adequate (10–200 mg L^{-1}, mg kg^{-1}, or ppm) because at high concentrations (>1000), as occurs with the application of other nanomaterials, plants show symptoms of toxicity [65]. Titanium evaluated with plants is the most commonly used titanium oxide (TiO_2) in anatase and rutile phases. Mattiello et al. [59] reported the effect of applications of TiO_2 NPs on edible plants. Promising effects were associated with root elongation, germination, and plant size. However, most of the reported studies were conducted under controlled conditions (laboratory), in contrast to in-field conditions. Therefore, various authors have reported that it is necessary to conduct evaluations under realistic conditions and at low concentrations, because this is where the best results have been observed [45].

TiO_2 NMs are effective in controlling disease-causing fungi and bacteria in edible crops. A study by [46] demonstrated that 40 mg L^{-1} TiO_2 NPs decreased the intensity of *Bipolaris sorokiniana*, which causes spot blotch disease in wheat. Additionally, researchers have observed an increase in morphological and biochemical parameters in wheat plants. In the document by [39, 104] can see more information about the effect of Ti NMs on various edible plants.

Other forms of NPs have been used in crop production and protection. A combination of two or more nanoparticles is known as a nanocomposite. Nanocomposites with nanomaterials have a core–shell structure with better optical, catalytic, and magnetic capabilities [50]. However, recent studies have shown that the combination of organic nanomaterials with NMs (Fe, Cu, Ag, Si, Zn, and Ti) promises better results than individual NMs. According to previous studies, simple NMs cause greater phytotoxicity. In such a way that for almost two decades, research has tried to demonstrate that biogenic materials are more compatible in crop production. Organic materials include chitosan (Ch), cellulose, starch, and carbon nanotubes [112].

Kaur et al. [46] demonstrated that compared to the control, the Ch-ZnO nanocomposite at a concentration of 100 ug ml reduced wilting in chickpea caused by Fusarium by almost 50%.

The use of Ch with other metals has been shown to improve crop growth because nanocomposites provide efficient delivery of minerals or participate indirectly in the absorption of nutrients. The combinations of Ch-Zn, Ch-Cu, Ch-K, Ch-Se, Ch-Fe, and Ch-Ag (applied foliar or soil) improved morphological characteristics and biochemical aspects in plants [112].

For the last decade, researchers have focused their attention on organic nanomaterials, mainly carbon nanotubes (CNTs). They are divided into two main groups: single-walled carbon nanotubes (SWCNTs) and multiwalled carbon nanotubes (MWCNTs). Owing to their particular effects on the growth of different crops and the promise of their potential use as intelligent distribution systems in plants, CNTs have been evaluated in laboratories, greenhouses, and crop fields, observing their effects as carriers of pesticides and fertilizers, or as simple nanomaterials that control pests and promise to be plant growth regulators. Hermes et al. [41] reported the effects caused by SWCNTs and MWCNTs. For example, compared to the control, CNTs at low concentrations (10–200 mg L^{-1}) increased the percentage of germination, fresh biomass, root size, and number of grains and fruits per plant in crops such as cucumber, tomato, corn, and soybean, among others. Other examples of positive effects of CNTs on plants have been found in recent reports. In addition, adverse effects at high concentrations have been reported [62].

In general, many nanomaterials continue to be evaluated for providing cleaner and more efficient agriculture for food production. It is expected that in the coming years, many nanomaterials and nanocomposites will have realistic applications in greenhouse and field production systems, demonstrating that they are non-toxic to plants, humans, and the environment, a topic that will be addressed in the next sections.

4 Food Quality

Nanotechnology serves as a pivotal tool for advancing and refining food quality, addressing aspects such as color, flavor, nutritional value, and texture, while mitigating the adverse effects of constituents such as excess fat, sugar, and salt in food products. This study delves into specific nanotechnological applications that demonstrate their influence on food characteristics.

Enzymes are frequently used to enhance the flavor and nutritional quality of foods. However, the effectiveness of certain enzymes may diminish because of their insolubility in water, resulting in nonuniform dispersion within the food matrix [82]. Nanotechnology offers a solution to this problem by providing suitable nanomaterials that can immobilize enzymes. This immobilization enhances dispersion within the food matrix, consequently amplifying the enzyme activity.

The development of nano-supplements, nano-additives, and nano-delivery systems represents a significant advancement in nanotechnology to improve food quality. When added to food products, nano-supplements enhance taste and texture while concurrently minimizing fat content, thereby yielding healthier food options [36].

Functional ingredients such as vitamins, antioxidants, flavors, preservatives, bioactive peptides, and probiotics contribute to the improvement of processed foods. These ingredients, which are often susceptible to degradation under processing conditions, are integrated into the delivery systems. Classification based on molecular and physical forms guides the application of strategies, such as immobilization, nanodispersion, association colloids, and nanoencapsulation, preserving the integrity of these ingredients and ensuring consistency in taste and texture [12].

Nano-colloidal and nanomicelle strategies have been adopted for the encapsulation and delivery of polar, nonpolar, and amphiphilic ingredients [36]. Additionally, sustained-release delivery systems, which are influenced by biological environments such as pH and temperature, employ various biopolymers and copolymers. Notable structures, such as cochleates, which are composed of a multilayer lipid bilayer in a spiral form, demonstrate efficacy in encapsulating bioactive materials with poor water solubility, hydrophilic molecules, drugs, peptides, and proteins [45].

Current research has also explored the application of graphene and metal chalcogenide nanoparticles as durable and efficient carriers of antioxidants, anti-browning agents, and enzymes, thereby contributing to the extension of the shelf life.

The convergence of lipids, proteins, and carbohydrate nanoparticles represents a next-generation delivery system. Recent investigations have suggested that nanomaterials derived from lipids, proteins, and carbohydrates can serve as efficient delivery systems for food ingredients [56, 85, 107]. Notably, ease of digestion in the gastrointestinal tract offers distinct advantages. Lipid nanoparticles, characterized by their optically transparent properties, can deliver hydrophobic bioactive ingredients, thereby acting as both carriers and sources of functional ingredients.

Natural protein nanoparticles, such as casein micelles, found in bovine milk and other dairy products, have been consumed since ancient times. Analogous to lipid nanoparticles, protein nanoparticles serve as delivery systems for color, flavor, preservatives, vitamins, and minerals. The fate of protein nanoparticles in the gastrointestinal tract is contingent upon their shape, size, and surface charge [67].

Finally, digestible and indigestible carbohydrate nanoparticles, structured as assembled polysaccharides, have been successfully employed in the development of delivery systems for functional ingredients. Synthesized from starch, cellulose, pectin, xanthan, etc., these nanoparticles can be prepared through top-down approaches involving the breakdown of larger polysaccharide structures or bottom-up approaches starting from polysaccharide molecules. The digestion of carbohydrate nanoparticles is contingent upon their source, with certain starch-derived carbohydrates readily hydrolyzed in the upper gastrointestinal tract, while others undergo fermentation in the lower gastrointestinal tract [69, 107].

4.1 Food Packaging

Nanocomposites composed of nanoparticles and polymers are used as packaging materials because of their enhanced resistance to heat, gas, and moisture. Food is directly contacted through packaging [87]. Maintaining an inert, low-oxygen environment that impedes microbial growth and spoilage is the primary method of preventing food contamination. As impermeable barriers, nanomaterials incorporated into the packaging polymer matrix serve as fillers, and the increased surface area of nanomaterials substantially enhances their impermeable characteristics [28]. Magnesium ash, specifically montmorillonite, is used to produce nanoclay, which is a polymer nanocomposite with complex metallic ores that exhibits exceptional gas permeability resistance [13]. This location is suitable for a variety of polymers including nylon, polyolefin, PET, epoxy resin, and polymethane.

4.2 Active Packaging

Active packaging functions as an inert barrier that separates food products from their external environment. By reacting with the food product, it prolongs shelf life and ensures favorable conditions [34].

Active food packaging consists of the following components: (1) nanocomposites comprising metal oxides of TiO_2 and MgO and silver, copper, and gold ions; (2) antimicrobial films comprising bacteriophages, gas scavengers, edible clove, pepper, cinnamon, coffee, chitosan, antimicrobial lysozyme, and bacteriophages; and (3) gas scavengers comprising titanium dioxide, iron powder, and silicate [16, 32, 71]. Enzymatic oxygen scavenging is accomplished through the action of alcohol oxidase enzymes on alcohols produced in food products.

The gas barrier properties of food packaging can be enhanced. The most frequently employed material is montmorillonite (also known as bentonite), which restricts the permeation of gases through its natural nanoscale layered structure when incorporated into a polymer [23].

Nanocomposites of polymers and inorganic materials (Ag, TiO_2, MgO, CuO, etc.) have been utilized in packaging systems [64]. Owing to their antimicrobial and antiodorant properties, Ag nanoparticles are extensively utilized in the production of dietary supplements and food packaging. SiO_2 nanoparticles are used in packaging, direct food contact applications, and clarification of beers and wines because of their biocompatibility [105]. Packaging materials frequently integrate nanoscale titanium dioxide (TiO_2) to facilitate photocatalytic sterilization [98]. An additional noteworthy accomplishment of nanotechnology is the integration of a highly effective nanosensor into a packaging system to identify instances of food spoilage caused by oxygen and moisture [101]. Packaging materials comprising silica gel and thiosulfate composites have been utilized because of their durable antibacterial properties [80].

To extend the shelf life of foods, various metal oxide-based antimicrobial materials and metals have been developed [2]. The packaging materials Ag/TiO_2 and Ag/carbon nanotube nanocomposites exhibited substantially enhanced antimicrobial activity against *Bacillus cereus* spores and *Escherichia coli*, respectively [61]. Antimicrobial activity against Bacillus cereus spores on aluminum- and polyester-coated surfaces was demonstrated using Ag-doped TiO_2 nanoparticles, which also eliminated a number of airborne bacteria [106].

4.3 Intelligent Packaging

Intelligent packaging is capable of scientific logic application, detection, sensing, recording, tracing, and communication to prolong shelf life; improve quality, safety, and information provision; and warn of potential problems [110]. Intelligent packages consist of the following components: (1) indicators (specifically, methyl blue, acid, antioxidants, mineral oil, and sugar) and elements (Na, K, Ca, Si, Al, and Mg). Responsive packaging technologies rely on the incorporation of a sensing interface into packaging films to enable continuous real-time quality monitoring [54]. This was achieved by implementing three critical design features: cross-contamination prevention, target analyte selection, and transduction system selection.

Stimulus-responsive materials, hydrogels, surfaces, particles, and supramolecules comprise the responsive packages [18]. By utilizing oxygen scavengers in intelligent packaging, the subsequent mechanism of iron oxidation prevents microbial growth, the development of off-flavors and odors, color changes, nutritional losses, and reduction in the shelf life of foods [24].

5 Environmental Impact and Safety Concerns

5.1 Environmental Impact

Globally, the environment receives thousands of tons of nanomaterials (NMs), particles less than 100 nm in diameter, annually [25]. These particles have a high active surface area and unique chemical properties, and can penetrate cells. The inherent potential toxicity of NMs is fundamentally conditioned by their ability to reach and invade different environmental systems by interacting with the soil, water, and air [33, 97]. This effect is directly related to the number of nanoparticles (NPs) released into the biosphere. Consequently, the risks associated with the use of nanomaterials are determined by all the processes that control their release into the environment, the shipment between facilities and areas or between organisms due to the trophic chain, and the transformations that can occur once they are released.

The released nanomaterials undergo various transformations, such as dissociation, dissolution, and aggregation, which affect the food chain. To understand the environmental impact of nanomaterials, it is necessary to understand their characteristics in detail, as well as their identification, physicochemical properties, release into the environment, and toxicity to living organisms [49]. Thus, the route taken by NMs to reach living cells provides information regarding their toxicity profiles.

In the environment, NMs interact with both biotic and abiotic components when they enter water bodies. Biological entities interact with nanomaterials and alter their fate by altering their original properties. Studies have shown that microorganisms interact with engineered nanomaterials and absorb large amounts of these materials. The presence of exopolymeric substances (EPS) in the outer membrane of cells favors the adsorption of engineered nanomaterials in aqueous media.

The interactions of NMs with abiotic factors, such as soil or clay particles, can influence the hetero-agglomeration and sedimentation of nanomaterials in aqueous environmental matrices [19, 43]. Moreover, biomacromolecules, acids, and other inorganic substances can alter the fate of engineered nanomaterials through various mechanisms such as adsorption or absorption, agglomeration or aggregation, sedimentation, speciation, ionization, plugging, and biotransfer. Importantly, transformations of engineered nanomaterials in the environment can also modify the bioavailability and toxicity potential of NMs to producers, consumers, and decomposers within an ecosystem [14].

In the same context, bioconcentration, bioaccumulation, and biomagnification are mechanisms that have been identified to determine the fate of nanomaterials (Fig. 1). In aquatic environments, the exposure of organisms to engineered nanomaterials occurs especially through the aqueous medium, which is referred to as bioconcentration [48, 52]. Passive absorption may occur in this process. Exposure to engineered nanomaterials through all possible pathways, including food contamination, is defined as bioaccumulation [102]. The increase in the contaminant concentration over time is also considered an important factor in the bioaccumulation process. The transfer of engineered nanomaterials from prey to predator with a magnitude more than one time higher is defined as biomagnification [38]. Ecotoxicological studies on established contaminants have suggested that the biomagnification of any contaminant is the most dangerous to human health and the environment.

Bacteria are a part of the basic food chain. Their importance, coupled with their ease of manipulation, makes them an essential starting point for studies on the potential environmental toxicity of NPs. Although the mechanisms of toxicity in bacteria are poorly understood, it has been shown that NPs affect cell walls, and with a probability of certainty that oxidative stress, metal ion release, and non-oxidative mechanisms are involved. One study showed that the effect of TiO_2 NPs (0, 0.5, 1.0, and 2.0 mg/g soil) and ZnO NPs (0.05, 0.1, and 0.5 mg/g soil) on bacteria was dose-dependent [35]. These studies proposed a dual mechanism of toxicity: direct toxicity due to cell damage, and indirect toxicity due to alterations in nutrients and soil properties. In the same context, another study analyzed the effects of TiO_2 and ZnO nanoparticles on soil bacteria and the enantioselective transformation of racemic metalaxyl (rac-metalaxyl) in agricultural soil with or without *Lolium perenne* in an

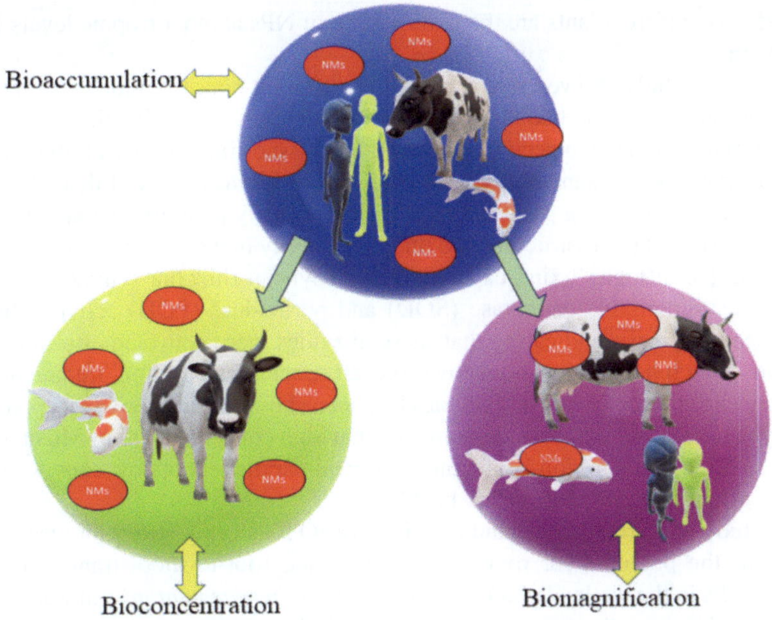

Fig. 1 Routes of engineered nanomaterials (NMs) in the food chain. Direct and indirect exposure to predators (human) and dietary (cow and fish) species

open-air greenhouse [115]. The results showed that the microbial carbon biomass decreased by 66% and the bacterial community composition changed significantly. While ZnO decreased the chlorophyll accumulation in *L. perenne* by 34%. ZnO also inhibited the enantioselective transformation of metalaxyl enantiomers and changed the fraction of metalaxyl enantiomers. TiO_2 showed similar effects, but to a lesser extent, *L. perenne* promoted rac-metalaxyl transformation and ingested both TiO_2 and ZnO. *L. perenne* changed bacterial co-occurrence networks and biomarkers in native soil and soil exposed to TiO_2 and ZnO. *L. perenne* reduced the inhibition effects of TiO_2 and ZnO on rac-metalaxyl transformation. These results demonstrate the existence of mutual interactions between the effects of engineered nanoparticles on different components (microbes, plants, and coexisting pollutants) in the terrestrial ecosystem.

Plants are living beings of great importance to study any source of environmental impact because of their interaction with all environmental compartments (soil, water, and air) that may contain NPs [58]. They constitute an essential part of the food chain and play an important role in NP transport to animals and humans. The most common mechanism used by plants to uptake NPs is through the leaves, flower surfaces, roots, or damaged areas of the plant. Most toxicity studies on plants that are usually used for human consumption (corn, wheat, and soybean) have focused on the analysis of parameters such as seed germination rate, root growth, and nitrogen fixation. Studies have shown that NPs concentrations in soil are higher than those in water and air

[6, 8, 96]. Therefore, plants are the main source of NPs at other trophic levels in the food chain.

A previous study showed that exposure to different concentrations and types of Mo-based nanomaterials had significant physiological and biochemical toxicity in soybean plants [111]. Mo accumulation in plant tissues increased with the concentration of nano-Mo in sand and vermiculite media. It was observed that Mo-based nanomaterials penetrating plant tissues affected not only plant growth and development, but also root microstructure and rhizobia activity in the symbiotic system, thus weakening the nitrogen-fixing capacity of the soybean-rhizobia symbiotic system. Meanwhile, superoxide dismutase (SOD) and peroxidase (POD) activity showed a significant increase, indicating that these enzymes were activated after nano-Mo exposure, which helps remove reactive oxygen species. In addition, treatment with high concentrations of Mo nanoparticles has been shown to have obvious epigenetic toxicity to plants. These results demonstrate that high concentrations of Mo nanoparticles can affect the agronomic and physiological parameters of soybeans, which in turn can affect human nutrition and health.

One study analyzed the fate and toxicity of ZnO NPs in aquatic plant mesocosms as well as the potential for root accumulation and root-to-shoot translocation of these Zn NPs in the wetland plant *Schoenoplectus tabernaemontani* exposed to ZnO NPs [114]. The growth of *S. tabernaemontani* in these hydroponic mesocosms was significantly inhibited by ZnO NPs (1000 mg L^{-1}) compared with the control. The distribution of ZnO NPs in *S. tabernaemontani* root tissues was confirmed by scanning electron microscopy (SEM). Transmission electron microscopy (TEM) images showed that ZnO NPs could pass through plant cell walls and were present inside *S. tabernaemontani* plant cells.

The toxicity effects generated by nanomaterials in animal models are essential for correlating the potential effects of nanoparticles in humans. Studies focusing on cytotoxicity, genotoxicity, tumorigenesis, and reproductive toxicity are of utmost importance [58]. One study investigated the fate, toxicity, and bioaccumulation of copper oxide nanoparticles (nCuO) in soil, with and without biosolid amendment, through chronic exposure to the earthworm *Eisenia Andrei* [103]. The fate of nCuO was evaluated by characterizing the dissolved and nanoparticle fractions, as well as the extractable Cu^{2+} over the duration of the selected trials. The effects on *E. andrei* reproduction were significant in all treatments (98–149 mg Cu kg^{-1}, dry soil). There were no significant differences in toxicity among Cu forms or in extractable Cu^{2+} activities, indicating that particle dissolution in soil and subsequent Cu^{2+} activity were probably the main modes of toxicity in nCuO exposure. The presence of biosolids affected the Cu^{2+} activity over time. Through dark-field hyperspectral imaging, the analysis showed nCuO accumulation was observed throughout the epidermis and gut of *E. andrei*, with traces observed in the muscle and chlorogenic tissue, providing evidence of nCuO translocation within the organism. Thus, the risk assessment of nanomaterials in the environment was demonstrated based on the solubility of the substance and the bioavailability of the dissolved free ions in exposure to nCuO from soil invertebrates.

Other nanomaterials of attention are those from ZnO, as they have been used in an uncontrolled and deliberate manner in industries as active additives in food processing, quality control, and nanofood/feed delivery systems, thus affecting food chains through soil and plant tissues [27, 93].

A study investigated the influence of different soil properties on the bioavailability and toxicity of zinc oxide (ZnO) nanoparticles and ionic Zn in the earthworm *Eisenia fetida* [52]. Earthworms were exposed to ZnO NPs and ionic Zn (100–4400 mg Zn/kg) in four different natural soils. Toxicity was also influenced by soil properties, with EC_{50}s based on total soil Zn ranging from 694 to > 2200 mg Zn/kg for ZnO NP, and 277 to 734 mg Zn/kg for ionic Zn. Earthworms exposed to ZnO NPs survived higher soil Zn concentrations and had higher tissue concentrations than those exposed to ionic Zn, particularly in calcareous soils with high organic content. The higher tissue concentrations in earthworms exposed to ZnO NPs could have implications for the persistence and trophic mobility of Zn in terrestrial systems.

An interesting paper cataloged 2102 experimental results on whole organisms for 22 classes of particles (mainly Ag, Zn, Ti, and Cu) to evaluate the biological responses, effective and lethal concentrations, and bioaccumulation of NMs [25].The analysis showed that smaller particles tended to be more toxic, aquatic organisms responded more negatively than terrestrial organisms, animals were more sensitive, and plants were less sensitive. Ag NMs generally had the most representative negative effects. Effective and lethal concentrations exceeded the concentrations modeled for the environment, and organisms did not accumulate or biomagnify to concentrations above those in their environment.

However, nanoplastics can adversely affect organisms of different phyla, with effects ranging from reproductive impairment to lethality [31, 78, 81]. However, no such information has been found in marine vertebrates (e.g., fish). These data show a high potential for bioaccumulation/biomagnification along marine food chains, as they can be easily retained within organisms. However, there is a lack of standardized methodologies for nanoplastic detection, and little or no legislation makes nanoplastics an environmental challenge [78].

5.2 Safety Concerns

The use of nanotechnology has been potentially increasing owing to its applications in various fields, such as the transformation and engineering industries (oil and gas, electronics, cosmetics, biomedicine, agriculture, medicine, public health, etc.), and even to address environmental problems [44]. However, the increasing use of nanomaterials has raised concerns about their impact on environmental health and safety,therefore, several studies have focused on nanotoxicity [37].

However, in the agricultural sector, the use of nanomaterials by farmers is still not encouraging [42]. Numerous studies have revealed that lower availability and higher prices compared to conventional products are the main causes. In addition, the safety of using NMs in agricultural production systems has played a synergistic role in their lower acceptance, while manufacturers are hesitant and remain hesitant about the large-scale production of nanomaterials in agriculture [44]. Hence, this calls for collaborative work between industries, regulatory organizations, and research institutes for the control and feasibility of the use of NMs in the agricultural sector [10].

In nanoremediation, the application of NMs that enhance the accumulation of a contaminant and increase its bioavailability should be reinforced along with the vigilant management of its transfer to higher levels of the food chain due to the biomagnification that could occur within the food chain [49]. Therefore, risk assessment of each new engineered nanomaterial on the environment and human health should be carried out along with the regulation of life cycle assessment (LCA) tools of nanomaterials used for nanoremediation, which can help to investigate potential environmental hazards.

The use of nanomaterials in food, especially insoluble and biopersistent nanoparticles, is safe for the health of consumers and the environment. There is a concern that nanomaterials with high external volatility may alter biological processes by reaching various parts of the body. Therefore, clear and adequate definitions of NMs are needed for their safe use [44]. At present, reliable approaches for the recognition and categorization of nanomaterials in complex food matrices are unavailable.

Uncertainty and limited knowledge about the effects of nanoparticles raise health and safety concerns, which can be addressed by well-defined regulations and legislation on a global scale. In 2003, the UK authorities commissioned the Royal Academy of Engineering and Royal Society to study the benefits, safety, and health issues arising from the use of nanomaterials. In 2004, the Royal Society published a report entitled "Nanoscience and Nanotechnologies: Opportunities and Uncertainties", which indicated that nanoparticles or nanotubes should be treated as new materials under the existing "Notification of New Substances" (NONS) and "Registration, Evaluation, Authorization and Restriction of Chemicals" (REACH) rules. For its part, the US Environmental Protection Agency (EPA), in its document EPA 100/B-07/001 (Nanotechnology White Paper), published in 2007, states that, "as the use of nanomaterials in society increases, it is reasonable to assume that their presence in the environment will increase proportionally, with consequences for human and environmental exposure"[44]. The Committees on Toxicity, Carcinogenicity and Mutagenicity of Chemicals in Food, Consumer Products and the Environment have also identified the risk assessment of nanomaterials as an area of interest in their "Joint Statement on Toxicology of Nanomaterials."

The European Commission's Scientific Committee on Emerging and Newly Identified Health Risks (CSCENIHR) also reviewed existing information/data and issues to be considered when performing risk assessments of NMs [22, 44, 90]. The Scientific Committee on Consumer Products (SCCP) of the European Commission published the document "Opinion on Safety of Nanomaterials in Cosmetic Products, " which focused on the safety of nanomaterials in cosmetic products. It expressed concern about the large data gaps, inadequacy of existing characterization techniques for risk assessment of nanomaterials, and inadequacy of available information.

The European Commission (EC), European Food Safety Agency (EFSA), Environmental Protection Agency (EPA), Organization for Economic Cooperation and Development (OECD), International Organization for Standardization (ISO), Food and Drug Administration (FDA), World Health Organization (WHO), and scientific committees and agencies are the main authorizing bodies in Europe and the United States, and direct regulatory issues and guidelines related to nanomaterials in food [100]. Common regulatory procedures indicate the identification of nanomaterials based on their size, surface characteristics, chemical composition, and stability, which are crucial for determining their possible interactions and persistence in an organism, thus providing data for risk assessment. Various analytical methods are widely used to detect and identify nanoparticles in food to assess their potential risks or hazards. In addition, in vitro and in vivo studies have provided significant data to determine the possible responses of the organism to nanoparticles.

In general, most countries lack defined regulations for the commercialization and use of industrial products derived from nanotechnology. Previous studies on the regulation and control of the use of nanomaterials have identified the inadequacy of existing regulatory processes and maintained the need for rapid policy and regulatory development to enter the market.

6 Socioeconomical Implications of the Use of Nanomaterials in Food Chains

Nanotechnology (NT) can potentially revolutionize the food chain, economy, and society. Because of its multidisciplinary nature, it can be applied across various technologies, leading to a significant transformation in the creation of materials. However, as an emerging technology, strategic management must ensure technological capabilities, market success, and social acceptance, which are critical to its overall success. The development of nanomaterials (NM) can enhance the solubility, absorption, and bioavailability of conventional products. However, the cellular-level interactions of nanoparticles (NP) need to be better understood. The demand for future food products should consider the ethical, legal, and socioeconomic factors related to both human and animal health concerns. There are significant differences in the governance of this technology between the European Union, the United States, and the rest of the world, which have a significant impact on socioeconomic and political conditions [11].

NT plays a crucial role in various fields, such as advanced material development and manufacturing techniques. These applications require specific skills to support national industrial systems to survive in the global context. However, strategies, innovation schemes, and learning dynamics in firms, academia, and government can vary from one region to another. Therefore, the government must establish laws and regulations before providing funding and infrastructure. NT has a significant potential for improving manufacturing processes in the food industry, particularly for the parts and components used. From a technological standpoint, nanotechnology is ready to be used in the food industry, with government support to increase its implementation. This can be achieved by facilitating the transition from laboratory research to prototype testing and manufacturing in the food industry, while promoting awareness and support for pathways to overcome challenges in nanotechnology research [11, 20].

To ensure equal access to the latest prototypes and testing platforms, it is important to encourage co-creation between public research agencies and businesses. We also need to launch targeted campaigns to educate and involve the public on scientific matters, particularly those related to nanotechnology and its applications. To convince people that nanotechnology is safe, we need to establish standards and certification processes for these products. Public policies should be put in place to create value based on economic, environmental, social, and technological factors.

It is widely acknowledged that raising public awareness of nanotechnology products is a significant challenge. There is a lack of understanding among people regarding the definition of these products, the risks involved in their handling and use, and the ethical considerations involved in their development processes. This lack of clarity is particularly evident in Latin America, where the concept of nanotechnology is just beginning to emerge. Laws and regulations in the region often rely on models from the United States or Europe, adding to the complexity of this issue [15, 57].

7 Future Trends and Research Directions

With the increasing incorporation of nanomaterials into the diverse facets of everyday life, the food industry has emerged as a leading innovator in the implementation of nanotechnology. The incorporation of nanomaterials into the food chain has introduced novel prospects including enhanced nutritional value and improved food safety. As one looks ahead, several emerging trends and research avenues have come into play, influencing the nanomaterials sector within the food industry.

1. Nanosensors for Real-time Monitoring: The development of nanosensors for real-time monitoring of food quality and safety is one of the most promising future trends. The development of nanoscale sensors capable of identifying contaminants, pathogens, or indicators of spoilage enables the prompt and precise evaluation of food freshness. The goal of this field is to develop scalable and economical nanosensor platforms that can be seamlessly incorporated into food processing and packaging facilities. This would enable the uninterrupted monitoring of food products across the entire supply chain.
2. The utilization of nanomaterials presents an opportunity to achieve precise nutrient delivery, thereby overcoming the obstacles associated with personalized nutrition and food fortification. Researchers are investigating the potential of nanocarriers to encapsulate and safeguard delicate nutrients, thereby enabling their regulated release throughout the digestion process. Through targeted nutrient delivery, bioavailability can be optimized, thereby guaranteeing

that consumers obtain the utmost nutritional benefits from the food they consume. Subsequent investigations will presumably focus on the fabrication of nanocarriers customized to deliver nutrients as well as the scrutiny of personalized nutrition approaches.

3. Sustainable Nanomaterials in Packaging: With the increasing emphasis on sustainability, there will be a corresponding transition towards environmentally friendly alternatives in the future of nanomaterials utilized in the food chain. The exploration of sustainable nanomaterials for food packaging, including biodegradable nanoparticles and nanocomposites fabricated from renewable resources, is a focal point of research. This phenomenon corresponds to the worldwide effort to implement environmentally sustainable methods in the food sector and tackle apprehensions regarding the ecological consequences of nanomaterials.
4. Risk Assessment and Regulation: A thorough comprehension of the potential hazards and regulatory structures associated with the extensive implementation of nanomaterials in the food chain is imperative. Further investigation is expected to concentrate on enhancing risk assessment methodologies tailored to nanomaterials, considering variables such as bioaccumulation and potential long-term impacts on human health. In addition, greater emphasis should be placed on the development of stringent regulations that strike a balance between the necessity to ensure consumer safety and the innovative potential of nanomaterials.
5. The importance of Consumer Perception and Education: With increasing prevalence of nanomaterials in the food industry, it is critical to address consumer perceptions and education. Further investigation is warranted to examine efficacious communication tactics that can enlighten consumers regarding the advantages and security of nanomaterials present in food products. Gaining insight into consumer sentiments and apprehensions will facilitate the formulation of educational initiatives and transparent labeling protocols, thereby promoting confidence and approval of nanotechnology throughout the food supply chain.

In summary, potential applications of nanomaterials in the food chain are highly promising, encompassing innovative packaging solutions and sophisticated monitoring systems. However, researchers must overcome the obstacles pertaining to consumer acceptance, safety, and sustainability. The optimization of nanomaterial integration into the food industry can be achieved by prioritizing these forthcoming trends and research avenues, thereby ensuring harmonious coexistence of innovation and responsible applications.

8 Conclusions

Studies on the impact of nanomaterials on the environment, and therefore on food chains, present several challenges. Most of these studies were carried out in vitro owing to the difficulties of the experimental models, as well as the periods of time that involve monitoring for months or years. In animal studies, ethical problems arise owing to the high number of animals sacrificed. Therefore, it is necessary to develop analytical methods that provide robust and orthogonal methods to analyze or quantify nanomaterials in abiotic and biotic samples from the environment or within the food chain.

In addition, numerous unique and untested NMs continue to enter the environment at an accelerated rate, potentially leading to negative effects. Therefore, the next challenge is the assessment and quantification of the number of nanomaterials released into the environment, which requires thorough investigation, starting with the production processes of nanomaterials and ending with the recycling and disposal procedures, considering how they are incorporated into the final products and how they are used.

Finally, to minimize the health and environmental risks associated with the use of nanomaterials in developed and developing countries, it is essential to recognize the toxicological, constructive, toxicodynamic, and toxicokinetic characteristics of NMs, which must be carefully clarified and emphasized. Therefore, studies on standards, policies, and guidelines focusing on NMs updates are necessary to safeguard the quality and safety of the environment, anthropogenic health, and environmental health.

References

1. Abutalib MM, Rajeh A (2021) Enhanced structural, electrical, mechanical properties and antibacterial activity of Cs/PEO doped mixed nanoparticles (Ag/TiO_2) for food packaging applications. Polym Testing 93:107013
2. Adeyemi JO, Fawole OA (2023) Metal-based nanoparticles in food packaging and coating technologies: a review. Biomolecules 13:1092
3. Adil M, Bashir S, Bashir S, Aslam Z, Ahmad N, Younas T, Asghar RMA, Alkahtani J, Dwiningsih Y, Elshikh MS (2022) Zinc oxide nanoparticles improved chlorophyll contents, physical parameters, and wheat yield under salt stress. Front Plant Sci 13:932861. https://doi.org/10.3389/fpls.2022.932861
4. Ahmad A, Hashmi SS, Palma JM, Corpas FJ (2022) Influence of metallic, metallic oxide, and organic nanoparticles on plant physiology. Chemosphere 290:133329. https://doi.org/10.1016/j.chemosphere.2021.133329
5. Ahmed R, Uddin MdK, Quddus MdA, Samad MYA, Hossain MAM, Haque ANA (2023) Impact of foliar application of Zinc and Zinc Oxide nanoparticles on growth, yield, nutrient uptake and quality of tomato. Horticulturae 9:162. https://doi.org/10.3390/horticulturae9020162
6. Akhtar O, Zoomi I, Pandey D, Kehri HK, Narayan RP (2020) Tripartite Interaction Among Nanoparticles, Symbiotic Microbes, and Plants: Current Scenario and Future Perspectives. Nanomaterials and Environmental Biotechnology, 55–64
7. Alghamdi HM, Abutalib M, Mannaa MA, Nur O, Abdelrazek E, Rajeh A (2022) Modification and development of high bioactivities and environmentally safe polymer nanocomposites doped by Ni/ZnO nanohybrid for food packaging applications. J Market Res 19:3421–3432
8. Ali S, Mehmood A, Khan N (2021) Uptake, translocation, and consequences of nanomaterials on plant growth and stress adaptation. J Nanomater 2021:1–17

9. Ambrosini VG, Rosa DJ, Bastos De Melo GW, Zalamena J, Cella C, Simão DG, Souza Da Silva L, Pessoa Dos Santos H, Toselli M, Tiecher TL, Brunetto G (2018) High copper content in vineyard soils promotes modifications in photosynthetic parameters and morphological changes in the root system of 'Red Niagara' plantlets. Plant Physiol Biochem 128:89–98. https://doi.org/10.1016/j.plaphy.2018.05.011
10. Arora S, Murmu G, Mukherjee K, Saha S, Maity D (2022) A comprehensive overview of nanotechnology in sustainable agriculture. J Biotechnol 355:21–41
11. Arumugam VR, Ng B-K, Thambiratnam K (2023) A delphi study on technical and socio-economic perspectives of nanotechnology and ICT industries relations. Systems 11:190. https://doi.org/10.3390/systems11040190
12. Assadpour E, Mahdi Jafari S (2019) A systematic review on nanoencapsulation of food bioactive ingredients and nutraceuticals by various nanocarriers. Crit Rev Food Sci Nutr 59:3129–3151
13. Bangar SP, Chaudhary V, Akhila PP, Sunooj KV (2023) Recent functionality developments in montmorillonite as a nanofiller in food packaging. Trends Food Sci Technol, 104148
14. Barhoum A, García-Betancourt ML, Jeevanandam J, Hussien EA, Mekkawy SA, Mostafa M, Omran MM, Abdalla S, M., Bechelany, M., (2022) Review on natural, incidental, bioinspired, and engineered nanomaterials: history, definitions, classifications, synthesis, properties, market, toxicities, risks, and regulations. Nanomaterials 12:177
15. Barragán-Ocaña A, Reyes-Ruiz G, Merritt H (2020) Scientific, technological, and innovation dynamics in nanotechnology for smart cities and villages: the OECD case and its implications for Latin America. Smart Village Technology: Concepts and Developments, pp 39–65
16. Becerril R, Nerín C, Silva F (2020) Encapsulation systems for antimicrobial food packaging components: An update. Molecules 25:1134
17. Bouwmeester H, Brandhoff P, Marvin HJ, Weigel S, Peters RJ (2014) State of the safety assessment and current use of nanomaterials in food and food production. Trends Food Sci Technol 40:200–210
18. Brockgreitens J, Abbas A (2016) Responsive food packaging: recent progress and technological prospects. Compr Rev Food Sci Food Saf 15:3–15
19. Buonomenna MG, Mousavi SM, Hashemi SA, Lai CW (2022) Water cleaning adsorptive membranes for efficient removal of heavy metals and metalloids. Water 14:2718
20. Cely-Bautista M-M, Castellar-Ortega G-C, Jaramillo-Colpas J-E, Higuera-Cobos O-F (2023) Global trends in normativity and regulatory issues on nanotechnology. Revista Facultad de Ingeniería 32:e16403–e16403
21. Choi SJ, McClements DJ (2020) Nanoemulsions as delivery systems for lipophilic nutraceuticals: strategies for improving their formulation, stability, functionality and bioavailability. Food science and biotechnology 29:149–168
22. Deng Y, White JC, Xing B (2014) Interactions between engineered nanomaterials and agricultural crops: implications for food safety. J Zhejiang Univ Sci A 15:552–572
23. Deshmukh RK, Hakim L, Gaikwad KK (2023) Active packaging materials. Curr Food Sci Technol Rep 1:123–132. https://doi.org/10.1007/s43555-023-00004-6
24. Dey A, Neogi S (2019) Oxygen scavengers for food packaging applications: a review. Trends Food Sci Technol 90:26–34
25. Dodds WK, Guinnip JP, Schechner AE, Pfaff PJ, Smith EB (2021) Fate and toxicity of engineered nanomaterials in the environment: a meta-analysis. Sci Total Environ 796:148843
26. Dola DB, Mannan MdA, Sarker U, Mamun MdAA, Islam T, Ercisli S, Saleem MH, Ali B, Pop OL, Marc RA (2022) Nano-iron oxide accelerates growth, yield, and quality of Glycine max seed in water deficits. Front Plant Sci 13:992535. https://doi.org/10.3389/fpls.2022.992535
27. Droepenu EK, Wee BS, Chin SF, Kok KY, Maligan MF (2022) Zinc oxide nanoparticles synthesis methods and its effect on morphology: a review. Biointerface Res. Appl. Chem 12:4261–4292
28. Ebner C, Morgan A, Manuel C (2021) Packaging of perishable food products. Food Safety and Quality-Based Shelf Life of Perishable Foods, pp 105–134

29. Fatima F, Hashim A, Anees S (2021) Efficacy of nanoparticles as nanofertilizer production: a review. Environ Sci Pollut Res 28:1292–1303. https://doi.org/10.1007/s11356-020-11218-9
30. Feng X, Tjia JYY, Zhou Y, Liu Q, Fu C, Yang H (2020) Effects of tocopherol nanoemulsion addition on fish sausage properties and fatty acid oxidation. Lwt 118:108737
31. Ferreira I, Venâncio C, Lopes I, Oliveira M (2019) Nanoplastics and marine organisms: what has been studied? Environ Toxicol Pharmacol 67:1–7
32. Firouz MS, Mohi-Alden K, Omid M (2021) A critical review on intelligent and active packaging in the food industry: research and development. Food Res Int 141:110113
33. Fukuto A, Kim S, Kang J, Gates BL, Chang MW, Pinkerton KE, Van Winkle LS, Kiuchi Y, Murphy CJ, Leonard BC (2021) Metal oxide engineered nanomaterials modulate rabbit corneal fibroblast to myofibroblast transformation. Transl Vis Sci Technol 10:23–23
34. Garba AI (2023) Food Preservation Packaging. In: Food Processing and Packaging Technologies-Recent Advances. Intech Open
35. Ge Y, Schimel JP, Holden PA (2012) Identification of soil bacteria susceptible to TiO_2 and ZnO nanoparticles. Appl Environ Microbiol 78:6749–6758
36. Ghosh C, Bera D, Roy L (2019) Role of nanomaterials in food preservation. In: Microbial nanobionics: Basic Research and Applications, vol 2, pp 181–211
37. Gottschalk F, Debray B, Klaessig F, Park B, Lacome J-M, Vignes A, Portillo VP, Vázquez-Campos S, Hendren CO, Lofts S (2023) Predicting accidental release of engineered nanomaterials to the environment. Nat Nanotechnol 18:412–418
38. Gupta GS, Shanker R, Dhawan A, Kumar A (2017) Impact of nanomaterials on the aquatic food chain. Nanoscience in Food and Agriculture 5:309–333
39. Haris M, Hussain T, Mohamed HI, Khan A, Ansari, Moh.S., Tauseef, A., Khan, A.A., Akhtar, N., (2023) Nanotechnology—A new frontier of nano-farming in agricultural and food production and its development. Sci Total Environ 857:159639. https://doi.org/10.1016/j.scitotenv.2022.159639
40. Hashem AH, El-Naggar ME, Abdelaziz AM, Abdelbary S, Hassan YR, Hasanin MS (2023) Bio-based antimicrobial food packaging films based on hydroxypropyl starch/polyvinyl alcohol loaded with the biosynthesized zinc oxide nanoparticles. Int J Biol Macromol 249:126011
41. Hermes PH, Gabriela MP, Ileana, VR, Fusaro C, Fernando LV, Mariana MA, Padilla-Rodríguez C, Fabián FL (2020) Carbon nanotubes as plant growth regulators: prospects. In: Patra JK, Fraceto LF, Das G, Campos EVR (eds) Green nanoparticles, nanotechnology in the life sciences. Springer International Publishing, Cham, pp 77–115. https://doi.org/10.1007/978-3-030-39246-8_4
42. Hossain A, Skalicky M, Brestic M, Mahari S, Kerry RG, Maitra S, Sarkar S, Saha S, Bhadra P, Popov M, Islam MT, Hejnak V, Vachova P, Gaber A, Islam T (2021) Application of nanomaterials to ensure quality and nutritional safety of food. J Nanomater 1:9336082. https://doi.org/10.1155/2021/9336082
43. Intisar A, Ramzan A, Sawaira T, Kareem AT, Hussain N, Din MI, Bilal M, Iqbal HM (2022) Occurrence, toxic effects, and mitigation of pesticides as emerging environmental pollutants using robust nanomaterials–a review. Chemosphere 293:133538
44. Jain A, Ranjan S, Dasgupta N, Ramalingam C (2018) Nanomaterials in food and agriculture: an overview on their safety concerns and regulatory issues. Crit Rev Food Sci Nutr 58:297–317
45. Jiang B, Zhao Y, Cao Y, Sun C, Lu W, Fang Y (2024) Advances in the interaction between food-derived nanoparticles and the intestinal barrier. J Agric Food Chem
46. Kaur P, Duhan JS, Thakur R (2018) Comparative pot studies of chitosan and chitosan-metal nanocomposites as nano-agrochemicals against fusarium wilt of chickpea (Cicer arietinum L.). Biocatal Agric Biotechnol 14:466–471. https://doi.org/10.1016/j.bcab.2018.04.014
47. Khan S, Hossain MK (2022) Classification and properties of nanoparticles. In: Nanoparticle-based polymer composites. Elsevier, pp 15–54
48. Kuehr S, Kaegi R, Maletzki D, Schlechtriem C (2021) Testing the bioaccumulation potential of manufactured nanomaterials in the freshwater amphipod Hyalella azteca. Chemosphere 263:127961

49. Kumar CV, Karthick V, Kumar VG, Inbakandan D, Rene ER, Suganya KU, Embrandiri A, Dhas TS, Ravi M, Sowmiya P (2022) The impact of engineered nanomaterials on the environment: release mechanism, toxicity, transformation, and remediation. Environ Res 212:113202
50. Kumar A, Kaur H, Choudhary A, Mehta K, Chattopadhyay A, Mehta S (2023) Role of nanocomposites in sustainable crop plants' growth and production. In: Engineered nanomaterials for sustainable agricultural production, soil improvement and stress management. Elsevier, pp 161–181. https://doi.org/10.1016/B978-0-323-91933-3.00002-7
51. Kumaraswamy RV, Saharan V, Kumari S, Chandra Choudhary R, Pal A, Sharma SS, Rakshit S, Raliya R, Biswas P (2021) Chitosan-silicon nanofertilizer to enhance plant growth and yield in maize (Zea mays L.). Plant Physiol Biochem 159:53–66. https://doi.org/10.1016/j.plaphy.2020.11.054
52. Lahive E, Matzke M, Svendsen C, Spurgeon D, Pouran H, Zhang H, Lawlor A, Pereira MG, Lofts S (2023) Soil properties influence the toxicity and availability of Zn from ZnO nanoparticles to earthworms. Environ Pollut 319:120907
53. Lazić V, Vivod V, Peršin Z, Stoiljković M, Ratnayake IS, Ahrenkiel PS, Nedeljković JM, Kokol V (2020) Dextran-coated silver nanoparticles for improved barrier and controlled antimicrobial properties of nanocellulose films used in food packaging. Food Packag Shelf Life 26:100575
54. Lee SJ, Rahman AM (2014) Intelligent packaging for food products. In: Innovations in food packaging. Elsevier, pp 171–209
55. Luo X, Han Y, Chen X, Tang W, Yue T, Li Z (2020) Carbon dots derived fluorescent nanosensors as versatile tools for food quality and safety assessment: a review. Trends Food Sci Technol 95:149–161
56. Luo Y, Wang Q, Zhang Y (2020) Biopolymer-based nanotechnology approaches to deliver bioactive compounds for food applications: a perspective on the past, present, and future. J Agric Food Chem 68:12993–13000
57. Martel-Estrada SA (2022) Ethical reflections related to the use and research of nanotechnolog. Instituto de Arquitectura Diseño y Arte
58. Martínez G, Merinero M, Pérez-Aranda M, Pérez-Soriano EM, Ortiz T, Villamor E, Begines B, Alcudia A (2020) Environmental impact of nanoparticles' application as an emerging technology: a review. Materials 14:166
59. Mattiello A, Lizzi D, Marchiol L (2018) Influence of Titanium Dioxide nanoparticles (TiO_2) on crop plants. In: Nanomaterials in plants, algae, and microorganisms. Elsevier, pp 277–296. https://doi.org/10.1016/B978-0-12-811487-2.00012-8
60. Mehmood A (2018) Brief overview of the application of silver nanoparticles to improve growth of crop plants. IET nanobiotechnol 12:701–705. https://doi.org/10.1049/iet-nbt.2017.0273
61. Mihaly Cozmuta A, Peter A, Mihaly Cozmuta L, Nicula C, Crisan L, Baia L, Turila A (2015) Active packaging system based on Ag/TiO_2 nanocomposite used for extending the shelf life of bread. chemical and microbiological investigations. Packag Technol Sci 28:271–284
62. Mohamed MA, Hashim AF, Alghuthaymi MA, Abd-Elsalam KA (2018) Nano-carbon: plant growth promotion and protection. In: Abd-Elsalam KA, Prasad R (Eds.), Nanobiotechnology applications in plant protection, nanotechnology in the life sciences. Springer International Publishing, Cham, pp 155–188. https://doi.org/10.1007/978-3-319-91161-8_7
63. Monte-Filho SS, Andrade SI, Lima MB, Araujo MC (2019) Synthesis of highly fluorescent carbon dots from lemon and onion juices for determination of riboflavin in multivitamin/mineral supplements. J Pharm Anal 9:209–216
64. Mulla MZ, Rahman MRT, Marcos B, Tiwari B, Pathania S (2021) Poly Lactic Acid (PLA) Nanocomposites: effect of inorganic nanoparticles reinforcement on its performance and food packaging applications. Molecules 26:1967
65. Mustafa N, Raja NI, Ilyas N, Abasi F, Ahmad MS, Ehsan M, Mehak A, Badshah I, Proćków J (2022) Exogenous application of green Titanium Dioxide nanoparticles (TiO_2 NPs) to improve the germination, physiochemical, and yield parameters of wheat plants under salinity stress. Molecules 27:4884. https://doi.org/10.3390/molecules27154884

66. Muñoz-Shuguli C, Vidal CP, Cantero-López P, Lopez-Polo J (2021) Encapsulation of plant extract compounds using cyclodextrin inclusion complexes, liposomes, electrospinning and their combinations for food purposes. Trends Food Sci Technol 108:177–186
67. Nahum V, Domb AJ (2021) Recent developments in solid lipid microparticles for food ingredients delivery. Foods 10:400
68. Van Nguyen D, Nguyen HM, Le NT, Nguyen KH, Nguyen HT, Le HM, Nguyen AT, Dinh NTT, Hoang SA, Van Ha C (2022) Copper nanoparticle application enhances plant growth and grain yield in maize under drought stress conditions. J Plant Growth Regul 41:364–375. https://doi.org/10.1007/s00344-021-10301-w
69. Nooshkam M, Varidi M (2020) Maillard conjugate-based delivery systems for the encapsulation, protection, and controlled release of nutraceuticals and food bioactive ingredients: a review. Food Hydrocolloids 100:105389
70. Ojagh SM, Hasani S (2018) Characteristics and oxidative stability of fish oil nano-liposomes and its application in functional bread. J Food Meas Charact 12:1084–1092
71. Omerović N, Djisalov M, Živojević K, Mladenović M, Vunduk J, Milenković I, Knežević NŽ, Gadjanski I, Vidić J (2021) Antimicrobial nanoparticles and biodegradable polymer composites for active food packaging applications. Compr Rev Food Sci Food Saf 20:2428–2454
72. Onyeaka H, Passaretti P, Miri T, Al-Sharify ZT (2022) The safety of nanomaterials in food production and packaging. Curr Res Food Sci 5:763–774
73. Osanlou R, Emtyazjoo M, Banaei A, Hesarinejad MA, Ashrafi F (2022) Preparation of solid lipid nanoparticles and nanostructured lipid carriers containing zeaxanthin and evaluation of physicochemical properties. Colloids Surf, A 641:128588
74. Patel DK, Kim H-B, Dutta SD, Ganguly K, Lim K-T (2020) Carbon nanotubes-based nanomaterials and their agricultural and biotechnological applications. Materials 13:1679
75. Pérez-Hernández H, García-Mayagoitia S, Torres-Gómez PA, Campos-Montiel RG, Fernández-Luqueño F (2022) Ecological effects of copper NPs: Advantages and drawbacks regarding current and potential applications. In: Copper nanostructures: next-generation of agrochemicals for sustainable agroecosystems. Elsevier, pp 719–750. https://doi.org/10.1016/B978-0-12-823833-2.00003-9
76. Pérez-Hernández H, Fernández-Luqueño F, Huerta-Lwanga E, Mendoza-Vega J, Álvarez-Solís José D (2020) Effect of engineered nanoparticles on soil biota: do they improve the soil quality and crop production or jeopardize them? Land Degrad Dev 31:2213–2230. https://doi.org/10.1002/ldr.3595
77. Pérez-Labrada F, López-Vargas ER, Ortega-Ortiz H, Cadenas-Pliego G, Benavides-Mendoza A, Juárez-Maldonado A (2019) Responses of tomato plants under saline stress to foliar application of copper nanoparticles. Plants 8:151. https://doi.org/10.3390/plants8060151
78. Petersen E, Barrios AC, Bjorkland R, Goodwin DG Jr, Li J, Waissi G, Henry T (2023) Evaluation of bioaccumulation of nanoplastics, carbon nanotubes, fullerenes, and graphene family materials. Environ Int 173:107650
79. Printz B, Guerriero G, Sergeant K, Audinot J-N, Guignard C, Renaut J, Lutts S, Hausman J-F (2016) Combining -omics to unravel the impact of copper nutrition on Alfalfa (*Medicago sativa*) stem metabolism. Plant Cell Physiol 57:407–422. https://doi.org/10.1093/pcp/pcw001
80. Priyadarshi R, Roy S, Ghosh T, Biswas D, Rhim J-W (2022) Antimicrobial nanofillers reinforced biopolymer composite films for active food packaging applications-a review. Sustain Mater Technol 32:e00353
81. Ramachandraiah K, Ameer K, Jiang G, Hong G-P (2022) Micro-and nanoplastic contamination in livestock production: entry pathways, potential effects and analytical challenges. Sci Total Environ 844:157234
82. Raveendran S, Parameswaran B, Ummalyma SB, Abraham A, Mathew AK, Madhavan A, Rebello S, Pandey A (2018) Applications of microbial enzymes in food industry. Food Technol Biotechnol 56:16
83. Riahi Z, Khan A, Rhim JW, Shin GH, Kim JT (2024) Carrageenan-based active and intelligent packaging films integrated with anthocyanin and TiO_2-doped carbon dots derived from sweet potato peels. Int J Biol Macromol, 129371

84. Rostamabadi H, Falsafi SR, Boostani S, Katouzian I, Rezaei A, Assadpour E, Jafari SM (2021) Design and formulation of nano/micro-encapsulated natural bioactive compounds for food applications. In: Application of nano/microencapsulated ingredients in food products. Elsevier, pp 1–41
85. Sahani S, Sharma YC (2021) Advancements in applications of nanotechnology in global food industry. Food Chem 342:128318
86. Salam A, Afridi MS, Javed MA, Saleem A, Hafeez A, Khan AR, Zeeshan M, Ali B, Azhar W, Sumaira U, Z., Gan, Y., (2022) Nano-priming against abiotic stress: a way forward towards sustainable agriculture. Sustainability 14:14880. https://doi.org/10.3390/su142214880
87. Sarfraz J, Gulin-Sarfraz T, Nilsen-Nygaard J, Pettersen MK (2020) Nanocomposites for food packaging applications: an overview. Nanomaterials 11:10
88. Satti SH, Raja NI, Javed B, Akram A, Mashwani Z-R, Ahmad MS, Ikram M (2021) Titanium dioxide nanoparticles elicited agro-morphological and physicochemical modifications in wheat plants to control Bipolaris sorokiniana. PLoS ONE 16:e0246880
89. Schmidt W, Thomine S, Buckhout TJ (2020) Editorial: Iron Nutrition and Interactions in plants. Front Plant Sci 10:1670. https://doi.org/10.3389/fpls.2019.01670
90. Schoonjans R, Castenmiller J, Chaudhry Q, Cubadda F, Daskaleros T, Franz R, Gott D, Mast J, Mortensen A, Oomen AG (2023) Regulatory safety assessment of nanoparticles for the food chain in Europe. Trends Food Sci Technol
91. Seleiman MF, Ahmad A, Battaglia ML, Bilal HM, Alhammad BA, Khan N (2023) Zinc oxide nanoparticles: a unique saline stress mitigator with the potential to increase future crop production. S Afr J Bot 159:208–218. https://doi.org/10.1016/j.sajb.2023.06.009
92. Shah B, Xu W, Mráz J (2021) Formulation and characterization of zein/chitosan complex particles stabilized pickering emulsion with the encapsulation and delivery of vitamin D3. J Sci Food Agric 101:5419–5428
93. Sharifan H, Moore J, Ma X (2020) Zinc oxide (ZnO) nanoparticles elevated iron and copper contents and mitigated the bioavailability of lead and cadmium in different leafy greens. Ecotoxicol Environ Saf 191:110177
94. Sharma AS, Ali S, Sabarinathan D, Murugavelu M, Li H, Chen Q (2021) Recent progress on graphene quantum dots-based fluorescence sensors for food safety and quality assessment applications. Compr Rev Food Sci Food Saf 20:5765–5801
95. Siddiqi KS, Husen A (2022) Plant response to silver nanoparticles: a critical review. Crit Rev Biotechnol 42:973–990. https://doi.org/10.1080/07388551.2021.1975091
96. Singh PM, Tiwari A, Maity D, Saha S (2022) Recent progress of nanomaterials in sustainable agricultural applications. J Mater Sci 57:10836–10862
97. Sun Y, Wang W, Zheng F, Zhang S, Wang F, Liu S (2020) Phytotoxicity of iron-based materials in mung bean: seed germination tests. Chemosphere 251:126432
98. Sungur Ş, Kaya P, Koroglu M (2020) Determination of titanium dioxide nanoparticles used in various foods. Food Addit Contam: Part B 13:260–267
99. Tang C-H, Chen H-L, Dong J-R (2023) Solid lipid nanoparticles (SLNs) and nanostructured lipid carriers (NLCs) as food-grade nanovehicles for hydrophobic nutraceuticals or bioactives. Appl Sci 13:1726
100. Tarhan Ö (2020) Safety and regulatory issues of nanomaterials in foods. Handb Food Nanotechnol 655–703
101. Tavker N, Sharma M (2022) Nanosensors for intelligent food packaging. In: Nanosensors for smart agriculture. Elsevier, pp 737–756
102. Uddin MN, Desai F, Asmatulu E (2020) Engineered nanomaterials in the environment: bioaccumulation, biomagnification and biotransformation. Environ Chem Lett. https://doi.org/10.1007/s10311-019-00947-0
103. Velicogna JR, Schwertfeger D, Jesmer A, Beer C, Kuo J, DeRosa MC, Scroggins R, Smith M, Princz J (2021) Soil invertebrate toxicity and bioaccumulation of nano copper oxide and copper sulphate in soils, with and without biosolids amendment. Ecotoxicol Environ Saf 217:112222

104. Verma SK, Das AK, Patel MK, Shah A, Kumar V, Gantait S (2018) Engineered nanomaterials for plant growth and development: a perspective analysis. Sci Total Environ 630:1413–1435. https://doi.org/10.1016/j.scitotenv.2018.02.313
105. Vieira IRS, de Carvalho APA, de, Conte-Junior, C.A., (2022) Recent advances in biobased and biodegradable polymer nanocomposites, nanoparticles, and natural antioxidants for antibacterial and antioxidant food packaging applications. Compr Rev Food Sci Food Saf 21:3673–3716
106. Vohra A, Goswami D, Deshpande D, Block S (2005) Enhanced photocatalytic inactivation of bacterial spores on surfaces in air. J Ind Microbiol Biotechnol 32:364
107. Wang C, McClements DJ, Jiao A, Wang J, Jin Z, Qiu C (2022) Resistant starch and its nanoparticles: recent advances in their green synthesis and application as functional food ingredients and bioactive delivery systems. Trends Food Sci Technol 119:90–100
108. Waqas Mazhar M, Ishtiaq M, Maqbool M, Akram R, Shahid A, Shokralla S, Al-Ghobari H, Alataway A, Dewidar AZ, El-Sabrout AM, Elansary HO (2022) Seed priming with iron oxide nanoparticles raises biomass production and agronomic profile of water-stressed flax plants. Agronomy 12:982. https://doi.org/10.3390/agronomy12050982
109. Yakdoumi FZ, Hadj-Hamou AS, Rahoui N, Rahman MM, Abetz V (2022) Polylactic acid nanocomposites containing functionalized multiwalled carbon nanotubes as antimicrobial packaging materials. Int J Biol Macromol 213:55–69
110. Yam KL, Takhistov PT, Miltz J (2005) Intelligent packaging: concepts and applications. J Food Sci 70:R1–R10
111. Yang J, Song Z, Ma J, Han H (2020) Toxicity of molybdenum-based nanomaterials on the soybean–rhizobia symbiotic system: implications for nutrition. ACS Appl Nano Mater 3:5773–5782
112. Yu J, Wang D, Geetha N, Khawar KM, Jogaiah S, Mujtaba M (2021) Current trends and challenges in the synthesis and applications of chitosan-based nanocomposites for plants: a review. Carbohyd Polym 261:117904. https://doi.org/10.1016/j.carbpol.2021.117904
113. Yuvaraj M, Sathya Priya R, Jagathjothi N, Saranya M, Suganthi N, Sharmila R, Cyriac J, Anitha R, Subramanian KS (2023) Silicon nanoparticles (SiNPs): challenges and perspectives for sustainable agriculture. Physiol Mol Plant Pathol 128:102161. https://doi.org/10.1016/j.pmpp.2023.102161
114. Zhang D, Hua T, Xiao F, Chen C, Gersberg RM, Liu Y, Stuckey D, Ng WJ, Tan SK (2015) Phytotoxicity and bioaccumulation of ZnO nanoparticles in Schoenoplectus tabernaemontani. Chemosphere 120:211–219
115. Zhou Q, Zhang X, Wu Z (2020) Impact of TiO_2 and ZnO nanoparticles on soil bacteria and the enantioselective transformation of Racemic-Metalaxyl in agricultural soil with lolium perenne: a wild greenhouse cultivation. J Agric Food Chem 68:11242–11252

Macro and Nanoplastics in Agriculture: Short Review

Esperanza Huerta Lwanga

Abstract There is a big need to restitute the agricultural soils. Consequences of decisions taken in the past, have given fingerprints in the agricultural sites. Among those fingerprints are the micro and nanoplastics. In the last decades, investigations of the presence and impact of micro/nanoplastics have been carried out in the agricultural soils. Polypropylene and polyethylene are the main plastic polymers prevailing in the agricultural soils, and there is awareness that agricultural soils are sink and source of micro and nanoplastics. The main aim of this short review is to demonstrate which is the impact of nanoplastics in the agricultural sites. For that, using the searching Scopus platform, a selection of papers has been carried out, where the words nanoplastics and agricultural sites were together, and the production of papers increases exponentially in the last years, having a total of 73 papers since 2017 to 2023. And most of the investigations have been performed in Asia. The impact of nanoplastics is recognized at the morphological, oxidative stress and genotoxicity on plants. Soil bacteria might shift their activities if micro/nanoplastics are present in soils. The gut bacteria microbiome diversity of earthworms is modified if soil micro/nanoplastics are present. The type of plastic, concentration and time of exposure play an important role on the impact that micro/nanoplastics might produce on plants, and soil organisms. It is evident that more studies are required since the agricultural soil is polluted with plastic debris. Human health relies on soil health. Therefore, the need of enhancing healthy soils, for healthy food and healthy humans.

Keywords Microplastics · Nanoplastics · Agriculture · Impact · Assessment

1 The Need of Healthy Soils

In agricultural lands, healthy soils are recognized, as those soils where soil invertebrates are active, where high diversity of micro, meso and macroinvertebrates is found, and the ecosystem services are ensured. In agricultural soil conditions,

E. H. Lwanga (✉)
Soil Physics and Land Management Group, Wageningen University, Wageningen, Netherlands
e-mail: esperanza.huertalwanga@wur.nl

© The Author(s), under exclusive license to Springer Nature Singapore Pte Ltd. 2025
A. Juárez-Maldonado et al. (eds.), *Plant Biostimulation with Nanomaterials*, Smart Nanomaterials Technology, https://doi.org/10.1007/978-981-96-4648-7_16

following the need of high production, different technologies have been used, among them the use of plastics. In this mini review it is presented how micro and nanoplastics are impacting soil conditions, and which are the repercussions in the agricultural systems with a small glance into humans.

Worldwide are different strategies trying to be implemented for ensuring soil health, soil health is translated into healthy food, and therefore in healthy humans. Agroecological systems are now more promoted worldwide, as those farming systems where diversity and food security are guaranteed. Within the European Union soil health strategy, enhancing soils free of pollutants remains one of the main goals, for enhancing healthy food.

2 Last Research on Nanoplastics in/and Agricultural Systems

To collect the highest number of papers, in this short review the assistance of the Scopus search platform was used. Nanoplastics & agricultural lands were the key words used for collecting the number of papers produced in relation to nanoplastics. Seventy-three papers were obtained, all of them have nanoplastics in the title and abstract. The number of papers per year is increasing exponentially since 2017, having the highest number in 2023 with 26 papers (Fig. 1).

From those papers, the article of Horton et al. [12] has been highly cited (>2000 papers), and the authors pointed out the high concern that represents the pollution of plastic debris in the aquatic and terrestrial environment. They highlighted that the emission of plastic debris into the land is 4 to 23 times higher than in the water. They pointed out through their review that plastic debris are considered as pollutants, and those pollutants accumulate in different compartments of the environment. In this review the impact of nanoplastics is described after meta-analysis performed previously. Therefore, the description is based per topic and per meta-analysis.

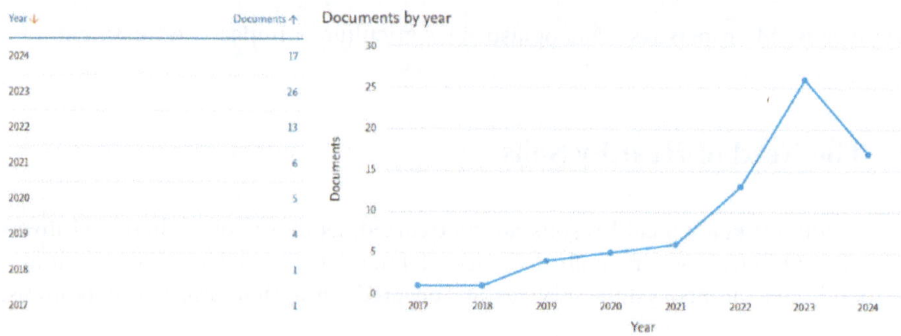

Fig. 1 Number of papers produced per year with the key words "nanoplastics and agriculture". Obtained from Scopus platform at June, 2024

3 Microplastics in the Agricultural Systems

In the last 2 decades it has been constated that microplastics are found in the terrestrial environment. Rillig et al. [41], manifested the great risk that microplastics are imposing into the agroecosystems. Even though plastic is an extraordinary and very flexible material, the misuse and overall, the waste mismanagement of plastics have produced an important concern in the aquatic and terrestrial environments. Koutnik et al. [24] analyze 196 studies informing over the concentration of microplastics in water and soils, being China the location with the highest production of studies [24], and findings [51].

And the concentration in soils is higher in the agricultural lands than in urban soils, existing a huge heterogeneity in the concentrations, after an analysis done by the authors with 117 studies in the terrestrial environment [24].

In agriculture, irrigation systems as certain pipes are made with plastic, slow-release fertilizers contain plastic coat, plastic mulches were maid originally with fossil based polyethylene, and the application of sewage sludge or compost (unfortunately due to inadequate waste separation by citizens plastic particles are also present in compost) polluted with microplastics have produced pollution in agricultural soils, making them rich in microplastic particles. Corradini et al. [5] found a direct correlation between the application of sewage sludge and the concentration of microplastics in agricultural soils. The use of plastic mulch also promotes the presence of plastic particles, not only in the soil but also in the surrounding ecosystems. Huerta Lwanga et al. [17] found after a study done in several agricultural sites, how microplastics can be transport from the agricultural soils to surrounding ditch sediments, and water after a strong correlation between those matrices.

Agricultural systems are sink and source of plastic pollution. Therefore, there is a primordial need to study the agricultural soils. According to Tajwar et al. [44], the agricultural soils with the highest abundance of microplastics are found in China. That might coincide with the fact that several studies were done likewise in China, and thus, more information is available, combined with the fact that the historical agricultural use, and condition of production in that country, together with extreme environmental factors, have influenced on the way of agricultural production in China. In the last 24 years there is an intense use of plastic mulches [44], and this situation coincides with high numbers of microplastic concentrations in soils; after a review, Tajwar et al. [44] affirm that there are in some places even above 1000 000 particles per kg, being Asia the continent with the highest concentrations reported from thousand to million particles per kg, following this analysis, in south America the concentration of microplastics are on from 30 to 2000 particles per kg, and in North America from 2000 to 14,000 particles per kg; in Europe the reported concentrations are varying from few items to thousand particles per gram of soil [34].

3.1 From Microplastics to Nanoplastics

According to the European Commission, a Nano Material is defined as "a natural, incidental or manufactured material within the size range of 1 nm–100 nm" [3]. Nanoplastics are the result of the fragmentation of microplastics, coming from places/areas high polluted with microplastics, and propense to emit high nanoplastic concentrations. Following this information, Asia is one of the continents with the highest risk and potential emission of nanoplastics. There is a big challenge to quantify nanoplastics from environmental samples. In the terrestrial environment, studies involving nanoplastics, are those with ecotoxicological risk assessments, evaluating the effects of nanoplastics on soil biota or/and plants. Even though the number of studies including nanoplastics is increasing, still exists different and important challenges in the use and assessment of nanoplastics.

3.2 Sources and Types of Micro and Nanoplastics

Nevertheless, mulching is one of the main sources of plastic pollution in the agricultural soils of China, with concentrations around 1075 items per gram with Polyethylene (PE) as main polymer, after 24 years of constant application of plastic mulch [4, 14], the use of sludge enhances even more the magnitude of microplastic pollution, also in China, Li et al. [28] report 56.4×10^3 items per kg with diverse polymers as Polyolefin, acrylic fibers, PE, and PA [4].

Fossil based polyethylene and polypropylene are plastic types that have dominated the agricultural soils due to their high persistence in the environment. Koutnik et al. [24], showed clearly how polypropylene and polyethylene are dominating not only the agricultural soils but also fresh water, coastal soils, urban soils, wetlands, and storm water.

Micro and nanoplastics are migrating from one compartment to another, from soil to water and from water to soil [25], this dispersion of plastic particles implies not only the transfer of polymers/oligomers but also of several chemicals that are part of the additives, present together with the plastic particles, as those mentioned by Kumar et al. [25] in their review, plasticizers, flame retardant, precursors/stabilizer, and or biocide such as, Bis (2-ethylhexyl) phthalate (DEHP), polybrominated diphenyl ethers (PBDEs), bisphenol A (BPA) and triclosan. Micro/nanoplastic pollution in soils jeopardize soil ecosystem services [29].

Experts emphasize that waste pollution is the main reason/source of plastic pollution in the environment, in 2016 approximately 2.7 millions of metric tons or plastic waste were emitted into the environment and from them around 27% is dumped again into the landfills [25, Plastic Europe 2018], in general from 20–42% of the amount of plastic that is produced finishes as waste in landfills [35]. Production and consumption of plastic wastes take place in the terrestrial environment; therefore, actions are needed before those materials are transformed into wastes. Migration of plastic wastes from land to rivers occurs mainly during runoff events, and the return of plastic particles from rivers to lands happens via flooding [25].

In 2019, a group of scientists worked together with different stakeholders, among them policymakers, they investigated and wrote the SAPEA (A scientific perspective on microplastics in nature and society) report, in that document several actions are introduced for solving the problem of plastic pollution, among them 9. Those actions are presented as the core of the report (https://sapea.info/topic/microplastics/), and they notify on the fact that more studies are required, in order to understand the effect of micro and nanoplastics in the environment. Here I mention 4 of them, those more related with the present short review, (i) improvement of risk assessment methods, (ii) reduction of the emission of plastic and enhancement of circular economy, (iii) to invoke the precautionary principle, and iv) to integrate societal dynamics and behavioral Science. It is evident the implication of societal actions for resolving plastic pollution problem. The pollution of micro and nanoplastics in the environment is the consequence of different actions and decisions taken by the society, from land to water together with monitoring actions of micro and nanoplastics it is important to work together with policymakers, farmers and plastic producers.

Agricultural lands are now situated worldwide in a specific -momentum- in which it is needed to take a turn, from the common ways of production or "business as usual" into sustainable agricultural actions. The common use of several inputs in agriculture, among them plastic products, has to be dosed. Agriculture has to change its paradigm; it is no longer possible to justify the use of high plastic inputs under the banner of high production for countries where there is no food. The reality is that agriculture is facing environmental challenges, given by climate change as well as the inherited fingerprints (pollutants in agricultural soils).

In the last two decades, extraordinary innovations have taken place in the production of environmentally safe plastic materials, more and more the industry is doing efforts for producing biodegradable plastic materials for different sectors of the economy. Among them, the agricultural sector. It is a strong need to replace persistent microplastics into non persistent ones. For decreasing the possibility of their accumulation in soils and through the terrestrial and aquatic food chain.

Soil conditions are affected when microplastics and nanoplastics are found in soils, plastic types and soil characteristics as texture, structure and porosity influence on the vertical and horizontal movement of micro and nanoplastics [44], the growth of the plants is also affected. In the following section the impact of micro and nanoplastics in agricultural systems is described.

4 Impact of Nanoplastics on Plants

Several studies have been performed, showing how microplastics and nanoplastics impact on plants [26, 29, 32, 47], and this fact represents a big treat to plant food [39]. In this mini review the selected studies are focused on nanoplastics. Dang et al.

[6], informed clearly after a meta-analysis carried out with 33 studies and 736 observations, how with physicochemical properties of nanoplastics, including nanoplastic size, mass concentration, type and surface charge, and plant traits, including plant species and life stages, likewise the exposure conditions including exposure time, medium, and pathways, and this meta-analysis confirmed that nanoplastics impact terrestrial plants negatively with 19%. Nevertheless, among the most important toxicity metrics, are the life history, morphological and physiological functions, as well the oxidative stress, and genotoxicity, the last one resulted as the toxicity metric with the highest negative percentage (highest impact) obtained after the meta-analysis. Around 150 observations reported high impact of nanoplastics on plant morphological functions and the oxidative stress with small nanoplastics $\leq$ 100 nm, and through low concentrations < 10 mg/kg [6].

Even though more attention is required for understanding how the presence of nanoplastics impact on plants at the gene level, still it is difficult to measure the uptake of nanoplastis in plants [50], where the type of method plays an important role.

More studies are required to understand the genotoxicity effect of nanoplastics including different types and sizes of plastics, including biodegradable and non-biodegradable plastics, with different plant species. Most of the studies are developed with model plants, it is important also to include no model plants for understanding more clearly different impacts on plants. Likewise, to include different environmental scenarios, due to the challenges provided by climate change, extreme scenarios including water scarcity or/and high temperatures are really needed.

5 Impact of Nanoplastics on Soil Fauna

Size and type of nanoplastics produce and impact [48] as well the feed strategy. Gomes et al. [9], after performing an interesting review and meta-analysis with 175 studies where the impact of mainly polyethylene (35% of the studies) and polystyrene (51% of the studies) on soil invertebrates was evaluated, observed that the size and type of nanoplastics play an important role on the impact to soil fauna. Most of the studies that the authors used in their analysis, have been carried out with spherical micro and nanoplastics because are easier to obtain in comparison with the fibers. Spherical microplastics are also easier to identify when using Spectro microscope methods. Among the assessed studies, the evaluated organisms were belonging to: Arthropoda (59 publications), followed by Chordata (41 publications), Mollusca (36 publications), Phytoplankton (25 publications), Annelida (16 publications), Cnidaria and Echinodermata (each 4 publications), Rotifera (3 publications) and finally Nematoda (1 publication). Gomes et al. [9] manifest a great concern related to the type of plastic used in the investigations, according to the review the type of polymer might influence the fate and behaviour of the particles, as well the additives present with the polymers. One of the observations of the study is also to carry out toxicological studies with additives, since most of the studies are done with pristine micro/

nanoplastics. Among the results it was observed a size dependent effect, this effect increases while the size decreases [9].

Studies implying complete life cycles of the organisms are scarce or absent. Most of the studies assess only adults or only juveniles, but both stadia are until now not assessed together in the same study. Gomes et al. [9] also include that the feeding strategy of the organisms implies a different reaction, in their review, 30% of the species used were filter feeders, followed by photosynthetic organisms (21%), predators (17%), detritivores (10%), grazers (9%), scavengers (8%) and deposit feeders (5%).

Micro and nanoplastics are transported along the terrestrial food chain, and more studies are needed to be carry out, with organisms that are present in the whole terrestrial food chain in tropics and template areas, with different terrestrial climate change scenarios. It is already proved if microplastics are present in soils, that soil invertebrates as earthworms try to adapt to the new conditions, their gut bacteria microbiome diversity and community shift, and this occurs according to the ecological category of the earthworms [36].

Microplastics affect also reproduction of the organisms as observed mainly with the aquatic beings. Ji et al. [20] after doing a meta-analysis found from 56 documents (1061 endpoints) how 49% of the studies show sever affectation of pristine micro/nanoplastics on organisms, and chlorine and phenyl groups produce strong impact on soil organisms.

6 Impact of Nanoplastics on Wildlife

In Europe as in the whole planet, agricultural areas are surrounded by natural areas, domestic animals, together with wildlife, are in contact with the agricultural systems. We know that microplastics can accumulate along the terrestrial trophic chain (ie. sheep, Beriot et al. [1], chicken, Huerta Lwanga et al. [16], Liu et al. [31]). It is important to recognize that polluted soils are sink and source of pollution, when microplastics are present in soils, the invertebrates and vertebrates that inhabit those soils or nearby, can transport the microplastics inside or on their bodies when grazing in the agricultural areas.

Since invertebrates and vertebrates transport microplastics in their bodies and in their biogenic structures. It is important to understand the relationship between the type and size of microplastics and the type and size of the invertebrates during the assessment of microplastics along the terrestrial food chain, and to test several hypotheses [15].

Nevertheless, small organisms are mostly impacted by nanoplastics, at the cellular level, also the vertebrates are impacted by the nanoplastics. Experiments with rats have shown, how nanoplastics remain in their intestine tract, once those vertebrates have ingested nanoplastics (60 nm plastic PS particles), 10% of the dose remains in the gut [11]. Microplastics have been found in the intestine tract of chicken [16], in feces of sheep [1] and in pellets of vultures [2]. There is evidence that plastic debris are transported from soil to vertebrates. Polystyrene nanoplastics produce effect on brain abnormality among rodent neonates, when these nanoplastics are transported by the milk of the rodents [19].

The agricultural systems are sources of micro/nanoplastics dissemination into surrounded natural areas or reserves. Different factors and interactions play a role, as the presence of plant protection products (PPP) and antibiotics, microplastics as emergent pollutants are not present alone in the agricultural soils, normally due to the industrialization of the agriculture, the agricultural soils have likewise other stressors for enhancing plant or cattle production. Those synergistic interactions are potentially more dangerous when plastic debris are small in size [10], studies are required to understand which are the impact of nanoplastics on wildlife with and without stressors. Already implications have been found in the aquatic systems, when nanoplastics are together with pesticides [22].

7 Possible Evolutive Selection by the Presence of Micro/ Nanoplastics

At the cellular and genetic level, it has been observed, a gene expression selection when PET microplastics are present. The etg gene cluster is enhanced within the protobacteria *Paracoccus denitrificans* when PET microplastics are abundant in its environment, and this occurs mainly because PET as other polyesters are covered with the ethylene glycol monomer, and the etg gene is expressed for using this ethylene glycol as resource source [40]. The main function of this protobacteria is inside the nitrogen cycle, but when microplastics are present in soils the function of this bacteria might shift instead of being more involved in the denitrification, it can be involved in using the ethylene glycol present in polyesters, of course this statement needs to be verified, and it is dependent on the concentration of microplastics in soils. Therefore, more studies are required. Thomas et al. [45] highlight several gaps in understanding the effect of nanoplastics, among them at the subcellular and molecular levels, it is still unknown which are the implications of nanoplastics on biota.

8 Impact of Nanoplastics on Humans Health

Nanoplastics are considered wastes in waster soil and air [38] and humans ingest accidentally micro and nanoplastics, it is in recent years that this ingestion is openly constated by different authors [21]. From the agricultural sites micro/nanoplastics can be transported by abiotic and biotic factors [15], and humans can be affected.

The existence of microplastics in drinking water has been studied [7], Koelmans et al [23] have demonstrated how microplastics are present in common aquatic animals, and it is constated that there is accumulation and translocation of microplastics in aquatic animal tissues (ie. mussels, Li et al. [27]), similar is observed in the terrestrial environment [42]. Hirt and Body-Malapel [11] emphasis on how microplastics have impacted the intestine mucuse of humans, and the implications of this situation on human immune system, even though the uptake by microplastics by the intestine it is not clear described, or indicated as inefficient, there is an immune system reaction, endocytosis, transcytosis, persorption, and inflammation. Although most of these results have been observed mainly in rodents [11].

In humans, it has been observed that there is stronger danger when humans are sick, human adenocarcinoma cells are more prone to be affected by microplastics with concentrations around 10 μg per militre (5–200 μm), those cells have less viability [8]. Other impact of microplastics on humans is via inhalation, there are found microplastics (3microm) in human lungs, where PET and PP are the most abundant polymers [18], and this situation produces an impact on human respiratory system [37]. Polypropylene being abundant in agricultural soils, have been found to produce a strong impact on humans [33].

9 Perspectives

In sustainable farming, soils play an important role, and microplastics have become an important fingerprint, due to inherited practices and decisions on the use of plastics that do not degrade at soil conditions. In many agricultural areas of the planet, actions are needed for stopping the emissions of micro/nanoplastics.

The need of re-connection between farmers and various stakeholders with soils, is imperative, in the aquatic system. Shen et al. [43] explain how the pollution found in water has a strong relation with the activities/behavior present in the fish sector.

Not only do farmers need to identify soil diversity as the path to soil health and healthy food, but the rest of society needs to acknowledge sustainable farming free of micro/nanoplastics as a societal goal. The future of new generations is in the hands of agroecological citizens realizing that healthy soils induce healthy humans, through above and below ground diversity, and stopping with the use of non-biodegradable plastics.

For the common good, society needs to trigger more sustainable plastic free farming, since society depends on farming and vice versa.

On the path towards sustainable plastic-free farming, recognizing that the synergy between societal and environmental pillars is needed, it is important to stress the need of soil biota enrichment in agricultural soils. Three steps are visualized for enhancing soil biota conditions in damaged agricultural sites:

- Identification at the farm level of soil conditions, assessing and enhancing above and below ground diversity in agricultural soils, resulting in the monitoring of soil ecosystem services; fingerprint monitoring, evaluating soil pollutants (ie. micro/

nanoplastics) at different compartments of the farming system (in soil, water, sediments, airborne sediment).

- Multidisciplinary participatory approach, for identifying problems and implementing new solutions. The farm, soil and surroundings are dependent on the society, on a local and regional scale, within each scale stakeholders should participate in giving ideas, and solutions for restoring the soil/field/area into a healthier plastic-free production place.
- The gut of the earthworms is interconnected with the gut of the humans, it has already been proved that human microbiome depends on soil microbiome [11]. Earthworm guts contain microorganisms together with nutrients and pollutants [46], those microorganisms are dependent on soil management, and therefore respond to soil conditions; high diversity of microorganisms inside the earthworms will at the long-term result in healthy and diverse food, which will be translated into rich diverse microbiome human guts, therefore plastic free food production is needed.

Micro/nanoplastics might be cleaned from agricultural soils, through different processes among them technics of polymer degradation, implying chemical, mechanical, UV, and microbial degradation [38].

References

1. Beriot N, Peek J, Zornoza R, Geissen V, Lwanga EH (2021) Low density-microplastics detected in sheep faeces and soil: a case study from the intensive vegetable farming in Southeast Spain. Sci Total Environ 755:142653
2. Borges-Ramírez MM, Escalona-Segura G, Huerta-Lwanga E, Iñigo-Elias E, Rendón-von Osten J (2021) Organochlorine pesticides, polycyclic aromatic hydrocarbons, metals and metalloids in microplastics found in regurgitated pellets of black vulture from Campeche, Mexico. Sci Total Environ 801:149674
3. Calzolai L, Gilliland D, Rossi F (2012) Measuring nanoparticles size distribution in food and consumer products: a review. Food Addit Contam Part A 29(8):1183–1193. https://doi.org/10.1080/19440049.2012.689777
4. Chai B, She Y, Wei Q, Lan W, Pan K (2022) Micro-and nanoplastic pollution in terrestrial ecosystems. In: Advances in toxicology and risk assessment of nanomaterials and emerging contaminants. Singapore: Springer Singapore, pp 207–226
5. Corradini F, Meza P, Eguiluz R, Casado F, Huerta-Lwanga E, Geissen V (2019) Evidence of microplastic accumulation in agricultural soils from sewage sludge disposal. Sci Total Environ 671:411–420
6. Dang F, Wang Q, Yan X, Zhang Y, Yan J, Zhong H, Wang Y (2022) Threats to terrestrial plants from emerging nanoplastics. ACS Nano 16(10):17157–17167
7. Danopoulos E, Twiddy M, Rotchell JM (2020) Microplastic contamination of drinking water: a systematic review. PLoS ONE 15(7):e0236838
8. Danopoulos E, Twiddy M, West R, Rotchell JM (2022) A rapid review and meta-regression analyses of the toxicological impacts of microplastic exposure in human cells. J Hazard Mater 427:127861
9. Gomes T, Bour A, Coutris C, Almeida AC, Bråte IL, Wolf R, Lusher AL (2022) Ecotoxicological impacts of micro-and nanoplastics in terrestrial and aquatic environments. Microplastic in the environment: Pattern and process, 199–260

10. Gupta C, Kaushik S, Jain S, Dhanwani I, Garg S, Paul A, Gupta N (2022) Bioaccumulation and toxicity of polystyrene nanoplastics on marine and terrestrial organisms with possible remediation strategies: A review. Environ Adv 8:100227
11. Hirt N, Body-Malapel M (2020) Immunotoxicity and intestinal effects of nano-and microplastics: a review of the literature. Part Fibre Toxicol 17:1–22
12. Horton AA, Walton A, Spurgeon DJ, Lahive E, Svendsen C (2017) Microplastics in freshwater and terrestrial environments: evaluating the current understanding to identify the knowledge gaps and future research priorities. Sci Total Environ 586:127–141
13. Huang D, Tao J, Cheng M, Deng R, Chen S, Yin L, Li R (2021) Microplastics and nanoplastics in the environment: macroscopic transport and effects on creatures. J Hazard Mater 407:124399
14. Huang Y, Liu Q, Jia W, Yan C, Wang J (2020) Agricultural plastic mulching as a source of microplastics in the terrestrial environment. Environ Pollut 260:114096
15. Huerta Lwanga EH, Beriot N, Corradini F, Silva V, Yang X, Baartman J, Geissen V (2022) Review of microplastic sources, transport pathways and correlations with other soil stressors: a journey from agricultural sites into the environment. Chem Biol Technol Agric 9(1):20
16. Huerta Lwanga E, Mendoza Vega J, Ku Quej V, Chi JDLA, Sanchez del Cid L, Chi C, Geissen V (2017) Field evidence for transfer of plastic debris along a terrestrial food chain. Sci Rep 7(1):14071
17. Huerta Lwanga E, Van Roshum I, Munhoz DR, Meng K, Rezaei M, Goossens D, Ritsema C (2023) Microplastic appraisal of soil, water, ditch sediment and airborne dust: the case of agricultural systems. Environ Pollut 316:120513
18. Jenner LC, Rotchell JM, Bennett RT, Cowen M, Tentzeris V, Sadofsky LR (2022) Detection of microplastics in human lung tissue using μFTIR spectroscopy. Sci Total Environ 831:154907
19. Jeong B, Baek JY, Koo J, Park S, Ryu YK, Kim KS, Lee DY (2022) Maternal exposure to polystyrene nanoplastics causes brain abnormalities in progeny. J Hazard Mater 426:127815
20. Ji Z, Huang Y, Feng Y, Johansen A, Xue J, Tremblay LA, Li Z (2021) Effects of pristine microplastics and nanoplastics on soil invertebrates: a systematic review and meta-analysis of available data. Sci Total Environ 788:147784
21. Jiang B, Kauffman AE, Li L, McFee W, Cai B, Weinstein J, Xiao S (2020) Health impacts of environmental contamination of micro-and nanoplastics: a review. Environ Health Prev Med 25:1–15
22. Junaid M, Abbas Z, Siddiqui JA, Liu S, Tabraiz S, Yue Q, Wang J (2023). Ecotoxicological impacts associated with the interplay between micro (nano) plastics and pesticides in aquatic and terrestrial environments. TrAC Trends Anal Chem, 117133
23. Koelmans AA, Nor NHM, Hermsen E, Kooi M, Mintenig SM, De France J (2019) Microplastics in freshwaters and drinking water: critical review and assessment of data quality. Water Res 155:410–422
24. Koutnik VS, Leonard J, Alkidim S, DePrima FJ, Ravi S, Hoek EM, Mohanty SK (2021) Distribution of microplastics in soil and freshwater environments: global analysis and framework for transport modeling. Environ Pollut 274:116552
25. Kumar M, Chen H, Sarsaiya S, Qin S, Liu H, Awasthi MK, Taherzadeh MJ (2021) Current research trends on micro-and nano-plastics as an emerging threat to global environment: a review. J Hazard Mater 409:124967
26. Larue C, Sarret G, Castillo-Michel H, Pradas del Real AE (2021) A critical review on the impacts of nanoplastics and microplastics on aquatic and terrestrial photosynthetic organisms. Small 17(20):2005834
27. Li J, Wang Z, Rotchell JM, Shen X, Li Q, Zhu J (2021) Where are we? towards an understanding of the selective accumulation of microplastics in mussels. Environ Pollut 286:117543
28. Li X, Chen L, Mei Q et al (2018) Microplastics in sewage sludge from the wastewater treatment plants in China. Water Res 142:75–85. https://doi.org/10.1016/j.watres.2018.05.034
29. Li T, Cui L, Xu Z, Liu H, Cui X, Fantke P (2023) Micro-and nanoplastics in soil: Linking sources to damage on soil ecosystem services in life cycle assessment. Sci Total Environ, 166925

30. Li X, Wang R, Dai W, Luan Y, Li J (2023) Impacts of Micro (nano) plastics on terrestrial plants: Germination, growth, and litter. Plants 12(20):3554
31. Liu Y, Shao H, Liu J, Cao R, Shang E, Liu S, Li Y (2021) Transport and transformation of microplastics and nanoplastics in the soil environment: A critical review. Soil Use Manag 37(2):224–242
32. Mateos-Cárdenas A, van Pelt FN, O'Halloran J, Jansen MA (2021) Adsorption, uptake and toxicity of micro-and nanoplastics: Effects on terrestrial plants and aquatic macrophytes. Environ Pollution 284:117183
33. McConnel G, Kasuske ZA, Mazariegos-Ortíz C, Muñoz SR, Cañas-Carrell JE (2024). A systems perspective of terrestrial, aquatic, and human health impacts of non-polystyrene-based nanoplastics. Curr Opin Environ Sci Health, 100557
34. Meixner K, Kubiczek M, Fritz I (2020) Microplastic in soil–current status in Europe with special focus on method tests with Austrian samples. AIMS Environ Sci 7(2)
35. Nizzetto L, Futter M, Langaas S (2016) Are agricultural soils dumps for microplastics of urban origin?
36. Papazlatani C, Garbeva P, Huerta-Lwanga E (2024) Effect of microplastics pollution on the gut microbiome of anecic and endogeic earthworms. FEMS Microbiol Lett, fnae040
37. Prata JC (2018) Airborne microplastics: consequences to human health? Environ Pollut 234:115–126
38. Ramasamy BSS, Palanisamy S (2021) A review on occurrence, characteristics, toxicology and treatment of nanoplastic waste in the environment. Environ Sci Pollut Res 28(32):43258–43273
39. Reddy AVK, Shankaraiah G, Kumar PS (2023) Toxicity Effects of Micro-and Nanoplastics in Terrestrial Environment. Micro and Nanoplastics in Soil: Threats to Plant-Based Food. Springer International Publishing, Cham, pp 191–220
40. Ren M, Li D, Addison H, Noteborn WEM, Andeweg EH, Glatter T, de Winde JH, Rebelein JG, Lamers MH, Schada von Borzyskowski L (2024) Promiscuous NAD-dependent dehydrogenases enable efficient bacterial growth on the PET monomer ethylene glycol
41. Rillig MC, Ingraffia R, de Souza Machado AA (2017) Microplastic incorporation into soil in agroecosystems. Front Plant Sci 8:1805
42. Rose PK, Yadav S, Kataria N, Khoo KS (2023) Microplastics and nanoplastics in the terrestrial food chain: Uptake, translocation, trophic transfer, ecotoxicology, and human health risk. TrAC Trends Anal Chem, 117249
43. Shen M, Zhang Y, Zhu Y, Song B, Zeng G, Hu D, Ren X (2019) Recent advances in toxicological research of nanoplastics in the environment: a review. Environ Pollut 252:511–521
44. Tajwar M, Shreya SS, Gazi MY, Hossain MB (2023) Abundance and distribution of MPs and NPs in soil: a global scenario. Micro and Nanoplastics in Soil: Threats to Plant-Based Food. Springer International Publishing, Cham, pp 35–57
45. Thomas PJ, Perono G, Tommasi F, Pagano G, Oral R, Burić P, Lyons DM (2021) Resolving the effects of environmental micro-and nanoplastics exposure in biota: A knowledge gap analysis. Sci Total Environ 780:146534
46. Vaccari F, Forestieri B, Papa G, Bandini F, Huerta-Lwanga E, Boughattas I, ... & Puglisi E (2022) Effects of micro and nanoplastics on soil fauna gut microbiome: an emerging ecological risk for soil health. Curr Opin Environ Sci Health 30:100402
47. Wang W, Yuan W, Xu EG, Li L, Zhang H, Yang Y (2022) Uptake, translocation, and biological impacts of micro (nano) plastics in terrestrial plants: Progress and prospects. Environ Res 203:111867
48. Wang W, Do ATN, Kwon JH (2022) Ecotoxicological effects of micro-and nanoplastics on terrestrial food web from plants to human beings. Sci Total Environ 834:155333
49. Yang S, Li M, Kong RYC, Li L, Li R, Chen J, Lai KP (2023) Reproductive toxicity of micro-and nanoplastics. Environ Int 108002
50. Yu Z, Xu X, Guo L, Jin R, Lu Y (2024) Uptake and transport of micro/nanoplastics in terrestrial plants: Detection, mechanisms, and influencing factors. Sci Total Environ 907:168155
51. Zeb A, Liu W, Shi R, Lian Y, Wang Q, Tang J, Lin D (2022) Evaluating the knowledge structure of micro-and nanoplastics in terrestrial environment through scientometric assessment. Appl Soil Ecol 177:104507

Correction to: Induction of Tolerance to Abiotic Stress by Applying Nanostimulants to Crops

Gonzalo Tortella Fuentes, Rodrigo Rodríguez, Paola Duran, Martín Fernández-Baldo, Mauricio Schoebitz, Ricardo Tighe, Joana Pieretti, and Paola Fincheira

Correction to: Chapter 8 in: A. Juárez-Maldonado et al. (eds.), *Plant Biostimulation with Nanomaterials*, Smart Nanomaterials Technology, https://doi.org/10.1007/978-981-96-4648-7_8

In the original version of the book, the author name in Chapter 8 has been updated as Martín Fernández-Baldo. The book and the chapter have been updated with the changes.

The updated version of this chapter can be found at
https://doi.org/10.1007/978-981-96-4648-7_8

© The Author(s), under exclusive license to Springer Nature Singapore Pte Ltd. 2025

A. Juárez-Maldonado et al. (eds.), *Plant Biostimulation with Nanomaterials*, Smart Nanomaterials Technology, https://doi.org/10.1007/978-981-96-4648-7_17

MIX
Papier aus verantwortungsvollen Quellen
Paper from responsible sources
FSC® C105338

If you have any concerns about our products, you can contact us on
ProductSafety@springernature.com

In case Publisher is established outside the EU, the EU authorized representative is:
Springer Nature Customer Service Center GmbH
Europaplatz 3, 69115 Heidelberg, Germany

Printed by Libri Plureos GmbH
in Hamburg, Germany